Biological Adhesives

AF832350

Andrew M. Smith
Editor

# Biological Adhesives

Second Edition

Springer

*Editor*
Andrew M. Smith
Department of Biology
Ithaca College
Ithaca, New York
USA

ISBN 978-3-319-83435-1 ISBN 978-3-319-46082-6 (eBook)
DOI 10.1007/978-3-319-46082-6

© Springer International Publishing Switzerland 2006, 2016
Softcover reprint of the hardcover 2nd edition 2016
This work is subject to copyright. All rights are reserved by the Publisher, whether the whole or part of the material is concerned, specifically the rights of translation, reprinting, reuse of illustrations, recitation, broadcasting, reproduction on microfilms or in any other physical way, and transmission or information storage and retrieval, electronic adaptation, computer software, or by similar or dissimilar methodology now known or hereafter developed.
The use of general descriptive names, registered names, trademarks, service marks, etc. in this publication does not imply, even in the absence of a specific statement, that such names are exempt from the relevant protective laws and regulations and therefore free for general use.
The publisher, the authors and the editors are safe to assume that the advice and information in this book are believed to be true and accurate at the date of publication. Neither the publisher nor the authors or the editors give a warranty, express or implied, with respect to the material contained herein or for any errors or omissions that may have been made.

Printed on acid-free paper

This Springer imprint is published by Springer Nature
The registered company is Springer International Publishing AG
The registered company address is: Gewerbestrasse 11, 6330 Cham, Switzerland

# Preface

The study of biological adhesives has advanced substantially since the first edition of this book. The use of a wider range of methods and a greater emphasis on an interdisciplinary approach have yielded dividends. Our ability to characterize the physical and biochemical structure of these glues and their resulting mechanical properties has grown in recent years, leading to much more detailed information about how they are put together and how they work. These new findings will help guide the development of novel biomimetic adhesives.

These new advances have come despite the inherent difficulty in studying biological adhesives. They are typically poorly soluble to insoluble, befitting the requirement of creating a bond that does not degrade under the reigning environmental conditions. Nevertheless, they are typically hydrated and much more deformable than common engineering materials. Many adhering organisms are so small that they provide little material for biochemists to analyze or materials scientists to test.

Despite these challenges, there has been substantial progress. One reason for this has been the success of an interdisciplinary approach driven by the growth of materials science. Our ability to analyze biochemical structure continues to advance, and it is now more often integrated with chemical and engineering approaches. Combining these approaches to analyze how structure gives rise to material properties and performance has led to a deeper understanding of biological adhesives. It is now common to characterize the secretory structures that synthesize and process a glue, the biochemical structures and chemical bonding involved in cohesion and adhesion, and the mechanical and functional properties that result. It is becoming more common for research groups to combine the expertise of scientists such as chemists, biochemists, molecular biologists, engineers, and physicists. These groups can be focused on the biological/biochemical end or the materials science end, but the ability to integrate approaches from different areas has led to many useful insights.

The ability to draw from different approaches is particularly important given the wide variety of structures, chemical interactions, and mechanical mechanisms involved in different biological adhesives. Adhesives are used for many different functions and environments. Microorganisms use adhesives to colonize and establish themselves on different surfaces. Larger organisms use them to ensconce themselves in favorable but often unstable environments. Adhesives can be used to capture prey or to foul potential predators. They can be used to build defenses. They can be used for rapid or slow locomotion. They may have to adhere to solid or flexible surfaces, hydrophobic or hydrophilic, and those surfaces depend on the environment in which they are found. Some are used in air and some in water, and the differences between dry and humid air or salt and freshwater will impact the way they work. Adhesives must be able to set at appropriate times, adhere to the desired surfaces, form a cohesive layer that resists separation, and provide adequate toughness.

Research over the 10 years since the publication of the first edition has advanced toward answers to how these different demands can be met. We have seen many different types of adhesives, including solid cements, fibrous holdfasts, tough gels, ultrafine, sticky tapes and threads, nanoparticle-based glues, hierarchically structured nanoarrays, and many different visco-elastic, sticky secretions. Molecules with specific biochemical features have been identified as being involved in interfacial adhesion, cross-linking, toughening, and protective coating or as providing some completely different function within the glue. While we continue to find further depths of complexity within biological glues, such complexity offers opportunity for those using a biomimetic approach. The structural and biochemical diversity of biological glues presents a diverse toolkit to inspire the development of novel biomimetic glues with tunable properties.

Thus, this book is aimed at researchers studying biological adhesives or synthesizing biomimetic or bioinspired versions of these adhesives. The goal is to bring together the most exciting current research in the field, including a wide variety of well-characterized biological adhesives. Such materials have the potential to inspire biomimetic or bioinspired materials that could have significant practical impacts. Biological adhesives are typically very different from common man-made adhesives. It is these differences that can provide inspiration for novel materials with desirable properties that are unlike those of current synthetic glues.

Ithaca, NY Andrew M. Smith

# Contents

# Chapter 1
# Adhesive Bacterial Exopolysaccharides

**Natalie C. Bamford and P. Lynne Howell**

**Abstract** Exopolysaccharides promote adhesion of bacteria to biotic and abiotic surfaces and are a key component of the extracellular matrix of many biofilms. Exopolysaccharides are chemically and structurally diverse and confer considerable advances to the bacteria that produce them. The increased tolerance to antibiotics and resistance to environmental changes that the polymers impart have significant consequences not only for human health but many industrial applications and processes. Herein we review current methods to analyze the composition and structure of exopolysaccharides and the molecular mechanisms used for their synthesis. We highlight the biosynthetic mechanisms and biological roles of four common bacterial adhesins: the Pel and Psl polysaccharides, poly-β-1,6-*N*-acetyl-D-glucosamine, and holdfast. Expanding our understanding of exopolysaccharide structure, production, and function will enhance current efforts to develop novel treatments for chronic bacterial biofilm-related infections.

## 1.1 Introduction

Bacteria live in diverse environments around the world in either a free-swimming planktonic state or in sessile biofilms. Biofilms are multicellular communities of bacteria encapsulated in a self-produced matrix. Biofilms form on biotic and abiotic surfaces and confer resistance to environmental changes. Once a biofilm is mature, the encapsulated bacteria have increased tolerance to desiccation, salinity, toxic environments, fluid force, and antibiotic treatment (Hall-Stoodley et al. 2004; Flemming and Wingender 2010). These advantages to bacteria can be detrimental

N.C. Bamford • P.L. Howell (✉)
Program in Molecular Structure & Function, The Hospital for Sick Children, 686 Bay St., Toronto, ON M5G 0A4, Canada

Department of Biochemistry, University of Toronto, 1 King's College Circle, Toronto, ON M5S 1A8, Canada
e-mail: howell@sickkids.ca

© Springer International Publishing Switzerland 2016 
A.M. Smith (ed.), *Biological Adhesives*, DOI 10.1007/978-3-319-46082-6_1

to human health and industrial processes and have led to increased research into biofilm formation and composition.

Biofilm formation is a multistage process and starts with bacterial attachment to a surface or imbedding in the mucosa. Attachment is initially reversible and mediated by bacteria-specific factors including proteinaceous adhesins and exopolysaccharides (Donlan 2002; Flemming and Wingender 2010; Kostakioti et al. 2013). After attachment becomes irreversible, monolayer and microcolonies form, and as the biofilm matures further, the bacteria become enveloped in a matrix of self-produced components. These components are variable and depend on the bacterial species and environmental conditions but most commonly include proteins, exopolysaccharides, extracellular DNA, and lipids (Kostakioti et al. 2013).

## 1.2 Characterization of Bacterial Exopolysaccharide Adhesins

Bacterial exopolysaccharides have diverse compositions and can carry out multiple functions. Monosaccharide composition, linkage, branching, and chemical modification all affect the function and properties of the polymer. Table 1.1 and Fig. 1.1 provide an overview of the bacterial exopolysaccharides that confer adhesive properties that will be discussed in this review. There are many more polysaccharides of bacterial origin that fulfill roles in adhesion, biofilm structure, water retention, and as nutrient sinks (Kostakioti et al. 2013).

Determining the composition and linkages present in bacterial exopolysaccharides can reveal common functions and helps to identify the biosynthetic pathways in their respective organism. There are many methods to determine polymer composition ranging from simple qualitative analyses to highly sensitive quantitative experiments. One barrier to all these methods is that purification of the polymer can be difficult. Exopolysaccharides are notoriously challenging to work with, as the long polymers are intrinsically insoluble and difficult to separate from contaminating polysaccharides.

### *1.2.1 Polysaccharide Purification*

Polysaccharides are often found in two different states: cell associated and cell free. For cell-free polysaccharides, ethanol precipitation of the culture supernatant is a common first step in the purification process. This is typically followed by digestion of the RNA, DNA, and protein contaminants. In the case of cell-bound exopolysaccharides, separation from the cell surface is a necessary first step. Purification of poly-β-1,6-*N*-acetyl-D-glucosamine (PNAG) from *Staphylococcus epidermidis* was done using sonication followed by pelleting of cellular debris

**Table 1.1** Adhesive bacterial exopolysaccharides and their properties

| Polysaccharide | Organisms studied | Composition | Modification | Charge |
|---|---|---|---|---|
| PEL | *Pseudomonas aeruginosa* | GalNAc[a] and GlcNAc | Deacetylation | + |
| PSL | *Pseudomonas aeruginosa* | Man, Glc, and Rha (3:1:1) | None | Neutral |
| PNAG | *Escherichia coli* | GlcNAc | Deacetylation | + |
| | *Bordetella bronchiseptica* | | | |
| | *Bordetella pertussis* | | | |
| | *Acinetobacter baumannii* | | | |
| | *Actinobacillus pleuropneumoniae* | | | |
| | *Yersinia pestis* | | | |
| | *Staphylococcus aureus* | | Deacetylation and Succinylation | |
| | *Staphylococcus epidermidis* | | | |
| Holdfast | *Caulobacter crescentus* | GlcNAc | Deacetylation | + |
| | *Asticcacaulis biprosthecum* | | | |
| | *Brevundimonas diminuta* | | | |
| | *Hirschia baltica* | | | |
| Cellulose | *Acetobacter xylinum* | Glc | None | Neutral |
| | *Escherichia coli* | | | |
| | *Rhizobium leguminosarum* | | | |
| | *Burkholderia* spp. | | | |
| | *Pseudomonas fluorescens* | | Acetylation | – |
| Cepacian | *Burkholderia cepacia* complex | Gal, Glc, GlcA, Man, and Rha (3:1:1:1:1) | Acetylation | – |
| *Listeria* polysaccharide | *Listeria monocytogenes* | ManNAc and Gal | Deacetylation | – |

[a]Abbreviations are as follows: *GalNAc* *N*-acetylgalactosamine, *GlcNAc* *N*-acetylglucosamine, *Man* mannose, *Glc* glucose, *Rha* rhamnose, *Gal* galactose, *GlcA* glucuronic acid, *ManNAc* *N*-acetylmannosamine

(Mack et al. 1996), while the cell-associated *Pseudomonas aeruginosa* Pel polysaccharide (PEL) has been isolated by treatment with EDTA (Jennings et al. 2015). Surface polysaccharide samples will often be heterogeneous and contain capsule forming polysaccharides and other surface antigens. In these cases analysis of the polysaccharide will require a bacterial strain that is incapable of making the target polymer as a comparator. Chromatographic purification can also be used to isolate

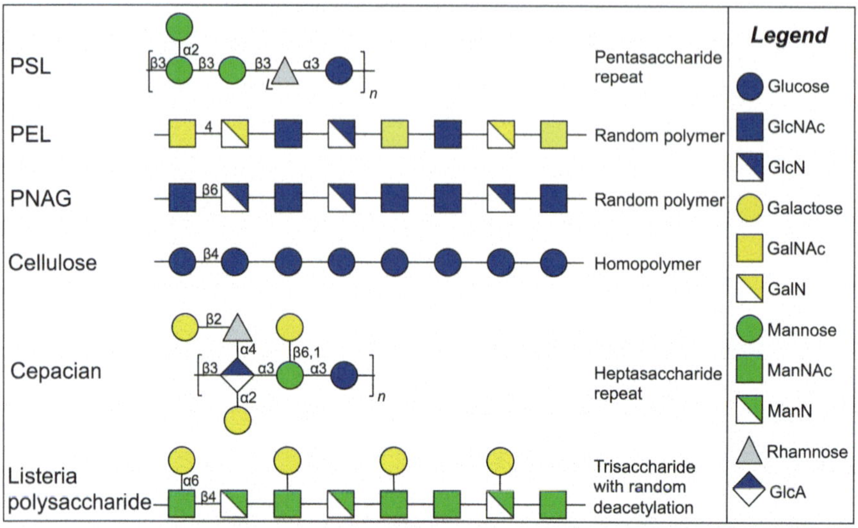


**Fig. 1.1** Schematic representation of adhesive exopolysaccharides of bacterial origin. The chemical composition of PSL, PEL, poly-β-1,6-*N*-acetyl-D-glucosamine (PNAG), cellulose, cepacian, and the *Listeria* polysaccharide is depicted. Known linkages are labeled and only once if repeating. The legend at the *right* displays the symbol notation consistent with the Consortium for Functional Glycomics (www.functionalglycomics.org)

different polymers, and cation-exchange (Jennings et al. 2015), anion-exchange (Mack et al. 1996; Cérantola et al. 1996), and size-exclusion (Byrd et al. 2009) chromatography have all been used successfully. For example, separation of *S. epidermidis* PNAG using ion-exchange chromatography yielded two unique fractions, one with net positive charge and one that was slightly electronegative (Mack et al. 1996). An alternative method to isolate exopolysaccharides is immunoprecipitation. This method requires an antibody specific to the polymer of interest. The *P. aeruginosa* PSL polysaccharide (PSL) has been isolated using this method (Borlee et al. 2010).

## *1.2.2 Polymer Length*

Once purified the exopolysaccharides can be analyzed to determine the monomer composition and polymer linkage, whether chemical modification has occurred, and polymer length. Polysaccharide size profiles are determined using size-exclusion chromatography with known standards such as dextran and cellobiose. The presence of carbohydrate in the elution fractions can be assayed using a phenol-sulfuric acid assay or UV absorption below 250 nm (Ovodova and Ovodov 1969). The amounts of specific polymers present can be determined using lectin staining and/or targeted antibodies.

### 1.2.3 Monosaccharide Composition

Monosaccharide composition is commonly determined using gas chromatography mass spectrometry (GC/MS). To enable this analysis, the polysaccharide must first be cleaved using acid hydrolysis or methanolysis. Acid hydrolysis will often remove sugar modifications such as *N*-acetyl groups and can also hydrolyze sugar monomers (Merkle and Poppe 1994). Methanolysis is therefore preferred as it is less harsh and is less likely to degrade the monosaccharides (Merkle and Poppe 1994). To detect amino sugars, the sample can be re-*N*-acetylated using acetic anhydride in methanol. Monosaccharides are subsequently derivatized using trimethylsilylate. This process increases the volatility of the sample allowing for effective gas-liquid chromatography and detection of amino, acetic, and neutral sugars (Merkle and Poppe 1994).

Acid hydrolysis can be used effectively if optimized (Manzi 1995). Incubation times and temperatures can be varied along with the strength of the acid. Treatment with trifluoroacetic acid is more predictable, and similar protocols can be used to hydrolyze linkages between pentoses, hexoses, amino sugars, and uronic acids. Milder acids are required for polymers containing sialic acids.

### 1.2.4 Alternative Composition Analysis Methods

Sugar binding lectins, specific antibodies, and glycoside hydrolases offer quick means to analyze exopolysaccharide composition and can be done in house at low cost. Treatment of unknown exopolysaccharides with a variety of glycoside hydrolases can reveal putative linkage and monosaccharide composition. The composition of the *P. aeruginosa* Pel polysaccharide (PEL) was initially determined to be glucose rich, based in part on the sensitivity of the biofilm formed to cellulase (Friedman and Kolter 2004a). DspB, a *N*-acetyl-β-glucosaminidase from *Actinobacillus actinomycetemcomitans*, has also been used to suggest the presence of PNAG in extracellular matrices of a number of bacteria including *Bordetella* spp. (Ciucanu 2006). Immunoblot assays with antibodies specific to polysaccharides have also been used to determine matrix composition. For example, Cywes-Bentley et al. used an antibody specific for PNAG to identify organisms producing PNAG-like polymers (Cywes-Bentley et al. 2013). Cross-reactivity is possible, and modifications to the same polysaccharide chain may give false negatives. Another method that exploits protein specificity is lectin-binding assays. Lectins fused to fluorophores can quickly determine the presence of specific motifs in the carbohydrate polymers. Ma et al. used lectins specific to galactose and mannose to support their compositional analysis of PSL (Ma et al. 2007), while Jennings et al. used a combination of antibody specificity, enzymatic digestion, and lectin staining to confirm the presence of *N*-acetylglucosamine (GlcNAc) and *N*-acetylgalactosamine (GalNAc) in PEL (Jennings et al. 2015). Neu et al. have also explored the use of fluorescently labeled lectins for imaging biofilm heterogeneity and component analysis (Neu et al. 2001; Neu and Lawrence 2014). Although these results are not definitive, they can be informative when the polysaccharide is difficult to isolate to high purity.

### *1.2.5 Linkage Analysis*

Polysaccharides can frequently have the same constituents but different linkages. For example, while chitosan and PNAG are both *N*-acetylglucosamine (GlcNAc) containing polymers, they have β-1,4 and β-1,6 linkages, respectively. A common method of linkage analysis uses per-*O*-methylation prior to depolymerization (Ciucanu 2006; Price 2008). After hydrolysis the samples are reduced and acetylated yielding monomers that have detectable acetyl groups at the linkage sites (Ciucanu 2006). The only definitive way to determine the anomeric configuration of a glycosidic bond between exopolysaccharide constituents is by nuclear magnetic resonance (NMR) spectroscopy. Partial degradation of complex branched polysaccharides may be required in order to resolve the structure from the spectra as found with cepacian polysaccharide (Cérantola et al. 2001).

### *1.2.6 Tertiary Structural Analysis*

Polysaccharides are often described by monosaccharide ratios and linkages, but increasing evidence suggests that these features and subsequent modification lead to tertiary structure. The high molecular weight, low solubility, and high flexibility of polysaccharides create a hurdle for some analytical methods (Larsen and Engelsen 2015). Combinations of polysaccharide staining, X-ray diffraction, NMR, atomic force spectroscopy (AFM), and molecular modeling have all been used to investigate polysaccharide conformation (Yu et al. 2010; Grachev et al. 2011; Larsen and Engelsen 2015). These experiments were first used on the exopolysaccharides of fungal origin scleroglucan and lentinan (Bluhm and Sarko 1977; Xu et al. 2010; Yu et al. 2010). These studies revealed that both linear and branched glucans form helical structures that are dependent on salinity and DMSO concentration. AFM studies of lentinan suggested long rods with a triple helical structure (Xu et al. 2010). NMR and molecular modeling of PNAG and the *Klebsiella pneumoniae* polysaccharide have also predicted helical conformations in solution (Guetta et al. 2003; Grachev et al. 2011).

## 1.3 Polysaccharide Biosynthesis Pathways

The synthesis of exopolysaccharides in Gram-negative bacteria has been classified into three different biosynthetic mechanisms (Fig. 1.2): the ATP-binding cassette (ABC) transporter, Wzx/Wzy-dependent, and synthase-dependent pathways. The ABC transporter and Wzx/Wzy-dependent pathways rely on building the initial chain on a lipid acceptor (Whitfield 2006; Whitney and Howell 2013). In contrast, while some exopolysaccharides produced using a synthase-dependent mechanism require a lipid acceptor most do not (Whitney and Howell 2013). Synthase-dependent systems are also typically regulated posttranslationally by bis-(3′–5′)-cyclic dimeric guanosine monophosphate (c-di-GMP) (Whitney and Howell 2013). Many heteropolymeric O-polysaccharides including those from *Salmonella enterica* serogroups and *P. aeruginosa* are produced using the Wzx/Wzy-dependent pathway (Whitfield 2006; Islam and Lam 2014). Homopolymers are commonly made using the ABC transporter or synthase-dependent mechanisms (Greenfield and Whitfield 2012; Whitney and Howell 2013).

### *1.3.1 Wzx/Wzy-Dependent Pathway*

Polysaccharides produced using the Wzx/Wzy-dependent pathway are created through the synthesis of repeat units in the cytoplasm (Drummelsmith and Whitfield 1999; Whitfield 2006; Islam and Lam 2014). These repeat units are generated by the attachment of sugar monomers to a lipid carrier by a series of specific glycosyltransferases. After assembly of the repeat unit, the lipid carrier is flipped to the periplasmic leaflet of the inner membrane. This step is unique to the Wzx/Wzy system and is carried out by a Wzx flippase (Whitfield 2006; Islam and Lam 2014). An inner-membrane-imbedded polymerase, Wzy, combines repeat units into a single chain in the periplasm (Woodward et al. 2010; Islam and Lam 2014). A polysaccharide copolymerase is required for chain length regulation but also appears to play a role in polysaccharide translocation (Cuthbertson et al. 2009; Woodward et al. 2010; Franklin et al. 2011; Islam and Lam 2014). Some polysaccharide copolymerases interact with the outer-membrane polysaccharide export protein creating a multi-protein complex within the periplasm (Cuthbertson et al. 2009). Modification of the polysaccharide can occur at different stages during the synthesis process, both pre- and postpolymerization (Whitfield 2006; Franklin et al. 2011; Wan et al. 2013; Whitfield et al. 2015).

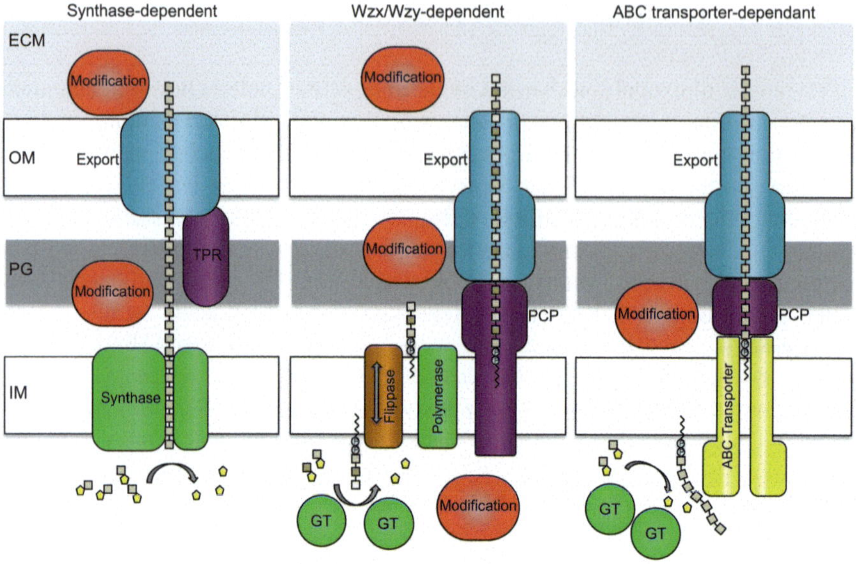


**Fig. 1.2** Exopolysaccharide biosynthesis pathways. Cartoon representations of synthase-, Wzx/Wzy-, and ABC-transporter-dependent pathways (*right to left*). The key feature of each is a system which is color coded and labeled in the representative Gram-negative cell membrane. The enzymes involved in building the polysaccharide; the inner-membrane synthase, glycosyltransferases (GT), and the polymerase that links the units in the Wzx/Wzy system are depicted in *green*. These enzymes take nucleotide di- or monophosphate (*yellow pentagon*)-activated sugar monomers (*beige squares*) and link them to produce random or repeating polymers. In the ABC transporter system the polysaccharide is assembled entirely in the cytoplasm on a lipid carrier before export through an ABC transporter (*yellow*). In *purple* are the tetratricopeptide repeat (TPR) and polysaccharide copolymerase proteins that are proposed to bind the sugar polymers during export. Modifying enzymes are displayed in *red*. These enzymes are not always required for synthesis. Export of the polymer through the outer membrane is carried out through a β-barrel porin in the synthase system or outer-membrane polysaccharide exporter (OPX) members for the other two systems (*blue*). *Abbreviations*: *IM* inner membrane, *PG* peptidoglycan, *OM* outer membrane, *ECM* extracellular matrix, *ABC* ATP binding cassette

### *1.3.2 ABC Transporter-Dependent Pathway*

Currently, the ABC transporter-dependent pathway has only been linked to the synthesis of linear O-polysaccharides. The first steps in this process are similar to those found in the Wzx/Wzy pathway where the sugar monomers are transferred onto a lipid acceptor on the cytoplasmic face of the inner membrane (Whitfield 2006). The major difference between the two mechanisms is that in the ABC transporter-dependent pathway the polymer is completed in the cytoplasm and then exported. Transport through the inner membrane is carried out by an ABC transporter and is driven by ATP hydrolysis (Whitfield 2006; Whitney and Howell 2013). The periplasmic and outer-membrane machinery have similarities to the Wzx/Wzy pathway as both mechanisms use proteins from the outer-membrane polysaccharide export and polysaccharide copolymerase families (Whitfield 2006; Whitney and Howell 2013).

### *1.3.3 Synthase-Dependent Pathway*

Synthase-dependent pathways are predicted to synthesize and translocate the polymer concomitantly (Whitney and Howell 2013). These systems are characterized by the presence of four protein domains: a glycosyltransferase responsible for polymerization; a cyclic-di-GMP-binding domain that posttranslationally regulates polymer synthesis, an outer-membrane beta-barrel porin, and finally a tetratricopeptide repeat (TPR) interaction domain. An inner-membrane complex consisting of a membrane-embedded glycosyltransferase and cyclic-di-GMP receptor produces and simultaneously translocates the polymer into the periplasm (Whitney and Howell 2013). The role of the TPR-containing protein is unknown but has been suggested to interact with the polymer and/or aid in creating a biosynthetic protein complex within the periplasm (Keiski et al. 2010; Whitney and Howell 2013).

## 1.4 Adhesive Exopolysaccharides

### *1.4.1 Pel Polysaccharide (PEL)*

*P. aeruginosa* is an opportunistic pathogen, which is often used as a model system for studying bacterial biofilm formation. *P. aeruginosa* is able to produce three distinct biofilm-related exopolysaccharides: alginate and the Psl (PSL) and Pel (PEL) polysaccharides (Friedman and Kolter 2004a). The Pel polysaccharide is characterized by the formation of a biofilm at the air–liquid interface termed a *pel*licle (Friedman and Kolter 2004a, b; Jennings et al. 2015). This polymer is composed of partially de-*N*-acetylated 1–4 linked *N*-acetylgalactosamine (GalNAc) and *N*-acetylglucosamine (GlcNAc) (Jennings et al. 2015). The *pel* locus was first identified in the non-mucoid *P. aeruginosa* PA14 strain using a transposon mutant library (Friedman and Kolter 2004b). This strain was obtained from a burn wound and lacks the ability to make PSL. Pellicle-deficient mutants led to the identification of a seven-gene operon that encoded carbohydrate-active enzymes (Friedman and Kolter 2004b). The genes were named *pelA* through *G* for *pel*licle producing. Based on bioinformatics analysis of the operon, the Pel polysaccharide is proposed to be produced via a synthase-dependent mechanism (Franklin et al. 2011).

#### 1.4.1.1 PEL Biosynthesis Pathway

Bioinformatics studies suggest that PelF is the only glycosyltransferase encoded in the *pel* operon (Friedman and Kolter 2004b). This protein is predicted to be cytoplasmic and uses UDP-activated sugar nucleotides as substrates for the production of the heteropolymer (Jennings et al. 2015). Earlier studies had suggested that PelF used UDP-Glc (Ghafoor et al. 2013), but the more recent results demonstrate that PEL is still produced in the absence of UDP-Glc (Jennings et al. 2015). PelD, PelE, and PelG are putative inner-membrane proteins (Franklin et al. 2011). PelD has a regulatory role and activates PEL production after binding of the secondary messenger cyclic-di-GMP (Lee et al. 2007; Whitney et al. 2012). It is predicted that PelD and some combination of the other inner membrane Pel proteins transport the nascent polymer into the periplasm (Vasseur et al. 2005). Once in the periplasm, the polymer is thought to interact with the periplasmic protein PelA (Colvin et al. 2013). PelA is a carbohydrate-modifying enzyme that has been demonstrated to have in vitro deacetylase and glycoside hydrolase activity (Colvin et al. 2013; Baker et al. 2016). Following modification by PelA, the nascent polymer is exported through the C-terminal beta-barrel porin domain of PelB (Franklin et al. 2011; Whitney and Howell 2013). The outer-membrane lipoprotein PelC is essential for polymer production and has also been suggested to be a part of the PEL export machinery (Vasseur et al. 2007; Kowalska et al. 2010). However, further experiments are required to determine PelC's exact function. In addition to its porin domain, PelB is predicted to have a large N-terminal TPR domain that may facilitate interactions between the polymer and or the PelC and/or PelA proteins (Vasseur et al. 2005). Periplasmic TPR domains are predicted to be present in many synthase-dependent systems, including alginate (Keiski et al. 2010), and while the function of the domain is not fully understood, this domain has been demonstrated to participate in protein interactions in the PNAG system (see below) (Forman et al. 2006). It has also been suggested that the TPR domain acts as a conduit to guide the polymer and protect it as it passages through the periplasm (Keiski et al. 2010; Whitney and Howell 2013).

#### 1.4.1.2 Modifications

The PEL polysaccharide is partially de-*N*-acetylated during synthesis by PelA (Colvin et al. 2013). PelA has been demonstrated to exhibit both de-*N*-acetylase and glycoside hydrolase activity in vitro (Colvin et al. 2013; Baker et al. 2016). To date, only the role of the deacetylase domain has been investigated in vivo (Colvin et al. 2013). The deacetylase domain is a member of the carbohydrate esterase 4 family (Colvin et al. 2013). Chromosomal mutation of active site residues revealed that de-*N*-acetylation is essential for pellicle formation (Colvin et al. 2013). These point mutants also resulted in no detectable PEL, although whether this is due to a lack of polymer production or whether the PEL-specific antisera do not recognize the unmodified polymer is unclear (Colvin et al. 2013).

#### 1.4.1.3 Interactions and Functions

Investigation into the role of the matrix components of *P. aeruginosa* biofilms is complicated by the many inherent redundancies in the system. Strains that overproduce components or have mutations in specific pathways are required in order to determine the traits of each polymers. Using an overproducing PEL strain, it was determined that PEL production leads to increased Congo red binding and cell aggregation in liquid culture (Colvin et al. 2011). PEL is also able to increase surface attachment in both PA14 and PAO1 strains, although PEL is not required for initial attachment (Colvin et al. 2011). Using the PSL-deficient strain, PA14, it was found that PEL is required for intercellular attachment and biofilm maturation after the monolayer stage. Initiation and maintenance of cell–cell interactions in the PA14 strain are also PEL dependent (Colvin et al. 2011), but these effects are masked in the PSL producing strain PAO1, as PSL is able to complement for the lack of PEL (Colvin et al. 2011). PEL was recently shown to cross-link to extracellular DNA and to bind in vitro host relevant polymers such as hyaluronan and mucin, suggesting that the ionic interaction mechanism of adhesion observed may have relevance in disease pathogenesis (Jennings et al. 2015).

### *1.4.2 Psl Polysaccharide*

The Psl polysaccharide (PSL) is the main biofilm polysaccharide component of most strains of *P. aeruginosa*. The *psl* operon was identified simultaneously by three different groups (Jackson et al. 2004; Matsukawa and Greenberg 2004; Friedman and Kolter 2004a). While looking for changes in auto-aggregation phenotypes using transposon mutagenesis, Friedman and Kolter found not only the *pel* operon (see above) but a second gene cluster with homology to polysaccharide processing genes (Friedman and Kolter 2004a). The annotated *P. aeruginosa* genome was used by two other groups to identify possible polysaccharide biosynthetic operons, which were then investigated experimentally (Jackson et al. 2004; Matsukawa and Greenberg 2004). This operon was termed *psl* for *p*olysaccharide *s*ynthesis *l*ocus (Jackson et al. 2004; Matsukawa and Greenberg 2004; Friedman and Kolter 2004a). Deletion and complementation studies have now demonstrated that 10 of the 15 genes in the *psl* operon are required for PSL production (Baker et al. 2015a; Cérantola et al. 2001; Friedman and Kolter 2004a; Byrd et al. 2009).

#### 1.4.2.1 PSL Biosynthetic Pathway

PSL is produced by a Wzx/Wzy-dependent pathway (Franklin et al. 2011). Synthesis of the PSL repeat unit requires three activated sugar nucleotides: guanosine diphosphate mannose (GDP-Man), uridine diphosphate glucose (UDP-Glu), and D-thymidine diphosphate rhamnose (dTDP-Rha) (Byrd et al. 2009). Only one of the proteins encoded by the *psl* operon, PslB, plays a role in precursor production. PslB is a bifunctional enzyme that is predicted to possess phosphomannose isomerase (PMI) and GDP-mannose pyrophosphorylase (GMP) activity (Byrd et al. 2009; Franklin et al. 2011). However, PslB is not essential for PSL production as WbpW, a protein involved in lipopolysaccharide (LPS) synthesis, can compensate for lack of PslB (Byrd et al. 2009). This overlap between PSL, LPS, and other cellular pathways means that no proteins within the *psl* operon are required for sugar-nucleotide precursor production. The *psl* operon encodes four predicted cytoplasmic glycosyltransferases (GT): PslC, PslF, PslH, and PslI (Franklin et al. 2011). PslF, PslH, and PslI are all predicted to be members of glycosyltransferase family 4 (GT-4), which transfer sugars by a retaining mechanism. PslC is predicted to belong to GT-2 family of enzymes; members of this family transfer sugars using an inverting mechanism. All four glycosyltransferases are required for PSL production, but their exact roles in assembling the pentasaccharide repeat have not been determined. In the Wzx/Wzy-dependent pathway, the repeat unit is built on a lipid acceptor and must be flipped across the inner membrane. The integral inner-membrane protein, PslA, is predicted to play a role in building the repeat unit onto the acceptor lipid similar to WbaP involved in *E. coli* capsule biosynthesis (Whitfield 2006; Franklin et al. 2011). The exact roles of the integral inner-membrane proteins PslJ, PslK, and PslL have not been demonstrated experimentally. PslJ and PslK are thought to be the polymerase and flippase machinery, respectively (Franklin et al. 2011). PslL has similarity to acyl transferases and may modify PSL during production (Franklin et al. 2011), although to date no modifications of PSL have been reported in the literature. PslE has similarity to the Wzc family proteins, which are polysaccharide copolymerases with C-terminal tyrosine autokinase domains (Cuthbertson et al. 2009; Franklin et al. 2011). Polysaccharide copolymerases play a role in length regulation of the secreted polymer (Islam and Lam 2014). Within the periplasm the nascent PSL polymer may be acted on by the glycoside hydrolase PslG. While it has been shown that PslG is not required for PSL production, it is thought that the protein may have a role in generating a low molecular weight cell-free PSL (Baker et al. 2015b). The final step in biosynthesis is secretion through the outer-membrane protein PslD (Cuthbertson et al. 2009; Franklin et al. 2011).

#### 1.4.2.2 Interactions and Functions

PSL increases cell–surface and cell–cell interactions in both PAO1 and ZK2870 strains of *P. aeruginosa* (Ryder et al. 2007). Attachment to both airway epithelial cells and mucin-coated surfaces was PSL dependent, thus connecting this polysaccharide to cystic fibrosis infections. In early stages of biofilm formation, PSL is found to form helical structures around the bacterial cells (Ma et al. 2009). This helical arrangement was visualized by confocal scanning laser microscopy using fluorescently labeled *Hippeastrum* hybrid (HHA) and *Marasmius oreades* agglutinin (MOA) lectins, which are specific for PSL. The helical arrangement of the polymer has been proposed to promote cell–cell interaction similar to how helical elements interact in other macromolecules. During the later stages of biofilm development, PSL covers the entire surface of the bacteria (Ma et al. 2009). In pellicle biofilms PSL has been found to colocalize with DNA in the center of the biofilm (Wang et al. 2013). PSL also interacts in vitro with DNA from diverse sources, including bacterial, salmon, and human samples, and modeling suggests that its repeat unit may bind to the minor groove of the DNA double helix (Wang et al. 2015). Fiber-like PSL strands have been visualized connecting microcolonies in the biofilms of laboratory strain PAO1 and at the base of mushroom-shaped biofilms grown in a flow-cell chamber (Wang et al. 2013). Fiber formation in both cases requires functional type 4 pili. Trails of PSL were found on surfaces after bacterial attachment and migration (Wang et al. 2013), and these trails were linked to migration of subsequent cells (Zhao et al. 2013). PSL production thus promotes microcolony formation by promoting colocalization of low and high producers by directing migration on surfaces (Zhao et al. 2013).

### *1.4.3 PNAG*

Poly-β-1,6-*N*-acetyl-D-glucosamine (PNAG) is produced by many Gram-negative and Gram-positive bacteria (Table 1.1) (Cywes-Bentley et al. 2013). PNAG was first identified over 20 years ago and originally termed polysaccharide intercellular adhesin (PIA) in Gram-positive systems (Mack et al. 1996). PNAG production has subsequently been identified in *E. coli*, *Bordetella* sp., *Actinobacillus pleuropneumoniae*, *Burkholderia cepacia* complex (Bcc), *Staphylococcus aureus*, *Staphylococcus epidermidis*, *Yersinia pestis*, and *Aggregatibacter actinomycetemcomitans*, among others (Cramton et al. 1999; Wang et al. 2004; Itoh et al. 2004; Vuong et al. 2004; Parise et al. 2007; Izano et al. 2007, 2008; Bobrov et al. 2008).

#### 1.4.3.1 Biosynthesis

PNAG biosynthesis has been investigated in both Gram-positive and Gram-negative species. The biosynthetic systems are different due to the difference in cell architecture, but some similarities exist. In both cases the polysaccharide is produced in a synthase-dependent manner with four proteins encoded within a single operon. The most well-studied systems are those of *E. coli* and *S. epidermidis*, which will be described here.

In *S. epidermidis* PNAG is synthesized by the products of *ica* operon. This operon encodes four genes, *icaADBC*. IcaA is an integral membrane protein with multiple transmembrane domains and a cytosolic domain belonging to glycosyltransferase family 2 (GT-2) (Heilmann et al. 1996; Gerke et al. 1998). IcaA is predicted to synthesize the polysaccharide and transport it through the membrane, potentially with the help of IcaD. Co-expression of IcaD with IcaA increases glycosyltransferase activity and may also aid in translocation (Gerke et al. 1998). Using *Staphylococcus carnosus*, a PNAG negative strain, transformed with ica genes it was shown that expression of *icaAD* alone were sufficient for PNAG synthesis. In this organism the polymer is 20 residues per chain or less, as opposed to over 130 residues/chain produced by strains expressing the whole operon (Mack et al. 1996; Gerke et al. 1998). Expression of *icaC* is required for production of longer PNAG (Gerke et al. 1998). Gerke et al. have proposed that IcaC completes the membrane complex and stabilizes the synthase. The final protein encoded in the operon, IcaB, is not required for the production of PNAG but for its modification (see below).

In *E. coli* the *pgaABCD* operon is responsible for PNAG synthesis (Wang et al. 2004). Homologous operons are found in other Gram-negative systems including the *Bordetella bpsABCD* (*Bordetella* polysaccharide) (Parise et al. 2007) and *Yersinia hmsHFRS* (hemin storage) operons, among others (Darby et al. 2002; Bobrov et al. 2008). Polysaccharide synthesis starts at the inner membrane with PgaCD. PgaC is the glycosyltransferase, with multiple transmembrane domains and a cytosolic GT-2 domain (Itoh et al. 2008). PgaC has a homologous function to IcaA. PgaD is a small integral membrane protein with two predicted transmembrane helices that does not belong to a known protein family (Steiner et al. 2013). PgaC and PgaD interact in a cyclic-di-GMP-dependent manner to create the active glycosyltransferase and inner-membrane transporter (Steiner et al. 2013). PgaB, similar to PelA (see above), is a periplasmic two-domain protein. However, in the case of PgaB, this protein is lipidated and attached to the inner leaflet of the outer membrane and contains a N-terminal deacetylase domain and a C-terminal PNAG-binding domain (Itoh et al. 2008; Little et al. 2012, 2014b). PgaA is the outer-membrane export protein with a C-terminal 16 stranded beta-barrel porin and an N-terminal TPR domain (Itoh et al. 2008; Wang et al. 2016). The TPR has recently been demonstrated to interact with PgaB (Wang et al. 2016). This interaction is necessary for PNAG secretion and has also been found in the *Y. pestis* system between the homologous proteins HmsHF (Abu Khweek et al. 2010). As it traverses the periplasm PNAG has been proposed to bind PgaB, which partially deacetylates the polymer (see below) and then guides it to the TPR domain of PgaA for subsequent export out of the cell via PgaA's beta-barrel porin (Little et al. 2014b; Wang et al. 2016).

#### 1.4.3.2 Modification

PNAG can be modified during synthesis by de-*N*-acetylation and *O*-succinylation (Mack et al. 1996; Heilmann et al. 1996; Whitfield et al. 2015). Deacetylation occurs in both Gram-positive and Gram-negative bacteria where as succinylation is specific to Gram-positive species (Whitfield et al. 2015). PNAG deacetylation is necessary for biofilm formation and gives PNAG its adhesive properties. The effects of succinylation have not been studied, but evidence suggests that succinylation may be used to modulate the anionic charge of PNAG leading to changes in fitness of *S. aureus* (Brooks and Jefferson 2014).

When PNAG was first purified from *S. epidermidis* cultures, it was isolated in two fractions (Mack et al. 1996). In the larger fraction, which represented over 80 % of the sample, the PNAG polymer was 15–20 % deacetylated. The second fraction was slightly anionic and was found to contain deacetylated PNAG with *O*-succinylation modifications. Atkin et al. have suggested that IcaC is responsible for the *O*-succinylation of PNAG due to its similarities to members of the *O*-acetyltransferase family of proteins and the absence of a homologous protein in Gram-negative PNAG systems (Atkin et al. 2014). In *S. epidermidis* deacetylation is required for PNAG adherence to the cell surface. PNAG produced by a Δ*icaB* strain is fully acetylated and does not bind to the cell. Deacetylation has also been shown to be necessary for surface colonization of polystyrene by *S. epidermidis* (Vuong et al. 2004). IcaB is a single-domain protein belonging to the carbohydrate esterase 4 (CE4) family of enzymes (Little et al. 2014a). The structure of IcaB from *Ammonifex degensii* revealed an active site groove that could potentially bind a trisaccharide (Little et al. 2014a). The structure was solved after the deletion of a hydrophobic loop that is not present in the CE4 domains of Gram-negative PNAG deacetylases. This loop as well as an electropositive patch on the protein has been suggested to play a role in cell–surface association.

Characterization of PNAG from *E. coli* revealed that this polymer is 3–5 % deacetylated and that the modification is required for export of the polymer and biofilm formation (Itoh et al. 2008). Although PgaB is a two domain protein, like IcaB, its N-terminal domain exhibits deacetylase activity and is a member of the CE4 family (Itoh et al. 2008; Little et al. 2012). Inactive point mutants of the CE4 domain or deletion of *pgaB* leads to a buildup of fully acetylated PNAG in the periplasm (Itoh et al. 2008). The C-terminal domain of PgaB has been shown to preferentially bind to deacetylated PNAG (Little et al. 2014b). This domain is proposed to guide the polymer to PgaA for export and suggests that the positive charge produced by deacetylation is required for trafficking. Recent structure–function studies of the PgaA porin have revealed that electronegative residues within the porin and on the periplasmic surface are important for efficient export

(Wang et al. 2016). Together these data suggest that in *E. coli* PNAG modification is necessary for polysaccharide recognition by the export machinery.

#### 1.4.3.3 Interactions and Functions

PNAG has been linked to antibiotic resistance, immune evasion, surface attachment, cellular cohesion, and detergent resistance (Heilmann et al. 1996; Itoh et al. 2004; Agladze et al. 2005; Izano et al. 2007, 2008). PNAG is required for adherence of *S. aureus* to human nasal epithelial cells. A *S. aureus* Δ*ica* strain had much lower adherence than the wild-type strain and was comparable to the adherence demonstrated by the *S. carnosus* PNAG-deficient strain (Lin et al. 2015). In a mouse lung infection model, bacterial burden is also significantly decreased in the *S. aureus* Δ*ica* strain compared to wild type (Lin et al. 2015). In *S. epidermidis*, PNAG extracellular matrix production increases resistance to human antibacterial peptides and phagocytic killing leading to heightened immune evasion (Vuong et al. 2004).

Although the production of PNAG and phenotypes of deficient strains has been investigated thoroughly, identification of specific interactions is not well understood. Investigation into cell–surface interactions between PNAG and *S. aureus* has determined that cell wall teichoic acid was not involved (Vergara-Irigaray et al. 2008). The same study also eliminated lipoteichoic acid and lysine–proline–X–threonine–glycine (LPXTG)-anchored proteins as being responsible for PNAG binding. In vitro analysis showed that PNAG binds purified peptidoglycan, but the importance of this interaction is not known (Vergara-Irigaray et al. 2008).

### *1.4.4 Holdfast*

The holdfast polymer is located at the poles of *Caulobacter crescentus* and other closely related alphaproteobacteria and is essential for surface attachment and biofilm formation in these species (Merker and Smit 1988; Ong et al. 1990; Mitchell and Smit 1990; Wan et al. 2013). The holdfast polymer contains partially deacetylated β-1,4 linked GlcNAc sugars, although the exact composition is unknown (Smith et al. 2003; Toh et al. 2008; Wan et al. 2013). It's suspected that there are other components since digestion with lysozyme weakens attachment but doesn't disrupt it fully (Li et al. 2005; Tsang et al. 2006). The polysaccharide is produced and secreted by the genes in the *hfsDABC* and *hfsEFGH* operons (Smith et al. 2003; Toh et al. 2008) and is bound to the cell surface by a protein complex, termed the holdfast anchor (Kurtz and Smith 1992; Kurtz et al. 1994). The anchor is encoded by the *hfaABDC* operon. Expression of *hfaABD* is required for holdfast attachment, but not *hfaC* (Cole et al. 2003). Both HfaA and HfaD proteins associate to make high molecular weight homomultimers that require HfaB for stability (Hardy et al. 2010). HfaA has been proposed to be the connection between the holdfast polymer and HfaD, which is situated in the cell membrane thus linking the polymer to the cell (Kurtz and Smith 1992; Kurtz et al. 1994). In the absence of the anchor holdfast doesn't adhere to the cell and is shed into the extracellular milieu (Kurtz et al. 1994; Cole et al. 2003; Hardy et al. 2010). Deletion of *hfaB* had the most dramatic holdfast-shedding phenotype leading to a model of holdfast anchoring that includes HfaB transferring both HfaA and HfaD across the outer membrane before anchor formation (Cole et al. 2003; Hardy et al. 2010).

#### 1.4.4.1 Biosynthesis

Holdfast synthesis is encoded on the *hfs* locus consisting of eight genes. The locus consists of two adjacent operons: *hfsDABC* and *hfsEFGH* (Toh et al. 2008). The holdfast genes encode proteins homologous to the Wzx/Wzy-dependent polysaccharide synthesis system. HfsE is a UDP-GlcNAc transferase that transfers the first monomer onto the lipid carrier. *hfsE* is not required for holdfast synthesis as PssY and PssZ are also capable of initiating synthesis of the repeat unit (Toh et al. 2008). PssY and PssZ were identified by screening the genome for HfsE homologues and may have roles in production of O-antigen. After transfer of the first GlcNAc to the lipid carrier, HfsG, the putative glycosyltransferase adds additional sugar residues elongating the repeat unit (Toh et al. 2008). The length of the repeating unit and exact structure of the polymer are not known. HfsH is predicted to be a deacetylase that is proposed to modify the repeat unit before it is flipped across the inner membrane by HfsF (Toh et al. 2008). Deletion of *hfsF* produces a strain with near wild-type adhesion and holdfast, suggesting that there is another flippase in the *C. crescentus* genome that can compensate. To date, the redundant flippase has not been identified as flippases being primarily alpha-helical and without any strongly conserved motifs are difficult to identify using bioinformatics. Once in the periplasm, the repeat unit is polymerized by either HfsC or HfsI. These genes are redundant. *hfsC* is located in the *hfs* gene cluster, while *hfsI* is located elsewhere on the chromosome. Deletion of both genes is required to abrogate adherence to a polystyrene surface; a phenotype that could be restored when the double mutant was complemented with *hfsC* (Toh et al. 2008). HfsDAB are responsible for the transport of the polymer (Toh et al. 2008). HfsA is an inner-membrane protein that is thought to function similarly to *E. coli* Wzc (Smith et al. 2003). HfsB is a tyrosine autokinase necessary for holdfast export and may function in concert with HfsA (Toh et al. 2008), while HfsD is the Wza equivalent, the outer-membrane auxiliary protein through which the polymer is exported (Smith et al. 2003; Toh et al. 2008).

#### 1.4.4.2 Modification

Mature holdfast is partially deacetylated by HfsH giving it a net positive charge. Holdfast deacetylation occurs in the cytoplasm before the repeat unit is flipped across the inner membrane (Toh et al. 2008; Wan et al. 2013). HfsH is a polysaccharide deacetylase belonging to the same CE4 family as PgaB/IcaB and PelA from the PNAG and PEL systems, respectively (Toh et al. 2008). Deletion or mutation of *hfsH* leads to a decrease in cellular adherence and coherence between cells (see below) (Toh et al. 2008; Wan et al. 2013). In these strains holdfast is still produced but does not anchor to the cell membrane (Wan et al. 2013). The holdfast polysaccharide produced by *hfsH* mutants also appears to be more diffuse and smaller than wild type when visualized using epifluorescence. Comparisons of purified holdfast from wild-type and Δ*hfsH* strains have also demonstrated the impact of deacetylation on adhesion (Berne et al. 2013). In vitro experiments with fully acetylated holdfast reveal 30–40 % lower binding affinity for both clean and coated glass. Interestingly, the binding of mature holdfast to abiotic surfaces was found to be modulated by pH and salinity, but when the polymer is fully acetylated, these trends are altered. The binding of acetylated holdfast is not affected by salinity, and the affinity was highest at slightly acidic pHs (pH 4–5) in contrast to the mature deacetylated holdfast where binding is strongest at neutral pH. Deacetylation of holdfast is thought to be required for proper tertiary structure formation of the sugar (Wan et al. 2013). Overexpression of *hfsH* creates a hyper-adhesive strain further supporting the importance of this modification in adhesion.

#### 1.4.4.3 Interactions and Functions

Holdfast is able to bind to a variety of both hydrophobic and hydrophilic surfaces. In vivo, holdfast contributes to the irreversible binding of cells during biofilm formation. Tsang et al. showed that single-cell adhesion was in the μN range when bound to borosilicate (Tsang et al. 2006). The strength of this binding is much greater than most biological adhesins. Although GlcNAc polymers are not the only component of holdfast, they play an integral role in adherence and structure as treatment of holdfast with lysozyme decreases the force constant measured by AFM by 90 % (Li et al. 2005). After initial binding holdfast attachment can strengthen, as the holdfast reaches maximum binding diameter and probably goes through a curing process in which the material becomes less fluid like (Li et al. 2005; Berne et al. 2013).

The unique anchoring of holdfast to the cell surface allowed for isolation of wild-type holdfast from the Δ*hfaB* strain (Berne et al. 2013). Isolation of holdfast in this manner led to studies of adhesion and kinetic binding parameters to multiple substrates (see above). Berne et al. used AFM to investigate the contact angle of holdfast on hydrophobic and hydrophilic surfaces. Initial binding to graphite, a hydrophobic surface, was found to be stronger than mica (hydrophilic) based on the lower contact angle measured. Holdfast binding was also lowered with increasing NaCl on clean glass but was unaffected by salt on treated glass. This difference in binding parameters suggests that there are multiple modes of holdfast binding through hydrophobic and electrostatic interactions.

## 1.5 Exopolysaccharide Adhesives in Infection

With the increased study of biofilms and their relevance in bacterial virulence, the mechanisms behind biofilm formation have become targets for new therapeutics and diagnostics. Exopolysaccharides have been shown to be essential for biofilm formation in many species where they play roles in biofilm initiation and structure. Strains deficient in exopolysaccharide production show decreased virulence in multiple animal models (Vuong et al. 2004; Kropec et al. 2005; Whitfield et al. 2015). Multiple different avenues are being explored therapeutically including carbohydrates for vaccine development, antibody therapy targeted against the polymers, and different methods to disrupt the polysaccharide component of the biofilm (Kostakioti et al. 2013; Haas 2013; Baker et al. 2016). The following are two specific examples of current developments for the polymers described herein that show promise as the treatment of biofilm-related infections. Polyclonal antibodies specific to PNAG have been used to show opsonic killing of multiple PNAG producing species suggesting that vaccination with pure PNAG oligomers may increase immune response to a wide range of pathogens (Cywes-Bentley et al. 2013). The use of a monoclonal antibody against PNAG as potential immunotherapy is currently being explored by Dr. Gerald Pier's group with Dr. Daniel Vlock (Cywes-Bentley et al. 2013; Haas 2013). Boosting immune response through biofilm disruption has also been explored recently (Baker et al. 2016). Glycoside hydrolases specific to PSL and PEL polysaccharides have been shown to disrupt *P. aeruginosa* biofilms in vitro and promote neutrophil killing as well as antibiotic susceptibility (Baker et al. 2016).

## 1.6 Conclusion

Exopolysaccharides play a key role in adhesion and virulence in many pathogenic and nonpathogenic organisms. As the pace of identifying and characterizing exopolysaccharides increases, combating chronic biofilm-related infections may become easier. Not only do multiple organisms produce similar polymers, but the mechanisms by which the various polymers are synthesized and modified share a great deal in common. Understanding these mechanism at the molecular level will be key to the design of novel therapeutics and may increase the effectiveness of current antibiotics.

**Acknowledgments** This work was supported in part by the grants from the Canadian Institutes of Health Research (CIHR) to PLH (#43998, #13337 and #81361). PLH is a recipient of Canada Research Chair. N.C.B. has been supported in part by graduate scholarships from the *Natural Sciences and Engineering Research Council of Canada* (NSERC), Mary H. Beatty, and Dr. James A. and Connie P. Dickson Scholarships from the University of Toronto, Cystic Fibrosis Canada, and The Hospital for Sick Children.

# References


Abu Khweek A, Fetherston JD, Perry RD (2010) Analysis of HmsH and its role in plague biofilm formation. Microbiology 156:1424–1438. doi:10.1099/mic.0.036640-0

Agladze K, Wang X, Romeo T (2005) Spatial periodicity of *Escherichia coli* K-12 biofilm microstructure initiates during a reversible, polar attachment phase of development and requires the polysaccharide adhesin PGA. J Bacteriol 187:8237–8246. doi:10.1128/JB.187.24.8237-8246.2005

Atkin KE, MacDonald SJ, Brentnall AS et al (2014) A different path: revealing the function of staphylococcal proteins in biofilm formation. FEBS Lett 588:1869–1872. doi:10.1016/j.febslet.2014.04.002

Baker P, Whitfield GB, Hill PJ, et al (2015a) Characterization of the *Pseudomonas aeruginosa* glycoside hydrolase PslG reveals that its levels are critical for Psl polysaccharide biosynthesis and biofilm formation. J Biol Chem 290:28374–28387

Baker P, Whitfield GB, Hill PJ et al (2015a) Characterization of the *Pseudomonas aeruginosa* glycoside hydrolase PslG reveals that its levels are critical for Psl polysaccharide biosynthesis and biofilm formation. J Biol Chem 290:28374–28387. doi:10.1074/jbc.M115.674929

Baker P, Hill PJ, Snarr BD et al (2016) Exopolysaccharide biosynthetic glycoside hydrolases can be utilized to disrupt and prevent *Pseudomonas aeruginosa* biofilms. Sci Adv 2:e1501632. doi:10.1126/sciadv.1501632

Berne C, Ma X, Licata NA et al (2013) Physiochemical properties of *Caulobacter crescentus* holdfast: a localized bacterial adhesive. J Phys Chem B 117:10492–10503. doi:10.1021/jp405802e

Bluhm TL, Sarko A (1977) The triple helical structure of lentinan, a linear β-(1→3)-D-glucan. Can J Chem 55:293–299. doi:10.1139/v77-044

Bobrov AG, Kirillina O, Forman S et al (2008) Insights into *Yersinia pestis* biofilm development: topology and co-interaction of Hms inner membrane proteins involved in exopolysaccharide production. Environ Microbiol 10:1419–1432. doi:10.1111/j.1462-2920.2007.01554.x

Borlee BR, Goldman AD, Murakami K et al (2010) *Pseudomonas aeruginosa* uses a cyclic-di-GMP-regulated adhesin to reinforce the biofilm extracellular matrix. Mol Microbiol 75:827–842. doi:10.1111/j.1365-2958.2009.06991.x

Brooks JL, Jefferson KK (2014) Phase variation of poly-N-acetylglucosamine expression in *Staphylococcus aureus*. PLoS Pathog 10:e1004292. doi:10.1371/journal.ppat.1004292

Byrd MS, Sadovskaya I, Vinogradov E et al (2009) Genetic and biochemical analyses of the *Pseudomonas aeruginosa* Psl exopolysaccharide reveal overlapping roles for polysaccharide synthesis enzymes in Psl and LPS production. Mol Microbiol 73:622–638. doi:10.1111/j.1365-2958.2009.06795.x

Cérantola S, Marty N, Montrozier H (1996) Structural studies of the acidic exopolysaccharide produced by a mucoid strain of *Burkholderia cepacia*, isolated from cystic fibrosis. Carbohydr Res 285:59–67. doi:10.1016/S0008-6215(96)90170-6

Cérantola S, Lemassu-Jacquier A, Montrozier H (2001) Structural elucidation of a novel exopolysaccharide produced by a mucoid clinical isolate of *Burkholderia cepacia*. Eur J Biochem 260:373–383. doi:10.1046/j.1432-1327.1999.00171.x

Ciucanu I (2006) Per-O-methylation reaction for structural analysis of carbohydrates by mass spectrometry. Anal Chim Acta 576:147–155. doi:10.1016/j.aca.2006.06.009

Cole JL, Hardy GG, Bodenmiller D et al (2003) The HfaB and HfaD adhesion proteins of *Caulobacter crescentus* are localized in the stalk. Mol Microbiol 49:1671–1683

Colvin KM, Gordon VD, Murakami K et al (2011) The pel polysaccharide can serve a structural and protective role in the biofilm matrix of *Pseudomonas aeruginosa*. PLoS Pathog 7: e1001264. doi:10.1371/journal.ppat.1001264

Colvin KM, Alnabelseya N, Baker P et al (2013) PelA deacetylase activity is required for Pel polysaccharide synthesis in *Pseudomonas aeruginosa*. J Bacteriol 195:2329–2339. doi:10.1128/JB.02150-12

Cramton SE, Gerke C, Schnell NF et al (1999) The intercellular adhesion (ica) locus is present in *Staphylococcus aureus* and is required for biofilm formation. Infect Immun 67:5427–5433

Cuthbertson L, Mainprize IL, Naismith JH, Whitfield C (2009) Pivotal roles of the outer membrane polysaccharide export and polysaccharide copolymerase protein families in export of extracellular polysaccharides in gram-negative bacteria. Microbiol Mol Biol Rev 73:155–177. doi:10.1128/MMBR.00024-08

Cywes-Bentley C, Skurnik D, Zaidi T et al (2013) Antibody to a conserved antigenic target is protective against diverse prokaryotic and eukaryotic pathogens. Proc Natl Acad Sci USA 110: E2209–E2218. doi:10.1073/pnas.1303573110

Darby C, Hsu JW, Ghori N, Falkow S (2002) *Caenorhabditis elegans*: plague bacteria biofilm blocks food intake. Nature 417:243–244. doi:10.1038/417243a

Donlan RM (2002) Biofilms: microbial life on surfaces. Emerg Infect Dis 8:881–890. doi:10.3201/eid0809.020063

Drummelsmith J, Whitfield C (1999) Gene products required for surface expression of the capsular form of the group 1 K antigen in *Escherichia coli* (O9a:K30). Mol Microbiol 31:1321–1332

Flemming H-C, Wingender J (2010) The biofilm matrix. Nat Rev Microbiol 8:623–633. doi:10.1038/nrmicro2415

Forman S, Bobrov AG, Kirillina O et al (2006) Identification of critical amino acid residues in the plague biofilm Hms proteins. Microbiology 152:3399–3410. doi:10.1099/mic.0.29224-0

Franklin MJ, Nivens DE, Weadge JT, Howell PL (2011) Biosynthesis of the *Pseudomonas aeruginosa* extracellular polysaccharides, alginate, Pel, and Psl. Front Microbiol 2:167. doi:10.3389/fmicb.2011.00167

Friedman L, Kolter R (2004a) Two genetic loci produce distinct carbohydrate-rich structural components of the *Pseudomonas aeruginosa* biofilm matrix. J Bacteriol 186:4457–4465. doi:10.1128/JB.186.14.4457-4465.2004

Friedman L, Kolter R (2004b) Genes involved in matrix formation in *Pseudomonas aeruginosa* PA14 biofilms. Mol Microbiol 51:675–690. doi:10.1046/j.1365-2958.2003.03877.x

Gerke C, Kraft A, Sussmuth R et al (1998) Characterization of the N-acetylglucosaminyltransferase activity involved in the biosynthesis of the *Staphylococcus epidermidis* polysaccharide intercellular adhesin. J Biol Chem 273:18586–18593

Ghafoor A, Jordens Z, Rehm BHA (2013) Role of PelF in pel polysaccharide biosynthesis in *Pseudomonas aeruginosa*. Appl Environ Microbiol 79:2968–2978. doi:10.1128/AEM.03666-12

Grachev AA, Gerbst AG, Gening ML et al (2011) NMR and conformational studies of linear and cyclic oligo-(1→6)-β-D-glucosamines. Carbohydr Res 346:2499–2510. doi:10.1016/j.carres.2011.08.031

Greenfield LK, Whitfield C (2012) Synthesis of lipopolysaccharide O-antigens by ABC transporter-dependent pathways. Carbohydr Res 356:12–24. doi:10.1016/j.carres.2012.02.027

Guetta O, Milas M, Rinaudo M (2003) Structure and properties of a bacterial polysaccharide from a Klebsiella strain (ATCC 12657). Biomacromolecules 4:1372–1379. doi:10.1021/bm030036u

Haas MJ (2013) PNAG: broadening infection protection. Sci-Bus eXchange. doi:10.1038/scibx.2013.566

Hall-Stoodley L, Costerton JW, Stoodley P (2004) Bacterial biofilms: from the natural environment to infectious diseases. Nat Rev Microbiol 2:95–108. doi:10.1038/nrmicro821

Hardy GG, Allen RC, Toh E et al (2010) A localized multimeric anchor attaches the Caulobacter holdfast to the cell pole. Mol Microbiol 76:409–427. doi:10.1111/j.1365-2958.2010.07106.x

Heilmann C, Schweitzer O, Gerke C et al (1996) Molecular basis of intercellular adhesion in the biofilm-forming *Staphylococcus epidermidis*. Mol Microbiol 20:1083–1091. doi:10.1111/j.1365-2958.1996.tb02548.x

Islam ST, Lam JS (2014) Synthesis of bacterial polysaccharides via the Wzx/Wzy-dependent pathway. Can J Microbiol 60:697–716. doi:10.1139/cjm-2014-0595

Itoh Y, Wang X, Hinnebusch BJ et al (2004) Depolymerization of -1,6-N-Acetyl-D-glucosamine disrupts the integrity of diverse bacterial biofilms. J Bacteriol 187:382–387. doi:10.1128/JB.187.1.382-387.2005

Itoh Y, Rice JD, Goller C et al (2008) Roles of pgaABCD genes in synthesis, modification, and export of the *Escherichia coli* biofilm adhesin poly-beta-1,6-N-acetyl-D-glucosamine. J Bacteriol 190:3670–3680. doi:10.1128/JB.01920-07

Izano EA, Sadovskaya I, Vinogradov E et al (2007) Poly-N-acetylglucosamine mediates biofilm formation and antibiotic resistance in *Actinobacillus pleuropneumoniae*. Microb Pathog 43:1–9. doi:10.1016/j.micpath.2007.02.004

Izano EA, Amarante MA, Kher WB, Kaplan JB (2008) Differential roles of poly-N-acetylglucosamine surface polysaccharide and extracellular DNA in *Staphylococcus aureus* and *Staphylococcus epidermidis* biofilms. Appl Environ Microbiol 74:470–476. doi:10.1128/AEM.02073-07

Jackson KD, Starkey M, Kremer S et al (2004) Identification of psl, a locus encoding a potential exopolysaccharide that is essential for *Pseudomonas aeruginosa* PAO1 biofilm formation. J Bacteriol 186:4466–4475. doi:10.1128/JB.186.14.4466-4475.2004

Jennings LK, Storek KM, Ledvina HE et al (2015) Pel is a cationic exopolysaccharide that cross-links extracellular DNA in the *Pseudomonas aeruginosa* biofilm matrix. Proc Natl Acad Sci USA 112:11353–11358. doi:10.1073/pnas.1503058112

Keiski C-L, Harwich M, Jain S et al (2010) AlgK is a TPR-containing protein and the periplasmic component of a novel exopolysaccharide secretin. Structure 18:265–273. doi:10.1016/j.str.2009.11.015

Kostakioti M, Hadjifrangiskou M, Hultgren SJ (2013) Bacterial biofilms: development, dispersal, and therapeutic strategies in the dawn of the postantibiotic era. Cold Spring Harb Perspect Med 3:a010306. doi:10.1101/cshperspect.a010306

Kowalska K, Soscia C, Combe H et al (2010) The C-terminal amphipathic α-helix of *Pseudomonas aeruginosa* PelC outer membrane protein is required for its function. Biochimie 92:33–40. doi:10.1016/j.biochi.2009.10.004

Kropec A, Maira-Litrán T, Jefferson KK et al (2005) Poly-N-acetylglucosamine production in *Staphylococcus aureus* is essential for virulence in murine models of systemic infection. Infect Immun 73:6868–6876. doi:10.1128/IAI.73.10.6868-6876.2005

Kurtz HD, Smith J (1992) Analysis of a *Caulobacter crescentus* gene cluster involved in attachment of the holdfast to the cell. J Bacteriol 174:687–694

Kurtz HD, Smit J, Smith J (1994) The *Caulobacter crescentus* holdfast: identification of holdfast attachment complex genes. FEMS Microbiol Lett 116:175–182. doi:10.1111/j.1574-6968.1994.tb06697.x

Larsen FH, Engelsen SB (2015) Insight into the functionality of microbial exopolysaccharides by NMR spectroscopy and molecular modeling. Front Microbiol 6:1374. doi:10.3389/fmicb.2015.01374

Lee VT, Matewish JM, Kessler JL et al (2007) A cyclic-di-GMP receptor required for bacterial exopolysaccharide production. Mol Microbiol 65:1474–1484. doi:10.1111/j.1365-2958.2007.05879.x

Li G, Smith CS, Brun YV, Tang JX (2005) The elastic properties of the *Caulobacter crescentus* adhesive holdfast are dependent on oligomers of N-acetylglucosamine. J Bacteriol 187:257–265. doi:10.1128/JB.187.1.257-265.2005

Lin MH, Shu JC, Lin LP et al (2015) Elucidating the crucial role of poly N-acetylglucosamine from *Staphylococcus aureus* in cellular adhesion and pathogenesis. PLoS One 10:e0124216. doi:10.1371/journal.pone.0124216

Little DJ, Poloczek J, Whitney JC et al (2012) The structure- and metal-dependent activity of *Escherichia coli* PgaB provides insight into the partial de-N-acetylation of poly-beta-1,6-N-acetyl-D-glucosamine. J Biol Chem 287:31126–31137. doi:10.1074/jbc.M112.390005

Little DJ, Bamford NC, Pokrovskaya V et al (2014a) Structural basis for the De-N-acetylation of Poly-β-1,6-N-acetyl-D-glucosamine in Gram-positive bacteria. J Biol Chem 289:35907–35917. doi:10.1074/jbc.M114.611400

Little DJ, Li G, Ing C et al (2014b) Modification and periplasmic translocation of the biofilm exopolysaccharide poly-beta-1,6-N-acetyl-D-glucosamine. Proc Natl Acad Sci USA 111:11013–11018. doi:10.1073/pnas.1406388111

Ma L, Lu H, Sprinkle A et al (2007) *Pseudomonas aeruginosa* Psl is a galactose- and mannose-rich exopolysaccharide. J Bacteriol 189:8353–8356. doi:10.1128/JB.00620-07

Ma L, Conover M, Lu H et al (2009) Assembly and development of the *Pseudomonas aeruginosa* biofilm matrix. PLoS Pathog 5:e1000354. doi:10.1371/journal.ppat.1000354

Mack D, Fischer W, Krokotsch A et al (1996) The intercellular adhesin involved in biofilm accumulation of *Staphylococcus epidermidis* is a linear beta-1,6-linked glucosaminoglycan: purification and structural analysis. J Bacteriol 178:175–183

Manzi AE (1995) Journal of Chromatography Library, Volume 58, Carbohydrate Analysis: High-performance liquid chromatography and capillary electrophoresis, Edited by Z. El Rassi, Elsevier Science, Amsterdam, The Netherlands, 1995, 672 p., $265.75. Anal Biochem 230:358. doi:10.1006/abio.1995.1490

Matsukawa M, Greenberg EP (2004) Putative exopolysaccharide synthesis genes influence *Pseudomonas aeruginosa* biofilm development. J Bacteriol 186:4449–4456. doi:10.1128/JB.186.14.4449-4456.2004

Merker RI, Smit J (1988) Characterization of the adhesive holdfast of marine and freshwater caulobacters. Appl Environ Microbiol 54:2078–2085

Merkle RK, Poppe I (1994) Carbohydrate composition analysis of glycoconjugates by gas-liquid chromatography/mass spectrometry. Methods Enzymol 230:1–15

Mitchell D, Smit J (1990) Identification of genes affecting production of the adhesion organelle of *Caulobacter crescentus* CB2. J Bacteriol 172:5425–5431

Neu TR, Lawrence JR (2014) Investigation of microbial biofilm structure by laser scanning microscopy. Adv Biochem Eng Biotechnol 146:1–51. doi:10.1007/10_2014_272

Neu T, Swerhone GD, Lawrence JR (2001) Assessment of lectin-binding analysis for in situ detection of glycoconjugates in biofilm systems. Microbiology 147:299–313. doi:10.1099/00221287-147-2-299

Ong CJ, Wong ML, Smit J (1990) Attachment of the adhesive holdfast organelle to the cellular stalk of *Caulobacter crescentus*. J Bacteriol 172:1448–1456

Ovodova RG, Ovodov YS (1969) The pectic substances of Zosteraceae. Carbohydr Res 10:387–390. doi:10.1016/S0008-6215(00)80898-8

Parise G, Mishra M, Itoh Y et al (2007) Role of a putative polysaccharide locus in Bordetella biofilm development. J Bacteriol 189:750–760. doi:10.1128/JB.00953-06

Price NPJ (2008) Permethylation linkage analysis techniques for residual carbohydrates. Appl Biochem Biotechnol 148:271–276. doi:10.1007/s12010-007-8044-8

Ryder C, Byrd M, Wozniak DJ (2007) Role of polysaccharides in *Pseudomonas aeruginosa* biofilm development. Curr Opin Microbiol 10:644–648. doi:10.1016/j.mib.2007.09.010

Smith CS, Hinz A, Bodenmiller D et al (2003) Identification of genes required for synthesis of the adhesive holdfast in *Caulobacter crescentus*. J Bacteriol 185:1432–1442. doi:10.1128/JB.185.4.1432-1442.2003

Steiner S, Lori C, Boehm A, Jenal U (2013) Allosteric activation of exopolysaccharide synthesis through cyclic di-GMP-stimulated protein-protein interaction. EMBO J 32:354–368. doi:10.1038/emboj.2012.315

Toh E, Kurtz HD, Brun YV (2008) Characterization of the *Caulobacter crescentus* holdfast polysaccharide biosynthesis pathway reveals significant redundancy in the initiating glycosyltransferase and polymerase steps. J Bacteriol 190:7219–7231. doi:10.1128/JB.01003-08

Tsang PH, Li G, Brun YV et al (2006) Adhesion of single bacterial cells in the micronewton range. Proc Natl Acad Sci USA 103:5764–5768. doi:10.1073/pnas.0601705103

Vasseur P, Vallet-Gely I, Soscia C et al (2005) The pel genes of the *Pseudomonas aeruginosa* PAK strain are involved at early and late stages of biofilm formation. Microbiology 151:985–997. doi:10.1099/mic.0.27410-0

Vasseur P, Soscia C, Voulhoux R, Filloux A (2007) PelC is a *Pseudomonas aeruginosa* outer membrane lipoprotein of the OMA family of proteins involved in exopolysaccharide transport. Biochimie 89:903–915. doi:10.1016/j.biochi.2007.04.002

Vergara-Irigaray M, Maira-Litran T, Merino N et al (2008) Wall teichoic acids are dispensable for anchoring the PNAG exopolysaccharide to the *Staphylococcus aureus* cell surface. Microbiology 154:865–877. doi:10.1099/mic.0.2007/013292-0

Vuong C, Kocianova S, Voyich JM et al (2004) A crucial role for exopolysaccharide modification in bacterial biofilm formation, immune evasion, and virulence. J Biol Chem 279:54881–54886. doi:10.1074/jbc.M411374200

Wan Z, Brown PJB, Elliott EN, Brun YV (2013) The adhesive and cohesive properties of a bacterial polysaccharide adhesin are modulated by a deacetylase. Mol Microbiol 88:486–500. doi:10.1111/mmi.12199

Wang X, Preston JF, Romeo T (2004) The pgaABCD locus of *Escherichia coli* promotes the synthesis of a polysaccharide adhesin required for biofilm formation. J Bacteriol 186:2724–2734

Wang S, Parsek MR, Wozniak DJ, Ma LZ (2013) A spider web strategy of type IV pili-mediated migration to build a fibre-like Psl polysaccharide matrix in *Pseudomonas aeruginosa* biofilms. Environ Microbiol 15:2238–2253. doi:10.1111/1462-2920.12095

Wang S, Liu X, Liu H et al (2015) The exopolysaccharide Psl-eDNA interaction enables the formation of a biofilm skeleton in *Pseudomonas aeruginosa*. Environ Microbiol Rep 7:330–340. doi:10.1111/1758-2229.12252

Wang Y, AndolePannuri A, Ni D et al (2016) Structural basis for translocation of a biofilm-supporting exopolysaccharide across the bacterial outer membrane. J Biol Chem 291:10046–10057. doi:10.1074/jbc.M115.711762

Whitfield C (2006) Biosynthesis and assembly of capsular polysaccharides in *Escherichia coli*. Annu Rev Biochem 75:39–68. doi:10.1146/annurev.biochem.75.103004.142545

Whitfield GB, Marmont LS, Howell PL (2015) Enzymatic modifications of exopolysaccharides enhance bacterial persistence. Front Microbiol 6:471. doi:10.3389/fmicb.2015.00471

Whitney JC, Howell PL (2013) Synthase-dependent exopolysaccharide secretion in Gram-negative bacteria. Trends Microbiol 21:63–72. doi:10.1016/j.tim.2012.10.001

Whitney JC, Colvin KM, Marmont LS et al (2012) Structure of the cytoplasmic region of PelD, a degenerate diguanylate cyclase receptor that regulates exopolysaccharide production in *Pseudomonas aeruginosa*. J Biol Chem 287:23582–23593. doi:10.1074/jbc.M112.375378

Woodward R, Yi W, Li L et al (2010) In vitro bacterial polysaccharide biosynthesis: defining the functions of Wzy and Wzz. Nat Chem Biol 6:418–423. doi:10.1038/nchembio.351

Xu X, Wang X, Cai F, Zhang L (2010) Renaturation of triple helical polysaccharide lentinan in water-diluted dimethylsulfoxide solution. Carbohydr Res 345:419–424. doi:10.1016/j.carres.2009.10.013

Yu Z, Ming G, Kaiping W et al (2010) Structure, chain conformation and antitumor activity of a novel polysaccharide from *Lentinus edodes*. Fitoterapia 81:1163–1170. doi:10.1016/j.fitote.2010.07.019

Zhao K, Tseng BS, Beckerman B et al (2013) Psl trails guide exploration and microcolony formation in *Pseudomonas aeruginosa* biofilms. Nature 497:388–391. doi:10.1038/nature12155

# Chapter 2
# Adhesion and Adhesives of Fungi and Oomycetes

**Lynn Epstein and Ralph Nicholson**

**Abstract** Many fungi adhere strongly to an outer surface of a host or a substratum before penetration; the ability to adhere is a virulence factor for plant and animal pathogens. Because substrata such as the plant cuticle are largely inert hydrophobic surfaces, plant pathogenic fungi often adhere to plastics. Strong adhesion is generally mediated by a secreted glue. Fungal glues are typically formed in an aqueous environment, and as such they may provide models for commercial glues, including for biomedical applications. The best-characterized fungal glue is a galactosaminogalactan polymer that serves as an adhesive to plastic, fibronectin, and epithelial cells in the mammalian pathogen *Aspergillus fumigatus*. However, most fungal glues are apparently mannoprotein(s). Despite our advances in fungal genomics, most fungal glues are uncharacterized, in contrast to the well-characterized fungal pathogen-mammalian host adhesins produced by *Candida albicans*; these adhesins, which bind to specific host receptors, are primarily large, glycosylphosphatidylinositol (GPI)-anchored cell wall mannoproteins that are posttranslationally covalently cross-linked to the cell wall β1,6-glucans. Adhesins typically have tandem repeats; adhesins such as the *C. albicans ALS* family bind to both specific cell types and to plastic. We postulate that the mannoprotein glues from plant pathogenic fungi are secreted in a more water-soluble form, are transiently sticky, and are cross-linked extracellularly. Anti-adhesive strategies may enable new plant disease control methods that are environmentally compatible because anti-adhesive strategies do not necessarily require any uptake of compounds into the fungus or disruption of a eukaryotic metabolic pathway.

---

**Ralph Nicholson** was deceased at the time of publication.

L. Epstein (✉)
Department of Plant Pathology, University of California, Davis, CA 95616-8680, USA
e-mail: lepstein@ucdavis.edu

© Springer International Publishing Switzerland 2016

A.M. Smith (ed.), *Biological Adhesives*, DOI 10.1007/978-3-319-46082-6_2

## 2.1 Introduction

The adhesion of fungi to host surfaces—both plant and animal—before penetration has been well documented (Mendgen et al. 1996; Epstein and Nicholson 1997, 2006; Hardham 2001; Osherov and May 2001; Tucker and Talbot 2001). Observational studies with microscopy indicate that many fungi adhere tenaciously onto inert surfaces such as polystyrene in addition to host substrata. Adhesion onto hydrophobic surfaces is not surprising since, for example, the aerial surfaces of plants are hydrophobic and relatively inert and aquatic fungi can adhere to rocks. Microscopy of fungi that are in the process of adhering also indicates that fungal-substratum adhesion is mediated by a glue, i.e., a secreted macromolecule that extends from the fungus onto the adjacent surface and binds to it in a relatively nonspecific manner. Here, we will primarily focus on fungal cell-substratum adhesion that is mediated by a glue. Some researchers differentiate between an initial and reversible attachment and a time-dependent and at least relatively irreversible adhesion. We will focus on adhesion.

We will use the term "adhesin" to indicate a molecule that mediates a comparatively specific attachment between a ligand on a fungus and a receptor on its host's substratum. The term adhesin is sometimes also used to indicate a molecule involved in cell-to-cell adhesion within a fungal biofilm, in which the ligand and the receptor are on cells of the same species; biofilms of the medically important *Candida* spp. are typically attached to either inert substrata such as implants or onto host cells. Whether an adhesin is somewhat nonspecifically "sticky" or has a conformational "good fit" for a particular substratum is generally unclear; *Candida albicans* cells, for example, adhere to inert substrata such as polystyrene (Masuoka et al. 1999) and to a variety of cell types. Consequently, while we will focus on examples of glue-mediated adhesion of fungi to substrata, we will also mention some of the best-characterized examples of cell-cell adhesion in which at least one of the cells is a fungus, and the adhesin may have glue-like properties. Flocculation as a cell-cell adhesion-mediated phenomenon in *Saccharomyces cerevisiae* has been reviewed relatively extensively and will only be mentioned briefly here (Brückner and Mösch 2012; Rossouw et al. 2015). Adhesins in biofilms produced by animal pathogens also will only be briefly mentioned (Ramage et al. 2009).

## 2.2 Prevalence and Importance of Adhesion in Fungi and Oomycetes

Cell-substratum adhesion is common in all taxonomic classes of microscopic fungi (Nicholson and Epstein 1991; Jones 1994). Cell-substratum adhesion also is common in the oomycetes, which include such economically important plant pathogens such as *Phytophthora* spp., *Pythium* spp., and *Plasmopara* spp. (Hardham 2001). Although oomycetes are stramenopiles, i.e., are phylogenetically distinct from fungi and are most closely allied to diatoms, brown algae, and some parasitic protozoans (Gunderson et al. 1987), we will include them here because they traditionally have been classified as fungi. More importantly, oomycetes and fungi share environments in which glues are secreted and function in similar ways.

### 2.2.1 Adhesion as Part of Many Stages of Morphogenesis in Many Fungi

In a single organism, multiple stages of development can be adherent and can involve differing mechanisms of adhesion. With wind-disseminated plant pathogens such as *Botrytis cinerea* (Doss et al. 1993) and *Erysiphe graminis* (see Sect. 2.2.3.2), hydrophobic interactions between a hydrophobic spore and the leaf cuticle can secure the initial contact. With water-disseminated pathogens such as *Colletotrichum graminicola* (Nicholson and Epstein 1991; Epstein and Nicholson 1997) and *Venturia inaequalis* (Schumacher et al. 2008), initial adhesion of conidia requires the release of a glue (Nicholson and Epstein 1991; Epstein and Nicholson 1997). Since most fungal spores require free water for germination, later stages of adhesion for both air- and water-disseminated spores typically occur in an aqueous environment. Regardless, less perturbable adhesion is temporally and spatially associated with de novo material on the cell surface (Hamer et al. 1988; Kwon and Epstein 1993; Braun and Howard 1994a). It is important to note that the fungal interface with its environment is biochemically complex and multifunctional, i.e., glues are only one component of surface-associated compounds and adhesion is only one function of the extracellular matrix (Nicholson and Epstein 1991). In addition, non-glue components in the extracellular matrix can alter the surface and increase the strength of attachment; *Uromyces viciae-fabae* urediniospores release cutinases and esterases that degrade the plant cuticle and enhance spore adhesion (Deising et al. 1992).

The strength of adhesion varies with the fungus and the cell type. For example, *Nectria haematococca* mating population I (synonym, *Fusarium solani* species complex 10) conidia and germlings adhere to polystyrene and to plant surfaces, but they do not adhere as tenaciously as spores or germlings of *Colletotrichum graminicola* or germlings of the rust fungus *Uromyces appendiculatus*. Unfortunately, relatively few papers quantify the strength of adhesion. Comparisons between studies would be facilitated by utilization of flow cells in which shear force can be determined (Li and Palecek 2003). Mechanical penetration from appressoria requires strong adhesion. Using a microscopic method, Bechinger et al. (1999) estimated that *Colletotrichum graminicola* appressoria had a turgor pressure of 5.4 MPa and exerted an approximate force of 17 μN. Using an indirect measure of turgor via osmotically induced collapse, Howard et al. (1991) estimated that *Magnaporthe oryzae* appressoria generated turgor pressures as high as

8.0 MPa. Consequently, the glue that affixes the appressorium to the plant surface must be able to withstand a very strong force.

### 2.2.2 *Functions of Adhesion*

Even if we limit our discussion to fungal-substratum adhesion that occurs on the plant host surface before penetration, adhesion serves multiple functions (Table 2.1). Most obviously, adhesion keeps a fungus from being blown or rinsed off a potentially suitable environment. Whether contact is required can depend upon the environmental conditions; many species of fungal spores germinate readily regardless of contact in a nutritive liquid medium, but in an oligotrophic environment, efficient germination requires contact with a substratum (Chaky et al. 2001). In *Phyllosticta ampelicida*, adhesion of pycnidiospores is required

**Table 2.1** Examples of documented functions of fungal glues on the surface of the plant host

| Benefits of adhesion for the fungus | Cell types | Examples of organisms[a] | References with experimental evidence |
|---|---|---|---|
| Prevents displacement by water and/or wind | Conidia, germlings, hyphae, appressoria | All | Reviewed in Epstein and Nicholson (1997) |
| Limits germination to potential host tissue (required for contact-stimulated germination) | Conidia | Pa | Kuo and Hoch (1996) |
| Increases the surface area of contact with its host | Germlings, hyphae | Cg | Apoga et al. (2004) |
| Facilitates chemical interaction between pathogen and host | Conidia, germlings | Nh | Jones and Epstein (1990) |
| Required for thigmotropism | Germlings | Ua | Epstein et al. (1987) |
| Required for thigmodifferentiation | Appressoria, hyphopodia | Cg | Chaky et al. (2001) |
| | | Eg | Yamaoka and Takeuchi (1999) |
| | | Ua | Epstein et al. (1985) |
| Required for host penetration via mechanical pressure | Appressoria, hyphopodia, cysts | Cg | Bechinger et al. (1999) |
| | | Mo | Howard et al. (1991) |

[a]This list provides examples and is a not comprehensive survey. All applies to all fungi in this list; Cg, *Colletotrichum graminicola*; Eg, *Erysiphe graminis*; Mo, *Magnaporthe oryzae*; Nh, *Nectria haematococca* mating population I (synonyms *Fusarium solani* species complex 10, *Fusarium solani* f. sp, *cucurbitae* race I); Pa, *Phyllosticta ampelicida*; Ua, *Uromyces appendiculatus*

for germination (Kuo and Hoch 1996). In fungi such as *Erysiphe graminis*, spore adhesion may facilitate "preparation of the infection court," i.e., alteration of the host surface so that it favors fungal development (see Sect. 2.2.3.1). That is, a tightly adherent spore or germling may be able to effectively maintain a higher concentration of its lytic enzymes at the interface between its wall and the substratum surface. *Nectria haematococca* mutants with conidia and germlings with reduced adhesiveness were less virulent than the wild type when deposited on the cucurbit fruit surface but were equally virulent when deposited within the fruit tissue (Jones and Epstein 1990). Adhesion of germlings to a substratum maximizes fungal reception of surface signals and absorption of nutrients from the host substratum because adherent germlings and hyphae are flattened on the bottom; in contrast, germlings and hyphae that have not developed attached to a substratum are round in cross section (Epstein et al. 1987).

Some fungi have contact-induced responses, and these generally require firm attachment. Rust fungi such as *Uromyces appendiculatus* display a type of thigmotropism, i.e., contact-dependent growth in which physical cues from leaf surface topography orient hyphal growth toward stomata (Hoch et al. 1987). Many fungi, including *U. appendiculatus* and *Colletotrichum* spp., display thigmodifferentiation, i.e., contact-dependent differentiation. Here, adhesion is required to induce formation of appressoria (Apoga et al. 2004); indeed the researchers demonstrated that to induce appressorial formation, *C. graminicola* germlings require 4 μm of continuous contact with a hydrophobic substratum. *U. appendiculatus* germlings that were rendered nonadherent with proteases continued to grow but were unable to establish pathogenicity because they were unable to grow thigmotropically or to undergo thigmodifferentiation (Epstein et al. 1985, 1987). Finally, since fungi mechanically penetrate the plant surface from appressoria (Howard et al. 1991), we can deduce that firm adhesion is required for appressorial function. Indeed attachment is so integral to appressorial function that appressoria are often defined as adherent.

Although most fungi are either saprophytic or associated with plants in either a parasitic or a mutualistic association, there are carnivorous fungi. The nematophagous fungi are the best studied of the carnivorous fungi; many of the nematophagous fungi produce adhesive hyphae or protuberances that are critical for nematode capture (Li et al. 2015; Nordbring-Hertz et al. 2006). More than 90 % of the carnivorous fungi are in a single monophyletic clade (the Orbiliomycetes) in the Ascomycota (Yang et al. 2012); additional nematophagous fungi are in the Zygomycotina and in the genus *Nematoctonus* in the Basidiomycota. Adhesive structures within the Orbiliomycetes include adhesive nets in *Arthrobotrys* spp., either adhesive columns or sessile adhesive knobs in *Gamsylella* spp., and stalked adhesive knobs in *Dactylellina* spp. (Yang et al. 2012).

Arbuscular mycorrhizal fungi in the Glomeromycota provide a critical ecosystem service by aggregating soil (Rillig 2004). Soil aggregation is apparently mediated by a highly expressed glycoprotein(s) which was named glomalin, or later, a glomalin-related soil protein (Gillespie et al. 2011). However, the glomalin-related soil protein(s) co-extracts with nonproteinaceous compounds and has never been identified. Nonetheless, the soil "super glue" apparently originates on fungal walls (Driver et al. 2005).

### 2.2.3 Selected Examples of Adhesiveness as a Part of a Developmental Sequence

Currently available sequences of whole fungal genomes make comparisons of the genes in adherent versus nonadherent fungi possible. The ascomycetous yeasts *Saccharomyces cerevisiae* and *Candida albicans* are comparatively nonadherent and adherent, respectively, for yeasts, but are far less adherent than many filamentous ascomycetes. The ascomycetes *Magnaporthe oryzae*, *Colletotrichum graminicola*, and *Botrytis cinerea* have adherent conidia, germlings, and appressoria (Hamer et al. 1988; Doss et al. 1993, 1995; Braun and Howard 1994b; Xiao et al. 1994a; Spotts and Holz 1996), but *Neurospora crassa* is nonadherent. Among the model basidiomycetes, *Ustilago maydis* has hyphae that grow attached to the leaf and then produces appressoria (Snetselaar and McCann 2001). The following examples illustrate that adhesion is just one component of development on a plant surface.

#### 2.2.3.1 *Colletotrichum graminicola*, Causal Agent of Anthracnose on Corn

The maize anthracnose pathogen, *C. graminicola*, produces conidia in acervuli. A complex mixture of high molecular weight glycoproteins surrounds the conidia. The glycoproteins have several roles. They act as an antidesiccant that allows the conidia to survive for an extended time during periods of drought and severe changes in temperature (Nicholson and Moraes 1980; Bergstrom and Nicholson 1999). In addition, the mucilaginous matrix contains mycosporine-alanine, a self-inhibitor of germination (Leite and Nicholson 1992). The mucilage also contains a cutinase and nonspecific esterase that assist in the process of adhesion and recognition of the infection court (Pascholati et al. 1993). Ungerminated conidia of *C. graminicola* adhere to plastic, but the same conidia bind less to hydrophilic surfaces such as glass (Mercure et al. 1994). Materials released by the conidia, including the adhesive material, are easily observed by scanning electron microscopy (Mercure et al. 1995) and contain a mixture of high molecular weight glycoproteins (Sugui et al. 1998). Most if not all *Colletotrichum* spp. produce adherent conidia, germ tubes, and appressoria (Sela-Buurlage et al. 1991). Glycoproteins in the ECM are considered responsible for host adhesion in *Colletotrichum* spp. (Pain et al. 1996).

#### 2.2.3.2 *Blumeria graminis* f. sp. *hordei* and f. sp. *tritici*, Causal Agent of Powdery Mildew of Barley and Wheat, Respectively

The powdery mildew pathogen of barley *B. graminis* produces conidia that release a liquid exudate on contact with the hydrophobic surface of the barley leaf (Kunoh et al. 1988). The conidial exudate contains a cutinase that degrades the host cuticle at the contact site (Nicholson et al. 1988; Pascholati et al. 1992). The substrata and the geometry of the interface between the conidium and the substratum affect the release of extracellular material by *B. graminis* conidia (Carver et al. 1999; Wright et al. 2002). On hydrophobic surfaces, a pad of extracellular material is released within 1 min after contact with the substratum. This phenomenon did not occur on a hydrophilic surface, suggesting that the hydrophobicity of the surface is critical for the initial host-pathogen interaction. The development of *B. graminis* f. sp. *tritici* is similar. A lipase, Lip1, is constitutively produced on the surface of the tips of mature conidia and, after surface contact, is induced on conidial germlings and secondary hyphae (Feng et al. 2009). Pretreatment of the wheat leaf with lipases reduced conidial adhesion, suggesting that the undigested cuticle is important for at least the initial attachment, which has been postulated to occur via hydrophobic interactions between the hydrophobic conidia, which are coated with long-chain fatty acids, and the hydrophobic leaf cuticle, which is also coated with long-chain fatty acids (Samuels et al. 2008). Feng et al. (2009) suggest that the Lip1 transesterification activity could result in the formation of covalent ester bonds between the fungus and the host cuticle. If so, this would result in an extremely strong attachment in the absence of a glue. However, such a covalent attachment would be limited to the plant cuticle and other lipase-sensitive surfaces and would not account for adhesion onto either inert substrata or root surfaces.

#### 2.2.3.3 *Magnaporthe oryzae*, Causal Agent of Rice Blast

In ungerminated *M. oryzae* conidia, after the conidia are hydrated, a "spore tip mucilage" is released from the periplasmic space at the spore apex; the mucilage attaches the spore to a hydrophobic surface (Hamer et al. 1988). The spore tip mucilage contains protein, an α-1,2 mannose, and a lipid (Howard and Valent 1996). The germ tubes are also adherent and function in sensing features of the surface; appressoria are typically induced at the juncture of two epidermal cells. Xiao et al. (1994b) inferred that hardness was a key feature of an appressorial-inducing surface. On an inductive surface, germ tubes produce an (adherent) appressorium. An adherent appressorium can withstand turgor pressures as high as 8.0 MPa (Howard et al. 1991; de Jong et al. 1997; Ebbole 2007), and consequently, the glue that attaches the appressorial wall to the substratum must withstand this force too. Antibody and enzyme treatments suggest that the appressorial glue is a glycoprotein that can be digested by gelatinase B, a collagenase (Inoue et al. 2007, 2011).

## 2.3 Challenges in Identifying Adhesives in Fungi

As indicated above, progress has been made in documenting the spatial and temporal expression of the development of adhesiveness and of cell-substratum adhesion. However, identification of the components of glues presents both biochemical and genetic challenges (Vreeland and Epstein 1996). Despite the recent advances in gene characterization of filamentous fungi, the most common fungal cell-inert substratum glue has still not been clearly identified.

### *2.3.1 Genetic "Knockout," "Knockin," and Overexpression Strategies*

Adhesion-reduced and nonadhesive mutants provide a superb tool to investigate the role of adhesion. Mutants have been used to formulate hypotheses about the adhesive compound(s); in the biocontrol yeast *Rhodosporidium toruloides*, the mannose-binding lectin concanavalin A (Con A) eliminated adhesion in the wild type and bound to the region of bud development in the wild type where the yeast adhered but did not bind to the same region in the mutant (Buck and Andrews 1999b).

Site-directed mutagenesis provides the most powerful technique to demonstrate that a specific gene is involved in the adhesive process. With the medically important fungi, there are many examples in which a putative adhesin gene has been knocked out, and there is a resultant loss of adhesiveness and virulence (e.g., Gale et al. 1998; Brandhorst et al. 1999; Loza et al. 2004; Zhao et al. 2004; de Groot et al. 2013); adhesion and virulence were restored after the mutant strain had the disrupted gene restored. Fungi produce both adhesives and anti-adhesives; *Candida albicans* with a deleted *Ywp1* had increased adherence to several surfaces and had enhanced biofilm formation (Granger et al. 2005; Granger 2012). Similarly, transmission electron microscopy of freeze-substituted conidia of *Fusarium solani* strains with reduced adhesiveness showed an increased amount of extracellular material (Caesar-TonThat and Epstein 1991).

Among the plant pathogenic fungi, researchers have focused more on genes that reduce pathogenicity than on adherence per se; in a few cases, these genes also affect fungal adhesion. However, the mutations have apparently been in genes that function in pathways upstream of adhesion and the mutants are pleiotropically affected. For example, disruption of the *Colletotrichum lagenarium* mitogen-activated protein (MAP) kinase *CMK1* does not affect mycelial growth but interrupts production of conidia, conidial adhesion, conidial germination, and appressorial formation (Takano et al. 2000). Somewhat similarly, disruption of a highly conserved transmembrane mucin (VdMsb) in the MAP kinase pathway in *Verticillium dahliae* resulted in a strain with reduced virulence, microsclerotial and conidial production, conidial adhesion, and invasive growth (Tian et al. 2014). The

*Aspergillus nidulans* signaling mucin MsbA regulates hyphal attachment onto low nutrient substrata in addition to starvation responses and cellulase secretion (Brown et al. 2014). Disruption of *Magnaporthe oryzae acr1* (Lau and Hamer 1998) affects conidiogenesis so that an Δ*acr1* mutant has conidiophores that produce conidia in a head-to-tail chain rather than sympodially. That is, a new conidium is produced from the tip of an older conidium rather than from a conidiophore. In the wild type, conidial adhesion is mediated by mucilage released from the spore tip. The Δ*acr1* conidia fail to produce the spore tip mucilage, are nonadherent, and are inefficient in forming appressoria.

Multiple authors have identified transcription factors that appear to activate genes involved in adhesion. Lin et al. (2015) demonstrated that the *Aspergillus fumigatus* transcription factor SomA, which is a homologue of *Saccharomyces cerevisiae* Flo8, is required for adhesion to host cells and for biofilm formation. Breth et al. (2013) demonstrated that mutants in the *M. oryzae* transcription factors tra1, TDG2, and to a lesser extent TDG6 had reduced conidial adhesion in addition to other phenotypes. Nonetheless, $tra1^{-}$ mutants appeared to form less spore tip mucilage, as determined by binding with the lectin concanavalin A-FITC. In *Colletotrichum lindemuthianum*, a mutant in the Ste12p transcription factor was reduced in its ability to adhere to artificial surfaces; an extracellular protein Clsp1 was absent from the mutant strain (Hoi et al. 2007). However, while mutants in regulatory genes are extremely informative about pathways (e.g., Liu and Kolattukudy 1999), they are not as useful for adhesion studies as knockouts of genes directly involved in glue production. Nonetheless, if multiple compounds can serve as building blocks of a glue, single-gene knockouts may only result in, at best, quantitatively discernible differences from the wild type.

Researchers have used "knockin" strategies to identify individual genes from *C. albicans* that transform the generally nonadhesive *Saccharomyces cerevisiae* into an organism that adheres to polystyrene (Barki et al. 1993; Fu et al. 1998; Li and Palecek 2003). These researchers screened libraries rather than specific genes, which resulted in identification of potentially new adhesins. Tran et al. (2014) transformed a non-flocculating *Saccharomyces cerevisiae Δflo8* strain with a cDNA library from the filamentous plant pathogen *Verticillium longisporum*, selected for adhesion onto agar and flocculation, and then identified 22 genes that conferred adhesiveness in *S. cerevisiae*. The majority of the identified genes are not glue components, but two potential GPI cell wall anchored candidates were identified: Fas1 and Wsc1. Successful utilization of this type of strategy to identify a glue depends upon there being a single or perhaps tandem genes that are expressed in a sticky organism that are not expressed in the nonadherent organism. Using a strategy similar to that described in Barki et al. (1993), the Epstein laboratory was not successful in identifying *Colletotrichum graminicola* genes that rendered *S. cerevisiae* transformants adhesive.

### 2.3.2 Biochemical Strategies

From a biochemical perspective, there are several reasons why identification of even a single fungal adhesive has remained a challenge. Particularly compared to *S. cerevisiae* and *C. albicans*, the cell surfaces of filamentous fungi are complex, and while advances have been made, they are still comparatively poorly characterized. As discussed in greater detail below (Sect. 2.4.2), many fungi probably produce glycoprotein- (or proteoglycan-) based glues. Perhaps not surprisingly, the external layer of fungal cell walls is largely composed of heavily glycosylated proteins (de Nobel et al. 2001). Indeed, all proteins in the wall may be glycosylated because glycosylation is part of the pathway of secreted proteins. Cell surface labeling of proteins by iodination and biotinylation indicates that there are many (glyco)proteins on the cell surfaces of fungi (Epstein et al. 1987; Apoga et al. 2001; de Nobel et al. 2001; de Groot et al. 2004). Adding to the complexity of identifying one or few proteins out of many, the macromolecules on the surface are highly cross-linked. For example, in *S. cerevisiae*, the glycosylphosphatidylinositol-dependent (GPI) cell wall proteins are bound to the plasma membrane and to β1,6-glucan; the β1,6-glucan is bound to chitin and to β1,3-glucan, which also is bound to chitin (de Nobel et al. 2001). Non-GPI cell wall proteins are also cross-linked to the wall. For example, in the animal pathogen *Blastomyces dermatitidis*, the proteinaceous adhesin is secreted through the wall, localizes on the cell surface and then reassociates with cell wall chitin (Brandhorst and Klein 2000). Similarly, the ALS proteins on *Candida albicans*, which adhere to both plastics and to host cells, are bound to the cell wall β1,6-glucan (de Groot et al. 2013). Moreover, *C. albicans* produces a family of eight ALS proteins that bind to plastic (Table 2.3). However, despite attachment by multiple proteins, single-deletion mutants still have discernible adhesion-reduced phenotypes.

Several properties of glues also make identification a challenge. Although assays, such as retention of microspheres, can be useful in identifying potential adhesives, there are limitations in that glues are transiently sticky (Vreeland and Epstein 1996). That is, a glue must be sufficiently nonsticky to be secreted to the wall surface, sticky during glue formation, and then typically, nonsticky after "curing" or "hardening." The ultimate loss of stickiness is accompanied by dramatic changes in solubility; fungal glues are extremely insoluble. Thus, there are always concerns that compounds that are not involved in adhesion are co-purified when the adhesive is sticky and polymerized with the glue after hardening.

## 2.4 Fungal and Oomycete Glues

### 2.4.1 Features

Conidia of many but not all fungal species must be alive in order for adhesion to occur (Slawecki et al. 2002). As indicated previously, although there may be some initial passive attachment of fungal spores to a surface, stronger adhesion seems mediated by secreted material on the cell surface (Epstein and Nicholson 1997; Apoga and Jansson 2000). Microscopic observations, including transmission electron microscopy with freeze-substituted specimens (Caesar-TonThat and Epstein 1991), are consistent with the interpretation that adhesives are liquid upon release and spread over the surface. The spreading of the adhesive obviously occurs underwater in aquatic fungi (Jones 1994) but also occurs underwater for most terrestrial fungi, which require "free water" to germinate. Thus, the glue must either displace or completely mix with the water molecules that are already on the surface. Microscopic observations are also consistent with the interpretation that the cell surface changes over time as the adhesive material is polymerized into a stronger, cross-linked compound (Caesar-TonThat and Epstein 1991; Apoga and Jansson 2000). Mature adhesives often appear to have a fibrous component (Watanabe et al. 2000; Nordbring-Hertz et al. 2006).

As indicated above, many fungi adhere well to inert surfaces including Teflon (Hamer et al. 1988). Overall, for the purpose of attachment, the predominant feature of a leaf surface appears to be its hydrophobicity. Many fungi adhere more tenaciously to hydrophobic than to hydrophilic surfaces (Epstein and Nicholson 1997). Terhune and Hoch (1993) found that the extent of adhesion of germlings of *Uromyces appendiculatus* to inert surfaces correlated closely with substratum hydrophobicity. However, other features on the leaf surface may also affect adhesion. The rust fungus *Uromyces viciae-fabae*, for example, secretes cutinases and esterases which change the potential binding sites on the surface and enhance adhesion (Deising et al. 1992). Buck and Andrews (1999b) utilized attachment mutants to conclude that localized positive charges, and not hydrophobic interactions, mediate attachment of the yeast *Rhodosporidium toruloides*. Adhesion to roots is not as well characterized as to leaves (Recorbet and Alabouvette 1997), and roots have potential carbohydrate receptors to which an adhesin could bind.

### 2.4.2 Postulated Composition of Glues

During the last three decades, we have learned much about why and when fungi adhere. We still know relatively little about how fungi adhere. Many researchers have deduced the composition of the glues by testing for the loss of adhesiveness with enzymes, lectins, antibodies, and inhibitors of specific metabolic pathways. Such indirect evidence indicates that most fungal adhesives are glycoproteins

(Chaubal et al. 1991; Jones 1994; Kuo and Hoch 1996; Pain et al. 1996; O'Connell et al. 1996; Bircher and Hohl 1997; Epstein and Nicholson 1997; Sugui et al. 1998; Hughes et al. 1999; Apoga et al. 2001; Rawlings et al. 2007). The carbohydrate moieties on glycoproteins seem particularly important in adhesion. In many fungi, including *Bipolaris sorokiniana*, *Colletotrichum graminicola*, *Magnaporthe oryzae*, *Nectria haematococca*, and *Phyllosticta ampelicida*, substratum adhesion is at least partially blocked by the lectin Con A (Hamer et al. 1988; Shaw and Hoch 1999; Apoga et al. 2001). Experimental evidence, including blockage of adhesion with the snowdrop lectin, indicates that mannose residues are involved in the adhesive mechanism (Kwon and Epstein 1997a). Importantly, Kahmann's and Corcelles' groups demonstrated that the yeast cells of the corn smut *Ustilago maydis* with a deletion in an intracellular o-mannosyltransferase *pmt4* had reduced adhesion onto starch agar compared to the wild type (Fernández-Álvarez et al. 2012). This supports the hypothesis that the *U. maydis* glue is a mannoprotein. In addition, the appressoria of the Δ*pmt4* strain were defective in penetration (Fernández-Álvarez et al. 2009), which could be due to defective appressorial adhesion, although this was not measured directly. Disruption of three *pmt* orthologs in the entomopathogen *Beauveria bassiana* was pleiotropically affected; adhesion of the defective strains was not examined (Wang et al. 2014).

Lectins, which are carbohydrate-binding proteins or glycoproteins, are candidates for cell-cell binding, such as on fungal-root interactions and on adhesive nets of nematophagous fungi (Li et al. 2015). Li et al. (2015) reviewed the recent literature on the role of lectins in the adhesiveness of traps of nematophagous fungi; based on a combination of lectin knockout and spatial/temporal expression of lectin genes, they suggest that lectins may not be involved in nematode trapping.

In the mammalian pathogen *Aspergillus fumigatus*, a galactosaminogalactan polymer serves as an adhesive to plastic, fibronectin, and epithelial cells (Gravelat et al. 2013); the polymer contains α-1,4-linked galactose and α-1,4-linked *N*-acetylgalactosamine. In addition to imparting adhesion, the polymer suppresses the host inflammatory response by preventing binding of the host dectin-1 to the fungal wall β-glucans (Gravelat et al. 2013). There have been a few other reports that fungal adhesives are carbohydrates (Pringle 1981), but in most cases, whether the adhesives are actually glycoproteins or proteoglycans was not determined. Buck and Andrews (1999a) reported that a mannose-containing compound, and possibly a mannoprotein, was involved in attaching the yeast *Rhodosporidium toruloides* to polystyrene and barley leaves. Howard et al. (1993) concluded that the adhesive *Magnaporthe oryzae* spore (conidial) tip mucilage is composed of a mixture of a polypeptide(s) with terminal mannose disaccharides and a lipid component. Determination of whether lipids are part of the *M. oryzae* appressorial adhesive (Ebata et al. 1998; Ohtake et al. 1999) requires additional experimental evidence.

### 2.4.3 *Secretion and Cross-Linking, with a Focus on Transglutaminases*

The precursors of fungal glues must be secreted in a water-soluble form so that the material migrates through the wall to the cell surface. Either in the wall or on the cell surface, the glue must polymerize into a high molecular mass. Theoretically, a variety of oxidases could polymerize the glue. A catechol oxidase polymerizes the glue in mussels, and a haloperoxidase may polymerize algal glues (Vreeland et al. 1998). Transglutaminases (EC 2.3.2–13) may be the polymerizing agents of fungal glues. Transglutaminases form a calcium-dependent interpeptidic cross-link between a glutamic acid and a lysine residue. Certainly transglutaminases serve functions in fungal cell walls that are distinct from adhesives; transglutaminases may cross-link proteins in ascomycete walls (Ruiz-Herrera et al. 1995; Iranzo et al. 2002) which would help to reduce the pore size of the fungal wall. In addition, transglutaminases in the oomycete *Phytophthora sojae* serve as elicitors of plant host defense (Brunner et al. 2002), which presumably is an indication that transglutaminases are present on the cell surface. Fabritius and Judelson (2003) identified a family of transglutaminases (M81, M81C, M81D, and M81E) in the plant pathogen *P. infestans*. Thus, transglutaminases are present, apparently on the surface, of fungal and oomycete walls. Adhesion of oomycete zoospores and some fungal spores requires $Ca^{++}$ (Shaw and Hoch 2000, 2001), as do transglutaminases. For example, adhesion of encysting zoospores of *P. cinnamomi* was enhanced with increasing $Ca^{++}$ concentration from 2 to 20 mmol/l; chelation of $Ca^{++}$ with EGTA eliminated adhesion (Gubler et al. 1989). Interestingly, with the *Candida albicans* adhesin Hwp1, the host transglutaminase covalently links the pathogen to the host epithelial cell (Staab et al. 1999).

### 2.4.4 *Cell Surface Macromolecules with Apparent Adhesive Properties*

Potential fungal glues are summarized in Table 2.2 and several are discussed below.

#### 2.4.4.1 PcVsv1, a Protein on Encysting Zoospores of *Phytophthora cinnamomi*

The oomycete *Phytophthora cinnamomi* causes serious diseases on a wide range of host plants. As with many oomycetes, *P. cinnamomi* are disseminated by motile zoospores. The zoospores contain an adhesive in secretory vesicles (Hardham and Gubler 1990). About 2 min after contact with a solid surface, the material in the vesicles is secreted onto the ventral surface of the zoospores, and the spores become sticky. The adhesive process is part of the process of transformation of zoospores

**Table 2.2** Fungal macromolecules that potentially function as fungal-substratum glues

| Organism | Glue component | Comments | Reference |
|---|---|---|---|
| *Candida albicans* | Eap1 | GPI-CWP[a] that is involved in adhesion to polystyrene and to epithelial cells | Li and Palecek (2003) |
| *Colletotrichum lindemuthianum* | 110-kDa glycoprotein on the conidial wall | Adhesion disrupted by monoclonal antibody UB20, which binds to several glycoproteins including the 110 kDa | Hughes et al. (1999) |
| *Metarhizium anisopliae* | MAD1 | GPI-CWP that is involved in conidial adhesion onto locust wings | Wang and St. Leger (2007) |
| *Nectria haematococca* mating population I | 90-kDa mannoprotein | Spatially and temporally associated with adhesion of macroconidia; specifically associated with conditions in which adhesion is induced | Kwon and Epstein (1997a, b) |
| *Rhodosporidium toruloides* | Mannose-containing compound | Involved in adhesion to polystyrene and barley leaves | Buck and Andrews (1999a, b) |
| *Phytophthora cinnamomi* | 220-kDa protein with thrombospondin type 1 repeats | Thrombospondin type 1 repeats are found in adherent cells in animals and in apicomplexan parasites | Robold and Hardham (2005) |
| *Saccharomyces cerevisiae* | Flo11p/Muc1 | GPI-CWP that is involved in adhesion to substrata (including polystyrene), cell-cell adhesion (flocculation), and filamentous growth | Reynolds and Fink (2001); Braus et al. (2003); Brückner and Mösch (2012) |

[a]GPI-CWP, glycosylphosphatidylinositol-dependent cell wall proteins

into adherent cysts. Within 20–30 min the encysted spores penetrate the underlying plant tissue. Adhesion is essential because a zoospore that is developing into a cyst must not be dislodged from a potential infection site and because a mature cyst must be sufficiently glued to the host surface so that the fungus can penetrate the host tissue.

Hardham and Gubler (1990) identified a monoclonal antibody that specifically bound material in the secretory vesicles and on the developing adhesive surface, i.e., the antigen was spatially and temporally consistent with being a component of the glue. Robold and Hardham (2004) screened a *P. cinnamomi* cDNA library with the antibody and then selected and cloned a gene that encodes for PcVsv1 (***P***. ***c****innamomi* **v**entral **s**urface **v**esicle) protein with an approximate mass of 220 kDa. Not surprisingly for an adhesive material, PcVsv1 contains a repetitive motif—47 copies of a ca. 50-amino-acid length segment that has homology to the thrombospondin type1 repeat. Thrombospondin motifs are found in proteins involved in attachment in animals and in apicomplexan parasites, including *Plasmodium*, *Cryptosporidium*, and *Eimeria* (Tomley and Soldati 2001; Deng et al. 2002; Tucker 2004). In animals, the thrombospondin motifs occur in proteins in the extracellular matrix, i.e., potentially in the location in which adhesion occurs. More experimental evidence is required before PcVsv1 can be declared the glue, or a component of the glue. However, even if thrombospondin is a glue component in oomycetes, it is probably not involved in adhesion of true fungi; based on a survey of sequences, Robold and Hardham (2005) found no evidence of thrombospondin motifs in either true fungi or in plants or green algae.

Other proteins of *Phytophthora* spp. may be involved in some aspects of adhesion, although probably not as components of the glue per se. Göernhardt et al. (2000) identified the car90 (for **c**yst germination-specific **a**cidic **r**epeat) protein on the surface of *Phytophthora infestans* germlings; the protein is transiently expressed during cyst germination and appressorium formation, i.e., during an adherent portion of the *P. infestans* lifecycle. The car90 protein contains 120 repeats of the consensus sequence TTYAPTEE. Because the repetitive sequence has homology with human mucins, the authors suggested that the car90 protein "may serve as a mucous cover protecting the germling from desiccation, physical damage, and adverse effects of the plant defense response." Because the protein is potentially viscous and sticky, the authors further speculated that car90 "may assist in adhesion to the leaf surface." Gaulin et al. (2002) identified the CBEL (for **c**ellulose-**b**inding **e**licitor **l**ectin) glycoprotein on the surface of hyphae of *Phytophthora parasitica* var. *nicotianae*. However, while transgenic strains lacking CBEL did not adhere to and develop on cellulose in vitro, they appeared to have normal adhesion and development on the plant surface and in planta.

#### 2.4.4.2 90-kDa Mannoprotein on Macroconidia of *Nectria haematococca* (Anamorph *Fusarium solani* f. sp. *cucurbitae*)

The Epstein laboratory selected *N. haematococca* as a model biochemical system because, unlike *C. graminicola* and *U. appendiculatus*, the *N. haematococca* glue components seem somewhat detergent soluble. When exposed to some but not all carbon sources, macroconidia of *N. haematococca* mating population I (anamorph, *Fusarium solani* f. sp. *cucurbitae* race I) become adhesive within 5 min and adhere at the spore apices to substrata and to other spores (Jones and Epstein 1989). At the same time, the macroconidia produce mucilage at the spore tips which binds Con A and the snowdrop lectin (SDL) and produce a 90-kDa glycoprotein which binds Con A and SDL; Con A and SDL inhibit macroconidial adhesion (Kwon and Epstein 1993, 1997a). When exposed to media that induce germination but not adhesiveness, neither the spore tip mucilage nor the 90-kDa glycoprotein is produced (Kwon and Epstein 1993). Adhesion of both ungerminated macroconidia and germlings was inhibited by a polyclonal IgG prepared against the 90-kDa glycoprotein (Kwon and Epstein 1997a). Anti-adhesive activity of the IgG was reduced by incubation of the antibodies with mannan; the mannan alone has no effect on adhesion. The anti-90-kDa IgG did not bind to the deglycosylated glycoprotein. Thus, we conclude that the 90-kDa glycoprotein and specifically its mannose residues are specifically involved in adhesion of *N. haematococca* macroconidia.

Visualization of the macroconidial tip mucilage and the detection of the 90-kDa glycoprotein are transient over time, consistent with compounds which become increasingly polymerized. Although *N. haematococca* macroconidia in zucchini extract are adherent for a 3-h incubation period, the macroconidial tip mucilage and the 90-kDa glycoprotein are primarily detectable only for the first hour. Macroconidia incubated in an adhesion-inducing medium with transglutaminase inhibitors (either 1 mM iodoacetamide or 10 mM cystamine) adhere significantly less, have less of the macroconidial tip mucilage that is spatially associated with adhesion, and have less recoverable 90-kDa glycoprotein that is temporally associated with adhesion; the inhibitors do not affect the presence of other extractable compounds visualized on the protein blot. In addition, the fact that two reducing agents do not affect adhesion suggests that the iodoacetamide and cystamine are interfering with the adhesion process and are not inhibiting adhesion by chemical reduction.

Thus the data are consistent with the following model. Within minutes of incubation in an adhesion-inducing medium, at the macroconidial apices, *N. haematococca* spores secrete a sticky lower-molecular-weight and more-water-soluble precursor of a 90-kDa glycoprotein (Kwon and Epstein 1993). At the spore apex, the glycoprotein is partially polymerized by a transglutaminase into a somewhat sticky 90-kDa form (Kwon and Epstein 1997a). After 1–2 h, the 90-kDa glycoprotein is extracellularly modified so that it is no longer sticky. After 2 h, adhesion is no longer localized at the spore apex; the macroconidia adhere along the entire lower spore surface and later along the germ tube substratum interface. Mutant analysis suggests that compounds other than the 90-kDa glycoprotein are involved in this later stage of adhesion (Epstein et al. 1994). However, inhibition of both spore and germling adhesion by anti-90-kDa IgGs suggests that related compounds may be involved in spore and germling adhesion (Kwon and Epstein 1997a).

The 90-kDa compound is hydrophobic, contains mannose, has N-linked carbohydrates, has sites for O-linked carbohydrates (18 % serine and threonine), and has an amino acid composition with ca. 38 % hydrophobic and 62 % hydrophilic residues (Kwon and Epstein 1997b). Genetic data and an in vitro system are required to demonstrate that the 90-kDa glycoprotein is a fungal glue per se.

Prados-Rosales et al. (2009) characterized the cell wall proteome from hyphae in adhesion-inducing conditions of the related fungus, *Fusarium oxysporum*. They compared cell wall proteins of the wild type with a nonadherent and nonpathogenic mutant with a deletion in a regulatory kinase, fmk1. Eighteen differentially expressed proteins were identified, but none to date have been identified as involved in adhesion.

#### 2.4.4.3 Hydrophobins: The Mannoprotein SC3, a *Schizophyllum commune* Hydrophobin; the Class I Hydrophobin BcHpb1 of *Botrytis cinerea*; and the HCf-6 Hydrophobin of *Cladosporium fulvum*

As indicated above, aerial plant surfaces are hydrophobic, and many fungi and oomycetes adhere more tenaciously to hydrophobic surfaces such as polystyrene than to hydrophilic surfaces such as glass (Sela-Buurlage et al. 1991; Terhune and Hoch 1993; Kuo and Hoch 1996; Wright et al. 2002). Fungal hydrophobins are small (approximately 100 amino acids), secreted, moderately hydrophobic proteins with eight conserved cysteine residues and a common hydropathy profile, but highly variable amino acid sequences (Kershaw and Talbot 1998). Despite the variability in sequence, complementation experiments in *Magnaporthe oryzae* with heterologous hydrophobin genes indicate that hydrophobins from different species encode for functionally related proteins (Kershaw et al. 1998).

Hydrophobins are interesting candidates as components of at least initial attachment because they are structural proteins on the fungal surface that self-assemble into insoluble aggregates (Wösten et al. 1994). Moreover, the self-assembly occurs at the interface between a hydrophobic and a hydrophilic surface, which includes the environment in which most foliar plant pathogens have to adhere, i.e., a leaf surface covered with water. Hydrophobins have the interesting property of forming amphipathic membranes (Scholtmeijer et al. 2002). That is, hydrophobins form a membrane with a hydrophobic and a hydrophilic side; the hydrophobic side binds to hydrophobic surfaces and the hydrophilic side binds to hydrophilic surfaces. In so doing, a hydrophobin changes the polarity of its substratum. In the context of fungal-plant interactions, *Colletotrichum graminicola*, *Erysiphe graminis*, and *Uromyces viciae-fabae* spores, for example, secrete material that increases the hydrophilicity of the surface (Nicholson et al. 1988; Deising et al. 1992; Pascholati et al. 1993). Films of class I hydrophobins form amyloid-like fibrils (Kwan et al. 2006), which are consistent with many microscopic observations of adhesion.

However, while the conidial rodlet hydrophobins, e.g., *Magnaporthe oryzae* class I hydrophobin Mpg1 (Kershaw et al. 1998) may allow for an initial hydrophobic interaction between a wind-blown spore and a leaf, it is a comparatively weak attachment. Indeed, a hydrophobin mutant of *M. oryzae Mpg1* adhered as well as the wild type, indicating that the MPG1 hydrophobin is not involved in adhesion (Talbot et al. 1996). In contrast, based on gene knockdown and knockout experiments, Inoue et al. (2016) concluded that *Mpg1* expression was correlated with adhesion, germling differentiation, and pathogenicity. Based on a knockout mutant in the class II hydrophobin mhp1, Inoue et al. (2016) concluded that mhp1 was not involved in adhesion, pathogenicity, or spore viability, although it affected hydrophobicity and appressorium differentiation on the artificial surface.

Although currently there is no compelling evidence that a hydrophobin is a component of a strong glue, two *Schizophyllum commune* hydrophobins are involved in attachment. SC4 coats the surface of hyphae within the fruiting body

(Schuren and Wessels 1990). As such, it serves in cell-cell adhesion. SC3 is involved in aerial hyphal formation. In comparison to the wild type, *sc3* mutants are reduced in adhesiveness to hydrophobic substrates (Wösten et al. 1994). We note that *S. commune* Sc3 is glycosylated with mannose residues (de Vocht et al. 1998), as many fungal glues appear to be. Similarly, deletion of the *Botrytis cinerea* class I hydrophobin BcHpb1 resulted in conidia that had reduced adhesion to a model hydrophobic surface, dimethylsilicated-coated glass (Izumitsu et al. 2010).

*Cladosporium fulvum* has six hydrophobins (Lacroix et al. 2008). Using epitope tagging, Lacroix et al. (2008) demonstrated that HCf-6 was produced during infection, is secreted, and coats surfaces under and around growing hyphae. Although a deletion mutant indicates that Hcf-6 **inhibits** adhesion of conidia and hyphae to glass slides, Lacroix et al. (2008) suggest that it is involved in attachment of hyphae to solid surfaces. Teertstra et al. (2006) demonstrated that the two hydrophobins in *Ustilago maydis* had no effect on hyphal adhesion to Teflon but that deletion of the repellent gene *rep1* (which is processed into 11 secreted peptides) increased hyphal wettability and decreased adhesion onto Teflon. While rep1-1 is not a hydrophobin, Teertstra et al. (2009) demonstrated that it formed amyloid-like filaments on the hyphal surface and concluded that it functioned as other hydrophobins.

#### 2.4.4.4 MAD1 and MAD2 (Metarhizium Adhesion-Like Proteins) in the Entomopathic Fungus *Metarhizium anisopliae*

Two putative adhesins were identified in an entomopathic fungus *Metarhizium anisopliae* (Wang and St. Leger 2007), which adheres to insect cuticles, a hydrophobic surface, and to plant root surfaces (Hu and St. Leger 2002), which are less hydrophobic. *M. anisopliae* MAD1 is present on the surface of conidia but interestingly plays a role in cytoskeletal organization and cell division. MAD1 knockouts had conidia with reduced adhesion onto locust wings and reduced virulence, but no reduction of adhesion onto onion epidermal peels, a more hydrophilic surface. Expression of *Mad1* in *Saccharomyces cerevisiae* cells transformed the wild type from nonadherent to adherent on insect cuticles but did not affect adhesion to onion peels. Conidia of MAD2 knockouts in *M. anisopliae* had wild-type adhesion on locust wings and wild-type virulence but reduced adhesion to onion peels. Expression of *Mad2* in *Saccharomyces cerevisiae* cells transformed the wild type from nonadherent to adherent on onion peels but did not affect adhesion on locust wings. Liang et al. (2015) disrupted the *A. oligospora* homologue AoMAD1, which was upregulated during trap formation of the nematophagous fungus. While the ΔAoMAD1 has an altered surface and wall structure, AoMAD1 does not appear to be the adhesive protein per se (Liang et al. 2015) because the ΔAoMAD1 strain formed more adhesive traps than the wild-type strain. Consequently, the authors suggested that AoMAD1 may have a negative regulatory role in the fungus in switching from a saprophytic to a pathogenic lifestyle.

#### 2.4.4.5 Selected Glycosylphosphatidylinositol-Dependent (GPI) Cell Wall Proteins

GPI cell wall proteins (GPI-CWP) are the most abundant proteins in fungal walls (de Groot et al. 2004). They have a GPI anchor to the plasma membrane at the C terminus and a signal peptide at the N terminus (de Nobel et al. 2001). GPI-CWP often have a repeated motif and a binding site that links the protein to the β1,6-glucan part of the wall. Here we will discuss GPI-CWP that are clearly adhesins but that may have more nonspecific glue-like properties.

The human pathogen *Candida albicans* adheres to polystyrene and to medical devices, in addition to human epithelial cells; *C. albicans* is more adherent than *Saccharomyces cerevisiae*. Li and Palecek (2003) expressed a library of *C. albicans* in a strain of the nonadherent *S. cerevisiae flo8*Δ and screened for adhesive cells on polystyrene. They isolated a gene *EAP1* (**e**nhanced **a**dherence to **p**olystyrene) which encodes for a GPI-CWP. Using expression in both *S. cerevisiae* and in a *C. albicans* mutant, they demonstrated that Eap1 affects adhesion on both polystyrene and on epithelial cells and biofilm formation (Li et al. 2007). In the diploid *C. albicans* SC5314, in one allele, the predicted protein product has 70 PATEST repeats in one allele (comprising 37 % of the protein) and 13 in the other (NCBI accession numbers EAK95619 and EAK95520). Following the PATEST repeat, there are nine or ten repeats of TPAAPGTPVESQP. Eap1 has homology to *C. albicans* Hwp1p and to *S. cerevisiae* Flo11p, which are discussed below. The *C. albicans eap1* complements *flo 11* mutations in *S. cerevisiae*.

In low glucose, *Saccharomyces cerevisiae* adheres to plastic and yeast cells agglutinate (Reynolds and Fink 2001). Genetic manipulation indicates that adhesion and agglutination require the cell surface glycoprotein Flo11p. Flo11p increases cell surface hydrophobicity and is expressed when *S. cerevisiae* grows as filaments, pseudohyphae, and invasively through agar but is not expressed in yeast cells (Braus et al. 2003). Based on the deduced amino acid sequence, Flo11p (also named MUC1 and YIR019C) contains 1367 amino acids of which 50 % are either serine or threonine and 10 % are proline. Starting at amino acid 333, there are 38 repeats of SSTTESS. This residue, which comprises 19 % of Flo11p, is contained within 22 copies of S/T SSTTESS S/V (A)P V/A PTP and 16 copies of S/T SSTTESSSAPV.

The ALS (agglutinin-like sequences) family of GPI-CWP is enriched in the *Candida* clade but absent from the *Saccharomyces* clade (Klis et al. 2010). As indicated in Table 2.3, different ALS may have different specificities to different cell types; mutational analysis indicates that Als1, Als2, Als3, Als4, Als5, Als7 and Als9 also adhere to plastic (Table 2.3). The ALS have repeat regions, and longer repeat regions are associated with greater adherence and shorter repeat regions are associated with less adherence (Verstrepen and Klis 2006); this may be because the longer protein extends further outside of the wall. Interestingly, within the threonine, isoleucine, and valine-rich "T" regions of the ALS proteins (de Groot

**Table 2.3** Examples of animal pathogenic fungi that produce adhesins and their receptors

| Fungus | Fungal adhesin | Apparent host-binding site | Reference |
|---|---|---|---|
| *Blastomyces dermatitidis* | BAD1 (=WI-1) protein | Binds yeast to CD11b/CD18 and CD14 receptors on human macrophages | Newman et al. (1995); Brandhorst et al. (1999); Brandhorst and Klein (2000) |
| *Candida albicans*[a] | Members of the ALS (agglutinin-like sequence) gene family (GPI-CWP[b]): Als3, Als1, Als4, Als5 (=Ala1), Als2, Als6, Als7, Als9 | Different ALS (in their N-terminal regions) have different specificities, e. g., for laminin, collagen, fibronectin, endothelial and/or epithelial cells. The indicated ALS also adhere to plastic | Fu et al. (1998); Hoyer (2001); de Groot et al. (2004); Loza et al. (2004); Zhao et al. (2004): Hoyer et al. (2008); Aoki et al. (2012); Alsteens et al. (2013); de Groot et al. (2013) |
| *Candida albicans* | Eap1 (enhanced adherence to polystyrene), a GPI-CWP | Epithelial cells; also polystyrene | Li and Palecek (2003); Li et al. (2007); Modrzewska and Kurnatowski (2015) |
| *Candida albicans* | Hwp1, a mannosylated, GPI-CWP | Epithelial cells | Tsuchimori et al. (2000); Staab et al. 2004 |
| *Candida albicans* | Iff4 | Epithelial cells; also plastic | Kempf et al. 2007; Fu et al. (2008) |
| *Candida albicans* | Int1, a RGD-protein | Epithelial cells | Gale et al. (1998) |
| *Candida albicans* | Mannan core and oligomannosyl side chains of the acid-stable portion of phosphormannoprotein | Marginal zone macrophages | Kanbe and Cutler (1998) |
| *Candida glabrata* | Epa1, a GPI-CWP | Epithelial cells | Cormack et al. (1999) |
| *Cryptococcus neoformans* | Cfl1, cell flocculation 1 (CNA07720) | Plastic biofilm | Wang et al. (2012) |
| *Metarhizium anisopliae* | MAD2, a GPI-CWP | Onion epidermal peels, presumably a model surface for plant roots | Wang and St. Leger (2007) |

[a]See de Groot et al. (2013) for more extensive documentation on adhesins on *Candida albicans* and other human pathogens
[b]GPI-CWP, glycosylphosphatidylinositol-dependent cell wall proteins

et al. 2013), Als1, Als3, and Als5, have sequences that form amyloid adhesion domains (Alsteens et al. 2010; de Groot et al. 2013). Fungal adhesins are discussed more below.

## 2.5 Fungal Adhesins

Attachment of fungal cells both within biofilms and to their host cells is a critical part of fungal morphogenesis and pathogenesis in animals (Sundstrom 2002; Wang et al. 2013). Adhesion of *Candida albicans* and other *Candida* spp. within biofilms, onto implanted medical devices and other model substrata, and onto host cells has been studied the most because invasive candidiasis is a common cause of nosocomial infections, and mortality, particularly with antibiotic-resistant isolates, has been correlated with a strain's ability to form biofilms and to adhere (Uppuluri et al. 2009). Examples of fungal adhesins that bind to animal cells are shown in Table 2.3; most are relatively well characterized (de Groot et al. 2013). Clf1 is a hyphal-specific adhesin from *Cryptococcus neoformans* involved in cell-cell adhesion in colonies and biofilms and in cell-substratum adhesion onto plastic (Wang et al. 2012); cells aggregated when CFL1 was overexpressed, but virulence was reduced.

Similar to the fungal glues, most adhesins are glycoproteins, and at least several are mannoproteins (Fukazawa and Kagaya 1997), with the mannan residues involved in adhesion (Kanbe and Cutler 1998). Timpel et al. (1998, 2000) demonstrated that strains of *Candida albicans* with either a disrupted mannosyltransferase Pmt1p or a Pmt6p adhered less to host cells and had reduced virulence; pmt1 mutants produced an Als1p adhesin with a lower molecular mass than the wild type. The results are consistent with a role of mannoproteins in cell-cell adhesion. Using *C. albicans* strains with deletions in the mannosyltransferases mnt1 and/or mnt2, Dutton et al. (2014) demonstrated that mannosylation is important for interkingdom (bacterial/*C. albicans*) biofilm formation.

Hazen and colleagues (Masuoka et al. 1999) have argued that cell surface hydrophobicity is an important factor in adhesion of *C. albicans* to host epithelial cells and to plastic, that adhesion is a virulence factor, and that hydrophobic glycoproteins on the surface of the cells contribute, if not determine, cell surface hydrophobicity. Based on 184 clinical isolates of *Candida* spp., Silva-Dias et al. (2015) provide evidence that cell surface hydrophobicity is a reasonably good predictor of biofilm formation, but not of adhesion. *C. tropicalis* and *C. parapsilosis* were the most adherent; *C. parapsilosis*, *C. krusei*, and *C. guilliermondii* were moderately adherent ; and *C. albicans* and *C. glabrata* were the least adherent.

As indicated above, several of the fungal adhesins, including Hwp1, Ala1p/Als5p, Als1p from *Candida albicans*, and Epa1p from *Candida glabrata*, are GPI cell wall proteins. The arginine-glycine-aspartic (RGD) tripeptide sequence appears to be a critical site in some fungal adhesins and in some plant and animal host receptors (Hostetter 2000; Senchou et al. 2004). Corrêa et al. (1996) concluded that thigmosensing of appressorium-inductive surfaces by the rust fungus *Uromyces appendiculatus* involved an RGD sequence, probably on a transmembrane, integrin-like glycoprotein.

## 2.6 Conclusions

Although many fungi produce glues, we have yet to characterize one definitively. Fungal glues are typically formed in an aqueous environment, and as such they may provide models for commercial glues, including for biomedical applications. The commercially available TISSEEL® (Baxter Corporation) is used during surgery as a sealant. Just before use, the TISSEEL® ingredients, fibrinogen and thrombin, are mixed; in mammals, thrombin activates fibrinogen which forms fibrin clots. The example is particularly relevant because many animal pathogenic fungi adhere to host fibrinogen and a putative adhesive from the oomycete *Phytophthora cinnamomi* is a protein with thrombospondin type 1 repeats (Robold and Hardham 2005).

Adhesion is a common feature of pathogenic fungi, regardless of whether they are pathogens of plants or animals. In addition, adhesion can be disrupted experimentally using enzymes that degrade the glue or compounds that block adhesion without directly killing the cell (Epstein et al. 1987). Consequently, anti-adhesives may provide an appealing disease control strategy because they do not necessarily require uptake into the fungus or disruption of a eukaryotic metabolic pathway. At least some of the antifungal activity of surfactants and detergents may be because they render plant surfaces less hydrophobic. Stanley et al. (2002) demonstrated that the phenolic zosteric acid (*p*-(sulfo-oxy) cinnamic acid) inhibits adhesion and infection by conidia of *Magnaporthe oryzae* and *Colletotrichum lindemuthianum*; concentrations that are effective as an anti-adhesive are not toxic to either the plant hosts or fungi.

Although it is perhaps surprising that more progress has not been made in the elucidation of fungal "superglues" in the past decade, we now have extremely powerful technologies with which to proceed– full genome assemblies in a variety of fungi which range from strongly adherent to weakly or essentially nonadherent, new tools in gene deletion and silencing using methods such as CRISPR/Cas (Schuster et al. 2016), and new technologies for imaging and analyzing adherent molecules (Alsteens et al. 2010; Beaussart et al. 2012). In combination, these tools should enable the fungal adhesion field to leap forward.

## References

Alsteens D, Garcia MC, Lipke PN, Dufrêne YF (2010) Force-induced formation and propagation of adhesion nanodomains in living fungal cells. Proc Natl Acad Sci USA 107:20744–20749

Alsteens D, Van Dijck P, Lipke PN, Dufrêne YF (2013) Quantifying the forces driving cell–cell adhesion in a fungal pathogen. Langmuir 29:13473–13480

Aoki W, Kitahara N, Miura N, Morisaka H, Kuroda K, Ueda M (2012) Profiling of adhesive properties of the agglutinin-like sequence (ALS) protein family, a virulent attribute of *Candida albicans*. FEMS Immunol Med Microbiol 65:121–124

Apoga D, Jansson H-B (2000) Visualization and characterization of the extracellular matrix of *Bipolaris sorokiniana*. Mycol Res 104:564–575

Apoga D, Jansson H-B, Tunlid A (2001) Adhesion of conidia and germlings of the plant pathogenic fungus *Bipolaris sorokiniana* to solid surfaces. Mycol Res 105:1251–1260

Apoga D, Barnard J, Craighead HG, Hoch HC (2004) Quantification of substratum contact required for initiation of *Colletotrichum graminicola* appressoria. Fungal Genet Biol 41:1–12

Barki M, Koltin Y, Yanko M, Tamarkin A, Rosenberg M (1993) Isolation of a *Candida albicans* DNA sequence conferring adhesion and aggregation on *Saccharomyces cerevisiae*. J Bacteriol 175:5683–5689

Beaussart A, Alsteens D, El-Kirat-Chatel S, Lipke PN, Kucharíková S, Van Dijck P, Dufrêne YF (2012) Single-molecule imaging and functional analysis of Als adhesins and mannans during *Candida albicans* morphogenesis. ACS Nano 6:10950–10964

Bechinger C, Giebel K-F, Schnell M, Leiderer P, Deising HB, Bastmeyer M (1999) Optical measurements of invasive forces exerted by appressoria of a plant pathogenic fungus. Science 285:1896–1899

Bergstrom GC, Nicholson RL (1999) The biology of corn anthracnose: knowledge to exploit for improved management. Plant Dis 83:596–608

Bircher U, Hohl HR (1997) Surface glycoproteins associated with appressorium formation and adhesion in *Phytophthora palmivora*. Mycol Res 101:769–775

Brandhorst T, Klein B (2000) Cell wall biogenesis of *Blastomyces dermatitidis*: evidence for a novel mechanism of cell surface localization of a virulence-associated adhesin via extracellular release and reassociation with cell wall chitin. J Biol Chem 275:7925–7934

Brandhorst T, Wüthrich M, Warner T, Klein B (1999) Targeted gene disruption reveals an adhesin indispensable for pathogenicity of *Blastomyces dermatitidis*. J Exp Med 189:1207–1216

Braun EJ, Howard RJ (1994a) Adhesion of *Cochliobolus heterostrophus* conidia and germlings to leaves and artificial surfaces. Exp Mycol 18:211–220

Braun EJ, Howard RJ (1994b) Adhesion of fungal spores and germlings to host surfaces. Protoplasma 181:202–212

Braus GH, Grundmann O, Brueckner S, Moesch H-U (2003) Amino acid starvation and Gcn4p regulate adhesive growth and *FLO11* gene expression in *Saccharomyces cerevisiae*. Mol Biol Cell 14:4272–4284

Breth B, Odenbach D, Yemelin A, Schlinck N, Schröder M, Bode M, Antelo L, Andresen K, Thines E, Foster AJ (2013) The role of the Tra1p transcription factor of *Magnaporthe oryzae* in spore adhesion and pathogenic development. Fungal Genet Biol 57:11–22

Brown NA, dos Reis TF, Goinski AB, Savoldi M, Menino J, Almeida MT, Rodrigues F, Goldman GH (2014) The *Aspergillus nidulans* signalling mucin MsbA regulates starvation responses, adhesion and affects cellulase secretion in response to environmental cues. Mol Microbiol 94:1103–1120

Brückner S, Mösch H-U (2012) Choosing the right lifestyle: adhesion and development in *Saccharomyces cerevisiae*. FEMS Microbiol Rev 36:25–58

Brunner F, Rosahl S, Lee J, Rudd JJ, Geiler C, Kauppinen S, Rasmussen G, Scheel D, Nuernberger T (2002) Pep-13, a plant defense-inducing pathogen-associated pattern from *Phytophthora* transglutaminases. EMBO J 21:6681–6688

Buck JW, Andrews JH (1999a) Attachment of the yeast *Rhodosporidium toruloides* is mediated by adhesives localized at sites of bud cell development. Appl Environ Microbiol 65:465–471

Buck JW, Andrews JH (1999b) Localized, positive charge mediates adhesion of *Rhodosporidium toruloides* to barley leaves and polystyrene. Appl Environ Microbiol 65:2179–2183

Caesar-TonThat TC, Epstein L (1991) Adhesion-reduced mutants and the wild type *Nectria haematococca*: an ultrastructural comparison of the macroconidial walls. Exp Mycol 15:193–205

Carver TLW, Kunoh H, Thomas BJ, Nicholson RL (1999) Release and visualization of the extracellular matrix of conidia of *Blumeria graminis*. Mycol Res 103:547–560

Chaky J, Anderson K, Moss M, Vaillancourt L (2001) Surface hydrophobicity and surface rigidity induce spore germination in *Colletotrichum graminicola*. Phytopathology 91:558–564

Chaubal R, Wilmot VA, Wynn WK (1991) Visualization, adhesiveness, and cytochemistry of the extracellular matrix produced by urediniospore germ tubes of *Puccinia sorghi*. Can J Bot 69:2044–2054

Cormack BP, Ghori N, Falkow S (1999) An adhesin of the yeast pathogen *Candida glabrata* mediating adherence to human epithelial cells. Science 285:578–582

Corrêa A Jr, Staples RC, Hoch HC (1996) Inhibition of thigmostimulated cell differentiation with RGD-peptides in *Uromyces* germlings. Protoplasma 194:91–102

de Groot PWJ, de Boer AD, Cunningham J, Dekker HL, de Jong L, Hellingwerf KJ, de Koster C, Klis FM (2004) Proteomic analysis of *Candida albicans* cell walls reveals covalently bound carbohydrate-active enzymes and adhesins. Eukaryot Cell 3:955–965

de Groot PW, Bader O, de Boer AD, Weig M, Chauhan N (2013) Adhesins in human fungal pathogens: glue with plenty of stick. Eukaryot Cell 12:470–481

de Jong JC, McCormack J, Smirnoff B, Talbot NJ (1997) Glycerol generates turgor in rice blast. Nature 389:244–245

De Nobel H, Sietsma JH, van den Ende H, Klis FM (2001) Molecular organization and construction of the fungal cell wall. In: Howard RJ, Gow NAR (eds) The Mycota, vol VIII, Biology of the fungal cell. Springer, Berlin, pp 181–200

De Vocht ML, Scholtmeijer K, van der Vegte EW, deVries OMH, Sonveaux N, Wösten HAB, Ruysschaert J-M, Hadziioannou G, Wessels JGH, Robillard GT (1998) Structural characterization of the hydrophobin SC3, as a monomer and after self-assembly at hydrophobic/hydrophilic interfaces. Biophys J 74:2059–2068

Deising H, Nicholson RL, Haug M, Howard RJ, Mengden K (1992) Adhesion pad formation and the involvement of cutinase and esterases in the attachment of uredospores to the host cuticle. Plant Cell 4:1101–1111

Deng MQ, Templeton TJ, London NR, Bauer C, Schroeder AA, Abrahamsen MS (2002) *Cryptosporidium parvum* genes containing thrombospondin type 1 domains. Infect Immun 70:6987–6995

Doss RP, Potter SW, Chastagner GA, Christian JK (1993) Adhesion of non-germinated *Botrytis cineria* conidia to several substrata. Appl Environ Biol 59:1786–1791

Doss RP, Potter SW, Soeldner AH, Christian JK, Fukunaga LE (1995) Adhesion of germlings of *Botrytis cinerea*. Appl Environ Microbiol 61:260–265

Driver JD, Holben WE, Rillig (2005) Characterization of glomalin as a hyphal wall component of arbuscular mycorrhizal fungi. Soil Biol Biochem 37:101–106

Dutton LC, Nobbs AH, Jepson K, Jepson MA, Vickerman MM, Alawfi SA, Munro CA, Lamont RJ, Jenkinson HF (2014) O-mannosylation in *Candida albicans* enables development of interkingdom biofilm communities. MBio 5:e00911–e00914

Ebata Y, Yamamoto H, Uchiyama T (1998) Chemical composition of the glue from appressoria of *Magnaporthe grisea*. Biosci Biotechnol Biochem 62:672–674

Ebbole D (2007) *Magnaporthe* as a model for understanding host-pathogen interactions. Annu Rev Phytopathol 45:437–456

Epstein L, Nicholson RL (1997) Adhesion of spores and hyphae to plant surfaces. In: Carroll G, Tudzynski P (eds) The Mycota, vol V, Plant relationships, Part A. Springer, Berlin, pp 11–25

Epstein L, Nicholson RL (2006) Adhesion and adhesives of fungi and oomycetes. In: Smith AM, Callow JA (eds) Biological adhesives. Springer, Berlin, pp 41–62

Epstein L, Laccetti L, Staples RC, Hoch HC, Hoose WA (1985) Extracellular proteins associated with induction of differentiation in bean rust uredospore germlings. Phytopathology 75:1073–1076

Epstein L, Laccetti LB, Staples RC, Hoch HC (1987) Cell-substratum adhesive protein involved in surface contact responses of the bean rust fungus. Physiol Mol Plant Pathol 30:373–388

Epstein L, Kwon YH, Almond DE, Schached LM, Jones MJ (1994) Genetic and biochemical characterization of *Nectria haematococca* strains with adhesive and adhesion-reduced macroconidia. Appl Environ Microbiol 60:524–530

Fabritius A-L, Judelson HS (2003) A mating-induced protein of *Phytophthora infestans* is a member of a family of elicitors with divergent structures and stage-specific patterns of expression. Mol Plant Microbe Interact 16:926–935

Feng J, Wang F, Liu G, Greenshields D, Shen W, Kaminskyj S, Hughes GR, Peng Y, Selvaraj G, Zou J, Wei Y (2009) Analysis of a *Blumeria graminis*-secreted lipase reveals the importance of host epicuticular wax components for fungal adhesion and development. Mol Plant Microbe Interact 22:1601–1610

Fernández-Álvarez A, Elías-Villalobos A, Ibeas JI (2009) The O-mannosyltransferase PMT4 Is essential for normal appressorium formation and penetration in *Ustilago maydis*. Plant Cell 21:3397–3412

Fernández-Álvarez A, Marín-Menguiano M, Lanver D, Jiménez-Martín A, Elías-Villalobos A, Pérez-Pulido AJ, Kahmann R, Ibeas JI (2012) Identification of O-mannosylated virulence factors in *Ustilago maydis*. PLoS Pathog 8:e1002563

Fu Y, Rieg G, Fonzi WA, Belanger PH, Edwards JE Jr, Filler SG (1998) Expression of the *Candida albicans* gene *ALS1* in *Saccharomyces cerevisiae* induces adherence to endothelial and epithelial cells. Infect Immun 66:1783–1786

Fu Y, Luo G, Spellberg BJ, Edwards JE, Ibrahim AS (2008) Gene overexpression/suppression analysis of candidate virulence factors of *Candida albicans*. Eukaryot Cell 7:483–492

Fukazawa Y, Kagaya K (1997) Molecular bases of adhesion of *Candida albicans*. J Med Vet Mycol 35:87–99

Gale CA, Bendel CM, McClellan M, Hauser M, Becker JM, Berman J, Hostetter MK (1998) Linkage of adhesion, filamentous growth, and virulence in *Candida albicans* to a single gene, *INT1*. Science 279:1355–1358

Gaulin E, Jauneau A, Villalba F, Rickauer M, Esquerre-Tugaye MT, Bottin A (2002) The CBEL glycoprotein of *Phytophthora parasitica* var. *nicotianae* is involved in cell wall deposition and adhesion to cellulosic substrates. J Cell Sci 115:4565–4575

Gillespie AW, Farrell RE, Walley FL, Ross ARS, Leinweber P, Eckhardt K-U, Regier TZ, Blyth RIR (2011) Glomalin-related soil protein contains non-mycorrhizal-related heat-stable proteins, lipids and humic materials. Soil Biol Biochem 43:766–777

Göernhardt B, Rouhara I, Schmelzer E (2000) Cyst germination proteins of the potato pathogen *Phytophthora infestans* share homology with human mucins. Mol Plant Microbe Interact 13:32–42

Granger BL (2012) Insight into the antiadhesive effect of yeast wall protein 1 of *Candida albicans*. Eukaryot Cell 11:795–805

Granger BL, Flenniken ML, Davis DA, Mitchell AP, Cutler JE (2005) Yeast wall protein 1 of *Candida albicans*. Microbiology 151:1631–1644

Gravelat FN, Beauvais A, Liu H, Lee MJ, Snarr BD, Chen D, Xu W, Kravtsov I, Hoareau CMQ, Vanier G, Urb M, Campoli P, Al Abdallah Q, Lehoux M, Chabot JC, Ouimet M-C, Baptista SD, Fritz JH, Nierman WC, Latgé JP, Mitchell AP, Filler SG, Fontaine T, Sheppard DC (2013) *Aspergillus* galactosaminogalactan mediates adherence to host constituents and conceals hyphal β-glucan from the immune system. PLoS Pathog 9:e1003575

Gubler F, Hardham AR, Duniec J (1989) Characterising adhesiveness of *Phytophthora cinnamomi* zoospores during encystment. Protoplasma 149:24–30

Gunderson JH, Elwood H, Ingold A, Kindle K, Sogin ML (1987) Phylogenetic relationships between chlorophytes, chrysophytes, and oomycetes. Proc Natl Acad Sci USA 84:5823–5827

Hamer JE, Howard RJ, Chumley FG, Valent B (1988) A mechanism for surface attachment in spores of a plant pathogenic fungus. Science 239:288–290

Hardham AR (2001) Cell biology of fungal infection of plants. In: Howard RJ, Gow NAR (eds) The Mycota: biology of the fungal cell, vol 7. Springer, Berlin, pp 91–123

Hardham A, Gubler F (1990) Polarity of attachment of zoospores of a root pathogen and pre-alignment of the emerging germ tube. Cell Biol Int Rep 14:947–956

Hoch HC, Staples RC, Whitehead B, Comeau J, Wolf ED (1987) Signaling for growth orientation and cell differentiation by surface topography in Uromyces. Science 235:1659–1662

Hoi JWS, Herbert C, Bacha N, O'Connell R, Lafitte C, Borderies G, Rossignol M, Rougé P, Dumas B (2007) Regulation and role of a STE12-like transcription factor from the plant pathogen *Colletotrichum lindemuthianum*. Mol Microbiol 64:68–82

Hostetter MK (2000) RGD-mediated adhesion in fungal pathogens of humans, plants and insects. Curr Opin Microbiol 3:344–348

Howard RJ, Valent B (1996) Breaking and entering: host penetration by the fungal rice blast pathogen *Magnaporthe grisea*. Annu Rev Microbiol 50:491–512

Howard RJ, Ferrari MA, Roach DH, Money NP (1991) Penetration of hard substrates by a fungus employing enormous turgor pressures. Proc Natl Acad Sci USA 88:11281–11284

Howard RJ, Sweigard JA, Hitz WD, Chumley FG, Valent BS (1993) Purified spore tip mucilage. US patent 5,264,569

Hoyer LL (2001) The ALS gene family of *Candida albicans*. Trends Microbiol 9:176–180

Hoyer LL, Green CB, Oh SH, Zhao X (2008) Discovering the secrets of the *Candida albicans* agglutinin-like sequence (ALS) gene family – a sticky pursuit. Med Mycol 46:1–15

Hu G, St. Leger RJ (2002) Field studies using a recombinant mycoinsecticide (*Metarhizium anisopliae*) reveal that it is rhizosphere competent. Appl Environ Microbiol 68:6383–6387

Hughes HB, Carzaniga R, Rawlings SL, Green JR, O'Connell RJ (1999) Spore surface glycoproteins of *Colletotrichum lindemuthianum* are recognized by a monoclonal antibody which inhibits adhesion to polystyrene. Microbiology 145:1927–1936

Inoue K, Suzuki T, Ikeda K, Jiang S, Hosogi N, Hyong GS, Hida S, Yamada T, Park P (2007) Extracellular matrix of *Magnaporthe oryzae* may have a role in host adhesion during fungal penetration and is digested by matrix metalloproteinases. J Gen Plant Pathol 73:388–398

Inoue K, Onoe T, Park P, Ikeda K (2011) Enzymatic detachment of spore germlings in *Magnaporthe oryzae*. FEMS Microbiol Lett 323:13–19

Inoue K, Kitaoka H, Park P, Ikeda K (2016) Novel aspects of hydrophobins in wheat isolate of *Magnaporthe oryzae*: Mpg1, but not Mhp1, is essential for adhesion and pathogenicity. J Gen Plant Pathol 82:18–28

Iranzo M, Aguado C, Pallotti C, Canizares JV, Mormeneo S (2002) Transglutaminase activity is involved in *Saccharomyces cerevisiae* wall construction. Microbiology 148:1329–1334

Izumitsu K, Kimura S, Kobayashi H, Morita A, Saitoh Y, Tanaka C (2010) Class I hydrophobin BcHpb1 is important for adhesion but not for later infection of *Botrytis cinerea*. J Gen Plant Pathol 76:254–260

Jones EBG (1994) Fungal adhesion. Mycol Res 98:961–981

Jones MJ, Epstein L (1989) Adhesion of *Nectria haematococca* macroconidia. Physiol Mol Plant Pathol 35:453–461

Jones MJ, Epstein L (1990) Adhesion of macroconidia to the plant surface and virulence of *Nectria haematococca*. Appl Environ Microbiol 56:3772–3778

Kanbe T, Cutler JE (1998) Minimum chemical requirements for adhesin activity of the acid-stable part of *Candida albicans* cell wall phosphomannoprotein complex. Infect Immun 66:5812–5818

Kempf M, Apaire-Marchais V, Saulnier P, Licznar P, Lefrançois C, Robert R, Cottin J (2007) Disruption of *Candida albicans IFF4* gene involves modifications of the cell electrical surface properties. Colloids Surf B Biointerfaces 58:250–255

Kershaw MJ, Talbot NJ (1998) Hydrophobins and repellents: proteins with fundamental roles in fungal morphogenesis. Fungal Genet Biol 23:18–33

Kershaw MJ, Wakley GE, Talbot NJ (1998) Complementation of the *Mpg1* mutant phenotype in *Magnaporthe grisea* reveals functional relationships between fungal hydrophobins. EMBO J 17:3838–3849

Klis FM, Brul S, De Groot PWJ (2010) Covalently linked wall proteins in ascomycetes. Yeast 27:489–493

Kunoh H, Yamaoka N, Yoshioka H, Nicholson RL (1988) Preparation of the infection court by *Erysiphe graminis*: I. Contact mediated changes in morphology of the conidium surface. Exp Mycol 12:325–335

Kuo K-C, Hoch HC (1996) Germination of *Phyllosticta ampelicida* pycnidiospores: prerequisite of adhesion to the substratum and the relationship of substratum wettability. Fungal Genet Biol 20:18–29

Kwan AH, Winefield RD, Sunde M, Matthews JM, Haverkamp RG, Templeton MD, Mackay JP (2006) Structural basis for rodlet assembly in fungal hydrophobins. Proc Natl Acad Sci USA 103:3621–3626

Kwon YH, Epstein L (1993) A 90-kDa glycoprotein associated with adhesion of *Nectria haematococca* macroconidia to substrata. Mol Plant Microbe Interact 6:481–487

Kwon YH, Epstein L (1997a) Involvement of the 90 kDa glycoprotein in adhesion of *Nectria haematococca* macroconidia. Physiol Mol Plant Pathol 51:287–303

Kwon YH, Epstein L (1997b) Isolation and composition of the 90 kDa glycoprotein associated with adhesion of *Nectria haematococca* macroconidia. Physiol Mol Plant Pathol 51:63–74

Lacroix H, Whiteford JR, Spanu PD (2008) Localization of *Cladosporium fulvum* hydrophobins reveals a role for HCf-6 in adhesion. FEMS Microbiol Lett 286:136–144

Lau GW, Hamer JE (1998) *Acropetal:* a genetic locus required for conidiophore architecture and pathogenicity in the rice blast fungus. Fungal Genet Biol 24:228–239

Leite B, Nicholson RL (1992) Mycosporine-alanine: a self-inhibitor of germination from the conidial mucilage of *Colletotrichum graminicola*. Exp Mycol 16:76–86

Li F, Palecek S (2003) *EAP1*, a *Candida albicans* gene involved in binding human epithelial cells. Eukaryot Cell 2:1266–1273

Li F, Svarovsky MJ, Karlsson AJ, Wagner JP, Marchillo K, Oshel P, Andes D, Palecek SP (2007) Eap1p, an adhesin that mediates *Candida albicans* biofilm formation in vitro and in vivo. Eukaryot Cell 6:931–939

Li J, Zou C, Xu J, Ji X, Niu X, Yang J, Huang X, Zhang KQ (2015) Molecular mechanisms of nematode-nematophagous microbe interactions: basis for biological control of plant-parasitic nematodes. Annu Rev Phytopathol 53:67–95

Liang L, Shen R, Mo Y, Yang J, Ji X, Zhang K-Q (2015) A proposed adhesin AoMad1 helps nematode-trapping fungus *Arthrobotrys oligospora* recognizing host signals for life-style switching. Fungal Genet Biol 81:172–181

Lin CJ, Sasse C, Gerke J, Valerius O, Irmer H, Frauendorf H, Heinekamp T, Straßburger M, Tran VT, Herzog B, Braus-Stromeyer SA (2015) Transcription factor SomA is required for adhesion, development and virulence of the human pathogen *Aspergillus fumigatus*. PLoS Pathog 11:e1005205

Liu ZM, Kolattukudy PE (1999) Early expression of the calmodulin gene, which precedes appressorium formation in *Magnaporthe grisea*, is inhibited by self-inhibitors and requires surface attachment. J Bacteriol 181:3571–3577

Loza L, Fu Y, Ibrahim AS, Sheppard DC (2004) Functional analysis of the *Candida albicans ALS1* gene product. Yeast 21:473–482

Masuoka J, Wu G, Glee PM, Hazen KC (1999) Inhibition of *Candida albicans* attachment to extracellular matrix by antibodies which recognize hydrophobic cell wall proteins. FEMS Immunol Med Microbiol 24:421–429

Mendgen K, Hahn M, Deising H (1996) Morphogenesis and mechanisms of penetration by plant pathogenic fungi. Annu Rev Phytopathol 34:367–386

Mercure EW, Leite B, Nicholson RL (1994) Adhesion of ungerminated conidia of *Colletotrichum graminicola* to artificial hydrophobic surfaces. Physiol Mol Plant Pathol 45:421–440

Mercure EW, Kunoh H, Nicholson RL (1995) Visualization of materials released from adhered, ungerminated conidia of *Colletotrichum graminicola*. Physiol Mol Plant Pathol 46:121–135

Modrzewska B, Kurnatowski P (2015) Adherence of *Candida* sp. to host tissues and cells as one of its pathogenicity features. Ann Parasitol 61:3–9

Newman SL, Chaturvedi S, Klein BS (1995) The WI-1 antigen on *Blastomyces dermatitidis* yeasts mediates binding to human macrophage CD18 and CD14 receptors. J Immunol 154:753–761

Nicholson RL, Epstein L (1991) Adhesion of fungi to the plant surface: prerequisite for pathogenesis. In: Cole GT, Hoch HC (eds) The fungal spore and disease initiation in plants and animals. Plenum Press, New York, pp 3–23
Nicholson RL, Moraes WBC (1980) Survival of *Colletotrichum graminicola*: importance of the spore matrix. Phytopathology 70:255–261
Nicholson RL, Yoshioka H, Yamaoka N, Kunoh H (1988) Preparation of the infection court by *Erysiphe graminis*. II. Release of esterase enzyme from conidia in response to a contact stimulus. Exp Mycol 12:336–349
Nordbring-Hertz B, Jansson H-B, Tunlid A (2006) Nematophagous fungi. In: Encyclopedia of life sciences. Wiley, New York, www.els.net
Ohtake M, Yamamoto H, Uchiyama T (1999) Influences of metabolic inhibitors and hydrolytic enzymes on the adhesion of appressoria of *Pyricularia oryzae* to wax-coated cover-glasses. Biosci Biotechnol Biochem 63:978–982
Osherov N, May GS (2001) The molecular mechanisms of conidial germination. FEMS Microbiol Lett 199:153–160
Pain NA, Green JR, Jones GL, O'Connell RJ (1996) Composition and organisation of extracellular matrices around germ tubes and appressoria of *Colletotrichum lindemuthianum*. Protoplasma 190:119–130
Pascholati S, Yoshioka H, Kunoh H, Nicholson RL (1992) Preparation of the infection court by *Erysiphe graminis* f. sp. *hordei*: cutinase is a component of the conidial exudate. Physiol Mol Plant Pathol 41:53–59
Pascholati SF, Deising H, Leite B, Anderson D, Nicholson RL (1993) Cutinase and non-specific esterase activities in the conidial mucilage of *Colletotrichum graminicola*. Physiol Mol Plant Pathol 42:37–51
Prados-Rosales R, Luque-Garcia JL, Martínez-López R, Gil C, DiPietro A (2009) The *Fusarium oxysporum* cell wall proteome under adhesion-inducing conditions. Proteomics 9:4755–4769
Pringle RB (1981) Nonspecific adhesion of *Bipolaris sorokiniana* sporelings. Can J Plant Pathol 3:9–11
Ramage G, Mowat E, Jones B, Williams C, Lopez-Ribot J (2009) Our current understanding of fungal biofilms. Crit Rev Microbiol 35:340–355
Rawlings SL, O'Connell RJ, Green JR (2007) The spore coat of the bean anthracnose fungus *Colletotrichum lindemuthianum* is required for adhesion, appressorium development and pathogenicity. Physiol Mol Plant Pathol 70:110–119
Recorbet G, Alabouvette C (1997) Adhesion of *Fusarium oxysporum* conidia to tomato roots. Lett Appl Microbiol 25:375–379
Reynolds TB, Fink GR (2001) Bakers' yeast, a model for fungal biofilm formation. Science 291:878–881
Rillig MC (2004) *Arbuscular mycorrhizae*, glomalin, and soil aggregation. Can J Soil Sci 84 (4):355–363
Robold AV, Hardham AR (2004) Production of monoclonal antibodies against peripheral vesicle proteins in zoospores of *Phytophthora nicotianae*. Protoplasma 223:121–132
Robold AV, Hardham AR (2005) During attachment *Phytophthora* spores secrete proteins containing thrombospondin type 1 repeats. Curr Genet 47:307–315
Rossouw D, Bagheri B, Setati ME, Bauer FF (2015) Co-flocculation of yeast species, a new mechanism to govern population dynamics in microbial ecosystems. PLoS One 10:e0136249
Ruiz-Herrera J, Iranzo M, Elorza MV, Sentandreu R, Mormeno S (1995) Involvement of transglutaminase in the formation of covalent cross-links in the cell wall of *Candida albicans*. Arch Microbiol 164:186–193
Samuels L, Kunst L, Jetter R (2008) Sealing plant surfaces: cuticular wax formation by epidermal cells. Annu Rev Plant Biol 59:683–707
Scholtmeijer K, Janssen MI, Gerssen B, de Vocht ML, van Leeuwen BM, van Kooten TG, Wősten HAB, Wessels JGH (2002) Surface modifications created by using engineered hydrophobins. Appl Environ Microbiol 68:1367–1373

Schumacher CFA, Steiner U, Dehne HW, Oerke E-C (2008) Localized adhesion of nongerminated *Venturia inaequalis* conidia to leaves and artificial surfaces. Phytopathology 98:760–768

Schuren FHJ, Wessels JGH (1990) Two genes specifically expressed in fruiting dikaryons of *Schizophyllum commune*: homologies with a gene not regulated by mating type genes. Gene 90:199–205

Schuster M, Schweizer G, Reissmann S, Kahmann R (2016) Genome editing in *Ustilago maydis* using the CRISPR-Cas system. Fungal Genet Biol 89:3–9

Sela-Buurlage MB, Epstein L, Rodriguez RJ (1991) Adhesion of ungerminated *Colletotrichum musae* conidia. Physiol Mol Plant Pathol 39:345–352

Senchou V, Weide R, Carrasco A, Bouyssou H, Pont-Lezica R, Govers F, Canut H (2004) High affinity recognition of a *Phytophthora* protein by *Arabidopsis* via an RGD motif. Cell Mol Life Sci 61:502–509

Shaw BD, Hoch HC (1999) The pycnidiospore of *Phyllosticta ampelicida*: surface properties involved in substratum attachment and germination. Mycol Res 103:915–924

Shaw BD, Hoch HC (2000) $Ca^{2+}$ regulation of *Phyllosticta ampelicida* pycnidiospore germination and appressorium formation. Fungal Genet Biol 31:43–53

Shaw BD, Hoch HC (2001) Ions as regulators of growth and development. In: Howard RJ, Gow NAR (eds) The Mycota, vol VIII, Biology of the fungal cell. Springer, Berlin, pp 73–89

Silva-Dias A, Miranda IM, Branco J, Monteiro-Soares M, Pina-Vaz C, Acácio G, Rodrigues AG (2015) Adhesion, biofilm formation, cell surface hydrophobicity, and antifungal planktonic susceptibility: relationship among *Candida* spp. Front Microbiol 6:205

Slawecki RA, Ryan EP, Young DH (2002) Novel fungitoxicity assays for inhibition of germination-associated adhesion of *Botrytis cinerea* and *Puccinia recondita* spores. Appl Environ Microbiol 68:597–601

Snetselaar KM, McCann MP (2001) From bud to appressorium: morphology of the *Ustilago maydis* transition from saprobic to parasitic growth. Phytopathology 91:S165 (abstract)

Spotts RA, Holz G (1996) Adhesion and removal of conidia of *Botrytis cinerea* and *Penicillium expansum* from grape and plum fruit surfaces. Plant Dis 80:691–699

Staab JF, Bradway SD, Fidel PL, Sundstrom P (1999) Adhesive and mammalian transglutaminase substrate properties of *Candida albicans* Hwp1. Science 283:1535–1538

Staab JF, Bahn YS, Tai CH, Cook PF, Sundstrom P (2004) Expression of transglutaminase substrate activity on *Candida albicans* germ tubes through a coiled, disulfide-bonded N--terminal domain of Hwp1 requires C-terminal glycosylphosphatidylinositol modification. J Biol Chem 279:40737–40747

Stanley MS, Callow ME, Perry R, Alberte RS, Smith R, Callow JA (2002) Inhibition of fungal spore adhesion by zosteric acid as the basis for a novel, nontoxic crop protection technology. Phytopathology 92:378–383

Sugui JA, Leite B, Nicholson RL (1998) Partial characterization of the extracellular matrix released onto hydrophobic surfaces by conidia and conidial germlings of *Colletotrichum graminicola*. Physiol Mol Plant Pathol 52:411–425

Sundstrom P (2002) Adhesion in *Candida* spp. Cell Microbiol 4:461–469

Takano Y, Kikuchi T, Kubo Y, Hamer JE, Mise K, Furusawa I (2000) The *Colletotrichum lagenarium* MAP kinase gene *CMK1* regulates diverse aspects of fungal pathogenesis. Mol Plant Microbe Interact 13:374–383

Talbot NJ, Kershaw M, Wakley GE, De Vries OMH, Wessels JGH, Hamer JE (1996) *MPG1* encodes a fungal hydrophobin involved in surface interactions during infection-related development of the rice blast fungus *Magnaporthe grisea*. Plant Cell 8:985–999

Teertstra WR, Deelstra HJ, Vranes M, Bohlmann R, Kahmann R, Kämper J, Wösten HA (2006) Repellents have functionally replaced hydrophobins in mediating attachment to a hydrophobic surface and in formation of hydrophobic aerial hyphae in *Ustilago maydis*. Microbiology 152:3607–3612

Teertstra WR, van der Velden GJ, de Jong JF, Kruijtzer JA, Liskamp RM, Kroon-Batenburg LM, Müller WH, Gebbink MF, Wösten HA (2009) The filament-specific Rep1-1 repellent of the phytopathogen *Ustilago maydis* forms functional surface-active amyloid-like fibrils. J Biol Chem 284:9153–9159

Terhune BT, Hoch HC (1993) Substrate hydrophobicity and adhesion of *Uromyces* urediniospores and germlings. Exp Mycol 17:241–252

Tian L, Xu J, Zhou L, Guo W (2014) VdMsb regulates virulence and microsclerotia production in the fungal plant pathogen *Verticillium dahliae*. Gene 550:238–244

Timpel C, Strahl-Bolsinger S, Ziegelbauer K, Ernst JF (1998) Multiple functions of Pmt1p-mediated protein *O*-mannosylation in the fungal pathogen *Candida albicans*. J Biol Chem 273:20837–20846

Timpel C, Zink S, Strahl-Bolsinger S, Schroppel K, Ernst J (2000) Morphogenesis, adhesive properties, and antifungal resistance depend on the Pmt6 protein mannosyltransferase in the fungal pathogen *Candida albicans*. J Bacteriol 182:3063–3071

Tomley FM, Soldati DS (2001) Mix and match modules: structure and function of microneme proteins in apicomplexan parasites. Trends Parasitol 17:81–88

Tran VT, Braus-Stromeyer SA, Kusch H, Reusche M, Kaever A, Kühn A, Valerius O, Landesfeind M, Aßhauer K, Tech M, Hoff K (2014) Verticillium transcription activator of adhesion Vta2 suppresses microsclerotia formation and is required for systemic infection of plant roots. New Phytol 202:565–581

Tsuchimori N, Sharkey LL, Fonzi WA, French SW, Edwards JE Jr, Filler SG (2000) Reduced virulence of *HWP1*-deficient mutants of *Candida albicans* and their interactions with host cells. Infect Immun 68:1997–2002

Tucker RP (2004) Molecules in focus. The thrombospondin type 1 repeat superfamily. Int J Biochem Cell Biol 36:969–974

Tucker SL, Talbot NJ (2001) Surface attachment and pre-penetration stage development by plant pathogenic fungi. Annu Rev Phytopathol 39:385–417

Uppuluri P, Pierce CG, Lopez-Ribot JL (2009) *Candida albicans* biofilm formation and its clinical consequences. Future Microbiol 4:1235–1237

Verstrepen KJ, Klis FM (2006) Flocculation, adhesion and biofilm formation in yeasts. Mol Microbiol 60:5–15

Vreeland V, Epstein L (1996) Analysis of plant-substratum adhesives. In: Jackson JF, Linskens H-F (eds) Modern methods of plant analysis, vol 17, Plant cell wall analysis. Springer, Berlin, pp 95–116

Vreeland V, Waite JH, Epstein L (1998) Polyphenols and oxidases in substratum adhesion by marine algae and mussels. J Phycol 34:1–8

Wang C, St. Leger RJ (2007) The MAD1 adhesin of *Metarhizium anisopliae* links adhesion with blastospore production and virulence to insects, and the MAD2 adhesin enables attachment to plants. Eukaryot Cell 6:808–816

Wang L, Zhai B, Lin XR (2012) The link between morphotype transition and virulence in *Cryptococcus neoformans*. PLoS Pathog 8:e1002765

Wang L, Tian X, Gyawali R, Lin X (2013) Fungal adhesion protein guides community behaviors and autoinduction in a paracrine manner. Proc Natl Acad Sci USA 110:11571–11576

Wang JJ, Qiu L, Chu ZJ, Ying SH, Feng MG (2014) The connection of protein O-mannosyltransferase family to the biocontrol potential of *Beauveria bassiana*, a fungal entomopathogen. Glycobiology 24:638–648

Watanabe K, Parbery DG, Kobayashi T, Doi Y (2000) Conidial adhesion and germination of *Pestalotiopsis neglecta*. Mycol Res 104:962–968

Wösten HAB, Schuren FHJ, Wessels JGH (1994) Interfacial self-assembly of a hydrophobin into an amphipathic membrane mediates fungal attachment to hydrophobic surfaces. EMBO J 13:5848–5854

Wright AJ, Thomas BJ, Kunoh H, Nicholson RL, Carver TLW (2002) Influences of substrata and interface geometry on the release of extracellular material by *Blumeria graminis* conidia. Physiol Mol Plant Pathol 61:163–178

Xiao J-Z, Ohshima A, Kamakura T, Ishiyama T, Yamaguchi I (1994a) Extracellular glycoprotein (s) associated with cellular differentiation in *Magnaporthe grisea*. Mol Plant Microbe Interact 7:639–644

Xiao J-Z, Watanabe T, Kamakura T, Ohshima A, Yamaguchi I (1994b) Studies on cellular differentiation of *Magnaporthe grisea*. Physicochemical aspects of substratum surfaces in relation to appressorium formation. Physiol Mol Plant Pathol 44:227–236

Yamaoka N, Takeuchi Y (1999) Morphogenesis of the powdery mildew fungus in water (4). The significance of conidium adhesion to the substratum for normal appressorium development in water. Physiol Mol Plant Pathol 54:145–154

Yang E, Xu L, Yang Y, Zhang X, Xiang M, Wang C, An Z, Liu X (2012) Origin and evolution of carnivorism in the Ascomycota (fungi). Proc Natl Acad Sci USA 109:10960–10965

Zhao X, Oh S-H, Cheng G, Green CB, Nuessen JA, Yeater K, Leng RP, Brown AJP, Hoyer LL (2004) ALS3 and ALS8 represent a single locus that encodes a *Candida albicans* adhesin; functional comparisons between Als3p and Als1p. Microbiology 150:2415–2428

# Chapter 3
# Diatom Adhesives: Molecular and Mechanical Properties

**Paul J. Molino, Anthony Chiovitti, Michael J. Higgins, Tony M. Dugdale, and Richard Wetherbee**

**Abstract** Diatoms are an incredibly diverse group of microalgae that are primarily characterized by their highly ornamented siliceous cell wall. Diatoms have long been a source of interest in the field of bioadhesion due to their ability to adhere and glide upon surfaces, as well as construct extracellular adhesive structures that can enable temporary or permanent adhesion to a surface. This chapter introduces the various adhesion and motility strategies employed by a number of benthic diatom species, describing the mechanisms that facilitate these processes. We review the chemical, molecular, and physical characterization of extracellular polymeric substances resident on the cell surface, as well as those secreted specifically for adhesion and motility from various structures in the cell wall. We highlight the significant work undertaken using atomic force microscopy to understand the nanomechanical properties of these adhesives, allowing sub-molecular structures to be identified that have allowed new insights into their role in enhancing diatom adhesion to surfaces even in high-energy aquatic environments. Finally we look at how new technological advances, such as the recent sequencing of several diatom genomes, together with an integrative research approach, will facilitate the further identification and characterization of diatom adhesives.

P.J. Molino (✉) • M.J. Higgins
Australian Institute for Innovative Materials, Intelligent Polymer Research Institute, University of Wollongong, Wollongong, NSW 2500, Australia
e-mail: pmolino@uow.edu.au

A. Chiovitti • T.M. Dugdale • R. Wetherbee
School of Botany, University of Melbourne, Melbourne, VIC 3010, Australia

© Springer International Publishing Switzerland 2016

A.M. Smith (ed.), *Biological Adhesives*, DOI 10.1007/978-3-319-46082-6_3

## 3.1 Diatoms and Adhesion

### 3.1.1 Diatom Morphology

Diatoms are a ubiquitous group of unicellular microalgae characterized by their highly ornate, silicified cell walls. They are the major primary producers in most aquatic habitats, responsible for ca. 20 % of global carbon fixation, and they are the dominant group of organisms involved in the global turnover of biogenic silica (Mann 1999). The cell wall, or "frustule," of diatoms comprises the silicified wall and associated organic polymers. The silicified portion of the frustule comprises two overlapping halves (or "valves") that fit together like a petri dish plus several encircling girdle bands that enclose the protoplast (Fig. 3.1i; for a review, see Pickett-Heaps et al. 1990). Diatoms are classified according to their valve symmetry as centric (radial) or pennate (bilateral) (Round et al. 1990). The centric forms are mostly planktonic, whereas the pennate forms are mostly benthic and capable of attachment and motility on natural or artificial substrata. This chapter focuses on the adhesives of benthic forms, which represent a small fraction of the total number of species.

### 3.1.2 Significance of Diatom Adhesion

Adhesion to a substratum is a well-documented strategy for growth and survival in a range of marine and freshwater organisms and provides photosynthetic protists with an optimal location to access soluble nutrients and light. The adhesion of benthic diatoms is associated with the secretion of mucilaginous material or extracellular polymeric substances (EPS) (Hoagland et al. 1993). An accumulation of diatom cells and their insoluble EPS (i.e., a biofilm) can beneficially stabilize benthic aquatic habitats by consolidating soft sediments (Paterson 1989; Underwood and Paterson 2003). However, diatoms also cause adverse economic impacts because they are the most frequent and successful microalgal foulers of submerged artificial structures (Bohlander 1991; Alberte et al. 1992). Even when diatoms are exposed to biocides that successfully discourage most organisms, such as copper or tributyl tin, surfaces frequently develop thick diatom slimes (Daniel et al. 1980; Callow 1986, 1996; Molino et al. 2008a). Owing to concerns about environmental toxicity, a worldwide ban has been placed on organotin antifouling paints by the International Maritime Organization and will become fully enforced in 2008 (Omae 2003). These biocidal antifouling paints are gradually becoming replaced by a new generation of environmentally benign antifouling paints based on silicone elastomers with low surface energy (Anderson et al. 2003), yet these paints are also relatively susceptible to fouling by diatoms (Clarkson and Evans 1995a, b; Holland et al. 2004; Molino et al. 2008a).

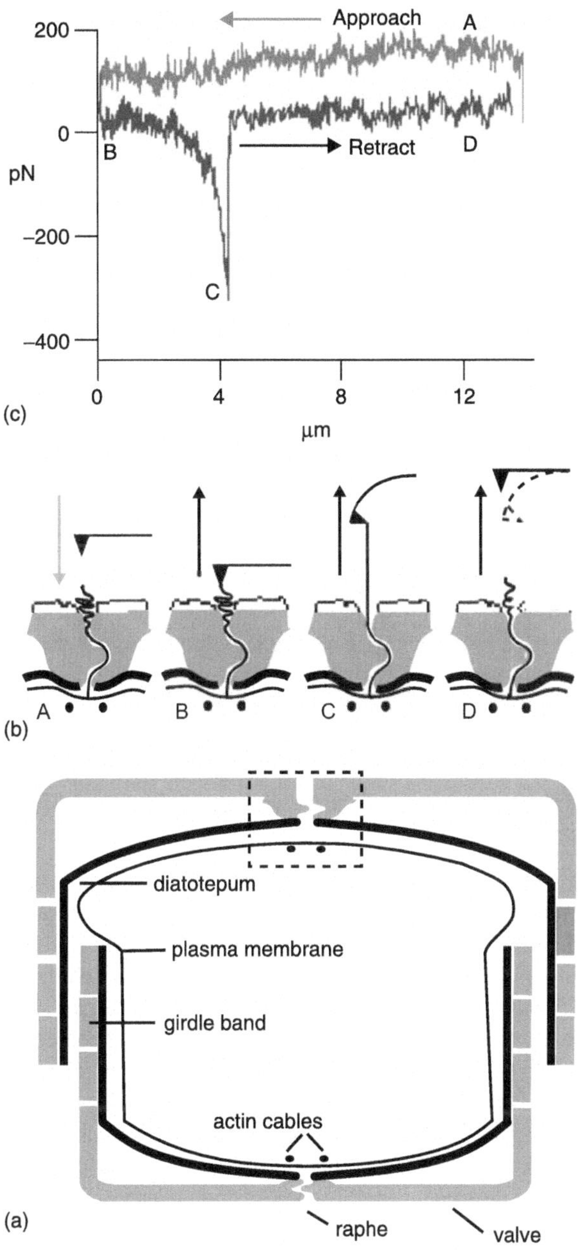


**Fig. 3.1** Interaction between the AFM probe and the raphe strands of a diatom. (i) Schematic diagram of a raphid diatom in transverse section. The non-driving raphe in the boxed region is expanded in (ii). (ii) Diagram depicting the interaction between the AFM cantilever tip and a single adhesive raphe strand. (iii) A force-separation curve recorded for the interactions depicted in (ii). Points A–D in (ii) correspond to those in (iii). During point A, the cantilever tip approaches the raphe without interactions occurring. At point B, the cantilever tip engages the raphe strand. Subsequent retraction of the cantilever tip extends the raphe strand. At point C, the raphe strand is fully extended causing the cantilever tip to deflect downward (ii) and is recorded as an increasingly applied force in the curve (iii). The raphe strand detaches from the tip at point D, allowing the cantilever tip to return to the zero deflection line. Adapted from Higgins et al. (2002)

### 3.1.3 Diatom Adhesion Strategies

Diatoms are passively carried to surfaces by the action of currents and localized water movement or by settling under gravity. With the exception of the motile male gametes, diatoms lack flagella throughout their life cycle (Round et al. 1990) and therefore are unable to actively seek a surface to attach to, although a surface-sensing mechanism in diatoms has been postulated (Cooksey and Wigglesworth-Cooksey 1992; Wigglesworth-Cooksey and Cooksey 1992) and may be activated after initial adhesion. Contact with the substratum initiates a set of processes that involve stabilization of initial attachment and reorientation of the cell (Wetherbee et al. 1998; Molino and Wetherbee 2008). Following initial adhesion, benthic diatoms generally adopt either of two lifestyle strategies. Many diatoms use attachment as traction for a form of cell motility termed "gliding" (Edgar and Pickett-Heaps 1984; Wetherbee et al. 1998; Molino and Wetherbee 2008). Motile diatoms may adjust their position on a substratum or in the sediment in response to a range of stimuli, including tidal fluctuation, diurnal rhythm, nutrient availability, and UV-B irradiation (Cooksey and Cooksey 1988; Serôdio et al. 1997; Smith and Underwood 1998; Moroz et al. 1999). Alternatively, diatoms become sessile, remaining attached to the substratum at one position for an extended period. For both motile and sessile adhesion strategies, cell-substratum adhesion is mediated by the secretion of EPS (Edgar and Pickett-Heaps 1984; Hoagland et al. 1993).

### 3.1.4 General Composition of Diatom Mucilages

Diatom mucilages are complex, multicomponent materials. Many early staining and compositional studies demonstrated that carbohydrates dominate diatom mucilages (for a review, see Hoagland et al. 1993), and during the last decade, this principle has shaped the direction and focus of most studies analyzing the EPS chemistry of unialgal diatom cultures (Wustman et al. 1997, 1998; McConville et al. 1999; Staats et al. 1999; Khandeparker and Bhosle 2001; de Brouwer and Stal 2002; Chiovitti et al. 2003b; Bellinger et al. 2005). The data in these studies generally indicate that the carbohydrates comprise complex, anionic polysaccharides with heterogeneous monosaccharide compositions, sulfate ester, and/or uronic acids. Neutral monosaccharides that have been identified include hexoses, pentoses, 6-deoxyhexoses, and *O*-methylated sugars. The co-occurrence of proteins and carbohydrates was noted in a number of studies (Smestad Paulsen et al. 1978; Bhosle et al. 1995; Lind et al. 1997; Wustman et al. 1997; McConville et al. 1999; Staats et al. 1999; Khandeparker and Bhosle 2001; Chiovitti et al. 2003a, b, 2008; Willis et al. 2013; Poulson et al. 2014), but the relationship between the protein and the carbohydrate has been rarely explored (Wustman et al. 1998; Chiovitti et al. 2003a). In contrast to most studies of diatom EPS, biochemical analyses of the extracellular mucilages of *Craspedostaouros australis* (Chiovitti et al. 2003a) and atomic force microscopy (AFM) of the adhesive pads of *Toxarium undulatum* (Dugdale et al. 2005; Chiovitti et al. 2008) demonstrated that protein was the fundamental component of the adhesives. For *Achnanthes longipes* stalks, the protein content of the EPS was relatively low, but its influence on EPS structure and properties was substantial (Wustman et al. 1998). A role for protein in *Navicula perminuta* adhesion was implied by reduction of adhesion strength by treatment with commercial proteases (Pettitt et al. 2004).

The extracellular location of components extracted from diatoms is often based on inference, and the inherent risk of misinterpreting the origin of nominal EPS was demonstrated for water-extractable intracellular glucans (Chiovitti et al. 2004). Localization of structural units in benthic diatom EPS has been probed by labeling with lectins or specific antibodies (Lind et al. 1997; Wustman et al. 1997, 1998; Wigglesworth-Cooksey and Cooksey 2005) or by assessing changes in the appearance and/or properties of diatom EPS by EM or AFM (McConville et al. 1999; Chiovitti et al. 2003b). The composition of various complex diatom EPS has been reviewed by Hoagland et al. (1993) and Underwood and Paterson (2003). The present discussion focuses on recent studies that relate mucilage chemistry to its fine structure and physical properties.

## 3.2 Adhesion and Gliding of Raphid Diatoms

### *3.2.1 Adhesion and Gliding Behavior*

In most benthic pennate diatoms, adhesive EPS is secreted from a longitudinal slit in the valve termed the "raphe" (Wetherbee et al. 1998). Most such "raphid" diatoms have a raphe in each valve, although members of the genus *Amphora* differ by having both raphes located on the ventral surface of the cell (Daniel et al. 1980; Round et al. 1990). When raphid diatoms settle from the water column onto the substratum, they typically land on their girdle surface. Initial adhesion is therefore mediated by cell surface mucilage and the substratum, and it is relatively weak since the diatom may be readily dislodged (Lind et al. 1997; Wang et al. 1997; Higgins et al. 2003a). Following initial adhesion, the cell may rock gently or remain stationary for a short period of time (e.g., 30–90 s in *C. australis*; Lind et al. 1997) before flipping onto one of its valves, bringing the raphe into contact with the substratum. Soon after, the cell typically commences gliding across the substratum parallel to the longitudinal axis of the frustule, simultaneously depositing a trail of adhesive mucilage (Molino and Wetherbee 2008).

Gliding speeds for raphid diatoms are in the range of 0.1–25 $\mu m\ s^{-1}$ but vary substantially depending upon the species, substratum type, and various environmental factors (Edgar and Pickett-Heaps 1984; Webster et al. 1985; Häder and Hoiczyk 1992; Cohn and Weitzell 1996; Lind et al. 1997; Poulsen et al. 1999; Cohn et al. 2003). Gliding diatoms are capable of abrupt reversals in direction, maintaining similar speeds in either direction for prolonged intervals. Rapid, alternating reversals in direction during which the diatom does not shift far from a fixed position have been described as “shunting” or “oscillating” (Moroz et al. 1999; Holland et al. 2004). Diatoms often veer from a linear course, forming paths that depict arcs of varying radii (Edgar and Pickett-Heaps 1984; Cohn and Weitzell 1996; Wigglesworth-Cooksey et al. 1999; Higgins et al. 2003a; Molino et al. 2006; Gutierrez-Medina et al. 2014). The path types engaged by gliding diatoms appear to be broadly characteristic for different species but are not apparently related to cell size or shape (Edgar and Pickett-Heaps 1984; Cohn and Weitzell 1996), except for a few asymmetrical species (Edgar and Pickett-Heaps 1984).

### 3.2.2 *Mechanism of Raphid Diatom Adhesion and Gliding*

Contemporary models for diatom adhesion and motility (Edgar and Pickett-Heaps 1984; Wetherbee et al. 1998) incorporate the elements of contact between the raphe and substratum, the secretion of adhesive mucilage, and the observation that a pair of actin filaments lies beneath the membrane at the raphe (Edgar and Zavortink 1983). By drawing analogies from animal and other protist systems, Wetherbee et al. (1998) elaborated on the earlier proposal of Edgar and Pickett-Heaps (1984) and suggested a detailed model for the adhesion complex (AC, Fig. 3.2) and gliding of raphid diatoms. The AC is a polarized continuum of connector molecules that extend from the actin filaments and associated intracellular motor proteins via transmembrane linker molecules to the extracellular adhesives attached to the

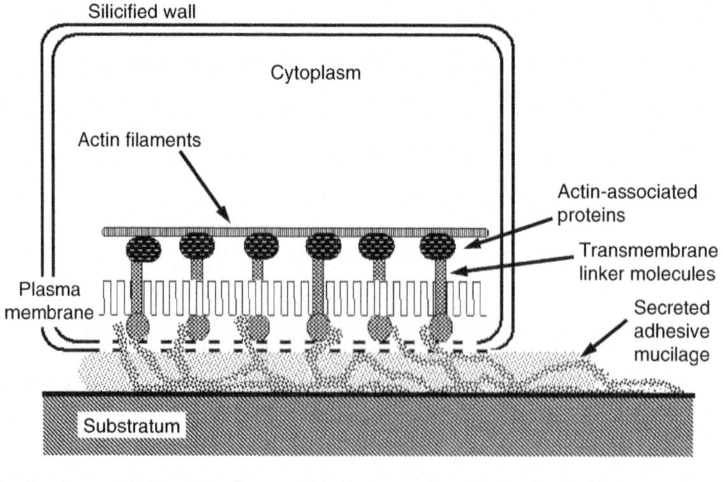


**Fig. 3.2** A schematic diagram of a raphid diatom cell with detail of the components of the proposed adhesion complex. Adapted from Wetherbee et al. (1998)

substratum. The force required to enable gliding is generated by myosin at the actin filaments. As the cell is propelled forward, adhesive EPS are secreted from the raphe, and the components of the AC continually assemble at the front boundary of the contact between the moving cell and the substratum. The fluidity of the plasma membrane allows the transmembrane components of the AC to remain fixed in position while the cell is gliding. The AC components continually disassemble at the rear boundary of cell-substratum contact, resulting in deposition of adhesive EPS in the mucilaginous trails. Although the available data on raphid diatom adhesion and gliding are consistent with the AC model, only the actin and extracellular adhesives have been identified so far. Furthermore, an authentic adhesive motif remains unknown.

The functional role of the actin filaments was demonstrated by reversible inhibition of diatom motility with an actin-destabilizing drug, latrunculin A, combined with in situ FITC-phalloidin labeling and Western blotting analysis (Poulsen et al. 1999). Latrunculin also prevented cells that had settled on their girdles from flipping onto their raphes, indicating the dependence of diatom adhesion on the actin filaments of the AC. An alternative hypothesis for raphid diatom locomotion proposes that the propulsive force is generated by the directional secretion and hydration of mucilage (Gordon 1987). However, when synthetic particles are added to the medium, the particles are observed to adhere to and stream back and forth along one of the raphes (Edgar and Pickett-Heaps 1984; Lind et al. 1997; Poulsen et al. 1999). This particle streaming is often bidirectional and reversible, and different particles can move at different speeds simultaneously. This activity is more sophisticated than can be explained by secretion and hydration of mucilage, which would merely repel the particles from the cell, but it can be rationalized as AC elements, including the extracellular adhesive, being driven independently along each of the actin filaments. TEM has shown that secretory vesicles appear near the raphe, apparently supplying the raphe adhesives (Daniel et al. 1980; Edgar and Pickett-Heaps 1982, 1983, 1984; Wang et al. 2000). Connections were observed between the vesicles and the actin at the raphe in *Navicula cuspidata* (Edgar and Pickett-Heaps 1984), indicating that the actin guides the vesicles into position at the raphe and possibly occurs concomitantly with assembly of the AC.

### 3.2.3 Fine Structure of Raphid Diatom Mucilages

Imaging by conventional SEM, FESEM, TEM, and AFM has all contributed to developing an understanding of the fine structure of diatom mucilages. The chief advantages of AFM over EM techniques are that it enables imaging of mucilages in their native, hydrated state, as well as measurements of the mechanical properties of mucilages at the molecular scale. The operating principles of the AFM (Binning et al. 1986) rely on there being contact between the AFM probe and the sample. The AFM probe comprises a fine silicon nitride tip (ca. 10–20 nm radius) mounted on a flexible cantilever. During imaging, the cantilever tip is scanned across the surface of the sample with either continuous (contact mode) or intermittent (tapping-mode) contact with the sample. The cantilever tip deflects as it encounters topographical changes in the sample, and these deflections are used to digitally construct an image.

#### 3.2.3.1 Cell Surface Mucilage

AFM has provided the most realistic images available of the cell surface mucilage of diatoms in its native, hydrated state (Crawford et al. 2001; Higgins et al. 2002, 2003a, b). Tapping-mode images of the girdle and valve regions of *C. australis* cells showed them to be covered in relatively thick mucilage that is mostly smooth and flat but with some regions showing irregularly striated patterns (Fig. 3.3i; Higgins et al. 2002, 2003b). Height images in tapping mode showed the topography of the *C. australis* mucilage varied up to 25 nm (Higgins et al. 2003b). By using the cantilever tip to scrape surface mucilage away from the siliceous wall, it was shown that mucilage was secreted from the girdle pores of stationary-phase cells but not from those of log-phase cells (Higgins et al. 2002). In contrast, the surface mucilage of *Pinnularia viridis* had the appearance of loosely packed humps which varied by up to 150 nm in height (Fig. 3.3ii; Higgins et al. 2003b). Essentially no surface mucilage was detected for log-phase *Nitzschia navis-varingica* cells, although mucilaginous strands were observed adhering to the valve but not the girdle (Higgins et al. 2003b). These observations demonstrate that the cell surface mucilage of diatoms is not merely an amorphous coating but has distinctive features that vary substantially between species.

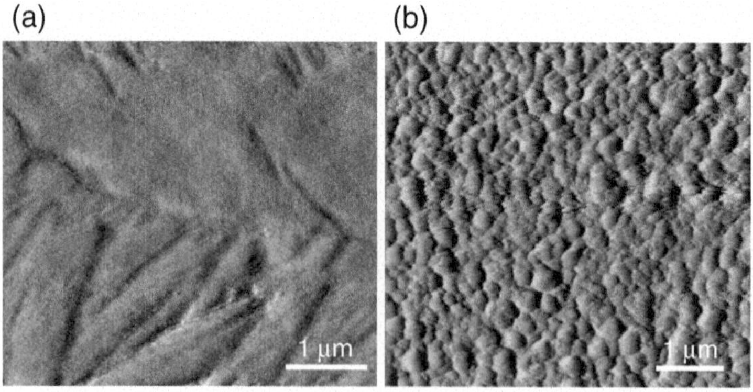


**Fig. 3.3** AFM tapping-mode images of the cell surface mucilage of raphid diatoms. (i) *Craspedostauros australis*. (ii) *Pinnularia viridis*. From Higgins et al. (2003b), reprinted with permission from the *Journal of Phycology*

#### 3.2.3.2 Raphe Mucilage

The adhesive raphe strands are yet to be imaged by AFM mainly because the strong binding of raphe strands to the cantilever tip impedes imaging (Higgins et al. 2003a) but probably also because the raphe strands are too mobile in fluid to be imaged. A description of the structure of this mucilage therefore necessarily relies on EM observations of chemically fixed and dehydrated diatom cells. However, the strand-like structure of raphe adhesives indicated by EM is supported by AFM analysis of the properties of native diatom raphe secretions (see Sect. 3.2.4).

SEM of *Navicula* spp., *Pinnularia* spp., and *A. coffeaeformis* showed that mucilage strands emerge from along the entire length of both the non-driving raphe and the driving raphe of inverted cells (Rosowski 1980, Edgar 1983; Rosowski et al. 1983; Webster et al. 1985; Higgins et al. 2003a). TEM of thin-sectioned *N. cuspidata* cells attached to a resin and prepared by critical point drying showed the cells to be adhered to the substratum by the raphe secretion (Edgar 1983). Longitudinal sections of these cells show the strands projecting as regular, fine bristles from the driving raphe, with the distal ends deformed as though the strands were under pressure. This suggests that the presentation of adhesive strands along the length of the raphe in vivo enables the diatom to anchor to and maintain relatively uniform adhesion over surfaces with nano- and microscale roughness since individual strands may be extended or compacted to varying degrees (Jagota and Bennison 2002). TEM of sections of freeze-fractured *N. cuspidata* cells that were rapidly frozen without pretreatment showed that the plasma membrane is closely appressed against the frustule throughout most of the cell, except along the raphe, where it is irregularly wrinkled, indicating that the proximal ends of the secreted adhesive strands retain contact with the plasma membrane (Edgar and Pickett-Heaps 1983, 1984).

SEM of chemically fixed and inverted *P. viridis* cells (exposing the driving raphe) showed distinct, thick mucilaginous strands (ca. 50–100 μm in diameter) protruding from the raphe, some of which were anastomosed (Fig. 3.4i,ii; Higgins et al. 2003a). *P. viridis* cells often had a blob of amorphous material at the central nodule of the non-driving raphe that was attached to collapsed raphe strands. In some cells, long tethers were observed extending 40 μm from the valve face to form an attachment with the substratum. The tethers were ca. 1–2 μm in diameter widening to 10–15 μm at the substratum interface (Fig. 3.4iii). Similar SEM observations of tethers and tangled bundles of strands (ca. 30 nm in diameter) at the central nodule of the non-driving raphe were made for *C. australis*. The invariant occurrence of adhesive EPS at both the driving and the non-driving raphes supports a model for constitutive, rather than induced, secretion of the raphe adhesive.

SEM of some *P. viridis* cells lying on their girdle region showed they bore mucilaginous sheets originating from the raphe of each valve and extending for 10 μm to the substratum (Fig. 3.4iv; Higgins et al. 2003a), an observation which underpins the occasional rocking motion of raphid cells following initial contact. Attachment at both raphes creates a tension between the two competing raphes,

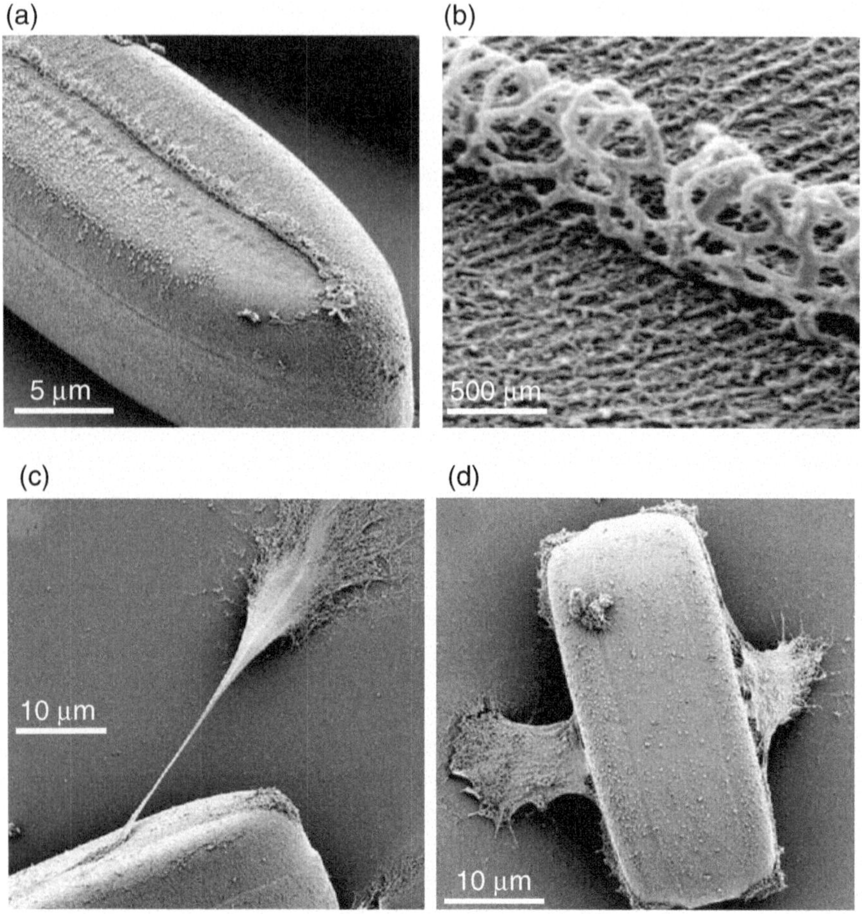


**Fig. 3.4** SEM images of the raphe mucilage of *Pinnularia viridis*. (i) The adhesive strands emerge from the driving raphe of a chemically fixed and inverted cell. (ii) Higher magnification image of entangled raphe strands. (iii) A tether formed between the elongated raphe adhesive of a cell and the substratum. (iv) A sheet of adhesive extends from each raphe and anchors the cell to the substratum. From Higgins et al. (2003a), reprinted with permission from the *Journal of Phycology*

essentially resulting in a tug-of-war (Wetherbee et al. 1998; Molino and Wetherbee 2008). The tension is broken when the force applied at one of the raphes is sufficient to break adhesion at the opposing raphe. However, SEM images, together with light microscopic observations of the occasional jerky movements and flip-flops between raphes of gliding *C. australis* and *P. viridis*, suggest that adhesion at the non-driving raphe is not always broken (Higgins et al 2003a). Instead, the tether at the non-driving raphe remains attached and is stretched as the cell moves on its driving raphe until it eventually detaches and becomes deposited on the substratum surface. Occasionally, detachment occurs at the driving raphe, so that the cell flips over.

#### 3.2.3.3 Secreted Trails

Freshly secreted trails deposited during diatom gliding are readily marked with tracer particles (Edgar and Pickett-Heaps 1984; Lind et al. 1997; Higgins et al. 2003b; Molino et al. 2006) or labeled by lectins or specific antibodies (Lind et al. 1997; Wustman et al. 1997; Wigglesworth-Cooksey and Cooksey 2005). However, the trails are difficult to label with conventional stains, and it has been suggested that this is probably because they are relatively soluble and disperse readily (Edgar and Pickett-Heaps 1984; Hoagland et al. 1993), although it is no less feasible that trails become cured and consequently resist staining.

Cryo-FESEM images of trails left after gliding *A. longipes* inoculated onto Formvar-coated grids revealed a tangle of linear and coiled strands (Wang et al. 2000). These strands covered the substratum surface, but "paths" leading to sessile *A. longipes* cells were cleared through them, indicating the relative ease with which the fresh trails could be displaced. FESEM observations of the trails of *P. viridis* left on glass coverslips showed they were ca. 3 μm wide and extended for several millimeters (Higgins et al. 2000). The trails often appeared as irregular ripples perpendicular to the direction of deposition and composed of short strands lying flat on the substratum. By contrast, tapping-mode AFM images demonstrated that the hydrated trails formed swollen, rounded, continuous ridges that were up to 300 nm high and were well adhered to the substratum along their entire length (Higgins et al. 2000). These ridges had no apparent substructure or specialized points of attachment, except for a thinner, more diffuse layer of hydrated mucilage either side of the main trail ridges. Some smearing was observed when the trails were imaged in tapping mode within an hour of deposition, but the trails were more stable if imaged 2–10 days later. These data indicated that the trail material was initially sticky but cured rapidly after deposition rather than dissolving (Higgins et al. 2000). The time-dependent deposition of mucilaginous trails by the diatoms *C. australis* and *A. coffeaeformis* on glass has been visualized using the stain Stains-All® (Molino et al. 2006). In this instance, the stain did not stain the organic trail material in the standard fashion (e.g., interacting with the biological material to generate a color change), rather the stain produced a blue precipitate when introduced into seawater, with this precipitate strongly binding to the trail material, revealing fine structural details. *A. coffeaeformis* was shown to produce discrete parallel trails on the glass surface, a signature of *A. coffeaeformis* cell morphology, with two raphes in contact with the surface during motility, as opposed to the single driving raphe observed in other diatom genera (Fig. 3.5). Additionally, reflection mode-scanning laser confocal microscopy images of the "stained" trails revealed *A. coffeaeformis* trails to be approximately 1 μm in height, allowing the cell to be slightly elevated from the underlying substratum. This study revealed that even at relatively low cell concentrations (~1631 cells/78.5 $mm^2$), motile cells can quickly (<20 h) cover the underlying substratum with adhesive trail material (Fig. 3.5).

The identity of the trail material and the secreted raphe adhesive was indicated by observations of *C. australis* (Lind et al. 1997). Silica microspheres added to the

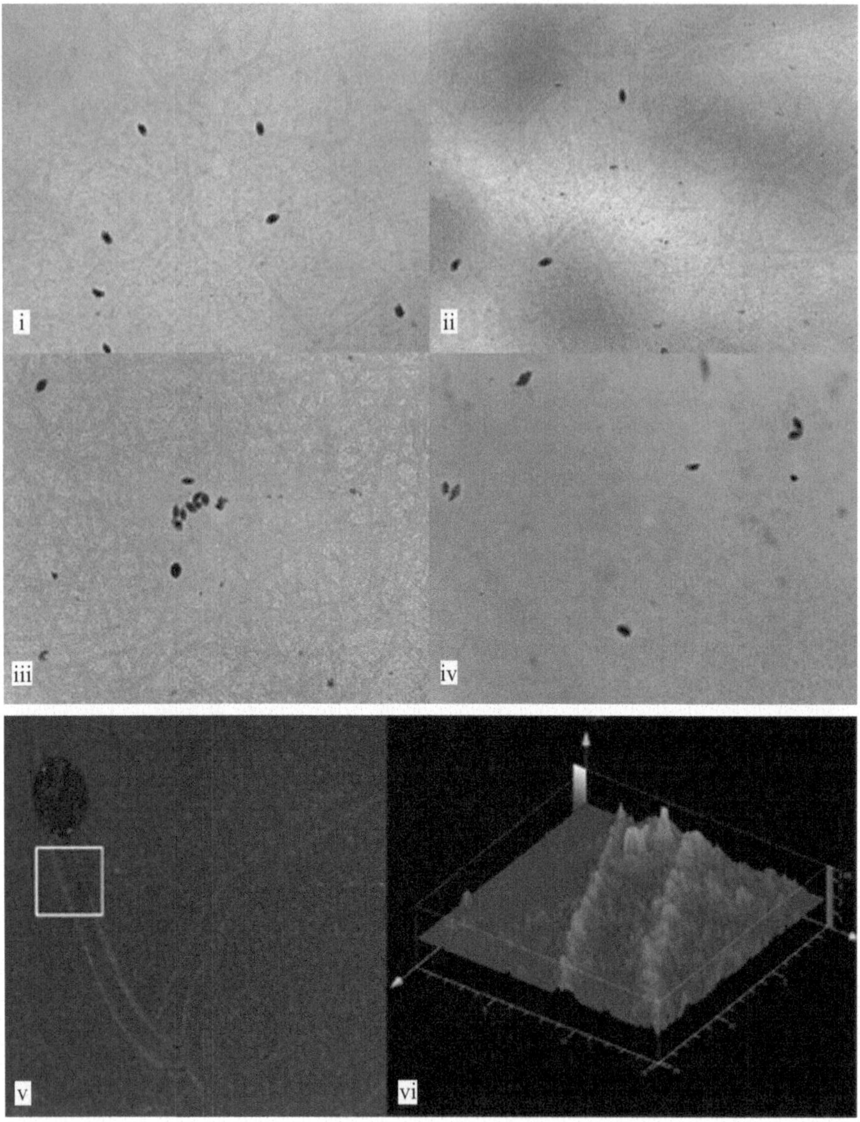


**Fig. 3.5** Stained mucilaginous trails of the diatom *Amphora coffeaeformis*. (i–iv) Time course study of the deposition of trails on glass after 30 min (i), 60 min (ii), 4 h (iii), and 20 h (iv) post-seeding. (v–vi) Reflection mode confocal laser scanning microscopy images of stained *A. coffeaeformis* trails on glass. From Molino et al. (2006), reprinted with permission from *Biomacromolecules*

medium became attached to and moved along the driving raphe until they reached the end and remained trapped in the trail secreted behind the cell. Microspheres were rarely seen to attach to any part of the cell other than the raphe. Occasionally, motionless cells with a taut trail of adhesive extending behind the cell recommenced gliding when the trail broke. Like the tethers characterized by SEM for *C. australis* and *P. viridis* (Higgins et al. 2003a), these observations suggest trails are sometimes cable-like and have sufficient tensile strength to arrest cell motility. Similarity between AFM images of the rounded filaments comprising *P. viridis* mucilage trails (Higgins et al. 2000) and SEM images of *P. viridis* tethers suggests they are the same thing (Higgins et al. 2003a). However, differential staining of tethers (positive) and trails (negative) with Stains-All® suggested there are materials in the raphe mucilage that are differentially incorporated into tethers and trails (Higgins et al. 2003a). Alternatively, the staining properties differ if the raphe adhesives coalesce into a tether rather than being deposited as a trail.

### *3.2.4 Nanomechanical Properties Determined by AFM and QCM-D*

#### 3.2.4.1 AFM

The mechanical properties of molecules can be measured by AFM in force mode. During one cycle of force mode, the cantilever tip approaches the sample and is then retracted (Fig. 3.1ii,iii). Deflections of the cantilever tip during the approach-retraction cycles are recorded and converted to force-distance curves. In conventional force mode, the cantilever tip approaches the sample until the resistance of the sample is detected. In "fly-fishing" force mode (Rief et al. 1997b), the AFM probe is allowed to closely approach the sample periphery enabling only the outermost polymers to adsorb to the cantilever tip.

##### Cell Surface Mucilage

AFM experiments showed that the cell surface mucilage of *P. viridis* was relatively soft, cohesive, and only weakly adhesive (Crawford et al. 2001). The mucilage was readily displaced by the sweeping action of the AFM cantilever tip in contact mode, and it remained piled up at the periphery of the scan area without deforming for a further 2 h. The extension phase of AFM force-distance curves recorded for the mucilaginous coating of the girdle and the valve regions was nonlinear, indicating that the mucilage was soft and compressible (Higgins et al. 2003b), and occasionally showed a peak that indicated the mucilage had a kind of skin which was punctured during the tip's approach to the cell surface (Higgins et al. 2002). The mucilage was therefore interpreted as having a degree of cross-linking (Crawford et al. 2001; Higgins et al. 2002), although the nature of this cross-linking remains undefined.

The retraction phase of AFM force curves taken on the cell surface mucilage of *C. australis* and *P. viridis* often displayed multiple detachment events and was interpreted as the interaction of many entangled polymers adsorbed to the cantilever tip, although single detachment events were also detected (Higgins et al. 2002, 2003b). The maximum force required to detach polymers from the cell surface mucilage varied with the loading force of the cantilever tip and was attributed to the depth with which the tip penetrated the mucilage layer. To achieve single binding events in the *C. australis* cell surface mucilage, Higgins et al. (2003b) used a loading force of 1–2 nN with a short contact time. The average maximum force required to detach polymers in the *C. australis* cell surface mucilage ($F_{max}$) was $203 \pm 9$ (SD) pN, and the average distance required to achieve detachment ($D_{max}$) was $106 \pm 5$ (SD) nm. Under the same conditions, no adhesive peaks were observed for *P. viridis*.

The relative elasticity of the surface mucilages of *C. australis* and *P. viridis* was characterized by force-volume mapping (Higgins et al. 2003b). The experimental data fitted the Hertz model well, and by making assumptions about tip radius (40 nm) and Poisson's ratio ($\mu = 0.25$), Young's modulus for elasticity was estimated as $0.25 \pm 0.01$ (SD) MPa in soft regions versus $0.45 \pm 0.01$ (SD) MPa in hard regions for *C. australis* and $0.54 \pm 0.02$ (SD) MPa versus $0.76 \pm 0.03$ (SD) MPa for *P. viridis*. Despite the uncertainties for precise measurement of Young's modulus, the data suggested the elastic properties of the mucilage of the two diatoms are similar.

### Raphe Adhesives

AFM force-separation curves recorded at the non-driving raphes of both *C. australis* and *P. viridis* were distinguishable from those recorded for cell surface mucilage and typically displayed multiple but irregular sawtooth patterns and significantly higher detachment forces (Higgins et al. 2003a). From these force curves, average $F_{max}$ for individual detachment events was measured as $4.5 \pm 3.6$ (SD) nN and average $D_{max}$ as $6.9 \pm 3.3$ (SD) μm for the raphe strands of *C. australis*, whereas average $F_{max}$ was $3.2 \pm 1.9$ (SD) nN, and average $D_{max}$ was $8.2 \pm 2.9$ (SD) μm for the raphe strands of *P. viridis* (Higgins et al. 2003a). However, a direct comparison between the average $F_{max}$ values estimated for the two species is not possible because $F_{max}$ is dependent on tip parameters that may differ between experiments and affect the number of strands adhering. A concern with recording conventional force-separation curves was that they measured $F_{max}$ over an area that was determined by the curvature of the cantilever tip, so that the number of strands binding could not be controlled. As a remedy for this, Higgins et al. (2003a) also sought to characterize binding of the raphe strands by the fly-fishing technique (Fig. 3.1ii,iii). The subsequent retraction phase of the cycle therefore ideally measured interaction with individual strands. Fly-fishing experiments above the raphe of *C. australis* gave single and/or multiple detachment events with average $F_{max}$ of $197 \pm 97$ (SD) pN, in the range for those of individual polymers (Zlatanova et al. 2000). However, the average $D_{max}$ of these adhesive strands was $3.5 \pm 1.2$ (SD) μm, demonstrating that they were highly extensible, even though the estimated value was regarded as less than the true $D_{max}$ because the tip would not necessarily contact the strand at its terminus (Higgins et al. 2003a). Based on these data, it appears that fly-fishing mode measured the force of detachment for individual adhesive strands that comprised single—or at most, a few—polymers. In contrast, the irregular sawtooth patterns of the curves obtained by conventional force mode are suggestive of successive unbinding of clumped adhesive strands from the cantilever tip and possibly also unbinding between strands and unfolding of domains within strands.

Less commonly, "incessant" S-shaped curves that did not return to the zero deflection line during successive extension-retraction cycles were obtained for the non-driving raphe of both *C. australis* and *P. viridis* (Higgins et al. 2002, 2003a). Although authentic $F_{max}$ and $D_{max}$ values could not be estimated, the energy required to extend and eventually detach these strands was up to 60 nN at distances of up to 15 μm. The attachment of raphe mucilage with the cantilever tip was visualized with Stains-All® and described as a kind of tether (Higgins et al. 2003a). Worm-like chain (WLC) modeling fitted poorly to the data in these incessant curves, but persistence length ($q$) was $0.006 \pm 0.005$ (SD) nm, about 100 times smaller than that of a simple monomer (0.1–0.5 nm; Higgins et al. 2002). Such persistence length values could arise from multiple chains extending in parallel, each with the same contour length and radius (Kellermeyer et al. 1997; Bemis et al. 1999; Dugdale et al. 2005).

#### 3.2.4.2 QCM-D

The quartz crystal microbalance with dissipation monitoring (QCM-D) is a highly sensitive surface-sensing technique that has been employed to study the mass and viscoelastic, or mechanical, properties of trail adhesives deposited on different surfaces during diatom locomotion (Molino et al. 2006; Molino et al. 2008b). The QCM-D allows the measurement of the mass adsorbed to an electrode surface on a quartz crystal sensor by measuring changes in the oscillating frequency ($f$) of the crystal, as well as the viscoelastic properties of the adsorbed mass through measurement of the dissipation ($D$) parameter (for review, see Marx 2003). The viscoelastic properties of the adsorbed mass can be characterized either through the application of complex viscoelastic models (e.g., Voigt model), or more simply defined as the ratio of the mass deposited on the sensor surface, $f$, or the amount of energy dissipated by that mass, $D$. QCM-D was used to study the adhesive mucilage deposited by *C. australis* and *A. coffeaeformis* on a gold-coated surface over a 20-h period, identifying differences both in the mass and viscoelastic properties of the material deposited for the two species (Molino et al. 2006). *C. australis* was shown to deposit far more material than *A. coffeaeformis* over the 20-h period, as evidenced through far larger $f$ and $D$ shifts. This was attributed to the faster motility rate of *C. australis*, thus allowing it to deposit trail mucilage on the electrode surface at a much faster rate. The study of the $f/D$ ratio of the adhesive mucilage revealed *A. coffeaeformis* to deposit a more rigid and less viscoelastic layer ($f/D$

ratio of ~6) compared to that of *C. australis* (ratio of ~2). Furthermore, there was no evidence of significant conformational or viscoelastic changes in the adhesive layers produced by either species over time, supporting the hypothesis that cross-linked-mediated adhesion, a process employed by other benthic marine organisms during adhesion, is not utilized by marine diatoms, at least during initial adhesion and colonization of a surface. This was the first study to identify bulk differences in the mechanical properties of the adhesive mucilage secreted by marine diatom species onto a surface, and given *A. coffeaeformis* has been identified as a much stronger fouling species than *C. australis* (Holland et al. 2004), differences into the viscoelastic properties of the adhesives may correlate with differences in fouling ability. Subsequent work investigating *A. coffeaeformis* and *C. australis* adhesion to surfaces of differing surface energies using QCM-D revealed the mucilaginous adhesives secreted by both species to be relatively soft and viscoelastic when deposited on a hydrophobic self-assembled monolayer surface (1-undecanethiol), compared to a hydrophilic surface (11-mercaptoundecanoic acid) which demonstrated a more rigid and less viscoelastic layer (Molino et al. 2008b). Interestingly, during initial interactions between *A. coffeaeformis* cells to the hydrophobic surface, a positive frequency shift was observed during the initial 20 min of adhesion to the surface, followed by a negative frequency shift for the following 20 h. This positive frequency shift was determined to result from the cell being tightly adhered to the hydrophobic surface, resulting in the cell adhesion complex "pulling" on the substratum surface. This process was proposed to increase the adhesion strength of adhesion between the cell and the substratum, and, given diatoms have been shown to adhere more strongly to hydrophobic, as opposed to hydrophilic, surfaces, provide a possible mechanism for these observations.

### 3.2.5 Molecular Composition

#### 3.2.5.1 *Craspedostauros australis*

Monoclonal antibodies (mAbs) were raised to four high molecular weight frustule-associated components ("FACs") from *C. australis* (Lind et al. 1997). These FACs are estimated to be 87 kDa, 112 kDa, and two >220 kDa and were visualized by SDS-PAGE with a glycan stain, indicating they were glycoconjugates. In immunolocalization studies, one mAb which had high affinity for all four FACs (StF.H4) labeled the cell surface, the raphe, and the diatotepum (a putative polysaccharide layer between the plasma membrane and the silica wall), whereas another mAb (StF.D5) bound with relatively high affinity to the two smaller FACs and labeled the cell surface only. In addition, cell adhesion and cell motility were inhibited in a concentration-dependent manner when cells were incubated with native StF.H4 or divalent $F(ab)_2$ fragments of StF.H4 but not by monovalent F (ab) StF.H4 fragments or by StF.D5. These observations indicated that inhibition was effected by cross-linking of the participating FACs recognized by the StF.H4 antibody. The contrasting inhibitory effects, affinities, and labeling patterns of the two mAbs suggested that the two larger FACs form the raphe and trail adhesive, whereas the two smaller FACs contributed to the cell surface mucilage. In contradiction to this interpretation, a third antibody (StF.F3) labeled the 112 kDa FAC with high affinity and immunolocalized to the raphe and girdle region (Lind et al. 1997). However, competitive inhibition assays, periodate treatment, and deglycosylation of the FACs (Lind et al. 1997; Chiovitti et al. 2003a) demonstrated that the antibodies were binding to carbohydrate. The localization data were therefore probably complicated by the antibodies cross-reacting with the same glycan epitopes on different FACs. The three largest FACs were subsequently shown to be high molecular weight glycoproteins decorated with a highly heterogeneous population of xylose-rich glycans, sulfate, and phosphate (Chiovitti et al. 2003a). The protein portions were relatively enriched in small nonpolar (Gly and Ala, 22 mol% of total amino acids) and hydroxylated amino acids (Ser, Thr, Tyr, Hyp, 22 mol% of total). The hydroxylated amino acids were proposed to be the main sites of linkage for the glycans. The glycans of the FACs located on the cell surface may function in a manner analogous to those of protistan and mammalian mucins, forcing the glycoproteins to adopt an extended conformation so that they would function as relatively nonadhesive lubricants (Chiovitti et al. 2003a).

Adhesives from *C. australis* deposited on petri dishes have more recently been isolated utilizing careful methodologies employed to minimize contamination from intracellular components, and their carbohydrate and amino acid composition determined and compared relative to another model biofouling species *A. coffeaeformis* (Poulson et al. 2014). Protein was identified as a significant component of the adhesive material, providing 30 % of the total mass, with polysaccharides contributing the other 70 %. For *C. australis*, glycine, serine, and threonine constituted about half of the amino acid residues for the adhesive material. This is in comparison to *A. coffeaeformis*, for which these residues only constituted about 20 mol%, with arginine dominating at 27.4 mol%. Carbohydrate analyses of the material revealed hexoses, pentoses, deoxyhexoses, and uronic acids to be present for both species. For *C. australis*, the relative content of pentoses and uronic acids was higher, while for *A. coffeaeformis*, relative amounts of hexoses and deoxyhexoses were greater.

#### 3.2.5.2 *Pinnularia viridis*

Biochemical analysis was combined with AFM to characterize the extracellular polysaccharides of *P. viridis* (Chiovitti et al. 2003b). Polysaccharides were successively extracted from the cells with water at 45 °C, a solution of sodium bicarbonate and EDTA at 95 °C, and hot alkali. AFM demonstrated that warm-water extraction removed an extracellular component that conferred a degree of softness and compressibility to the cell surface mucilage, leaving behind material that was more rigid and less elastic than the original mucilage layer. The rigid material was subsequently extracted with bicarbonate/EDTA, exposing the siliceous cell wall underneath. Carbohydrate analyses detected 19 sugars and 65 linkage and substitution patterns. The warm-water fraction was relatively enriched in fucose residues (and intracellular glucans), whereas the bicarbonate/EDTA fraction was relatively enriched in rhamnose residues, indicating that polymers enriched in fucose or rhamnose, respectively, contributed to the observed changes in the properties of the modified mucilage layer. However, many of the same sugars and a proportion of the linkage patterns were observed in all the fractions, indicating that the *P. viridis* polysaccharides comprised a spectrum of highly heterogeneous but structurally related sulfated polysaccharides. It was proposed that polydisperse molecular weight of the polysaccharides, distribution of residues within the polysaccharides, and covalent and noncovalent intermolecular associations with protein also affected the properties of the mucilage (Chiovitti et al. 2003b).

#### 3.2.5.3 *Amphora coffeaeformis*

*A. coffeaeformis* cells have a tightly bound organic sheath and a loosely associated mucilaginous capsule (Daniel et al. 1980; Wustman et al. 1997). Early histochemical studies strongly indicated that *A. coffeaeformis* mucilage was composed of sulfated polysaccharides and uronic acids (Daniel et al. 1980). Mechanically isolated capsules from *A. coffeaeformis* contained 13 % sulfate, 6 % protein, and a heterogeneous sugar composition consisting of glucose (40 %) and nearly equal proportions of glucuronic acid, galactose, mannose, xylose, fucose, and rhamnose (7–13 % each; Wustman et al. 1997). Sequential extraction with hot water and hot bicarbonate solution enriched the proportions of fucose and galactose (up to 32 % each; Wustman et al. 1997). Surprisingly perhaps, concanavalin A labeled mainly the organic sheaths associated with the frustule, whereas *Abrus precatorius* agglutinin (specific for D-galactose) labeled the loosely associated capsule material (Wustman et al. 1997). However, in a later study, polymers associated with the cell walls, trails, footpads, and intercellular biofilm were all stained with lectins recognizing glucose, mannose, and galactose (Wigglesworth-Cooksey and Cooksey 2005). The apparent differences in staining in the two studies may be due to variations in interpretation, diatom physiology, or strain types. The height of *A. coffeaeformis* footpads estimated from confocal microscopy of FITC-concanavalin A-labeled footpads is 5–7 μm, indicating the cell is substantially raised above the substratum (Wigglesworth-Cooksey and Cooksey 2005). Apart from an intracellular role in facilitating diatom cell motility (Cooksey 1981; Cooksey and Cooksey 1988), calcium ions apparently also mediate cross-linking of *Amphora* adhesives. Treatment of attached *A. coffeaeformis* with 10 mM EGTA causes "cohesive breaks," with the effect that cells can be freely washed away and the adhesive pads are left behind on the substratum (Cooksey and Cooksey 1980). Similar effects have been described for other diatom species (Geesey et al. 2000). Calcium ions in the pads can be exchanged with strontium ions without any apparent effect on adhesion, although motility is affected (Cooksey and Cooksey 1980).

#### 3.2.5.4 *Phaeodactylum tricornutum*

*Phaeodactylum tricornutum* is a pennate diatom that possesses ovoid and fusiform morphotypes. Ovoid cells have been demonstrated to adhere more strongly to surfaces than fusiform cells, with ovoid cells capable of forming a confluent adhering biofilm (Willis et al. 2013). Willis et al. (2013) attempted to identify whether differences in the protein or carbohydrate composition of the adhesives of the respective morphotypes could be linked to their variable adhesion ability. Staining of the extracellular mucilage produced by a 7-day-old biofilm for monosaccharides using lectins did not reveal any discernible spatial pattern in carbohydrate distribution. Constituent monosaccharide profiles and linkages of fractions taken from the extracellular mucilage were broadly similar for both morphotypes, indicating subtle variation in mucilage structure and composition, as opposed to substantial changes in bulk composition, which likely guide the variable adhesion capacity of the two morphotypes (Willis et al. 2013). Principal component analysis supported this, indicating the monosaccharide composition of the fractions to be similar but to differ significantly in their linkage composition (i.e., 4-linked fucose and 3,4-linked rhamnose were only found in secreted fraction for the ovoid morphotype).

The sequencing of the *P. tricornutum* genome has allowed a bioinformatics approach to the study of uncharacterized or novel cell adhesion molecules through genomic identification by amino acid profiling (Willis et al. 2014). Therein ten candidate putative diatom cell-substratum adhesion molecules (PDCs) were identified based on the amino acid profile predicting involvement in cell adhesion, the highest numbers of expressed sequence tags, homology to known cell adhesion molecules from other organisms, presence of signal peptides, and whether they were predicted to be secreted. Transgenic cell lines for each PDC were then created, and their adhesive properties assayed using a flow chamber apparatus and their extracellular adhesives studied using AFM and QCM-D. Three transgenic lines demonstrated increased cell adhesion to glass (~50 % of cells detached) when tested against a water shear force on the flow chamber, compared to control cells (~80 % cell detachment). AFM was used to characterize the nanomechanical properties of the adhesives. The percentage of adhesion events, when probing the adhesive with the AFM cantilever tip in force mode, varied between 11 % and 45 % for the transgenic cell lines compared to 24 % for control cells. The increase in the number of adhesion events was proposed to result from an increase in the number of adhesion molecules exposed at the surface of the mucilage adhered to the substratum. Maximum adhesion strength was also greater for all transgenic cell lines (290–660 pN) relative to control cells (206.5 pN). This study identified five PDCs that have homology to genes previously identified to participate in cell adhesion, along with three novel genes, and provides a valuable proof of principle for the application of genome-based methods for the identification of proteins involved in diatom adhesion.

## 3.3 Sessile Adhesion

Often, after a period of gliding, benthic diatoms settle and form sessile adhesion structures. These adhesive structures are described primarily as pads and stalks (Daniel et al. 1987; Hoagland et al. 1993). The structures may be intergraded but are differentiated on the basis of relative length (Hoagland et al. 1993). However, the pads and the stalks of stalk-forming diatoms, such as *A. longipes*, differ in ultra-structure and composition (Daniel et al. 1987; Wang et al. 1997; Wustman et al. 1997). The site of stalk secretion varies for different species and may be the raphe, the mantle edge, or an apical region of the frustule bearing numerous small pores (apical pore field) (Gibson 1979; Pickett-Heaps et al. 1991; Hoagland et al. 1993; Kooistra et al. 2003).

### *3.3.1 Physical Properties of Adhesive Pads with AFM*

Adhesion and motility are normally associated with raphid diatoms, but some diatoms that lack a raphe, such as *Ardissonea crystallina* and *T. undulatum*, are also capable of gliding (Pickett-Heaps et al. 1991; Kooistra et al 2003). This mode of motility is apparently simpler than that of raphid diatoms, and propulsion is probably achieved by swelling and hydration of secreted mucilage (Pickett-Heaps et al. 1991). Adhesive mucilage in both species is apparently secreted at the cell apices in the region of the first girdle band. Sessile adhesion is attained by formation of an adhesive pad, described as a "stipe" for *A. crystallina*.

#### 3.3.1.1 *Toxarium undulatum*

The properties of the *T. undulatum* adhesive pad have been investigated in substantial detail by AFM (Dugdale et al. 2005). Fly-fishing experiments on native pads with the cells still attached produced force curves displaying regular sawtooth patterns. Like *C. australis* raphe strands (Higgins et al. 2003a, see Sect. 3.2.4.2), but in contrast to most polymers and proteins studied by AFM (Hugel and Seitz 2001), the *T. undulatum* mucilage adsorbed to the cantilever tip without applying detectable force ($<5$ pN) or pausing. This observation underscores the role of the mucilage as an effective adhesive. The sawtooth pattern was still retained on pads from which cells had been removed for up to 12 h. However, after 48–89 h, these curves were obtained with reduced frequency, and the curves sometimes had indistinct peaks or no peaks, indicating that the adhesive slowly underwent degradation and/or curing.

The sawtooth curves of the adhesive nanofibers had regular features (Dugdale et al. 2005). The average distance to the first peak was $201 \pm 10.6$ (SE) nm, the average distance between each peak was $34.3 \pm 1.5$ (SE) nm, and the average peak force was $0.801 \pm 0.071$ (SE) nN. The distance to the last peak varied and was dependent upon the total number of peaks in the curve, which varied from 3 to 27. These data showed that the first ca. 200 nm portion of the adhesive nanofiber invariably unraveled without restriction. This portion must be proximal to the anchor point of the adhesive, probably enabling the adhesive to trawl away from the parent surface. The remainder of the adhesive nanofiber comprises a backbone of up to 27 domains, each successively unfolding to increase overall length by ca. 34 nm. Most of the peaks were relatively accurately fitted with the WLC model giving a persistence length ($q$) of $0.036 \pm 0.004$ (SE) nm. The low average persistence length, together with the high average force for unfolding (0.8 nN) indicated that the adhesive nanofibers are composed of multiple polymer chains that must be aligned in parallel, unfolding and refolding synchronously in order to achieve the observed regular sawtooth patterns.

The adhesive nanofiber of *T. undulatum* also showed very efficient self-healing properties (Dugdale et al. 2005). The regular sawtooth pattern was reproducibly recorded for up to 666 successive approach-retraction cycles of adhesive nanofibers that remained bridged between the pad and the cantilever tip to a distance of 800 nm. The average proportion of domains that refolded decreased as the scan rate increased (from 92 % at 0.8 μm/s down to 53 % at 7.4 μm/s). However, all the domains refolded if the tip was paused in the extended position for 2 s. These data clearly demonstrated that strong, reversible intradomain bonding was involved in refolding the stretched adhesive nanofiber. It was suggested that hydrogen bonding, which has extremely low activation energies, must facilitate the refolding (Dugdale et al. 2005). The strength and flexibility of the adhesive makes the cells difficult to detach, even in fast-moving currents.

The sawtooth pattern of the *T. undulatum* adhesive nanofibers was the characteristic AFM fingerprint for unfolding of modular proteins (Fisher et al. 2000), but this was previously only recorded for purified or recombinant proteins (Rief et al. 1997a; Oroudjev et al. 2002). The multi-cycle sawtooth patterns for bridged adhesive nanofibers were abolished by incubating the *T. undulatum* pads with protease, demonstrating that the backbone of the adhesive nanofiber was composed predominantly of protein (Dugdale et al. 2005). During these experiments, the sawtooth pattern of the initial extension and retraction cycles was the same as those recorded for native bridged adhesive nanofibers. However, the retraction phase of the second and subsequent cycles gave no cantilever deflection. These data indicated that the protease did not interfere with the binding or initial unfolding of the adhesive nanofiber but, when the adhesive nanofiber was extended, it became susceptible to hydrolysis by the protease, resulting in cleavage and rendering it undetectable by AFM.

Subsequent chemical characterization of the *T. undulatum* adhesives using energy dispersive X-ray analysis (EDXA) and Fourier transform infrared (FTIR) revealed the adhesives to mainly contain protein, carbohydrate, sulfate, calcium, and magnesium (Chiovitti et al. 2008). The extract has a protein-to-carbohydrate-to-sulfate weight ratio of 1.0:0.2:0.9 that contained a single high molecular weight (>220 kDa) anionic macromolecule. The most abundant amino acids included glycine (22 mol%), aspartic acid/aspartame (14 mol%), and histidine (11 mol%). Eleven neutral sugars were identified and were dominated by mannose (50 mol%) and xylose (29 mol%). AFM analyses of adhesive pads treated with EDTA resulted in the rapid loss of the signature sawtooth pattern, and thus the adhesive structure, of the nanofibers. Subsequent treatment with solutions containing divalent cations ($Ca^{2+}$ and $Mg^{2+}$) saw the signature adhesive nanofiber structure restored. Thus the adhesive pads of *T. undulatum* were determined to be composed of sulfated high-molecular-mass glycoproteins that are cross-linked with magnesium and calcium ions, with the cross-linking allowing adjacent protein backbones to unfold and refold in register (Chiovitti et al. 2008).

#### 3.3.1.2 *Eunotia sudetica*

Irregular unbinding events have been recorded on the adhesive pads of the freshwater diatom, *E. sudetica*. To examine the adhesive, *E. sudetica* cells were dislodged from a glass slide, and the adhesive remaining on the substratum was probed with AFM for several hours. Multiple detachment events were recorded in the retraction phase of the cycle with a separation distance of ca. 600 nm (Gebeshuber et al. 2003). The self-healing properties of the *E. sudetica* adhesive were not as efficient as those of the *T. undulatum* adhesive. An interval of 30 s was required between successive scanning cycles for the *E. sudetica* adhesive to recover progressively diminished proportions of its relaxed-state structure until, after five scans, the adhesive was undetectable (Gebeshuber et al. 2002).

### 3.3.2 *Molecular Composition and Chemical Properties of Stalks:* Achnanthes longipes

#### 3.3.2.1 Stalk Formation

The best studied example of stalk formation in a diatom is that of *A. longipes*. Cell density is a key stimulator for stalk production in *A. longipes*, interpreted as a response to crowding by raising cells above the biofilm and possibly triggered by a diffusible autoinducer molecule (Lewis et al. 2002). The adhesion process of *A. longipes* has been described as occurring in four key stages (Wang et al. 1997). In stage 1, cells actively adhere to the substratum via the raphe and glide. Cell motility ceases after ca. 6 h. In stage 2, small globular pad structures are produced from the terminal nodule of the raphe. In stage 3, a flexible, elongated shaft is secreted and elevates the cell above the substratum. The shaft connects the cell to the pad, and adhesion of residual pad material to the cell results in formation of a collar at the top of the shaft (Daniel et al. 1987; Wang et al. 1997). Stalk secretion is associated with active vesicle traffic (Daniel et al. 1987; Wang et al. 2000). In stage 4, successive cell divisions lead to formation of filamentous colonies.

Stalk synthesis and cell gliding in *A. longipes*, but not cell division, were reversibly inhibited in a dose-dependent manner by the herbicide, 2,6-dichlorobenzonitrile (DCB), and four DCB analogues, indicating that events in gliding and stalk secretion are related (Wang et al. 1997). A fluorescent analogue of DCB bound an 18 kDa protein associated with the membrane fraction obtained by detergent extraction of cells that were actively synthesizing adhesives, leading Wang et al. (1997) to postulate that the effect may be sited at the plasma membrane, although the identity of the 18-kDa protein and the mode of DCB action remain unknown.

#### 3.3.2.2 Composition and Properties

The length and diameter of *A. longipes* stalks generally varies depending on the amount of time elapsed since the cells were inoculated (Wang et al. 1997), but, when grown under ideal conditions, stalks are 200–500 μm long and 7–9 μm wide (Johnson et al. 1995). The stalks are composed of four layers identified by TEM of chemically fixed cells (Wang et al. 1997, 2000) or by LM of stained thin sections (Daniel et al. 1987). The core of the shaft comprised a central ribbon of densely packed fibers surrounded by one layer of perpendicularly arranged fibers, a second layer of fibers arranged parallel to the stalk axis, and a relatively diffuse, outermost layer. The pad and collar are comparatively amorphous.

Based on available data (discussed below), Wustman et al. (1998) proposed a model for *A. longipes* cell adhesion. During sessile adhesion, highly sulfated fucoglucuronogalactans (FGGs) are secreted from the raphe and expand to form a cylinder comprising fibrils that are arrayed perpendicular to the axis of the stalk. This material is assumed to be related to the adhesive used by the cell for motility. The low-sulfate FGGs are secreted from the apical pore field surrounding the raphe pole and form the outer layer of fibrils that run parallel to the stalk axis. Wustman et al. (1998) suggested that the adhesives putatively cure by cross-linking proteins (presumably glycoproteins), polysaccharides, and phenolics through covalent *O*-linkages.

Mechanically isolated stalks (MIS) of *A. longipes* were composed of carbohydrate (40 %), uronic acid (10 %, mainly as glucuronic acid), and small amounts of protein (5 %) (Wustman et al. 1997). The carbohydrates of MIS and water-insoluble/base-soluble (WIBS) fractions from the stalks were dominated by fucose and galactose (comprising 50–70 % of total sugars) with a complex mix of multiple linkages and substitution patterns. $^{13}$C-NMR indicated that the fucose was present in the furanosyl form. Differential cytochemical staining and labeling with FITC-lectin conjugates demonstrated that the composition of the pads and collars differed from that of the shaft of the stalk (Wustman et al. 1997). However, a continuity of labeling with some lectins, such as concanavalin A, also showed that related glucose and mannose structural units occurred throughout the stalk.

The WIBS was fractionated by size-exclusion chromatography into three distinct size classes: $F_1 \geq 20{,}000$ kDa, $F_2 \approx 100$ kDa, and $F_3 < 10$ kDa (Wustman et al. 1998). $F_1$ and $F_2$ were the major fractions, composed mainly of carbohydrate with small amounts of protein (5 % in $F_1$ and < 1 % in $F_2$). $F_3$ was almost entirely protein and ran as a smear on SDS-PAGE, indicating it was still bound to carbohydrates. Cold NaOH treatment of WIBS led to a decrease in $F_1$ and an increase in $F_2$ and $F_3$, whereas trypsin treatment eliminated $F_3$, demonstrating that $F_1$ was a complex of $F_2$ (mainly polysaccharide) and $F_3$ (mainly protein). The complexes were likely stabilized by covalent *O*-linkages between carbohydrates and proteins because chelating agents and imidazole buffers, which disrupt ionic interactions, did not solubilize *A. longipes* stalks (Wustman et al. 1998). CPC-soluble material from WIBS was enriched in protein and contained relatively more terminal and 4-linked mannopyranose and less galactopyranose residues than unfractionated WIBS, suggesting the occurrence of mannoprotein (Wustman et al. 1998). The major amino acids in MIS were the small nonpolar residues (Ala 17 mol%, Gly 13 mol%), together with Ser, Val, and Asx (13 mol% each, Wustman et al. 1997).

High- and low-sulfate FGG fractions of WIBS were isolated by anion-exchange chromatography (Wustman et al. 1998). Both fractions had similar carbohydrate (80–84 %), protein (7–10 %), and uronic acid content (18–19 %) but differed substantially in sulfate content (11 % versus 2 %). Wustman et al. (1998) raised five antibodies that recognized a family of related epitopes with antigenicity dependent upon fucose-containing side chains present on both high- and low-sulfate FGGs. The antibodies differentially localized the high-sulfate FGGs to the shaft core and the low-sulfate FGGs to the outer layers. These observations corroborated earlier cytochemical staining and X-ray microanalysis of sectioned *A. longipes* stalks that indicated that the levels of sulfation were higher in the central core than in the surrounding layers (Daniel et al. 1987; Johnson et al. 1995). Cells grown in media containing methionine in place of sulfate appeared healthy but lacked adhesion and had reduced rates of cell division (Johnson et al. 1995), indicating a crucial role for sulfate in the formation of the stalks. The sulfate may effect cross-linking by divalent cations (Johnson et al. 1995) but may also promote flexibility within the stalk by mutual charge repulsion (Daniel et al. 1987). The mAb with the highest affinity for high-sulfate FGGs (AL.C1) was the only mAb that labeled the inner core of the shaft (Wustman et al. 1998). Interestingly, AL.C1 also labeled the raphe region of motile *A. longipes*. Wustman et al. (1998) concluded that the central core region is raphe derived and is related to the material exuded during cell gliding.

Bromide is essential for stalk formation in *A. longipes*, with an optimum concentration of 30 mM for maximal stalk production, although bromide was not essential for optimal growth rates (Johnson et al. 1995; Lewis et al. 2002). As bromide concentrations were decreased, stalk morphology became progressively more distorted until only pads were produced (Johnson et al. 1995). Insoluble EPS was not produced in the absence of bromide, but cells secreted relatively high levels of soluble carbohydrates, indicating that the EPS was incompletely cross-linked. Increasing iodide concentrations also inhibited the extent and morphological integrity of stalk formation, and only soluble EPS was produced at 67 μM of iodide, even with optimum levels of bromide. Based on their observations, Johnson et al. (1995) proposed that *A. longipes* stalk formation requires the action of a haloenzyme, possibly a vanadate-dependent bromoperoxidase, that stabilizes extracellular matrix components by oxidizing phenolic compounds and effecting their cross-linking to carbohydrates in polysaccharides and glycoproteins. Bromoperoxidase activity has been detected in other chromophytes, mainly representatives of the Phaeophyceae, such as *Ascophyllum* and *Fucus* species (Vilter 1984; Vreeland et al. 1998). Cross-linking of *Fucus serratus* polyphenolics was achieved with an exogenous vanadate-dependent bromoperoxidase from *Ascophyllum nodosum*, and bromide was demonstrated to be essential for the intermolecular cross-linking (Berglin et al. 2004). *A. longipes* stalks autofluoresce when excited at 360 nm, suggesting phenolics are present (Wustman et al. 1997). The inhibitory effect of elevated iodide concentrations on *A. longipes* stalk curing may be due to competitive inhibition of the bromide in the putative bromoperoxidase, with the implication that the binding pocket of the enzyme is equally accessible to both ions, despite the relative difference in their sizes (Colin et al. 2005).

## 3.4 Concluding Remarks

The structure, properties, and composition of diatom mucilages vary according to function and include adhesion for motility, sessile adhesion, cell-surface lubrication, and cell elevation. The composition and properties of the adhesives also vary between diatom species and at different sites on the same diatom cell. Disparate data sets for the properties and composition derived for different diatom species could be consolidated by standardizing some model species for complementary studies. The field would benefit, for example, from detailed AFM studies on *A. longipes* stalks, which are well characterized biochemically (Sect. 3.3.2). However, some tentative principles may be emerging. For example, the very low persistence lengths estimated by AFM for both the *C. australis* raphe strands (Sect. 3.2.4.2) and the *T. undulatum* adhesive nanofibers (Sect. 3.3.1.1) suggest that the adhesive properties of these mucilages derive from the coordinated activity of amalgamated polymers. AFM studies on more diatom species would assist to determine whether these represent isolated observations or a genuine principle underpinning diatom adhesives. Knowledge of the distribution of specific sugar moieties in diatom EPS would enhance understanding of the physical properties of EPS. However, the occurrence of protein in various diatom adhesives highlights a need to focus biochemical studies on characterizing these protein components. With the availability of diatom genomes (Armbrust et al. 2004; Montsant et al. 2005; Bowler et al. 2008), protein sequences would enable dissection of diatom adhesives with an integrated approach incorporating molecular biology, immunolocalization, and AFM, targeting such goals as determining how adhesives are secreted from the frustule without occluding the secretion sites, characterizing cross-linking processes, and, ultimately, identifying adhesive domains.

## References

Alberte RS, Snyder S, Zahuranec BJ, Whetsone M (1992) Biofouling research needs for the United States Navy: program history and goals. Biofouling 6:91–95

Anderson C, Atlar M, Callow M, Candries M, Townsin RL (2003) The development of foul-release coatings for seagoing vessels. J Mar Des Oper 84:11–23

Armbrust EV, Berges JA, Bowler C, Green BR, Martinez D, Putnam NH, Zhou S, Allen AE, Apt KE, Bechner M, Brzezinski MA, Chaal BK, Chiovitti A, Davis AK, Demarest MS, Detter JC, Glavina T, Goodstein D, Hadi MZ, Hellsten U, Hildebrand M, Jenkins BD, Jurka J, Kapitonov VV, Kröger N, Lau WWY, Lane TW, Larimer FW, Lippmeier JC, Lucas S, Medina M, Montsant A, Obornik M, Schnitzler Parker M, Palenik B, Pazour GJ, Richardson PM, Rynearson TA, Saito MA, Schwartz DC, Thamatrakoln K, Valentin K, Vardi A, Wilkerson FP, Rokhsar DS (2004) The genome of the diatom *Thalassiosira pseudonana*: ecology, evolution, and metabolism. Science 306:79–86

Bellinger BJ, Abdullahi AS, Gretz MR, Underwood GJC (2005) Biofilm polymers: relationship between carbohydrate biopolymers from estuarine mudflats and unialgal cultures of benthic diatoms. Aquat Microb Ecol 38:169–180

Bemis JE, Akhremitchev BB, Walker GC (1999) Single polymer chain elongation by atomic force microscopy. Langmuir 15:2799–2805

Berglin M, Delage L, Potin P, Vilter H, Elwig H (2004) Enzymatic cross-linking of a phenolic polymer extracted from the marine alga *Fucus serratus*. Biomacromolecules 5:2376–2383

Bhosle NB, Sawant SS, Garg A, Wagh A (1995) Isolation and partial chemical analysis of exopolysaccharides from the marine fouling diatom *Navicula subinflata*. Bot Mar 38:103–110

Binning G, Quate CF, Gerber CH (1986) Atomic force microscope. Phys Rev Lett 56:930–933

Bohlander GS (1991) Biofilm effects on drag: measurements on ships. In: Long DM, Bufton R, Yakimiuk P, Williams K (eds) Polymers in a marine environment. Institute of Marine Engineering, London, pp 135–138

Bowler C, Allen AE, Badger JH, Grimwood J, Jabbari K, Kuo A et al (2008) The Phaeodactylum genome reveals the evolutionary history of diatom genomes. Nature 456(7219):239–244

Callow ME (1986) Fouling algae from "in-service" ships. Bot Mar 29:351–357

Callow ME (1996) Ship-fouling: the problem and method of control. Biodeter Abst 10:411–421

Chiovitti A, Bacic A, Burke J, Wetherbee R (2003a) Heterogeneous xylose-rich glycans are associated with extracellular glycoproteins from the biofouling diatom *Craspedostauros australis* (Bacillariophyceae). Eur J Phycol 38:351–360

Chiovitti A, Higgins MJ, Harper RE, Wetherbee R, Bacic A (2003b) The complex polysaccharides of the raphid diatom *Pinnularia viridis* (Bacillariophyceae). J Phycol 39:543–554

Chiovitti A, Molino P, Crawford SA, Teng R, Spurck T, Wetherbee R (2004) The glucans extracted with warm water from diatoms are mainly derived from intracellular chrysolaminarin and not extracellular polysaccharides. Eur J Phycol 39:117–128

Chiovitti A, Heraud P, Dugdale TM, Hodson OM, Curtain RCA, Dagastine RR, Wood BR, Wetherbee R (2008) Divalent cations stabilize the aggregation of sulfated glycoproteins in the adhesive nanofibers of the biofouling diatom *Toxarium undulatum*. Soft Matter 4:811–820

Clarkson N, Evans LV (1995a) Further studies investigating a potential non-leaching biocide using the marine biofouling diatom *Amphora coffeaeformis*. Biofouling 9:17–30

Clarkson N, Evans LV (1995b) Raft trial experiments to investigate the antifouling potential of silicone elastomer polymers with added biocide. Biofouling 9:129–143

Cohn SA, Weitzell RE (1996) Ecological considerations of diatom motility. I. Characterization of adhesion and motility in four diatom species. J Phycol 32:928–939

Cohn SA, Farrell JF, Munro JD, Ragland RL, Weitzell RE, Wibisono BL (2003) The effect of temperature and mixed species composition on diatom motility and adhesion. Diatom Res 18:225–243

Colin C, Leblanc C, Gurvan M, Wagner E, Leize-Wagner E, Van Dorsselaer A, Potin P (2005) Vanadium-dependent iodoperoxidases in *Laminaria digitata*, a novel biochemical function diverging from brown algal bromoperoxidases. J Biol Inorg Chem 10:156–166

Cooksey KE (1981) Requirement for calcium in adhesion of a fouling diatom to glass. Appl Environ Microbiol 41:1378–1382

Cooksey B, Cooksey KE (1980) Calcium is necessary for motility in the diatom *Amphora coffeaeformis*. Plant Physiol 65:129–131

Cooksey B, Cooksey KE (1988) Chemical signal-response in diatoms of the genus *Amphora*. J Cell Sci 91:523–529

Cooksey KE, Wigglesworth-Cooksey B (1992) The design of antifouling surfaces: background and some approaches. In: Melo LF, Bott TR, Fletcher M, Capdeville B (eds) Biofilms – science and technology. Kluwer Academic, Dordrecht, pp 529–549

Crawford SA, Higgins MJ, Mulvaney P, Wetherbee R (2001) Nanostructure of the diatom frustule as revealed by atomic force and scanning electron microscopy. J Phycol 37:543–554

Daniel GF, Chamberlain AHL, Jones EBG (1980) Ultrastructural observations on the marine fouling diatom *Amphora*. Helgoländer Meeresunters 34:123–149

Daniel GF, Chamberlain AHL, Jones EBG (1987) Cytochemical and electron microscopical observations on the adhesive materials of marine fouling diatoms. Br Phycol J 22:101–118

de Brouwer JF, Stal LJ (2002) Daily fluctuations of exopolymers in cultures of the benthic diatoms *Cylindrotheca closterium* and *Nitzschia* sp. (Bacillariophyceae). J Phycol 38:464–472

Dugdale TM, Dagastine R, Chiovitti A, Mulvaney P, Wetherbee R (2005) Single adhesive nanofibers from a live diatom have the signature fingerprint of modular proteins. Biophys J 89 (6):4252–4260

Edgar LA (1983) Mucilage secretions of moving diatoms. Protoplasma 118:44–48

Edgar LA, Pickett-Heaps JD (1982) Ultrastructural localization of polysaccharides in the motile diatom *Navicula cuspidata*. Protoplasma 113:10–22

Edgar LA, Pickett-Heaps JD (1983) The mechanism of diatom locomotion. I. An ultrastructural study of the motility apparatus. Proc R Soc Lond B 281:331–343

Edgar LA, Pickett-Heaps JD (1984) Diatom locomotion. Prog Phycol Res 3:47–88

Edgar LA, Zavortink M (1983) The mechanism of diatom locomotion. II. Identification of actin. Proc R Soc Lond B 218:345–348

Fisher TE, Marszalek PE, Fernandez JM (2000) Stretching single molecules into novel conformations using the atomic force microscope. Nat Struct Biol 7:719–723

Gebeshuber IC, Thompson JB, Del Amo Y, Stachelberger H, Kindt JH (2002) In vivo nanoscale atomic force microscopy investigation of diatom adhesion properties. Mater Sci Technol 18:763–766

Gebeshuber IC, Kindt JH, Thompson JB, Del Amo Y, Stachelberger H, Brzezinski MA, Stucky GD, Morse DE, Hansma PK (2003) Atomic force microscopy study of living diatoms in ambient conditions. J Microsc 212:292–299

Geesey GG, Wigglesworth-Cooksey B, Cooksey KE (2000) Influence of calcium and other cations on surface adhesion of bacteria and diatoms. Biofouling 15:195–205

Gibson RA (1979) Observations of stalk production by *Pseudohimantidium pacificum* Hust. & Krasske (Bacillariophyceae: Protoraphidaceae). Nova Hedwigia 31:899–915

Gordon R (1987) A retaliatory role for algal projectiles, with implications for the mechanochemistry of diatom gliding motility. J Theor Biol 126:419–436

Gutierrez-medina B, Guerra AJ, Maldonado AIP, Rubio YC, Meza JVG (2014) Circular random motion in diatom gliding under isotropic conditions. Phys Biol 11:066006

Häder D-P, Hoiczyk E (1992) Gliding motility. In: Melkonian M (ed) Algal cell motility. Chapman and Hall, New York, pp 1–38

Higgins MJ, Crawford SA, Mulvaney P, Wetherbee R (2000) The topography of soft, adhesive diatom 'trails' as observed by atomic force microscopy. Biofouling 16:133–139

Higgins MJ, Crawford SA, Mulvaney P, Wetherbee R (2002) Characterization of the adhesive mucilages secreted by live diatom cells using atomic force microscopy. Protist 153:25–38

Higgins MJ, Molino P, Mulvaney P, Wetherbee R (2003a) The structure and nanomechanical properties of the adhesive mucilage that mediates diatom substratum adhesion and motility. J Phycol 39:1181–1193

Higgins MJ, Sader JE, Mulvaney P, Wetherbee R (2003b) Probing the surface of living diatoms with atomic force microscopy: the nanostructure and nanomechanical properties of the mucilage layer. J Phycol 39:722–734

Hoagland KD, Rosowski JR, Gretz MR, Roener SC (1993) Diatom extracellular polymeric substances: function, fine structure, chemistry, and physiology. J Phycol 29:537–566

Holland R, Dugdale TM, Wetherbee R, Brennan AB, Finlay JA, Callow JA, Callow ME (2004) Adhesion and motility of fouling diatoms on a silicone elastomer. Biofouling 20:323–329

Hugel T, Seitz M (2001) The study of molecular interactions by AFM force spectroscopy. Macromol Rapid Commun 22:989–1016

Jagota A, Bennison SJ (2002) Mechanics of adhesion through a fibrillar microstructure. Integr Comp Biol 42:1140–1145

Johnson LM, Hoagland KD, Gretz MR (1995) Effects of bromine and iodine on stalk secretion in the biofouling diatom *Achnanthes longipes* (Bacillariophyceae). J Phycol 31:401–412

Kellermeyer MSZ, Smith SB, Granzier HL, Bustamante C (1997) Folding-unfolding transitions in single titin molecules characterized with laser tweezers. Science 276:1112–1116

Khandeparker RD, Bhosle NB (2001) Extracellular polymeric substances of the marine fouling diatom *Amphora rostrata* Wm.S. Biofouling 17:117–127

Kooistra WHCF, de Stafano M, Mann DG, Salma N, Medlin LK (2003) Phylogenetic position of *Toxarium*, a pennate-like lineage within centric diatoms (Bacillariophyceae). J Phycol 39:185–197

Lewis R, Johnson LM, Hoagland KD (2002) Effects of cell density, temperature, and light intensity on growth and stalk production in the biofouling diatom *Achnanthes longipes* (Bacillariophyceae). J Phycol 38:1125–1131

Lind JL, Heimann K, Miller EA, van Vliet C, Hoogenraad NJ, Wetherbee R (1997) Substratum adhesion and gliding in a diatom are mediated by extracellular proteoglycans. Planta 203:213–221

Mann DG (1999) The species concept in diatoms. Phycologia 38:437–495

Marx K (2003) Quartz crystal microbalance: a useful tool for studying thin polymer films and complex biomolecular systems at the solution-surface interface. Biomacromolecules 4 (5):1099–1120

McConville MJ, Wetherbee R, Bacic A (1999) Subcellular location and composition of the wall and secreted extracellular sulphated polysaccharides/proteoglycans of the diatom *Stauroneis amphioxys* Gregory. Protoplasma 206:188–200

Molino PJ, Wetherbee W (2008) The biology of biofouling diatoms and their role in the development of microbial slimes. Biofouling 24(5):365–379

Molino PJ, Hodson OM, Quinn JF, Wetherbee R (2006) Utilizing QCM-D to characterize the adhesive mucilage secreted by two marine diatom species in-situ and in real-time. Biomacromolecules 7:3276–3282

Molino PJ, Campbell E, Wetherbee R (2008a) Development of the initial diatom microfouling layer on antifouling and fouling-release surfaces in temperate and tropical Australia. Biofouling 25(8):685–694

Molino PJ, Hodson OM, Quinn JF, Wetherbee R (2008b) The quartz crystal microbalance: a new tool for the investigation of the bioadhesion of diatoms to surface of differing surface energies. Langmuir 24(13):6730–6737

Montsant A, Jabbari K, MaheswariU BC (2005) Comparative genomics of the pennate diatom *Phaeodactylum tricornutum*. Plant Physiol 137:500–513

Moroz AL, Ehrman JM, Clair TA, Gordon RJ, Kaczmarska I (1999) The impact of ultraviolet-B radiation on the motility of the freshwater epipelic diatom *Nitzschia lineariz*. Global Change Biol 5:191–199

Omae I (2003) Organotin antifouling paints and their alternatives. Appl Organomet Chem 17:81–105

Oroudjev E, Soares J, Arcidiacono S, Thompson JB, Fossey SA, Hansma HG (2002) Segmented nanofibers of spider dragline silk: atomic force microscopy and single-molecule force spectroscopy. Proc Natl Acad Sci USA 99:6460–6465

Paterson DM (1989) Short-term changes in the erodibility of intertidal cohesive sediments related to the migratory behaviour of epipelic diatoms. Limnol Oceanogr 43:223–234

Pettitt ME, Henry SL, Callow ME, Callow JA, Clare AS (2004) Activity of commercial enzymes on settlement and adhesion of cypris larvae of the barnacle *Balanus amphitrite*, spores of the green alga *Ulva linza*, and the diatom *Navicula perminuta*. Biofouling 20:299–311

Pickett-Heaps JD, Schmid A-M, Edgar LA (1990) The cell biology of diatom valve formation. Prog Phycol Res 7:1–168

Pickett-Heaps JD, Hill DRA, Blaze KL (1991) Active gliding in an araphid marine diatom, *Ardissonea* (formerly *Synedra*) *crystallina*. J Phycol 27:718–725

Poulsen NC, Spector I, Spurck TP, Schultz TF, Wetherbee R (1999) Diatom gliding is the result of an actin-myosin motility system. Cell Motil Cytoskeleton 44:23–33

Poulson N, Kroger N, Harrington MJ, Brunner E, Paasch S, Buhmann MT (2014) Isolation and biochemical characterization of underwater adhesives from diatoms. Biofouling 30 (4):513–523

Rief M, Gautel M, Oesterhelt F, Fernandez JM, Gaub HE (1997a) Reversible unfolding of individual titin immunoglobulin domains by AFM. Science 276:1109–1112

Rief M, Oesterhelt F, Heymann B, Gaub HE (1997b) Single molecule force spectroscopy on polysaccharides by AFM. Science 275:1295–1297

Rosowski JR (1980) Valve and band morphology of some freshwater diatoms. II. Integration of valves and bands in *Navicula confervacea* vr. confervacea. J Phycol 16:88–101

Rosowski JR, Hoagland KD, Roemer SC (1983) Valve and band morphology of some freshwater diatoms. IV. Outer surface mucilage of *Navicula confervacea* var. confervacea. J Phycol 19:342–347

Round FE, Crawford RM, Mann DG (1990) The diatoms. Cambridge University Press, Cambridge, 747 pp

Serôdio J, da Silva JM, Catarino F (1997) Nondestructive tracing of migratory rhythms of intertidal benthic microalgae using in vivo chlorophyll *a* fluorescence. J Phycol 33:542–553

Smestad Paulsen B, Haug A, Larsen B (1978) Structural studies of a carbohydrate-containing polymer present in the mucilage tubes of the diatom *Berkeleya rutilans*. Carbohydr Res 66:103–111

Smith DJ, Underwood GJC (1998) Exopolymer production by intertidal epipelic diatoms. Limnol Oceanogr 43:1578–1591

Staats N, de Winder B, Stal LJ, Mur LR (1999) Isolation and characterization of extracellular polysaccharides from the epipelic diatoms *Cylindrotheca closterium* and *Navicula salinarum*. Eur J Phycol 34:161–169

Underwood GJC, Paterson DM (2003) The importance of extracellular carbohydrate production by epipelic diatoms. In: Callow JM (ed) Advances in botanical research, vol 40. Elsevier Academic Press, Oxford, pp 183–240

Vilter H (1984) Peroxidases from Phaeophyceae – a vanadium (V)-dependent peroxidase from *Ascophyllum nodosum*. 5. Phytochemistry 23:1387–1390

Vreeland V, Waite JH, Epstein L (1998) Polyphenols and oxidases in substratum adhesion by marine algae and mussels. J Phycol 34:1–8

Wang Y, Lu J, Mollet J-C, Gretz MR, Hoagland KD (1997) Extracellular matrix assembly in diatoms (Bacillariophyceae). II. 2,6-dichlorobenzonitrile inhibition of motility and stalk production in the marine diatom *Achnanthes longipes*. Plant Physiol 113:1071–1080

Wang Y, Chen Y, Lavin C, Gretz MR (2000) Extracellular matrix assembly in diatoms (Bacillariophyceae). IV. Ultrastructure of *Achnanthes longipes* and *Cymbella cistula* as revealed by high-pressure freezing/freeze substitution and cryo-field emission scanning electron microscopy. J Phycol 36:367–378

Webster DR, Cooksey KE, Rubin RW (1985) An investigation of the involvement of cytoskeletal structures and secretion in gliding motility of the marine diatom, *Amphora coffeaeformis*. Cell Motil Cytoskeleton 5:103–122

Wetherbee R, Lind JL, Burke J, Quatrano RS (1998) The first kiss: establishment and control of initial adhesion by raphid diatoms. J Phycol 34:9–15

Wigglesworth-Cooksey B, Cooksey KE (1992) Can diatoms sense surfaces?: state of our knowledge. Biofouling 5:227–238

Wigglesworth-Cooksey B, Cooksey KE (2005) Use of fluorophore-conjugated lectins to study cell–cell interactions in model marine biofilms. Appl Environ Microbiol 71:428–435

Wigglesworth-Cooksey B, van der Mei H, Busscher HJ, Cooksey KE (1999) The influence of surface chemistry on the control of cellular behaviour: studies with a marine diatom and a wettability gradient. Colloids Surf B Biointerfaces 15:71–79

Willis A, Chiovitti A, Dugdale TM, Wetherbee R (2013) Characterisation of the extracellular matrix of Phaeodactylum tricornutum (Bacillariophyceae): structure, composition, and adhesive characteristics. J Phycol 49:937–949

Willis A, Eason-Hubbard M, Hodson O, Maheswari U, Bowler C, Wetherbee R (2014) Adhesion molecules from the diatom *Phaeodactylum tricornutum* (Bacillariophyceae): genomic identification of amino-acid profiling and in vivo analysis. J Phycol 50:837–849

Wustman BA, Gretz MR, Hoagland KD (1997) Extracellular matrix assembly in diatoms (Bacillariophyceae). I. A model of adhesives based on chemical characterization and localization of polysaccharides from the marine diatom *Achnanthes longipes* and other diatoms. Plant Physiol 113:1059–1069

Wustman BA, Lind J, Wetherbee R, Gretz MR (1998) Extracellular matrix assembly in diatoms (Bacillariophyceae). III. Organization of fucoglucuronogalactans within the adhesive stalks of *Achnanthes longipes*. Plant Physiol 116:1431–1441

Zlatanova J, Lindsay SM, Leuba SH (2000) Single molecule force spectroscopy in biology using atomic force microscopy. Prog Biophys Mol Biol 74:37–61

# Chapter 4
# Progress in the Study of Adhesion by Marine Invertebrate Larvae

**Nick Aldred and Luigi Petrone**

**Abstract** This chapter summarises recent progress towards the characterisation of bioadhesives secreted by the larvae of marine invertebrates, with reference also to their subsequent developmental stages. These adhesives vary structurally and biochemically between species and between life stages to satisfy the requirements of the particular organism for permanent or temporary/reversible adhesion, often under hostile conditions. To date, a small number of bioadhesives have been described for the adult forms of marine invertebrates, while a functional understanding of larval adhesives remains elusive. Progress is essential, however, since the larval forms perform a key role in the fouling of marine structures, and their adhesives may have characteristics of interest for development of synthetic, bio-inspired glues. Despite recent advances in the fields of proteomics and genomics, major obstacles exist in the isolation and analysis of tiny quantities of larval adhesives, which have largely precluded these approaches. Further challenges relate to the in situ detection of larval adhesive materials, being usually secreted underwater and buried at the interface between a solid substrate and the organism's body. Here, we discuss a range of novel experimental approaches that have surmounted these technical issues and provided useful insight into the morphology and composition of larval bioadhesives in situ. These involve imaging and spectroscopic approaches as well as nano-/micromechanical and surface-sensitive techniques that have enabled quantification of adhesion forces and surface adsorption of purified adhesive proteins.

N. Aldred (✉)
School of Marine Science and Technology, Newcastle University, Newcastle Upon Tyne NE1 7RU, UK
e-mail: nicholas.aldred@ncl.ac.uk

L. Petrone
Brookes Bell Hong Kong Ltd, Room 2207, Tower 2, Lippo Centre, 89, Queensway, Hong Kong

© Springer International Publishing Switzerland 2016

A.M. Smith (ed.), *Biological Adhesives*, DOI 10.1007/978-3-319-46082-6_4

## 4.1 Introduction

Marine invertebrate larvae are of fundamental interest for their ability to attach rapidly, tenaciously and, sometimes, reversibly to a wide range of immersed surfaces. They also represent the key stage in surface colonisation by marine fouling species, and characterisation of their surface attachment is thus of significant practical and commercial value. At present, however, there are no examples of full, functional understanding of adhesion in larval systems. This is due primarily to the challenges associated with collecting and analysing such tiny quantities of material (on the order of nanograms), and it is therefore timely to review the recent advances in larval bioadhesion that have been facilitated, in large part, by improved accessibility and sensitivity of analytical techniques essential for their study.

Historically, the barriers to progress in this field have been predominantly technical; invertebrate larvae present particular challenges to the study of adhesion and adhesives. They are small, typically less than 1 mm, and consequently produce limited quantities of adhesive that are insufficient for proteomics or biochemical characterisation. Often, the larvae of marine invertebrates constitute the dispersal stage of the life cycle, and the adhesive(s) they produce may therefore only be available once, as they initiate settlement. Simultaneous attachment of many larvae may be required to procure a sufficient sample for analysis; however, this may be difficult if the larvae are selective about the surfaces they colonise (Clare and Aldred 2009). Advances in surface imaging and analytical and spectroscopic methods have provided additional means to interrogate some larval adhesion mechanisms that have, to date, proven refractory to analysis by traditional methods that require collection of material. The advancing nature of these techniques has added a new dynamism to the study of larval adhesion and promises to advance our understanding significantly in the coming years. It is important to note, however, that no single imaging/spectroscopic technique can provide the full suite of data necessary to elucidate a given adhesion system. Multiple complementary approaches are usually required.

Perhaps the most fundamental question pertains to the degree of similarity between larval and adult adhesion systems in the same species. In general, larval adhesives are poorly understood compared to their counterparts in the adult life stages, and in recognition of the scarcity of information for diverse species, this review will focus primarily on barnacles for which there has been notable progress in recent years. The cypris larva of the barnacle (Fig. 4.1) is arguably the most comprehensively studied marine invertebrate larva in terms of surface selection and adhesion, but the experimental approaches described herein for investigating its adhesion/adhesives are broadly applicable. Surprisingly, despite the thorough descriptions of adhesive proteins in adult barnacles (Kamino 2013), studies of the larval adhesives remain insufficiently detailed to allow meaningful comparison between life stages (e.g., He et al. 2013). Data for other well-studied species is similarly scant, although in the Crinoid and Asteroid classes of the Echinodermata, the larval adhesive organs apparently have no equivalent in the adult forms (Strathmann 1978). While the temporary adhesive of asteroid larvae appears to be biochemically similar to that of the adult (Haesaerts et al. 2005), these organisms

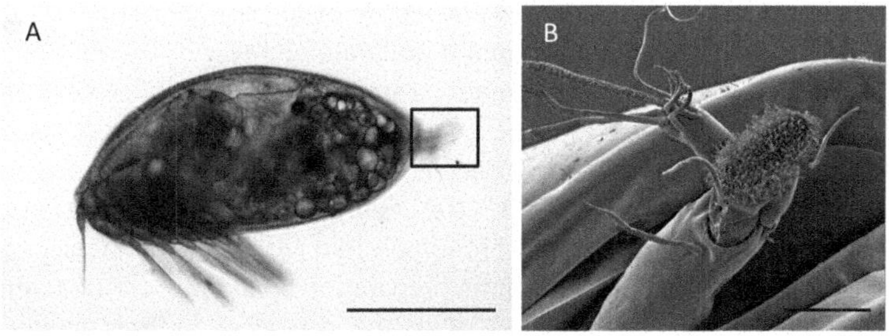


**Fig. 4.1** (**a**) A cyprid of the barnacle *Balanus amphitrite*, imaged using transmitted light through a ×20 objective lens on a Leica DMi8 inverted microscope (scale bar = 200 μm). (**b**) Scanning electron micrograph of an antennule of *B. amphitrite* [*boxed area* in (**a**)] terminating in the adhesive disc (scale bar = 20 μm). For a full description of this figure, the reader is referred to Aldred and Clare (2006) [reproduced with permission from Aldred and Clare (2006)]

also have a permanent adhesive which is probably unique to the larval stage. Petrone et al. (2008) conducted the first detailed investigation into the larval adhesive of marine mussels (*Perna canaliculus* veligers) and concluded that the material secreted by larvae for initial surface attachment was distinct from the well-studied byssus system of the adult. Other model 'biofouling' invertebrates such as the serpulid polychaete *Hydroides* spp. (Tanur et al. 2009), the bryozoan *Bugula neritina* (Dahms et al. 2004), the solitary ascidian *Ciona intestinalis* (Aldred and Clare 2014) and the oysters *Crassostrea* spp. (Burkett et al. 2010), while used routinely in assays of settlement, removal and presettlement behaviour, have larval adhesives about which practically nothing is known.

A deeper understanding of larval adhesion processes would benefit not only antifouling technologies that are designed to prevent larval attachment but also provide means to investigate the development, adaptation and evolution of adhesion in biological systems. As for all adhesion systems, the physical and biochemical nature of the materials from marine larvae are of paramount importance. In particular, the phenomena occurring at the interface between the adhesive and the attachment surface are fundamental to understanding the process of adhesion. The following discussion highlights recent advances in the study of larval bioadhesives that were enabled by the adoption of imaging, spectroscopic and micro-/nanomechanical methods. We also discuss several techniques that may be well suited to the study of adhesives from invertebrate larvae, but which have not yet been applied and therefore constitute exciting avenues for future work.

## 4.2 In Situ Imaging for Morphological and Compositional Studies of Larval Adhesives

Observational techniques can provide significant insight into the composition and mechanism of action of biological adhesives. Historically, such studies were limited to low-magnification transmitted light microscopy with histochemistry and higher-resolution methods such as scanning electron microscopy (SEM) that required potentially destructive sample pretreatment (e.g., dehydration, gold-coating; Fig. 4.2). The technical limitations imposed upon early studies of larval adhesion by the available techniques were therefore significant. Although the temporary adhesive secretion of barnacle cyprids was correctly identified by Walker and Yule (1984a) using SEM, for example, further investigations of the material beyond its mere existence were impossible with the available technology, despite the authors' experimental sample preparation techniques (Walker and Yule 1984a; Fig. 4.2). It was not until the 1990s that advances in high-magnification imaging such as cryo-SEM and environmental-SEM (ESEM) began to allow for direct, three-dimensional observation of adhesive structures in an unmodified form (Callow et al. 2003). These techniques allowed for improved visualisation of adhesive structures under more natural conditions and, in the case of ESEM, for experimental procedures involving the modulation of vapour pressure in the vacuum chamber during sample observation (Callow et al. 2003). The use of ESEM to perform in situ experiments enabled determination of the hygroscopic nature of larval mussel adhesives (Petrone et al. 2009) and marked a shift in the use of microscopy techniques from observational tools to advanced methods for experimental and investigative research. These advances continue with modern high-power (up to 300 kV) cryo-SEM facilities revolutionising the fields of structural biology and 3D tomography of nanoscale cellular structures (Callaway 2015). While these techniques remain prohibitively expensive to date, it is likely that in time they will contribute substantially to our understanding of larval adhesion.

Early techniques such as energy-dispersive X-ray spectroscopy (EDX) provided compositional information that was limited to elemental profiles. To understand better the biochemical profile of adhesives both pre- and post-release by larvae, parallel advances in visible light microscopies and spectroscopies proved to be crucial. The potential impact of these advances on our understanding of larval adhesion processes is best exemplified in the case of barnacle cyprid permanent adhesive. Prior to settlement, barnacle cypris larvae (Fig. 4.1) explore immersed surfaces using a temporary adhesion mechanism involving both viscoelastic and behavioural components (reviewed by Aldred et al. 2013a). Once a suitable surface for settlement has been identified, the larva will attach permanently in a process that involves explosive release of an adhesive that is stored, prior to settlement, in two large kidney-shaped glands located posterior to the compound eyes and connected to the exterior by ducts that travel down the length of the paired antennules (Nott and Foster 1969). The exact composition of the cyprid permanent adhesive and its relatedness to that of the adult barnacle remain unclear. Attempts to purify and analyse the expressed material biochemically failed due to the small volume of adhesive produced, so an alternative approach involving imaging spectroscopy was pursued. Using selected fluorophores and careful sample preparation, Gohad et al. (2014) used confocal laser-scanning microscopy (CLSM) to confirm the identity of two distinct strata (Walker 1971) in naturally secreted permanent adhesive. While the innermost material stained strongly for phosphorylated amino acid residues, moieties common in adhesive proteins, the outermost layer

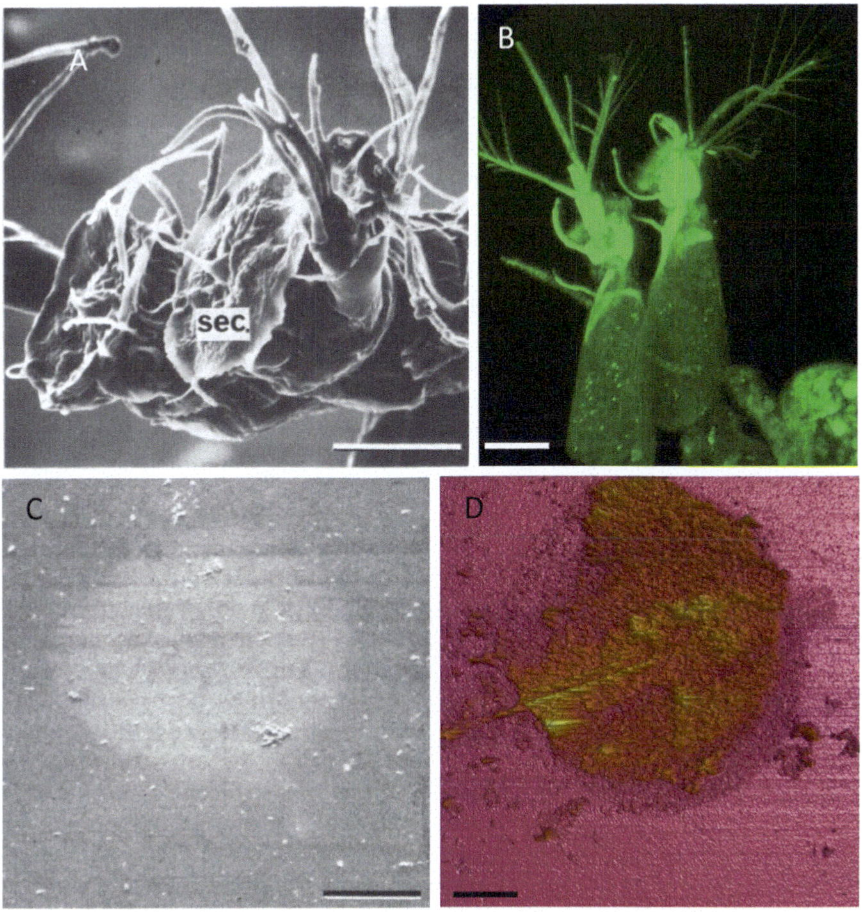


**Fig. 4.2** A comparison of imaging techniques available at the time of identification of the cyprid temporary adhesive and those now available for the study of such materials. (**a**) A scanning electron micrograph (SEM) of the antennules of a barnacle cyprid (*Semibalanus balanoides*) prepared by freeze-drying in order to retain the temporary adhesive in situ ([sec.] see Fig. 4.1 for a similar specimen prepared using traditional methods) (Walker and Yule 1984a). (**b**) Antennules of *Balanus amphitrite* imaged using confocal laser-scanning microscopy, demonstrating the ability of this technique to produce 3D-like images of the delicate structures that collapsed in (**a**) as a result of the freeze-drying procedure and high vacuum conditions in the SEM (Gohad et al. 2012). (**c**) A putative 'footprint' of temporary adhesive deposited on a surface by a cyprid of *S. balanoides*, imaged using SEM (Walker and Yule 1984a). (**d**) A footprint deposited by *B. amphitrite* and imaged using atomic force microscopy, a scanning-probe technique that allows for both 3D imaging and interaction force measurements of biological materials (all scale bars = 20 μm) [reproduced with permission from Walker and Yule (1984a) and Gohad et al. (2012)]

(released first) stained weakly for proteins and strongly for lipids. The identification of a lipidic precursor to cyprid permanent adhesion significantly altered the conceptual model of surface wetting by hydrophilic adhesive proteins, as is thought to be common in biological systems. Displacement of water from a hydrated interface is a major challenge for adhesion underwater, and, while mussels (Lee et al. 2011; Petrone 2013; Wei et al. 2014) and tubeworms (Hennebert et al. 2015) may avoid competition of their adhesives with water by physically excluding it or incorporating it into a coacervate-based adhesive, it may be that cyprids have evolved an alternative approach, using a hydrophobic primer to drive water from the interface.

After identifying two phases in the secreted adhesive, Gohad et al. (2014) were able to trace these precursors back to the cement glands and, using two-photon microscopy, localise them, respectively, into the two cell types (α and β) described previously by Walker (1971) (Fig. 4.3). Two-photon microscopy allows optical sectioning similar to CLSM but with greater sample penetration and reduced out of plane fluorescence. With a modified two-photon system, it is also possible to conduct in situ imaging using coherent anti-Stokes Raman scattering (CARS) which, in this case, provided compositional data for unstained adhesive plaques to support the conclusions of the CLSM-based histology. CARS is a technique

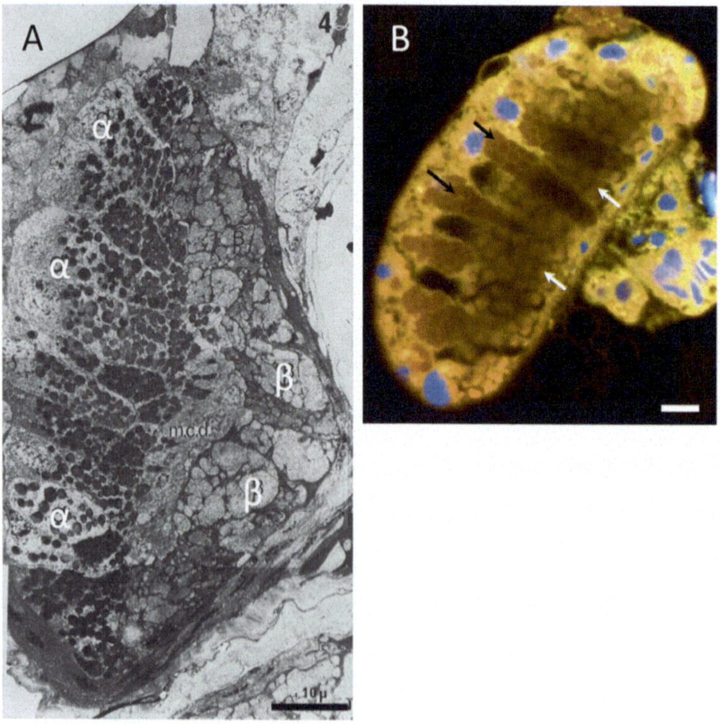


**Fig. 4.3** (**a**) An oblique section through a cyprid (*Semibalanus balanoides*) cement gland, imaged using transmission electron microscopy (Walker 1971), indicating the presence of α and β cell types and (**b**) the same from *Balanus amphitrite* using confocal laser-scanning microscopy (Gohad et al. 2014), showing clearly the spatial relationship of the α (*black arrows*) and β (*white arrows*) cells as well as providing compositional data via fluorescent labelling of lipids (*white arrows*) (scale bars = 10 μm) [reproduced with permission from Walker (1971) and Gohad et al. (2014)]

underutilised in the study of bioadhesives and, in particular, those of marine invertebrate larvae where it allows spatially resolved compositional data to be collected without the requirement for sample manipulation. The resolution and precision of the CARS data collected, confirming the presence of lipids, also highlighted the advancing nature of this technology when compared to the more ambiguous results of a similar study performed 5 years earlier by Schmidt et al. (2009), using confocal Raman scattering imaging. Building on these findings, Senkbeil et al. (2016) recently confirmed anecdotal reports of significant quantities of bromine being present in the cyprid adhesive. While such observations were previously believed to be erroneous, the synchrotron-based X-ray fluorescence study of Senkbeil et al. conclusively identified a strong bromine signal in the cyprid adhesive. The most plausible explanation for its presence was proposed to be in its role as a cross-link promoter during arthropod cuticle sclerotisation. Thus, the presence of lipids secreted from modified epithelial cells within the cyprid cement glands [as occurs in the adhesion of many terrestrial arthropods (Federle et al. 2002)] and the possibility of a sclerotisation-like cross-linking mechanism in the permanent adhesive present the possibility of adhesion via a modified cuticle secretion mechanism. Far from providing a 'starting point' for investigation, therefore, advanced imaging methods have potential to provide significant compositional and mechanistic information, identifying new and promising avenues for larval adhesive research.

## 4.3 In Situ Spectroscopic Investigations of Larval Adhesive Secretions

Noninvasive, in situ spectroscopic techniques lend themselves to the investigation of the properties of biointerfaces without altering their biological structure (Berglin et al. 2005; Barlow and Wahl 2012; Petrone 2013). However, the physicochemical characterisation of biological interfaces by conventional infrared (IR) spectroscopy in aqueous media has historically proven to be a challenge, due to the strong absorbance of water in the mid-IR range. The introduction of the attenuated total reflection IR (ATR-IR) method presented new opportunities for the investigation of phenomena at liquid/solid interfaces by circumventing issues related to the presence of water (McQuillan 2001). Briefly, an IR beam travelling into a high refractive index prism in contact with a liquid medium of lower refractive index is totally internally reflected above a critical angle of incidence at the prism/liquid interface. Beyond the reflecting prism and within the liquid, an evanescent wave is generated whose intensity decays exponentially away from the prism (Goormaghtigh et al. 1999) with a penetration depth of ~1–3 μm (Fig. 4.4). The bulk of the solvent is effectively absent and can be subtracted conveniently by recording an appropriate background spectrum. Only the molecules in close contact with the prism surface are probed by ATR-IR spectroscopy, which has a detection

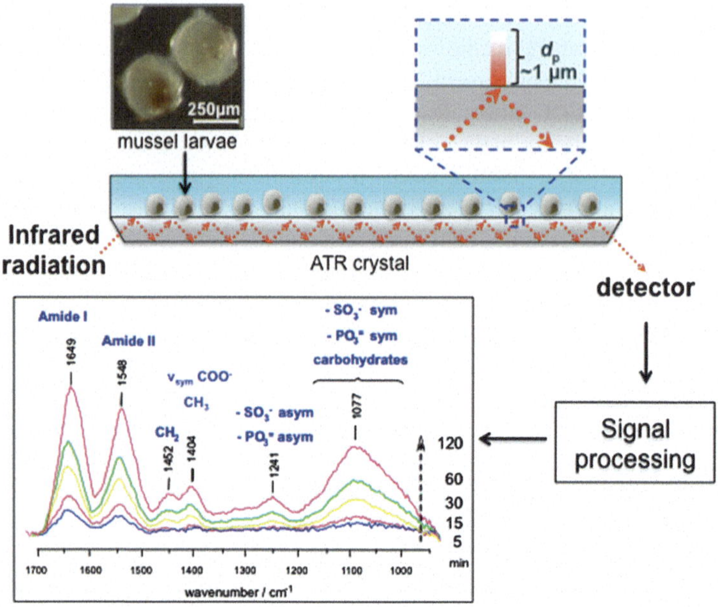


**Fig. 4.4** Schematic diagram of an ATR-IR setup for the in situ detection of the adhesive secretion from mussel larvae (*Perna canaliculus*). This configuration has ~1 μm penetration depth ($d_p$), the intensity decaying exponentially away from the multiple internal reflection ATR crystal's surface into the aqueous medium. After obtaining the appropriate IR background from mussel shell, ATR-IR spectra may be recorded over the course of larval settlement. The identity of each absorbance peak is also reported on the ATR-IR spectrum [reproduced with permission from Petrone et al. (2008)]

limit of a few micrograms. In addition, the deposition of thin nanoparticle films (a few hundred nm) on the ATR prism expands the array of surface chemistries available for spectroscopic investigations. Nanoparticle films possess a considerably larger surface area than the flat prism surface, thus greatly enhancing the sensitivity of the ATR method towards adsorbed molecules (Sperline et al. 1992; Hug and Sulzberger 1994; Connor et al. 1995). Notably, changes in the IR spectra of molecules in solution occur upon adsorption due to the formation of new chemical bonds, ionic interactions and inevitable structural rearrangements (Hug 1997; Peak et al. 1999; Parikh and Chorover 2006; Petrone and McQuillan 2011). Of particular relevance to this chapter, ATR-IR spectroscopy has recently expanded our knowledge of the chemical composition of marine bioadhesives, including binding modes and adsorption mechanisms of their reactive functionalities to wet mineral surfaces. Marine invertebrate larvae generally encounter and settle on mineral and metal oxide surfaces; hence, the use of metal oxide nanoparticles as thin films in ATR-IR spectroscopy studies to replicate the molecular interactions on naturally occurring substrata.

Following initial studies on catecholate ligands in siderophores secreted by bacteria (McWhirter et al. 2003), the McQuillan group exploited the surface sensitivity of the ATR mode in the context of wet bioadhesion, revealing proteins and polysaccharides secreted by the pediveliger larvae of mussels [*Mytilus galloprovincialis* and *Perna canaliculus* (Gao et al. 2007)] on a bare ATR germanium (Ge) prism. The experimental setup was then improved with a cooling system underneath the Ge prism to reproduce the seawater temperatures for each target fouling species in their natural environment. Living larvae of *P. canaliculus* measuring ~250 μm in length settled under seawater flow at 16 °C and were interrogated by in situ ATR-IR spectroscopy (Petrone et al. 2008). ATR-IR spectra, in combination with EDX spectra and ESEM observations (Petrone et al. 2009), demonstrated the presence of proteins and polysaccharides along with carboxylate, monoester-sulphate and monoester-phosphate functionalities held together via ionic interactions with calcium ions. In a subsequent study, in situ ATR-IR investigations on a bare Ge prism conducted with flowing seawater at 14 °C revealed a carboxylated glycoproteinaceous glue secreted by kelp zoospores (*Undaria pinnatifida*) of ~4 μm in diameter (Petrone et al. 2011a, b). Moreover, the use of a thin (~100 nm) $TiO_2$ particle film allowed the identification of monoester-phosphate groups in the algal spore glue, as revealed by the appearance of characteristic absorbances due to the formation of phosphate-Ti covalent bonds.

Although not yet applied to the study of barnacle larvae, ATR-IR spectroscopy has been used to investigate the chemical composition of the adult barnacle cement. Barlow et al. (2009) recorded a spectroscopic proteinaceous fingerprint of barnacle primary cement in vivo by allowing an adult barnacle (*B. amphitrite*) to settle directly onto a Ge wafer, which was subsequently brought into contact with a Ge ATR prism. It was discovered (Barlow et al. 2010) that both primary (original cement secreted by the barnacle) and secondary (used for reattachment) cements were predominantly composed of proteins with cross β-sheet secondary structures. This structure is characteristic of amyloid-like proteins, a hypothesis confirmed by the presence of amyloid nanofibrils imaged by atomic force spectroscopy. More recently, Burden et al. (2012) revealed that barnacles (*B. amphitrite*) adhere via a stepwise cementation process, whereby one of the two types of secretion presents a larger quantity of amyloid-like structures, as observed from the enhanced cross β-sheet peak in transmission IR spectra. This process has since been observed in real time, using an imaging surface plasmon resonance method (Golden et al. 2016) discussed in Sect. 4.4. Recently, Nakano and Kamino (2015) have demonstrated that a series of tyrosine-rich peptides, designed based on a bulk 52 kDa barnacle cement protein, self-assembled into fibrillar microstructures with amyloid-like motifs rich in β-sheets. These studies demonstrate the potential of ATR-IR spectroscopy for investigating the composition and conformation of adhesive secretions, which could reasonably be extended to the study of invertebrate larvae.

Although similar in some respects to IR spectroscopy, Raman spectroscopy has the advantage of being relatively insensitive to the presence of water at visible wavelengths, thus allowing the analysis of hydrated biological samples. Mostaert et al. (2009) used Raman spectroscopy to detect proteins forming predominantly amyloid-like β-sheet structures in the adhesive from two green unicellular microalgae (Chlorophyta), while Harrington et al. (2010) used in situ resonance Raman spectroscopy to demonstrate that the cuticle of the mussel byssus of *M. californianus* and *M. galloprovincialis* consists of highly cross-linked granules containing catecholato (Dopa)-iron chelation complexes embedded in a collagenous matrix. Confocal resonance Raman microspectroscopy was applied to the investigation of the spatial distribution of carotenoids in the diffusion boundary layer of marine green (*Ulva* spp.) and brown (*Fucus vesiculosus*) algae (Grosser et al. 2012). To date, however, Raman-based techniques have been sparsely applied to studies of adhesion by invertebrate larvae (Schmidt et al. 2009; Gohad et al. 2014). Techniques with higher surface sensitivity are generally required to probe the relatively small deposits of larval adhesives. Looking forward, spectroscopic techniques with signal enhancement, such as surface-enhanced IR absorption spectroscopy (SEIRAS) and surface-enhanced Raman spectroscopy (SERS), could help to resolve physicochemical make-up of such micro-scale adhesive deposits as the footprints of barnacle or echinoderm larvae, using specially prepared metal substrata.

## 4.4 Quantifying the Interaction of Naturally Secreted Adhesives with Surfaces

The characterisation of cyprid permanent adhesive (Sect. 4.2) demonstrated that imaging-based studies can be illuminating beyond the provision of micro-scale structural details. Indeed, in the context of adhesion, interactions with surfaces are all important. When bioadhesion imaging is combined with careful selection of the substrate for larval attachment, further information can be gathered regarding the dynamics of adhesion at the surface interface. For adult mussels, this approach was elegantly demonstrated in a recent study by Martinez Rodriguez et al. (2015). In this series of experiments, the authors employed surfaces functionalised with a lipid bilayer and tethered fluorochrome (Oregon Green 448 DHPE), which reported the acidic interfacial pH beneath the foot of attaching mussels during byssus deposition. Although this approach has not yet been attempted at the scale of settling larvae, experimental use of well-characterised surfaces to report on adhesive physicochemical properties has been achieved. Aldred et al. (2013b) used hydrophobic and hydrophilic self-assembled monolayers (SAMs) of alkanethiols on gold-coated microscope slides to investigate, using CLSM, the interaction of cyprid permanent adhesive with surfaces differing in wettability. The authors observed that spreading of the adhesive was more restricted on the hydrophobic SAM. This finding was contrary to conventional wetting theory, presuming a hydrophilic proteinaceous adhesive (Callow et al. 2005), but consistent with observations of adhesives from other species (Callow et al. 2005; Aldred et al. 2006). It was also consistent with the presence of a hydrophobic outer phase in the adhesive, as later described by Gohad et al. (2014). In their study, Aldred et al. (2013b) clearly discerned two distinct strata in the adhesive deposits and noted a difference in contact angle between the inner phase and outer phase in contact with the hydrophobic surface, suggesting significant differences in the composition of the two layers that were later confirmed (Gohad et al. 2014). The authors suggested that this confocal microscopy-based goniometry method could be used to measure the interaction of cyprid adhesives with surfaces as a proxy measurement of affinity between the adhesive and a surface of interest.

The Aldred et al. (2013b) study of cyprid permanent adhesive was not, however, the first example of this approach towards gathering information about the physical properties of larval adhesive deposits by proxy. Aldred et al. (2011) had previously developed a method drawing upon series of SAMs and an imaging surface plasmon resonance system (Andersson et al. 2009) to collect comparable surface interaction data relating to cyprid temporary adhesion. Before committing to irreversible settlement, barnacle cyprids explore surfaces, often extensively, using temporary adhesion. The viscoelastic adhesive component of this system, the temporary adhesive, is a proteinaceous material often deposited as footprints on explored surfaces (Aldred et al. 2013a). Aldred et al. (2011) used a series of SAMs including hydrophilic and hydrophobic, positively and negatively charged chemistries to observe the frequency of deposition of temporary adhesive footprints. They also investigated poly(ethylene glycol) (PEG) and PEGMA/hydroxyethylmethacrylate (HEMA) hydrogels which were reported to demonstrate anti-adhesion efficacy. Using this technique, surface contacts (termed 'touchdowns') could be recorded as bright flashes in the surface plasmon field (Fig. 4.5), an approach later applied by Golden et al. (2016) to the investigation of growth and adhesion by adult barnacles. Deposited footprints remained as persistent bright spots, and the ratio between these values provided a measure of adhesion success for each surface, i.e., whether the temporary adhesive was more likely to stick to the surface or remain on the cyprid's adhesive disc. The number of footprints remaining on the negatively charged carboxyl SAM was significantly higher than on any other surface tested, implying stronger adhesion to that surface. Further, footprints were thicker on that surface than on the other SAMs, and the measurement of footprint thickness highlighted the PEG and PEGMA/HEMA surfaces as retaining very little temporary adhesive. Settlement data published later by Petrone et al. (2013) also demonstrated the highest settlement of cyprids on negatively charged SAMs, broadly supporting a correlation between perceived adhesion strength and settlement site selection by barnacle cyprids (Aldred et al. 2011). The authors concluded that an adhesive optimised for attachment to negatively charged surfaces would make sense in the context of an ocean environment dominated by surfaces carrying net-negative charges.

Finally, it must be recognised that although larval surface selection, adhesion strength and survival to adulthood are relevant measures of efficacy in the development of novel fouling-resistant materials, the relationship between them is complex and requires further investigation. Although a conceptual framework exists for determining the likelihood of cyprid settlement by observation of pre-settlement behaviour (Crisp 1976), for example, in reality this is simplistic. Similarly, by measuring adhesion strength and settlement rate independently, it is

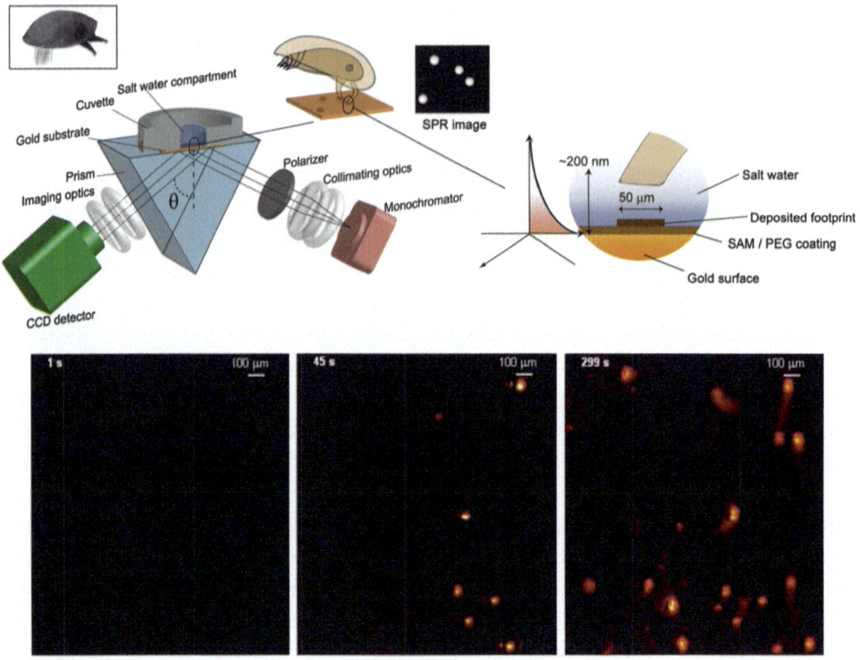


**Fig. 4.5** A schematic diagram of the imaging surface plasmon resonance technique (Aldred et al. 2011) used to image the temporary adhesive interactions of cyprids with surfaces. The time series of images demonstrates the accumulation of footprints on a surface during an experiment [reproduced with permission from Aldred et al. (2011)]

challenging to argue a causal influence of the former on the latter. To address this challenge, the field requires new approaches to combine long-term measures of presettlement behaviour (e.g., Maleshlijski et al. 2012), up to final settlement of larvae, with simultaneous measurement of adhesive interactions. The first attempt to achieve this has recently been published (Maleshlijski et al. 2015) in a study that combined 3D stereoscopic tracking of barnacle cyprids with iSPR quantification of adhesive deposition during surface exploration on well-characterised SAMs. Although preliminary, this study highlights the necessary direction for systematic determination of surface characteristics on larval adhesion and the influence of the latter on surface selection and settlement.

## 4.5 Measuring Surface Adsorption of Purified Larval Adhesive Proteins

Once proteins and other functional biomolecules are isolated from the bioadhesives of marine invertebrate larvae, their role as adhesive primers and/or bulk contributions to cohesion can be ascertained more precisely. There exists a range of surface-sensitive techniques that are applicable to this field of research. These include SPR as discussed, but also quartz crystal microbalance (QCM), ellipsometry and surface forces apparatus (SFA) techniques which have been used previously, for example, in the study of mussel adhesive proteins (Höök et al. 2001; Yu et al. 2011; Wei et al. 2014; Petrone et al. 2015a).

The findings of Aldred et al. (2011) and Petrone (2013) discussed in Sect. 4.4 demonstrated an effect of surface chemistry on the efficacy of cyprid temporary adhesion and implied a relationship between this and the propensity of larvae to settle on a given surface. It was thus of interest to determine whether the same surface specificity would be observed in experiments using a purified form of the adhesive versus the naturally secreted form. Although the exact composition of cyprid temporary adhesive is unknown, due to incompatibility of analytical techniques with the small quantity of material available for study, one component has been identified. The settlement-inducing protein complex (SIPC) of barnacles is the conspecific cue that mediates the settlement of cyprids close to adults of the same species. It is a large cuticular glycoprotein with significant sequence homology to the α-2-macroglobulin family of blood complement protease inhibitors (Dreanno et al. 2006a) and has been identified in footprints of temporary adhesive deposited onto surfaces (Dreanno et al. 2006b). While we do not know the exact proportion of the temporary adhesive that SIPC comprises, immunohistochemistry suggests that it is a significant component and raised the possibility of a dual role for SIPC, as both a pheromone and an adhesive protein.

SPR provides a convenient and highly sensitive method for estimating mass accumulation on surfaces. While this technique has usually been employed for the investigation of specific interactions, it is equally applicable to the study of nonspecific protein adsorption, the process that underpins the initiation of bioadhesive bonds. Petrone et al. (2015b) investigated, using SPR, the adsorption of SIPC to a range of SAMs and compared the adsorption of this putative bioadhesive protein to both α-2-macroglobulin as a negative control (similar structure, no role in adhesion) and fibrinogen as a positive standard. Fibrinogen is widely recognised to be a protein with high potential for nonspecific adsorption to most surfaces. The mass uptake of SIPC to all surfaces was remarkable and comparable to that of fibrinogen. Notably, SIPC adsorbed in measurable quantities to a protein-resistant oligo(ethylene glycol) SAM and such a propensity for broad-spectrum nonspecific surface adsorption is unlikely to have evolved in a molecule whose function is not one of adhesion. Adsorption of α-2-macroglobulin was weak by comparison, with notable accumulation recorded only on the positively charged amine-terminated SAM where adsorption was presumably driven electrostatically. This series of experiments provided an evidence base from which to further investigate the role of SIPC in cyprid temporary adhesion and demonstrated a tractable method to determine the adsorption characteristics of putative adhesive proteins from invertebrate larvae.

## 4.6 Micromechanical Measurement of Adhesion Force in Larval Adhesives

Mechanical testing has been the cornerstone of bioadhesive research throughout the history of the field; however at the scale of invertebrate larvae, experiments can be less than straightforward. Early work successfully measured the adhesion of whole organisms to surfaces by a combination of micromanipulation and tensile testing (Walker and Yule 1984b). However, it is impossible in such an experimental design to eliminate the behavioural component that is now known to be central to the process of temporary adhesion in barnacle larvae (Aldred et al. 2013a) and, in all likelihood, the larvae of other marine invertebrates. Mechanical testing of the adhesive in situ and in an unmodified state, but in the absence of the larva, is the ideal solution to this technical challenge but was impossible until the relatively recent introduction of biophysical techniques to the field.

Atomic force microscopy (AFM) and related scanning-probe techniques have had arguably the greatest impact in advancing the field of bioadhesion research. Fundamentally, AFM is a nanomechanical method rather than an optical microscopy method and can thus provide simultaneous topological and mechanical data (force mapping), allowing researchers to produce high-resolution 3D renderings of adhesive deposits at the micro-scale while also collecting force data relating directly to the material under observation. AFM experiments can be performed in ambient conditions including under liquid and without pretreatment of samples. Crucially, careful experimentation can allow measurement of interaction forces between molecules or between an adhesive molecule and an AFM probe, providing physical as well as morphological/topological information.

Higgins et al. (2000) were among the first to apply AFM to the study of whole-organism adhesion, producing images of adhesive trails deposited on surfaces by pennate diatoms, unicellular algae comprising the vast majority of marine fouling diatom species. This study was followed by a series in which the application of AFM advanced from an imaging tool to a force-data gathering technique (Higgins et al. 2002, 2003) using both diatoms and unicellular algal spores as models (Callow et al. 2000). Direct measurements of biological adhesion at the nanoscale were shown to be achievable, even in a liquid medium, and these early studies paved the way for ever-more precise and informative experiments. In a series of papers, Dugdale et al. (2005, 2006a, b) investigated the so-called adhesive nanofibres (ANFs) of pennate diatoms, which are used for adhesion and gliding propulsion on surfaces. Dugdale presented evidence that modular proteins were present and played a functional role in adhesion, contrary to the consensus that carbohydrates were the predominant functional component of diatom adhesive mucilage (Chiovitti et al. 2006). The evidence was the presence of distinctive 'sawtooth' profiles in force-distance curves recorded while pulling individual adhesive strands. These profiles were shown to be diagnostic of the regular unfolding of modular proteins and have been identified throughout biological systems that rely on efficient force dissipation and 'self-healing', including the adhesives of some invertebrate larvae. Soon after this pioneering work, the same methods were applied for the first time to marine invertebrate larvae (Aldred and Clare 2008). Previously, the physical nature of the footprints of temporary adhesive deposited onto surfaces by exploring cyprids had been a mystery. AFM provided a unique opportunity to probe the footprints, generating topological maps, 3D images and simultaneous force measurements (Phang et al. 2008). Again, sawtooth force-distance curves were recorded (Phang et al. 2010), and, in a further advance, cantilevers were functionalised with different chemical terminations to observe the effects of hydrophilic/hydrophobic characteristics on the adhesion/adsorption of footprint proteins to surfaces. In these experiments, adhesion of footprint proteins at the molecular level was found to be stronger to hydrophobic surfaces, and this was later born out in the iSPR study of Aldred et al. (2011) where footprint deposition and footprint thickness values were higher on a hydrophobic $CH_3$ SAM than the very hydrophilic PEG surface. Evidence was also found for a peeling detachment mechanism that was, again, supported by future observations using high-speed microscopy (Aldred et al. 2013a). Phang et al. (2006) also applied AFM to the permanent adhesive of barnacle cyprids and identified modular proteins via their characteristic unfolding profile, with the reduction in 'pull off length' of the proteins with time providing the first direct measurement of cyprid adhesive curing.

## 4.7 Conclusions

Research into the adhesion of marine invertebrate larvae has, by necessity, embraced techniques that would previously have been considered the purview of the physical sciences and must continue to do so. These advanced techniques make possible compositional studies of trace quantities of larval adhesives and also allow investigation of the deposition of adhesive material in situ and in real time, allowing quantitative studies of the attachment process to a variety of different surfaces. Most notable, perhaps, has been a shift from simple adoption of surface-science techniques to the process of developing combinations of approaches to realise their maximum potential in the study of adhesion. In some cases, for example, CARS imaging, these advances have been made outside of the bioadhesion field but have potential to contribute enormously to investigations of spatial/temporal release of adhesives, adhesive mixing/phase separation and the identification of nonprotein components. CARS combines chemical characterisation with spatially resolved imaging capability, bridging the gap between the traditional approaches used to judge biochemical composition and diagnostic imaging methods to confirm the presence of chemical moieties. Secondly, there are techniques that have been developed primarily for investigations of adhesion. These include single instruments like the surface forces apparatus but also combinations of methods such as stereoscopic video tracking and iSPR, the results of which can marry descriptions of the behaviour of larvae on surfaces with direct observation and measurement of their adhesive bonds.

The investigation of adhesion by marine invertebrate larvae will continue to present challenges; however it is hoped that the pace of technological development will gradually diminish the obstacles set in place by sample size, availability and heterogeneity. These are challenges worth addressing, because not only are larval adhesives fascinating biologically and ecologically, fulfilling a crucial role in the life cycle of countless benthic organisms, but they are also of clear practical relevance in combatting marine biofouling and developing bio-inspired adhesives and sealants for a variety of purposes.

**Acknowledgements** N. Aldred acknowledges funding support from the Office of Naval Research award number N00014-13-1-0634 to A. S. Clare and N. Aldred. N. Aldred receives additional support from a Newcastle University SAgE Faculty Research Fellowship. Both authors would like to extend their gratitude to Prof. A. M. Smith for the invitation to contribute a chapter to this edited volume.

## References

Aldred N, Clare AS (2008) The adhesive strategies of cyprids and development of barnacle-resistant marine coatings. Biofouling 24:351–363

Aldred N, Clare AS (2014) Mini-review: impact and dynamics of surface fouling by solitary and compound ascidians. Biofouling 30:259–270

Aldred N, Ista LK, Callow ME et al (2006) Mussel (*Mytilus edulis*) byssus deposition in response to variations in surface wettability. J R Soc Interface 3:37–43

Aldred N, Ekblad T, Andersson O et al (2011) Real-time quantification of microscale bioadhesion events in situ using imaging surface plasmon resonance (iSPR). Appl Mater Interfaces 3:2085–2091

Aldred N, Høeg JT, Maruzzo D et al (2013a) Analysis of the behaviours mediating barnacle cyprid reversible adhesion. PLoS One 8:e68085

Aldred N, Gohad NV, Petrone L et al (2013b) Confocal microscopy-based goniometry of barnacle cyprid permanent adhesive. J Exp Biol 216:1969–1972

Andersson O, Ekblad T, Aldred N et al (2009) Novel application of imaging surface plasmon resonance for in situ studies of the surface exploration of marine organisms. Biointerphases 4:65–68

Barlow DE, Wahl KJ (2012) Optical spectroscopy of marine bioadhesives interfaces. Annu Rev Anal Chem 5:229–251

Barlow DE, Dickinson GH, Orihuela B et al (2009) In situ characterization of the primary cement interfaces of the barnacle *Balanus amphitrite*. Biofouling 25:359–366

Barlow DE, Dickinson GH, Orihuela B et al (2010) Characterization of the adhesive plaque of the barnacle *Balanus amphitrite*: Amyloid-like nanofibrils are a major component. Langmuir 26:6549–6556

Berglin M, Hedlund J, Fant C et al (2005) Use of surface-sensitive methods for the study of adsorption and cross-linking of marine bioadhesives. J Adhes 81:805–822

Burden DK, Barlow DE, Spillmann CM et al (2012) Barnacle *Balanus amphitrite* adheres by a stepwise cementing process. Langmuir 28:13364–13372

Burkett JR, Hight LM, Kenny P et al (2010) Oysters produce an organic–inorganic adhesive for intertidal reef construction. J Am Chem Soc 132:12531–12533

Callaway E (2015) The revolution will not be crystallized: A new method sweeps through structural biology. Nature 525:172–174

Callow JA, Crawford SA, Higgins MJ et al (2000) The application of atomic force microscopy to topographical studies and force measurements on the secreted adhesive of the green alga *Enteromorpha*. Planta 211:641–647

Callow JA, Osborne MP, Callow ME et al (2003) Use of environmental scanning electron microscopy to image the spore adhesive of the marine alga *Enteromorpha* in its natural hydrated state. Colloid Surf B 27:315–321

Callow JA, Callow ME, Ista LK et al (2005) The influence of surface energy on the wetting behaviour of the spore adhesive of the marine alga *Ulva linza* (synonym *Enteromorpha linza*). J R Soc Interface 2:319–325

Chiovitti A, Dugdale TM, Wetherbee R (2006) Diatom adhesives: molecular and mechanical properties. In: Smith AM, Callow JA (eds) Biological adhesives. Springer, Heidelberg, pp 79–103

Clare AS, Aldred N (2009) Surface colonisation by marine organisms and its impact on antifouling research. In: Hellio C, Yebra D (eds) Advances in marine antifouling coatings and technologies. Woodhead Publishing, Oxford, pp 46–79

Connor PA, Dobson KD, McQuillan AJ (1995) New sol–gel attenuated total reflection infrared spectroscopic method for analysis of adsorption at metal oxide surfaces in aqueous solutions. chelation of $TiO_2$, $ZrO_2$ and $Al_2O_3$ surfaces by catechol, 8-quinolinol, and acetylacetone. Langmuir 11:4193–4195

Crisp DJ (1976) Two settlement responses in marine organisms. In: Newell RC (ed) Adaptation to environment. Essays on the physiology of marine animals. Butterworths, London, pp 83–124

Dahms HU, Dobretsov S, Qian PY (2004) The effect of bacterial and diatom biofilms on the settlement of the bryozoan *Bugula neritina*. J Exp Mar Biol Ecol 313:191–209

Dreanno C, Matsumura K, Dohmae N (2006a) An $\alpha_2$-macroglobulin-like protein is the cue to gregarious settlement of the barnacle. Proc Natl Acad Sci USA 103:14396–14401

Dreanno C, Kirby RR, Clare AS (2006b) Smelly feet are not always a bad thing: The relationship between cyprid footprint protein and the barnacle settlement pheromone. Biol Lett 2:423–425

Dugdale TM, Dagastine R, Chiovitti A (2005) Single adhesive nanofibres from a live diatom have the signature fingerprint of modular proteins. Biophys J 89:4252–4260

Dugdale TM, Dagastine R, Chiovitti A et al (2006a) Diatom adhesive mucilage contains distinct supramolecular assemblies of a single modular protein. Biophys J 90:2987–2993

Dugdale TM, Willis A, Wetherbee R (2006b) Adhesive modular proteins occur in the extracellular mucilage of the motile, pennate diatom *Phaeodactylum tricornutum*. Biophys J 90:L58–L60

Federle W, Riehle M, Curtis ASG et al (2002) An integrative study of insect adhesion: mechanics and wet adhesion of pretarsal pads in ants. Integr Comp Biol 42:1100–1106

Gao Z, Bremer PJ, Barker MF et al (2007) Adhesive secretions of live mussels observed in situ by attenuated total reflection-infrared spectroscopy. Appl Spectrosc 61:55–9

Gohad NV, Aldred N, Orihuela B et al (2012) Observations on the settlement and cementation of the barnacle (*Balanus amphitrite*) cyprid larvae after artificial exposure to noradrenaline and the locations of adrenergic-like receptors. J Exp Mar Biol Ecol 416–417:153–161

Gohad NV, Aldred N, Hartshorn CM et al (2014) Synergistic roles for lipids and proteins in the permanent adhesive of barnacle larvae. Nat Commun 5:4414

Golden JP, Burden DK, Fears KP et al (2016) Imaging active surface processes in barnacle adhesive interfaces. Langmuir 32(2):541–550. doi:10.1021/acs.langmuir.5b03286

Goormaghtigh E, Raussens V, Ruysschaert J-M (1999) Attenuated total reflection infrared spectroscopy of proteins and lipids in biological membranes. Biochim Biophys A 1422:105–185

Grosser K, Zedler L, Schmitt M et al (2012) Disruption-free imaging by Raman spectroscopy reveals a chemical sphere with antifouling metabolites around macroalgae. Biofouling 28:687–696

Haesaerts D, Jangoux M, Flammang P (2005) The attachment complex of brachiolaria larvae of the sea star Asterias rubens (Echinodermata): an ultrastructural and immunocytochemical study. Zoomorphology 124:67–78

Harrington MJ, Masic A, Holten-Andersen N et al (2010) Iron-clad fibers: a metal-based biological strategy for hard flexible coatings. Science 328:216–220

He L-S, Zhang G, Qian P-Y (2013) Characterization of two 20kDa-cement protein (cp20k) homologues in *Amphibalanus amphitrite*. PLoS One 8(5), e64130

Hennebert E, Maldonado B, Van De Weerdt C et al (2015) From sand tube to test tube: the adhesive secretion from Sabellariid tubeworms. In: Bianco-Peled H, Davidovich-Pinhas M (eds) Bioadhesion and biomimetics: from nature to applications. CRC Press, Boca Raton, pp 109–127

Higgings MJ, Molino P, Mulvaney P et al (2003) The structure and nanomechanical properties of the adhesive mucilage that mediates diatom substratum adhesion and motility. J Phycol 39:1181–1193

Higgins MJ, Crawford SA, Mulvaney P et al (2000) The topography of soft, adhesive diatom 'trails' as observed by atomic force microscopy. Biofouling 16:133–139

Higgins MJ, Crawford SA, Mulvaney P et al (2002) Characterization of the adhesive mucilage secreted by live diatom cells using atomic force microscopy. Protist 153:25–38

Höök F, Kasemo B, Nylander T et al (2001) Variations in coupled water, viscoelastic properties, and film thickness of a Mefp1 protein film during adsorption and cross-linking: a quartz crystal microbalance with dissipation monitoring, ellipsometry, and surface plasmon resonance study. Anal Chem 73:5796–5804

Hug SJ (1997) In situ Fourier transform infrared measurements of sulfate adsorption on hematite in aqueous solutions. J Colloid Interface Sci 188:415–422

Hug SJ, Sulzberger B (1994) In situ Fourier transform infrared spectroscopic evidence for the formation of several different surface complexes of oxalate on $TiO_2$ in the aqueous phase. Langmuir 10:3587–3597

Kamino K (2013) Mini-review: Barnacle adhesives and adhesion. Biofouling 29:735–749

Lee BP, Messersmith BP, Israelachvili JN et al (2011) Mussel-inspired adhesives and coatings. Annu Rev Mater Res 41:99–132

Maleshlijski S, Sendra GH, Di Fino A et al (2012) Three dimensional tracking of exploratory behaviour of barnacle cyprids using stereoscopy. Biointerphases 7:50

Maleshlijski S, Sendra GH, Aldred N et al (2015) Imaging SPR combined with stereoscopic 3D tracking to study barnacle cyprid-surface interactions. Surf Sci 643:172–177

Martinez Rodriguez NR, Das S, Kaufman Y et al (2015) Interfacial pH during mussel adhesive plaque formation. Biofouling 31:221–227

McQuillan AJ (2001) Probing solid-solution interfacial chemistry with ATR-IR spectroscopy of particle films. Adv Mater 13:1034–1038

McWhirter MJ, Bremer PJ, Lamont IL et al (2003) Siderophore-mediated covalent bonding to metal (oxide) surfaces during biofilm initiation by *Pseudomonas aeruginosa* bacteria. Langmuir 19:3575–3577

Mostaert AS, Giordani C, Crockett R et al (2009) Characterisation of amyloid nanostructures in the natural adhesive of unicellular subaerial algae. J Adhes 85:465–483

Nakano M, Kamino K (2015) Amyloid-like conformation and the interaction for the self-assembly in barnacle underwater cement. Biochemistry 54:826–835

Nott JA, Foster BA (1969) On the structure of the antennular attachment organ of the cypris larva of *Balanus balanoides* (L.). Philos Trans R Soc B 256:115–134

Parikh SJ, Chorover J (2006) ATR-FTIR spectroscopy reveals bond formation during bacterial adhesion to iron oxide. Langmuir 22:8492–500

Peak D, Ford RG, Sparks DL (1999) An in Situ ATR-FTIR investigation of sulfate bonding mechanisms on goethite. J Colloid Interface Sci 218:289–299

Petrone L (2013) Molecular surface chemistry in marine bioadhesion. Adv Colloid Interface Sci 195–196:1–18

Petrone L, McQuillan AJ (2011) Alginate ion adsorption on a $TiO_2$ particle film and interactions of adsorbed alginate with calcium ions investigated by Attenuated Total Reflection Infrared (ATR-IR) spectroscopy. Appl Spectrosc 65:1162–1169

Petrone L, Ragg NLC, McQuillan AJ (2008) In situ infrared spectroscopic investigation of *Perna canaliculus* mussel larvae primary settlement. Biofouling 24:405–413

Petrone L, Ragg NLC, Girvan L (2009) Scanning electron microscopy and energy dispersive X-ray microanalysis of *Perna canaliculus* mussel larvae adhesive secretion. J Adhes 85:78–96
Petrone L, Easingwood R, Barker MF et al (2011a) In situ ATR-IR spectroscopic and electron microscopic analyses of settlement secretions of *Undaria pinnatifida* kelp spores. J R Soc Interface 8:410–422
Petrone L, DiFino A, Aldred N et al (2011b) Effects of surface charge and Gibbs surface energy on the settlement behaviour of barnacle cyprids (*Balanus amphitrite*). Biofouling 27:1043–1055
Petrone L, Kumar A, Sutanto CN et al (2015a) Mussel adhesion is dictated by time-regulated secretion and molecular conformation of mussel adhesive proteins. Nat Commun 6:8737
Petrone L, Aldred N, Emami K et al (2015b) Chemistry-specific surface adsorption of the barnacle settlement-inducing protein complex. Interface Focus 5:20140047
Phang IY, Aldred N, Clare AS et al (2006) An in situ study of the nanomechanical properties of barnacle (*Balanus amphitrite*) cyprid cement using atomic force microscopy (AFM). Biofouling 22:245–250
Phang IY, Aldred N, Clare AS et al (2008) Towards a nanomechanical basis for temporary adhesion in barnacle cyprids (*Semibalanus balanoides*). J R Soc Interface 5:397–402
Phang IY, Aldred N, Ling XY et al (2010) Atomic force microscopy of the morphology and mechanical behaviour of barnacle cyprid footprint proteins at the nanoscale. J R Soc Interface 7:285–296
Schmidt M, Cavaco A, Gierlinger N et al (2009) In situ imaging of barnacle (*Balanus amphitrite*) cyprid cement using confocal Raman microscopy. J Adhes 85:139–151
Senkbeil T, Mohamed T, Simon R et al (2016) In vivo and in situ synchrotron radiation μ-XRF reveals elemental distributions during the early attachment phase of barnacle larvae and juvenile barnacles. Anal Bioanal Chem 408(5):1487–1496
Sperline RP, Song Y, Freiser H (1992) Fourier transform infrared attenuated total reflection spectroscopy linear dichroism study of sodium dodecyl sulfate adsorption at the alumina/water interface using alumina-coated optics. Langmuir 8:2183–2191
Strathmann R (1978) Length of pelagic period in echinoderms with feeding larvae from the Northeast Pacific. J Exp Mar Biol Ecol 34:23–27
Tanur AE, Gunari NA, Sullan RM et al (2009) Biomineralisation by the marine tubeworm *Hydroides dianthus* and composition of the adhesive cement. Biophys J 96:640–641
Walker G (1971) A study of the cement apparatus of the cypris larva of the barnacle *Balanus balanoides*. Mar Biol 9:205–212
Walker G, Yule AB (1984a) Temporary adhesion of the barnacle cyprid: The existence of an antennular adhesive secretion. J Mar Biol Assoc UK 64:679–686
Walker G, Yule AB (1984b) The temporary adhesion of barnacle cyprids: effects of some differing surface characteristics. J Mar Biol Assoc UK 1984(64):429–439
Wei W, Tan Y, Martinez Rodriguez NR et al (2014) A mussel-derived one component adhesive coacervate. Acta Biomater 10:1663–1670
Yu J, Wei W, Danner E et al (2011) Effects of interfacial redox in mussel adhesive protein films on mica. Proc Natl Acad Sci USA 23:2362–2366

# Chapter 5
# The Adhesive Tape-Like Silk of Aquatic Caddisworms

**Nicholas N. Ashton, Ching-Shuen Wang, and Russell J. Stewart**

**Abstract** Aquatic caddisfly larva spin a sticky silk tape used underwater to construct a protective composite stone case. Caddisworm silk fibers are drawn on-demand from fluid precursors stored in the posterior region of the silk gland. Fibers begin to form in the gland at a cuticular narrowing at the entrance into the short (2–3 mm) anterior conducting channel leading to the spinneret. The caddisworm silk comprises a thin adhesive peripheral coating on a tough viscoelastic core fiber. The thin adhesive layer contains glycoproteins and a heme-peroxidase in the peroxinectin subfamily (Pxt). Pxt catalyzes dityrosine cross-linking in the fiber periphery and may catalyze covalent adhesive cross-links to surface-active natural polyphenolic compounds. The major component of the silk core, H-fibroin, contains around 13 mol% phosphoserines (pS) in repeating $(pSX)_n$ motifs, wherein X is usually hydrophobic, and $n$ is 4 or 5. The $(pSX)_n$ motifs form β-domains crossbridged and stabilized by multivalent metal ions, predominantly $Ca^{2+}$ in natural fibers. During loading, the $Ca^{2+}/(pSX)_n$ β-domains reversibly rupture to reveal hidden length and dissipate strain energy. The tough fibers can be strained to more than 100 % of their initial length before fracture. The work of extension to failure, ~$17.3 \pm 6.2$ MJ/m$^3$, is higher than articular cartilage. Silk fibers cycled to 20 % elongation completely recover their initial stiffness, strength, and hysteresis within 120 min as an elastic covalent network guides the post-yield recovery of the $Ca^{2+}/(pSX)_n$ β-domains.

## 5.1 Introduction

Caddisflies are the seventh largest insect order, and larvae of their species are the second most abundant aquatic insect, represented by over 14,000 species worldwide in freshwater and several marine environments (Malm et al. 2013; Riek 1976). Caddisflies (Trichoptera) are a sister order to terrestrial moths and butterflies

N.N. Ashton • C.-S. Wang • R.J. Stewart (✉)
Department of Bioengineering, University of Utah, Salt Lake City, UT 84112, USA
e-mail: russell.stewart@utah.edu

© Springer International Publishing Switzerland 2016 
A.M. Smith (ed.), *Biological Adhesives*, DOI 10.1007/978-3-319-46082-6_5

(Lepidoptera), together comprising the insect superorder Amphiesmenoptera. Trichoptera and Lepidoptera branched ~234 million years ago (Malm et al. 2013). Like moth and butterfly larvae, caddisfly larvae (caddisworms) pull silk fibers from fluid precursors stored in a pair of large modified salivary glands.

Caddisflies live most of their year-long lifecycle feeding underwater in five larval stages. After pupating, the adult flies emerge for a brief 1–2 week period of mating and egg-laying. The successful radiation of the order into diverse habitats has been attributed, in part, to inventive uses of its underwater silk (Wiggins 2004). Caddisworms use their silks to build shelters and procure food. Three suborders of caddisflies are distinguished by the types of silken structures they create. Integripalpia assemble portable protective cases, usually incorporating adventitiously gathered materials like stones, plant matter, or mollusk shells (Fig. 5.1).

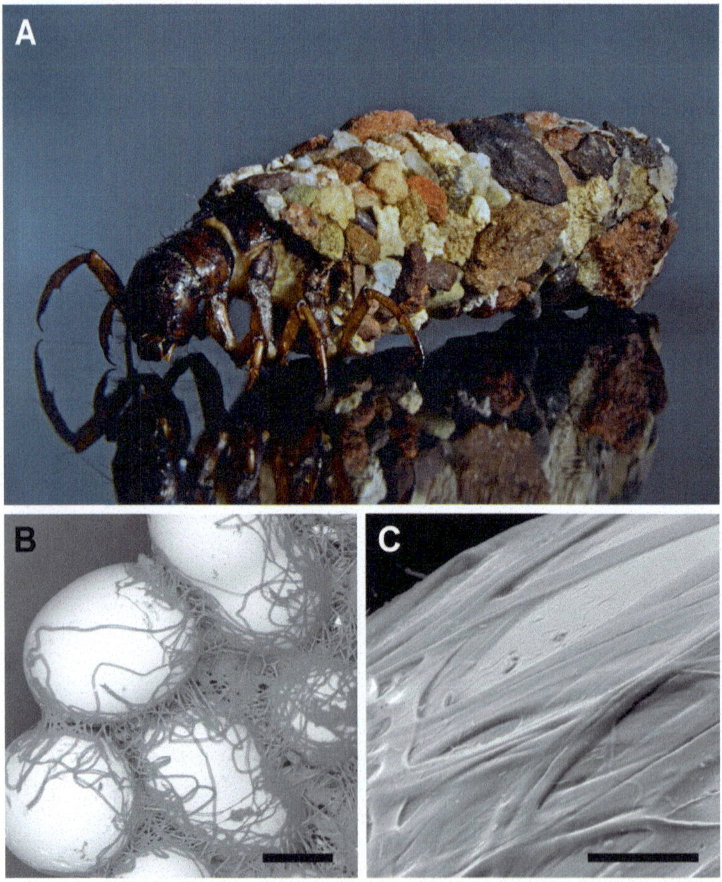


**Fig. 5.1** *Casemaker caddisfly larvae, Hesperophylax occidentalis.* (**a**) The casemaking larva uses adhesive silk to build portable protective cases with adventitiously gathered stones. (**b**) An inside view of the composite silk case. Scale bar = 250 μm. (**c**) Higher magnification of silk threads on the substrate. Scale bar = 20 μm

Annulipalpia build stationary retreats accompanied by suspended capture nets for filtering materials and food from flowing water. Spicipalpia are free-living and, except for occasional anchoring tethers, spin silk only to construct a semipermeable cocoon just before pupation (Malm et al. 2013; Wiggins 2004; Merritt and Cummins 2008).

The caddisfly suborders branched early in their evolutionary history; all three were distinct by ~203 million years ago (Malm et al. 2013). The different uses of silks across the three suborders suggest there may be significant variations in their molecular structure and mechanical properties. The sequences of H-fibroin, the major silk structural protein, differ significantly between suborders. For example, a repetitive phosphorylated molecular sequence that toughens Integripalpia silk is less precise in Annulipalpia silk, and Spicipalpia H-fibroin contains ~7 mol% histidine not found in the other suborders (Yonemura et al. 2009). The independent adaptation of the silks means that progress in understanding the mechanochemistry of silks in one caddisfly suborder does not necessarily translate to the silks in the other orders. This chapter is about the silks of casemaker *Hesperophylax* species, *H. occidentalis* and *H. consimilis*, found in the western United States. The focus is on the adhesive mechanism and mechanical properties of the casemaker silk. Eventual comparisons with silks from other suborders and the mechanical adaptation of their silks to fit their lifestyle will provide an opportunity to further connect caddisworm silk molecular structure to mechanical properties. The long-term motivation for these studies is to discover design principles that may guide future development of soft but tough materials.

## 5.2 Multi-scale Structure of Casemaker Caddisworm Silk

The silk of casemaking caddisworms is a sticky tape with an adhesive coating on a tough, viscoelastic backing material. The silk comprises two ~5 μm diameter core fibers fused together and encased by a thin fuzzy coating of adhesive (Fig. 5.2a) (Engster 1976a). The caddisworm silk core is an elliptical bundle of 80–100 nm diameter, axially aligned subfibrils (Fig. 5.2b) (Engster 1976a; Wang et al. 2014). The nanoscale subfibrils of casemaker silks are evenly distributed in the core compared to the silks of retreat-building caddisflies (Annulipalpia) in which the subfibrils are inhomogeneously distributed into two concentric morphological regions (Ashton et al. 2012; Hatano and Nagashima 2015). Fibers fuse together into cohesive bundles, adding strength to the adhesive joints (Ashton et al. 2012; Stewart and Wang 2010). Atomic force microscopy of fibers adhered to a surface showed that outer nanofibrils splayed apart and conformed to the surface to increase contact area and interfacial adhesive strength (Ashton et al. 2012).

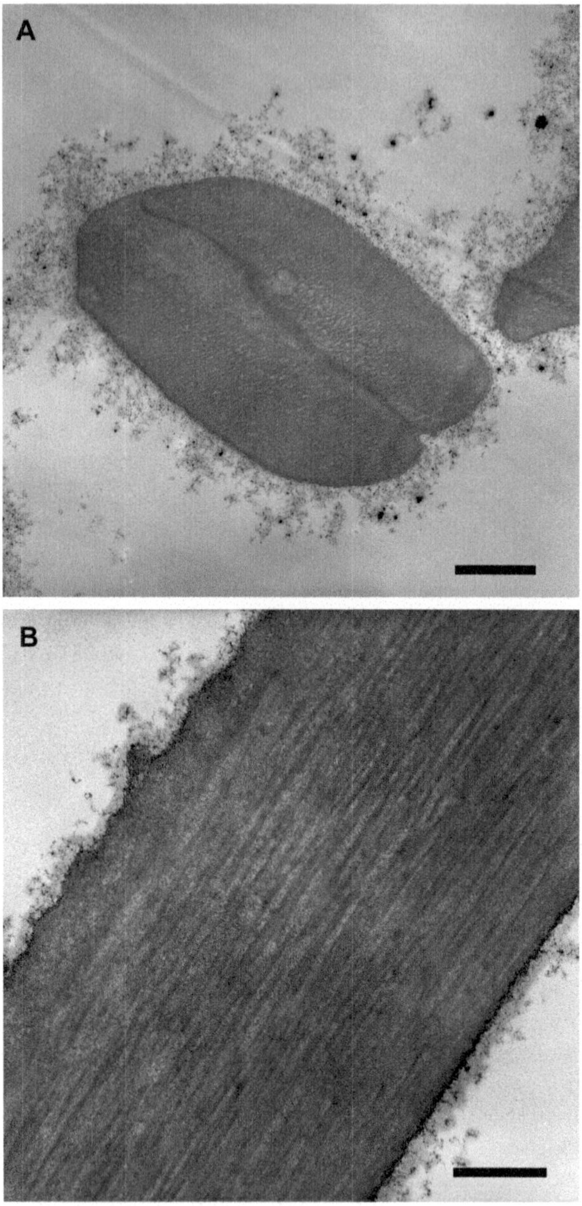


**Fig. 5.2** *Transmission electron micrographs of caddisworm silk sections.* (**a**) Cross-section through the two silk subfibers. Scale Bar = 1 μm. (**b**) Longitudinal section through a single silk subfiber. Scale bar = 0.5 μm. The fibers have a fuzzy peripheral coating on a fibrous core. Reproduced with permission from Wang et al. (2014)

## 5.3 The Peripheral Adhesive Coating

### *5.3.1 Composition of the Silk Coating*

Engster demonstrated that the adhesive peripheral layer of the casemaking caddisworms *Pycnopsyche guttifer* and *Neophylax concinnus* is comprised of neutral and acidic glycoproteins (Engster 1976a, 1976b). Similarly, glycoprotein-staining of *H. occidentalis* silk sections demonstrated that carbohydrates are predominantly distributed in the peripheral layer but may also be present in the core (Wang et al. 2015). The thin carbohydrate-rich coating on caddisworm silk is functionally analogous to the sticky gum coating of the silk of the domesticated silkworm (*B. mori*). The coating on *B. mori* silk is much thicker than that observed on caddisworm silk making up 19–28 % of the silk mass (Naskar et al. 2014). In contrast, the sugar content of *P. guttifer* case silk was only 1.2 wt% (Engster 1976b). The peripheral coating is 0.3–0.5 μm in *H. occidentalis* silk sections (Fig. 5.2a). The silk gum on silkworm silk comprises almost entirely the glycoprotein sericin, which can be removed with warm alkaline water. A sericin homologue has not been found in caddisworms despite extensive searches in several species in different suborders (Yonemura et al. 2006, 2009; Wang et al. 2014, 2015; Ashton et al. 2013). In contrast to the silkworm sericin coating, it is difficult to remove the peripheral coating of caddis silk fibers even with strong detergents, denaturants, reducing agents, elevated temperatures, chelating agents, and extremes in pH (Engster 1976a; Wang et al. 2014). The caddisworm silk gland contents, on the other hand, are easy to solubilize (Yonemura et al. 2009; Wang et al. 2015; Ohkawa et al. 2012), which suggests that the silk gland proteins are covalently cross-linked after they are drawn. The case fibers from *H. occidentalis* gradually redden over 24 h post-draw. Rudall observed that some caddisworm cases appeared "like insect-tanned material" (Engster 1976a). Enzymes and phenols are co-drawn into the silk of terrestrial saturniid cocoons, which covalently cross-link the silks (Mitchinson 1974). The insolubility and reddening of *H. occidentalis* silk is evidence of a cross-linking enzyme co-drawn with caddisworm silk fibers.

### *5.3.2 Enzyme-Catalyzed Cross-Linking in the Peripheral Coating*

A heme-peroxidase in the peroxinectin (Pxt) subfamily was identified by searching a de novo assembled *H. occidentalis* silk gland transcriptome with peptide sequences identified by LC-MS/MS of trypsin digested silk gland proteins. Heme-peroxidases catalyze the oxidation of various substrates by reducing $H_2O_2$ to water. Peroxidase activity in caddisworm silk was demonstrated with Amplex® Red, a peroxidase substrate that is converted to fluorescent resorufin in the presence of $H_2O_2$. Resorufin fluorescence in caddisworm silk did not require the addition of

$H_2O_2$, although florescence was greatly enhanced, which provided evidence that endogenous $H_2O_2$ is produced in the fibers (Wang et al. 2014). A superoxide dismutase enzyme (SOD3) identified in the silk gland transcriptome may be the source of endogeneous $H_2O_2$. Superoxide dismutases are a broad category of enzyme, ubiquitous in living cells and tissues that convert reactive superoxides into more benign $H_2O_2$. Both csPxt and csSOD3 were shown to be glycosylated by lectin column chromatography (Wang et al. 2014, 2015). The SOD3 enzyme may be co-drawn into the silk fiber where it uses natural reactive oxygen species (ROS) in natural water to generate $H_2O_2$ for the csPxt. There is a steady state reservoir of ROS in natural surface waters, including singlet dioxygen, superoxide, hydroxyl radicals, and numerous peroxyl radicals, maintained by photochemical reactions of sunlight with naturally occurring chromophores, like humic acid (Blough and Ziepp 1995).

To identify potential targets of caddisworm silk peroxinectin (csPxt), the free amino acids Tyr, Arg, His, Lys, and Gly were added to compete with the Amplex® Red® assay (Wang et al. 2014). Only Tyr competed appreciably, decreasing fluorescence by ~75 %. Fluorescent imaging of silk fiber cross-sections revealed strong dityrosine fluorescence (415 nm) in the peripheral coating, confirming intrinsic tyrosine as a target for csPxt (Fig. 5.3a). CsPxt itself is perhaps the most likely target for dityrosine cross-linking in the peripheral coating. Peroxinectins lack strong substrate specificity and instead achieve specificity through localization in proximity to their substrate (Soudi et al. 2012). Of the 24 tyrosines in csPxt, five occur in the N-terminus outside of the active site. In some peroxidases, this region contains recognition-binding sequences that localize the enzyme near their substrates, such as collagens in the extracellular matrix, basement membrane, or to integrins on the surface of cells (Soudi et al. 2012). Dityrosine cross-links with core components may covalently couple the peripheral coating to the fiber core. Dityrosine cross-linking occurs during or after the fibers are drawn since dityrosine fluorescence was not detected in the silk gland proteins within the silk gland (Wang et al. 2014).

### 5.3.3 *Interfacial Adhesion Mechanisms*

Successful surface attachment of caddisworm silk fibers likely involves multiple adhesion mechanisms. The neutral and acidic polysaccharides in the fuzzy peripheral adhesive coating are efficient adhesion promoters, as evident from their prevalence in aquatic bioadhesives. Examples include the glycosylated silk proteins of the aquatic midge larva (Hertner et al. 1983), sulfated polysaccharides in the sandcastle worm adhesive (Wang and Stewart 2013), glycosylated anchoring plaques of sessile barnacles (Jonker et al. 2015), and anionic exopolysaccharide bacterial biofilms, to name a few (Cooksey and Wigglesworth-Cooksey 1995). The carboxylate anion in acidic polysaccharides can bond electrostatically or H-bond with metal hydroxide surfaces of minerals (Stewart et al. 2011). Caddisworm silk

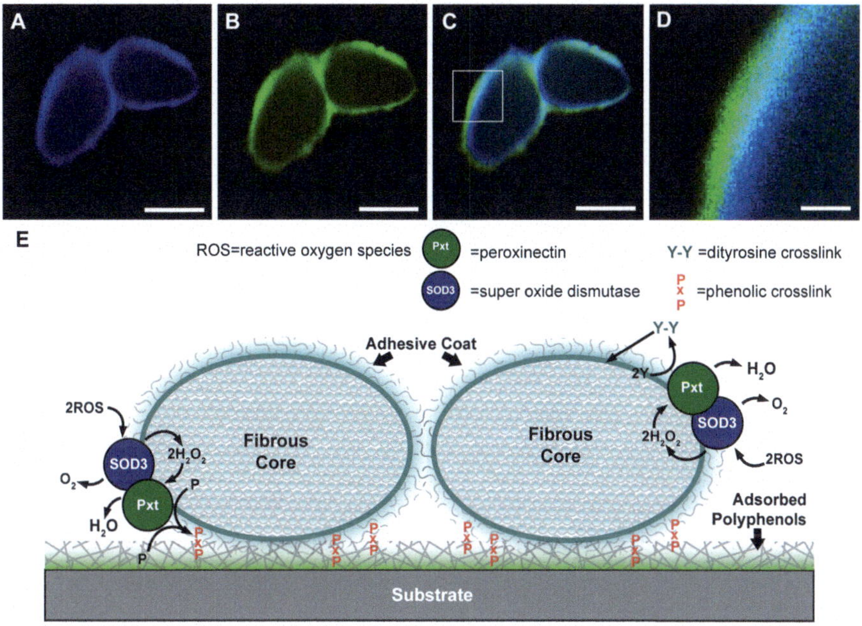


**Fig. 5.3** *CsPxt-catalyzed cross-linking in the silk peripheral layer and to*. (**a**) Intrinsic dityrosine fluorescence, *Blue*, Em: 415 nm. (**b**) Humic acid fluorescence of a silk fiber incubated in humic acid and $H_2O_2$, *Green*, Em: 518 nm. (**c**) Overlay of intrinsic dityrosine (*Blue*) and covalently bound humic acid penetrating the peripheral coating (*Green*). (**d**) Higher magnification of the *boxed* region in (**c**). (**e**) Schematic of csPxt-catalyzed phenolic cross-linking to the surface. The glycosylated csPxt is embedded in the silk peripheral adhesive coating where it catalyzes cross-linking between tyrosines within the peripheral coating and between the peripheral coating and external phenolic compounds that foul submerged surfaces. The $H_2O_2$ substrate of csPxt may be generated by glycosylated csSOD3 from ROS. Panels (**a**)–(**c**) were reproduced with permission from Wang et al. (2015)

must adhere to fouled surfaces since pristine surfaces do not exist in natural waters. Rather, polyphenolic macromolecules, like humic and fulvic acid, or bacterial biofilms are adsorbed to natural surfaces. Soluble humic acid, which fluoresces green (518 nm), became covalently bound within the peripheral coating of silk fibers in the presence of $H_2O_2$, providing evidence that csPxt can catalyze covalent interfacial cross-linking to external polyphenolic macromolecules (Fig. 5.3b–d). A model of the distribution csSOD3 and csPxt in the peripheral coating and their role in dityrosine cross-linking are diagrammed in Fig. 5.3e. In the peripheral coating, csSOD3 may generate $H_2O_2$ from environmental ROS that in turn is used by csPxt to catalyze dityrosine cross-linking within the peripheral coating and with surface-active polyphenolic compounds (Wang et al. 2014, 2015).

## 5.4 The Tough Fibrous Silk Core

### 5.4.1 Composition of the Silk Core

The primary structural component of both aquatic trichopteran and terrestrial lepidopteran silks is H- and L-fibroin, >350 kDa and ~25 kDa, respectively (Yonemura et al. 2009). In Lepidoptera, the L-fibroin is linked by a single disulfide bridge to the H-fibroin C-terminus in a 1:1 ratio (Tanaka et al. 1999). Formation of this disulfide cross-link is essential for proper silk secretion (Mori et al. 1995). Yonemura and colleagues showed that the positions of the cysteines are conserved in the H- and L-fibroins in all three caddisfly suborders and therefore suggested that disulfide cross-linking is also conserved in caddisflies (Yonemura et al. 2009). H-fibroins are the major silk fiber structural proteins. Although there are general structural similarities between the H-fibroins of Trichoptera and Lepidoptera, highly repetitive central regions flanked by short non-repetitive N- and C-termini, for example, the primary sequences and amino acid composition are not similar. In addition to the fibroins, a novel repetitive structural protein, rich in Pro, Glu, Val, and Lys (12.2, 19.5, 16.3, and 11.3 mol%, respectively), was also identified in *H. occidentalis* silk by LC-MS/MS. It was called PEVK-like because its amino acid content and primary sequence are reminiscent of the entropic spring-like spacer domains of the giant muscle protein, titin (Wang et al. 2014). Immunostaining with monoclonal antibodies raised against synthetic peptide epitopes of PEVK-like labeled the fiber core rather than the peripheral coating (Wang et al. 2014). Its function in the silk fiber is not known but likely structural.

### 5.4.2 H-Fibroin Structure

Yonemura and colleagues recognized imperfectly alternating repeating sequence blocks in a casemaker caddisworm H-fibroin, which they designated as D, E, and F repeats (Fig. 5.4a) (Yonemura et al. 2006). The homologous H-fibroin D, E, and F repeats of *H. occidentalis*, and their relation to the non-repetitive termini, is depicted in Fig. 5.4b. Analysis of trypsin digests of *H. occidentalis* silk proteins by LC-MS/MS revealed that nearly all of the ~15 mol% serines in the H-fibroin repeats are phosphorylated (Fig. 5.4, underlined sequences) (Stewart and Wang 2010). The extensive phosphorylation of caddisworm H-fibroin serines has been confirmed by several methods: labeling with anti-phosphoserine antibodies (Ashton et al. 2012; Stewart and Wang 2010), Pro-Q® fluorescent phosphopeptide stain (Wang et al. 2015), infrared spectroscopy (Ashton et al. 2013; Ashton and Stewart 2015), and $^{31}$P solid-state NMR (Addison et al. 2013, 2014). Silks from species in each one of the three caddisfly suborders contain phosphoserines, an indication that H-fibroin phosphorylation is conserved across the Trichoptera order (Ashton

A

```
nT:MRAILLLIVFCSWQVHRTGSSAVSNKVQDLFKHGRHLDSDGLH
   ERILEEEDIIETNSKGEIIEKIISRKEIISDDASNSWSDSDSS
   ESGSTEQIIKQIIIVQERPKHGHHHAKEKIYEEEIVIKKIGDV
   VHGVDVPIDTYGGIGRGR
D: RRGPNISISRSISIEQVVRPRVYTKISRSSSVSVEGG
E: RRHGPWGHGKGAWGPLGDLSSSGDLDGLGGLGSIGG
F: RRGPWGRSVSASLSVSVERGI
D: RRGPWGRRGKVSISRSVSIER ......

E:                    ...... DLDGYGGLGGYGGLGGL
                       SGSGDLDGLGGYGGLGGYGGLGGYGGLGG
F: RRGPWGRSVSSSLSVSVERGI
D: RRGPWGRRGKVSISRSVSIERIVTPGVYTKISRSSSVSIEGG
E: RR PWGYGRGLGGLSGSGDYDGLGGYGGLGGYGGLGG
F: RRGPWGRSVSVSLSVSVEGG
E: RR PWGYGRGLGGLSGSGDLDGLGGYGGLGGYGGLGG
F: RRGPWGRSVSSSLSVSVERGI
D: RRGPWGRRGKVSISRSVSIERIVTPGVYTKISRSSSVSVEGG
E: RR PWGYGRGLGGLSGSGDLDGLGGLGGLGGYGGLGG
E: RRGPWGRRGLGGLSGSGDLDGYGGLGGYGGLGGL
                  SGSGDLDGLGGYGGLGGYGGLGGYGGLGG
F: RRGPWGRSVSASLSVSVERGI
D: RRGPWGRRGKVSISRSVSIERIVTPGVYTKISRSSSVSVEGG
E: RR PWGYGRGLGGLSGSGDLDGLGGLGGLGGYGGLGGL
                     SGSGDLDGLGGYGGLGGYGGLGGYGGLGG
F: RRGPWGRSVSSSLSVSVERGI
D: RRGPWGRRGKVSISRSVSIERIVTPGVYTKISRSSSVSVEGG
E: RR PWGYGRGLGGLSGSGDLDGLGGYGGLGGYGGLGG
F: RRGPWGRSVSASLSVSVERGI
D: RRGPWGRRGKVSISRSVSIERIVTPSVYSKISRSSSVSVEGGI
   RRGPWG
cT:SFGGYDVSASSSLESVRPGRVLAGLGRVGALGRSGAVPVPSVY
TNHPGINSVKSPCKLSDFNLLVKVGNVRKSNGNC*
```

B

**Fig. 5.4** *Partial sequence and organization of H-fibroin from Hesperophylax occidentalis.* (**a**) Partial H-fibroin sequence aligned to highlight the *D*, *E*, and *F* subrepeats. The positively charged Arg blocks are highlighted in *gray*. $(pSX)_n$ repeats are *underlined*. Conserved Pro are *bolded*. The secretion signal peptide is in *bold italics*. (*asterisk*) = stop codon. (**b**) A schematic representation of the H-fibroin imperfectly repeating pattern of conserved subrepeats. Reproduced with permission from Ashton et al. (2013)

et al. 2012; Stewart and Wang 2010; Ohkawa et al. 2013). Silk of the casemaker *H. occidentalis* contains multivalent metal ions, mostly $Ca^{2+}$ and small amounts of $Mg^{2+}$, $Fe^{2+}$, $Zn^{2+}$, and $Mn^{2+}$, in roughly equimolar ratio to pS (Stewart and Wang 2010; Ashton et al. 2013). Charged residues comprise ~35 mol% of caddisworm H-fibroin residues. For comparison, the H-fibroin repetitive core of silkworm silk has less than 1 % charged residues, few multivalent metal ions, and is relatively hydrophobic (Lucas et al. 1960; Tulachan et al. 2014). In addition to silk proteins, caddisworm silk (*H. occidentalis*) comprises 60–70 wt% water.

Except for the short termini, caddisworm H-fibroin is ampholytic with alternating blocks of acidic and basic residues. Each D, E, and F repeats contains both a positively charged block of 3–4 Arg and a negatively charged block of at least one phosphorylated $(pSX)_n$ motif. In both D and F repeats, X is usually Val, Leu, Iso, sometimes Arg, and *n* is 4 or 5, respectively. In E repeats, X is always Gly and *n* is 2. The D repeats contain two $(pSX)_4$ domains equidistant from a central Gly-Pro pair (Figs. 5.4 and 5.5a).

Pro-Gly pairs are often the middle two residues of a tight turn (Zimmerman and Scheraga 1977). The two $(pSX)_4$ domains in the D repeats were hypothesized to form a $Ca^{2+}$-stabilized β-hairpin secondary structure motif, designated $[(pSX)_4]_2$ (Fig. 5.5b) (Stewart and Wang 2010). The basic Lys and acidic Glu residues that flank the $(pSX)_4$ blocks may promote the proper self-assembly of the $[(pSX)_4]_2$ β-hairpins by correctly registering the antiparallel strands, as depicted in Fig. 5.5a by the + and – symbols. In support of the proposed structure, steered molecular dynamic simulations to explore the stability of phosphorylated and non-phosphorylated D-repeat secondary structures demonstrated that the unfolding force was dramatically higher in $Ca^{2+}$-stabilized β-sheets than other structures (Strzelecki et al. 2011).

Higher order structures can be envisioned in which the $[(pSX)_4]_2$ β-hairpin modules associate laterally into β-sheets (Fig. 5.5c). The $[(pSX)_4]_n$ sheets can stack into β-domains with the aliphatic X residues and dianionic phosphates of the $(pSX)_n$ motifs on opposite sides of the sheet (Fig. 5.5d). Hydrophobic association of the aliphatic residues would provide a driving force for formation of the proposed β-structure (Dobson et al. 1998). Charge repulsion between the dianionic phosphate side groups would destabilize the structure unless the charges were balanced by bonds with multivalent metal ions. Indeed, native caddisworm silk has roughly equimolar ratios of metal ions to phosphates. The 600-fold decrease in the solubility of dibasic phosphate salt with $Ca^{2+}$ vs. $Na^{+}$ reflects the potential of multivalent metal ions to stabilize the pS-face of the proposed $[(pSX)_4]_n$ β-sheets (Fig. 5.5b, c). Participation in $Ca^{2+}$-stabilized β-sheets may not be limited to the D repeats. An F repeat $(pSX)_5$ block can be paired with a D-repeat hairpin to create a $Ca^{2+}$-stabilized three-stranded antiparallel β-sheet (Ashton et al. 2013). The proposed $(pSX)_n$ β-domains are stabilized by three reversible interactions: hydrophobic association of the aliphatic X residues, bridging of pS groups through multivalent metal ion complexes, and H-bonds between adjacent antiparallel peptide backbones. The caddisworm silk $Ca^{2+}$/phosphate-stabilized β-domains are structurally analogous to the stacked β-sheets comprising the poly(GA) repeats in

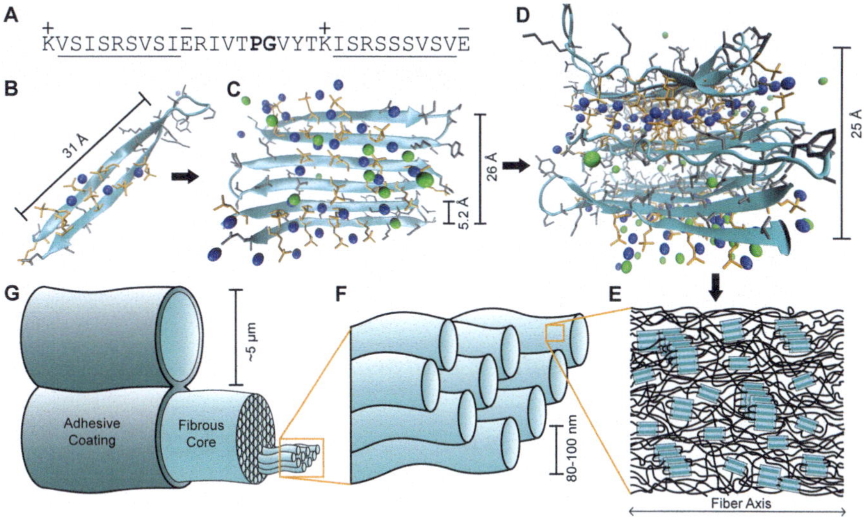


**Fig. 5.5** *Hierarchical structural model of caddisworm silk.* (**a**) The sequence of the D repeats used to create the structural models in **b**–**d**. Serines (S) in the *underlined* segments are phosphorylated. The *plus* and *minus* symbols depict the charges on the Lys and Glu residues, respectively. The Pro-Gly bend is *bold*. (**b**) A snapshot of the structure of a $Ca^{2+}$-stabilized D-repeat during a 40 ns molecular dynamic simulation. $Ca^{2+}$ = *blue spheres*, $Cl^{-}$ = *green spheres*. (**c**) Snapshot during a 10 ns molecular dynamic simulation of an extended β-sheet structure built with three D-repeat β-hairpins. (**d**) Representative structure of a three sheet β-domain model stabilized through alternating $Ca^{2+}$-pS-bridged interfaces and hydrophobic interfaces. (**e**) Macroscale subfibril model. The $Ca^{2+}$-stabilized $[(pSX)_4]_n$ β-domains (*turquoise*) form physical cross-links within an amorphous silk matrix (*black*). (**f**) The silk core comprises ~80–100 nm axially aligned subfibrils. (**g**) The caddisworm silk comprises two core subfibers fused together and encased in an adhesive coating. Adapted with permission from Ashton et al. (2013)

silkworm silk that have inter-sheet spacing between alternating Gly and Ala faces of 3.70 and 5.27 Å, respectively (Warwicker 1960).

Corroborating evidence for the $Ca^{2+}$-stabilized $(pSX)_n$ β-sheet structural model includes X-ray diffraction analysis of the case silk of *Olinga feredayi*, which showed a β-sheet structure with a 17 Å inter-sheet spacing. Treatment with the chelating agent EDTA greatly reduced the β-structure, prompting the authors to conclude that within some of the sheets "there are acidic groups binding an ion such as calcium" (Rudall and Kenchington 1971). Likewise, IR spectroscopy revealed a decrease in β-sheet content when multivalent metal ions were removed from *H. occidentalis* silk with $Na^{+}$-EDTA (Ashton et al. 2013). Removal of $Ca^{2+}$ caused the silk fibers to swell to more than twice their original diameter, further evidence of a $Ca^{2+}$-dependent structure. The IR absorption band corresponding to dianionic phosphoserine symmetric stretching modes when complexed with $Ca^{2+}$ was split into bands at 975 and 1007 $cm^{-1}$. When $Ca^{2+}$ is exchanged with $Na^{+}$-EDTA there was a single 976 $cm^{-1}$ peak (Ashton et al. 2013). Solid-state $^{31}P$ NMR showed that H-fibroin native $Ca^{2+}$/phosphates were in a rigid environment but became solvated and highly mobile when $Ca^{2+}$ was exchanged with $Na^{+}$-EDTA (Addison et al. 2014). Furthermore, $^{13}C$ solid-state NMR analysis of isotopically enriched *H. occidentalis* silk showed that the chemical shifts of Val residues, which occur almost exclusively in the D and F $(pSX)_{4–5}$ repeats, were consistent with a rigid β-structure (Addison et al. 2013).

## 5.5 Mechanical Properties of Caddisworm Silk

### *5.5.1 Mechanically Probing Caddisworm Silk Structure*

Single-fiber mechanical analysis provided additional evidence that $Ca^{2+}$-phosphate complexes are critical mechanical elements of caddisworm silk (Ashton et al. 2013; Ashton and Stewart 2015). Single silk fibers from the casemaker, *H. occidentalis,* were initially stiff (80–140 MPa) but dramatically strain softened (yield) at ~3 % elongation. Beyond 3 % elongation, the force profile plateaued then gradually increased (strain hardens) until failure at 90–120 % elongation and 25–40 MPa (Ashton et al. 2013). The silk fibers were tough, requiring twice the energy (~$17.3 \pm 6.2$ $MJ/m^3$) to break as tendon collagen (Ashton et al. 2013; Pollock and Shadwick 1994). The shape of a representative force-elongation profile of *H. occidentalis* silk (Fig. 5.6) was qualitatively similar to spider dragline silk, albeit slightly more extensible and lower in magnitude (Elices et al. 2011a). In spider silks, yield is the manifestation of sacrificial H-bonds in β-domains breaking at a critical force to dissipate mechanical energy and reveal hidden length (Elices et al. 2011a; Du et al. 2011). When submerged, water plasticizes the energy-dissipating H-bonded domains, and the dragline silk behaves like an elastomer with a progressive spring-like force profile (Elices et al. 2011a, 2011b). Analogously, when multivalent metal ions are removed from caddisworm silk with $Na^{+}$-EDTA, the apparent yield behavior is lost, and the silk behaves like a weak elastomer (Fig. 5.6) (Ashton et al. 2013).

The contribution of phosphate/$Ca^{2+}$ complexes to caddisworm silk mechanical properties was confirmed by comparing the IR spectra of the phosphate groups to the yield stress of cyclically strained silk fibers as the pH was progressively decreased from pH 8.0 to 3.0 (Fig. 5.7) (Ashton and Stewart 2015). The sigmoidal decrease of the combined intensity of two IR absorption bands, corresponding to phosphate symmetric stretching modes at 975 and 1007 $cm^{-1}$, (Fig. 5.7a, b) was nearly superimposable on the sigmoidal decrease in the fiber yield stress (Fig. 5.7c, d). The inflection points of the sigmoidal curves were at pH 4.4 and 4.6, respectively, providing direct evidence that protonation of the phosphate groups disrupted the $Ca^{2+}$/phosphate complex, with a coincident loss of stiffness, strength, and yield behavior in the fibers. A small decrease in yield force was observed from the pH 3.5

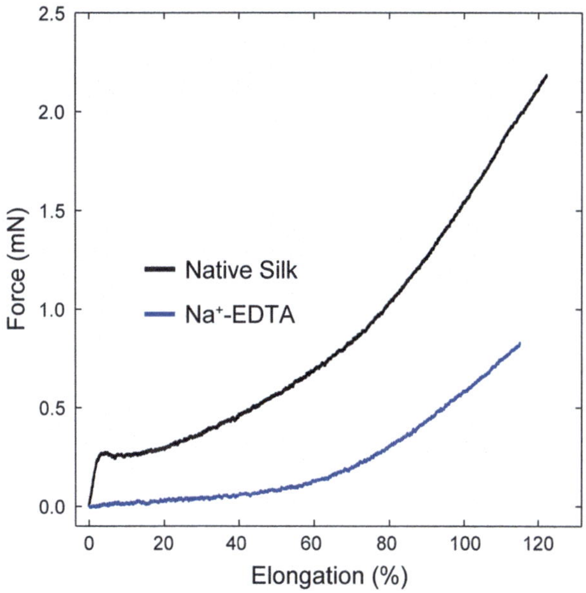


**Fig. 5.6** *$Ca^{2+}$-dependent caddisworm silk stiffness and strength.* Representative force-elongation profiles of a native silk fiber (*Black*) and a $Na^+$-EDTA-exchanged silk fiber (*Blue*) pulled until failure. Fibers were tested submerged at pH 7.0. The force profiles were created using methods described in (Ashton et al. 2013) and (Ashton and Stewart 2015)

to 2.0, coinciding with the pKa of carboxylate side chains, suggesting that a network of $Ca^{2+}$/carboxylate complexes may also contribute to strengthening caddisworm silk (Figs. 5.7c, d). With 20 % of its amino acids having carboxylate side chains, the caddisworm silk PEVK-like protein may account for at least part of the $Ca^{2+}$/carboxylate network (Wang et al. 2014). The similar PEVK regions of titin bind $Ca^{2+}$, which stiffens the domains (Lee et al. 2007; Nagy et al. 2005).

A major distinction between caddisworm silk and terrestrial silks is that the post-yield softening of caddisworm silk is transient; single fibers from *H. occidentals* recover their initial stiffness and strength within 120 min during cyclical strains to 20 % elongation (Fig. 5.8) (Ashton and Stewart 2015). The hysteresis in the loading and unloading profiles demonstrated that caddisworm silk can efficiently and repeatedly dissipate strain energy. In this regard, caddisworm silk is similar to a group of natural and synthetic materials that display yield-like behavior during tensile loading and spontaneous, hysteretic self-recovery of initial strength when the load is removed. Natural examples include the distal portion of the marine mussel byssal threads (Harrington et al. 2009), mammalian hair (Hearle 2007), and the whelk snail egg capsule (Rapoport and Shadwick 2007). Some synthetic double network hydrogels also demonstrate hysteretic self-recovery (Haque et al. 2011; Sun et al. 2012; Henderson et al. 2010), including a caddisworm silk-inspired hydrogel comprising a metal ion-cross-linked polyphosphate network within an elastic covalent polyacrylamide matrix (Lane et al. 2015). The caddisworm silk-inspired hydrogel qualitatively replicated the yield, efficient self-recovery, and energy-dissipating properties of the natural silk.

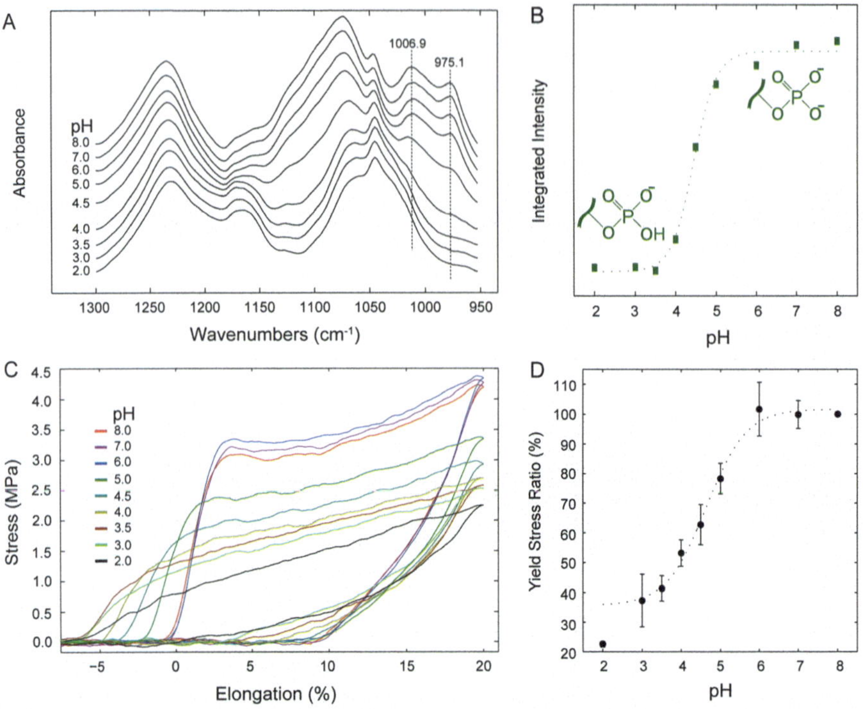


**Fig. 5.7** *pH dependence of caddisworm silk*. (**a**) Bundles of silk fibers were analyzed by ATR-FTIR as the pH was decreased from 8.0 to 2.0. The dianionic phosphate symmetric stretching modes are indicated at 975 and 1007 $cm^{-1}$. (**b**) The integrated intensity under the two dianionic phosphate symmetric stretching mode fitted with a logistic function. (**c**) A single silk fiber mechanically cycled as the pH was decreased from 8.0 to 2.0 with a 1 h recovery period between cycles. (**d**) The average percent change in yield force of fibers cycles at descending pH as in (**c**). Error bars = standard deviation, $n = 3$. (**a–d**) were adapted with permission from Ashton and Stewart (2015)

## 5.5.2 A Mechanical Model of Caddisworm Silk

A multi-network model of the caddisworm silk is presented in Fig. 5.9 (Ashton and Stewart 2015). Two parallel networks of $Ca^{2+}$-stabilized sacrificial domains act in parallel with an elastic network. The predominating $Ca^{2+}$-dependent network corresponds to the H-fibroin serial $[(pSX)_n]_n$ domains. The second $Ca^{2+}$-dependent network corresponds to $Ca^{2+}$/carboxylate complexes of unknown structure and organization. The third network is an elastic network with a force-elongation profile that may be similar to the profile of $Na^+$-EDTA-treated silk fibers (Fig. 5.6). When the fiber is unloaded, recoil of the elastic network may assist the mechanical recovery of the $Ca^{2+}$-dependent networks. Components of the elastic network may include amorphous regions formed by the YGGLGG regions of the H-fibroin E repeats. Solid-state $^{13}C$ NMR of isotopically enriched *H. consimilis* silk was

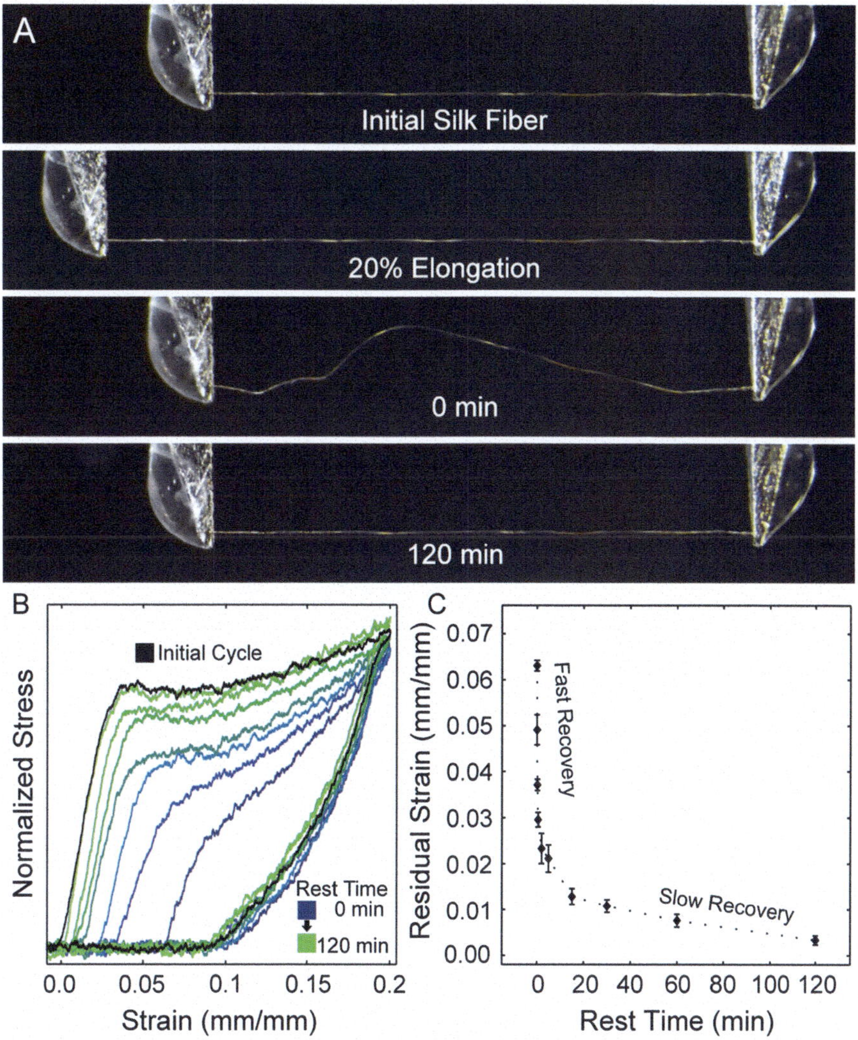


**Fig. 5.8** *Post-yield recovery of silk mechanical properties.* (**a**) A caddisworm silk fiber submerged at pH 8.0 was held taut at its initial length, cycled to 20 % elongation, returned to 0 % elongation, and allowed to recover for 120 min (**b** and **c**). In a similar experiment, three fibers were cycled to 20 % elongation and then allowed to recover for either 0, 0.5, 1, 2, 5, 10, 15, 30, 60, or 120 min before the next cycle. (**b**) Representative normalized stress–strain curves for the second cycle after recovery periods of 0, 0.5, 5, 15, 30, 60, and 120 min. (**c**) Recovery of caddisworm silk is biphasic. (**b** and **c**) were adapted with permission from Ashton and Stewart (2015)

consistent with Gly and Leu occupying a random coil configuration. Other components of the covalent elastic network may include the csPxt-catalyzed dityrosine cross-linked network, as well as a disulfide cross-linked network.

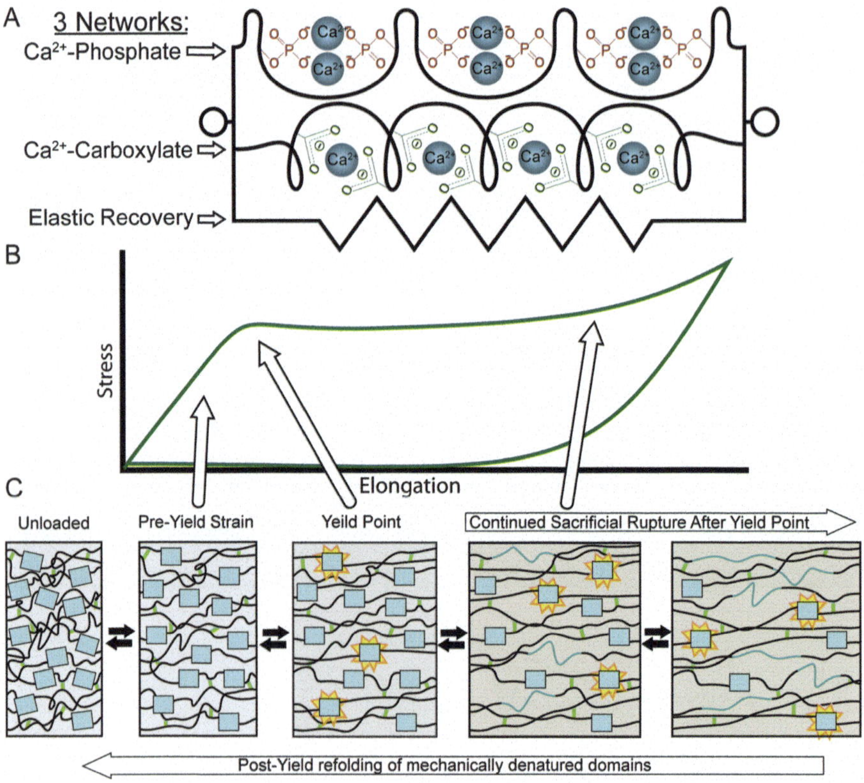


**Fig. 5.9** *Caddisworm silk mechanical model*. (**a**) Multi-network model of caddisworm silk with two parallel $Ca^{2+}$-cross-linked sacrificial networks and an elastic covalent network. $Ca^{2+}$-phosphate and $Ca^{2+}$-carboxylate complexes rupture under mechanical load to dissipate mechanical energy. During unloading, an elastic network guides refolding of the mechanically denatured $Ca^{2+}$-dependent domains. (**b**) An idealized stress–strain plot during a load cycle. (**c**) Hypothetical molecular events occurring at the indicated regions of the stress–strain curve. *Blue rectangles* represent $Ca^{2+}$-stabilized sacrificial domains. *Yellow stars* represent reversible rupture of the $Ca^{2+}$-stabilized domains. *Blue lines* represent unraveled sacrificial domains that reveal hidden length. *Black lines* represent amorphous silk components. *Green lines* represent disulfide and dityrosine covalent cross-links that may contribute to silk elasticity

The pre-yield region of the stress–strain curve represents the elasticity of the intact sacrificial domains (Fig. 5.9b, c). At a critical stress—the yield stress—individual sacrificial domains begin to unravel and lengthen resulting in a stress plateau. The stress plateau limits the load on the interfacial adhesive bonds protecting them from irreversible damage. The effects of the weaker second $Ca^{2+}$/carboxylate network are hidden within the first strain cycle but become apparent when a single fiber is cyclically loaded with a progressive increase in strain with no recovery time between cycles (Ashton and Stewart 2015). When unloaded, fiber recovery is biphasic. The $Ca^{2+}$/carboxylate network may recover more quickly than the $Ca^{2+}$/phosphate network (Fig. 5.8c). The slow recovery phase may be due to temporarily reformed $Ca^{2+}$ crossbridges. Dissipation of strain energy and self-recovery maintains the integrity of the case without material reprocessing and rebuilding.

## 5.6 The Silk Spinning System

### *5.6.1 Silk Gland Anatomy*

The caddisworm's two identical silk glands fold several times along the length of the ventral side of the alimentary canal, filling much of the abdominal cavity. Unfolded, the silk glands are about twice the length of the larva (Fig. 5.10). The posterior and anterior segments of the glands are separated by a bulbous region (Fig. 5.10, arrows). Silk components are secreted into and stored in the larger posterior silk gland. The anterior silk gland is a 2–4 mm long, narrow silk conducting channel (Fig. 5.10, inset). The anterior lumens of the paired silk glands fuse into a common duct at the silk press just before the spinneret located ventral to the mouthparts on the underside of the head (Fig. 5.10, arrowhead) (Tszydel et al. 2009; Spanhoff et al. 2003). Accessory silk glands, also known as Filippi's or Lyonet's glands (Patra et al. 2012), which are connected to the anterior conducting channel in lepidopteran silk glands, have not been identified in caddisworm silk glands. The interior lumen of the anterior conducting channels is lined with chitin. The silk press is controlled by four muscles that connect the silk press to the exoskeleton of the head (Glasgow 1936). In the relaxed state, the silk press is closed. When the muscles are contracted, the press opens and fibers can be drawn from the gland (Glasgow 1936). By modulating the contractile strength of the silk press muscles, the larva may have control of the silk fiber tension during spinning (Glasgow 1936).

The posterior silk gland is covered in large hexagonal cells with interdigitated membranes and large, extensively branched nuclei (Engster 1976b; Glasgow 1936; Packard 1898). A basement membrane lines the lumen of the gland (Engster 1976b). The degree of nuclei branching in caddisworm glands was reported to be related to the secretory activity of the cell; more active cells had more extensively branched nuclei (Beams and Sekhon 1966). Engster observed a peripheral coating along the entire length of the silk gland lumen, increasing gradually in thickness from ~0.25 μm at the extreme posterior to 3.0 μm in the anterior region. She suggested the components of the fiber core and peripheral coating may be co-secreted and phase separate once in the lumen (Engster 1976b).

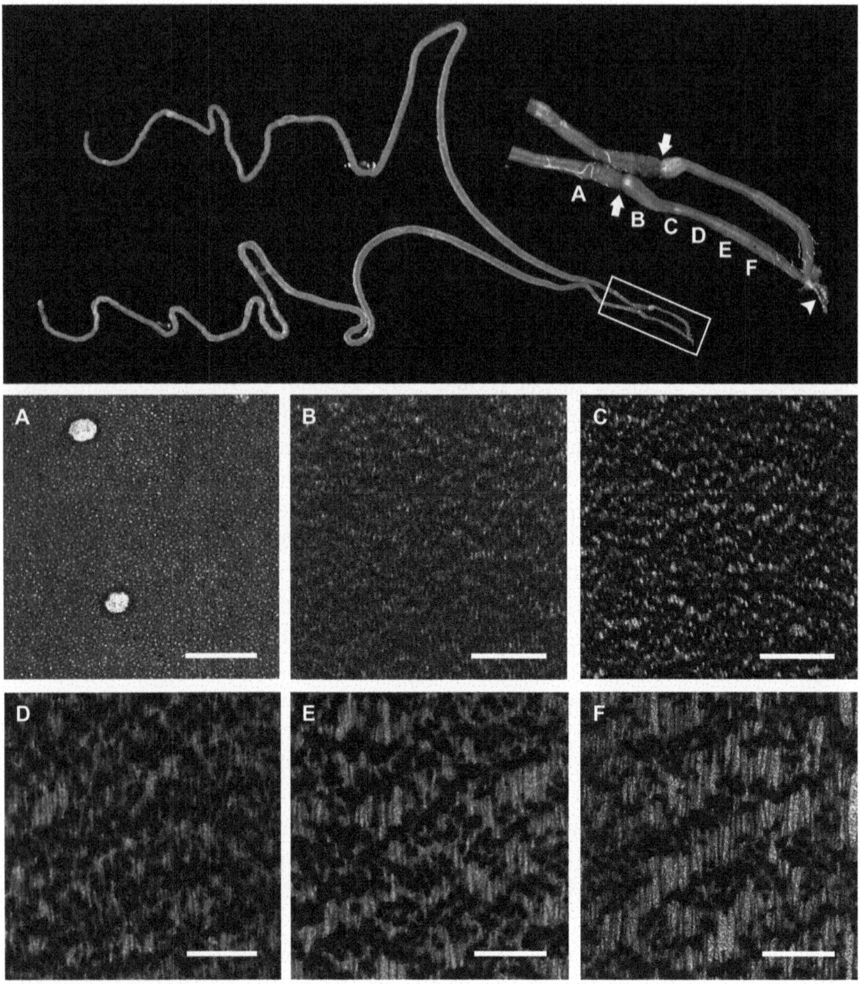


**Fig. 5.10** *Silk fiber morphology in the anterior silk gland lumen.* (*Top*) A pair of intact silk glands dissected from a fifth instar larva of *H. occidentalis*. *Inset*: higher magnification of the anterior silk gland. The *arrows* mark the bulbous feature at the transition from the posterior to anterior silk glands. The *arrowhead* marks the silk press and spinneret. The approximate locations of the TEM images in panels (**a**)–(**f**) are lettered. (**a**) Silk dope stored in the posterior silk gland lumen with spherical, bubble-like inclusions. (**b**)–(**f**) Progressive formation of fibrous silk as precursors are draw out of the gland. The axial direction is downward in the micrographs. Scale bars = 1 μm. Methods: Silk glands were dissected from fifth instar *H. occidentalis* larvae collected in Red Butte Creek, Salt Lake County, UT, USA. The glands were fixed in 2 % glutaraldehyde and 1 % paraformaldehyde in 0.1 M cacodylate buffer (pH 7.4) for 24 h. Silk samples were treated with 2 % $OsO_4$, rinsed with DI-$H_2O$, and stained with 4 % uranyl acetate. After dehydration in a graded series of ethanol, the samples were embedded in Embed-812 (Electron Microscopy Sciences). *Blocks* were sectioned to approximately 70 nm thickness with a diamond ultra 45° blade (Diatome). Sections on copper grids were stained with saturated uranyl acetate, rinsed with water, stained with Reynold's lead citrate, and imaged in a Tecnai 12 TEM (FEI)

### 5.6.2 *Fiber Formation in the Anterior Silk Gland*

Silk gland contents become anisotropic, and a silk fiber begins to form as they are drawn through the rigid, cup-shaped, chitinous structure at the beginning of the anterior conducting channel in the bulbous region (Fig. 5.11). In earlier work, the silk components became refringent in the same region when viewed under polarized light (Ashton et al. 2012). It appears that the cup-shaped structure is designed to shear the silk dope during the spinning process. From a material processing standpoint, this structure is the true spinneret, as opposed to the anatomical fiber exit point below the mouth (Tszydel et al. 2009; Spanhoff et al. 2003). The process of caddisfly silk fiber formation is similar to industrial spinning of polymeric fibers that involve mechanically shearing fluid precursors accompanied by a liquid or gel to solid phase transition (Chawla 1998). It may also be similar to the spinning of lepidopteran and spider silks (Vollrath and Knight 2001). The components of the terrestrial silks are stored in the silk gland in a liquid crystalline state. Mechanical shear is thought to induce molecular alignment of β-sheet forming domains in adjacent H-fibroin molecules that zip together into interstrand cross-links. In the case of caddisworm silk, shearing could align H-fibroin molecules and promote formation of interstrand $[(pSX)_n]_n$ domains.

TEM images of longitudinal sections, taken at successive 150–250 μm lengths along the anterior silk gland, demonstrate the progressive transition of isotropically distributed ~30 nm granules in the stored silk dope into aligned ~100 nm subfibrils (Fig. 5.10b–f). The bands of globular structures separated by thin strands (Fig. 5.10f) are not apparent in the TEM image of a longitudinal section through

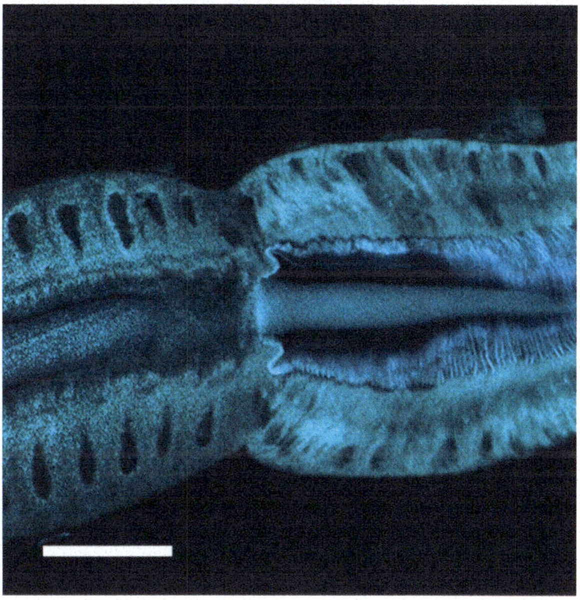

**Fig. 5.11** *The silk gland transition region of Hesperophylax consimilis.* Autofluorescent image of a 5 μm thick longitudinal section through the transition region. The chitin-lined anterior conducting channel is to the right. Scale bar = 50 μm. Adapted from Ashton et al. (2012)

a final, fully extruded fiber (Fig. 5.2). It is not known when $Ca^{2+}$ is added to the caddisworm silk. Experiments are in progress to determine if it is added in the silk gland before or during fiber formation or if it is absorbed from the environment after the fiber is drawn into natural waters.

**Acknowledgments** Funding from the Army Research Office is gratefully acknowledged. We thank Nancy Chandler for the assistance with TEM and Daniel Taggart for the confocal microscopy.

## References

Addison JB et al (2013) β-sheet nanocrystalline domains formed from phosphorylated serine-rich motifs in caddisfly larval silk: a solid-state NMR and XRD study. Biomacromolecules 14:1140–1148

Addison JB et al (2014) Reversible assembly of β-sheet nanocrystals within caddisfly silk. Biomacromolecules 15:1269–1275

Ashton NN, Stewart RJ (2015) Self-recovering caddisfly silk: energy dissipating, $Ca^{2+}$-dependent, double dynamic network fibers. Soft Matter 11:1667–1676

Ashton NN, Taggart DS, Stewart RJ (2012) Silk tape nanostructure and silk gland anatomy of trichoptera. Biopolymers 97:432–445

Ashton NN, Roe DR, Weiss RB, Cheatham TE, Stewart RJ (2013) Self-tensioning aquatic caddisfly silk: Ca2+-dependent structure, strength, and load cycle hysteresis. Biomacromolecules 14:3668–3681

Beams HW, Sekhon SS (1966) Morphological studies on secretion in the silk glands of the caddis fly larvae Platyphylax designatus Walker. Zeitschrift für Zellforsch und Mikroskopische Anat 72:408–414

Blough NV, Ziepp RG (1995) Reactive oxygen species in natural waters, 2nd edn. Chapman & Hall, London

Chawla KK (1998) Fibrous materials. Cambridge University Press, New York City

Cooksey KE, Wigglesworth-Cooksey B (1995) Adhesion of bacteria and diatoms to surfaces in the sea: a review. Aquat Microb Ecol 9:87–96

Dobson CM, Šali A, Karplus M (1998) Protein folding: a perspective from theory and experiment. Angew Chem Int Ed 37:868–893

Du N, Yang Z, Liu XY, Li Y, Xu HY (2011) Structural origin of the strain-hardening of spider silk. Adv Funct Mater 21:772–778

Elices M, Plaza GR, Pérez-Rigueiro J, Guinea GV (2011a) The hidden link between supercontraction and mechanical behavior of spider silks. J Mech Behav Biomed Mater 4:658–669

Elices M, Guinea GV, Pérez-Rigueiro J, Plaza GR (2011b) Polymeric fibers with tunable properties: lessons from spider silk. Mater Sci Eng C 31:1184–1188

Engster MS (1976a) Studies on silk secretion in the Trichoptera (F. Limnephilidae): II. Structure and amino acid composition of the silk. Cell Tissue Res 169:77–92

Engster M (1976b) Studies on silk secretion in the trichoptera (F. Limnephilidae): I. Histology, histochemistry, and ultrastructure of the silk glands. J Morphol 150:183–212

Glasgow JP (1936) Internal anatomy of Caddis (Hydropsyche colonica). Q J Microsc Sci 79:151–179

Haque MA, Kurokawa T, Kamita G, Gong JP (2011) Lamellar bilayers as reversible sacrificial bonds to toughen hydrogel: hysteresis, self-recovery, fatigue resistance, and crack blunting. Macromolecules 44:8916–8924

Harrington MJ, Gupta HS, Fratzl P, Waite JH (2009) Collagen insulated from tensile damage by domains that unfold reversibly: in situ X-ray investigation of mechanical yield and damage repair in the mussel byssus. J Struct Biol 167:47–54
Hatano T, Nagashima T (2015) The secretion process of liquid silk with nanopillar structures from Stenopsyche marmorata (Trichoptera: Stenopsychidae). Sci Rep 5:9237
Hearle JWS (2007) Protein fibers: structural mechanics and future opportunities. J Mater Sci 42:8010–8019
Henderson KJ, Zhou TC, Otim KJ, Shull KR (2010) Ionically cross-linked triblock copolymer hydrogels with high strength. Macromolecules 43:6193–6201
Hertner T, Eppenberger HM, Lezzi M (1983) The giant secretory proteins of Chironomus tentans salivary glands: the organization of their primary structure, their amino acid and carbohydrate composition. Chromosoma 88:194–200
Jonker JL, Morrison L, Lynch EP, Grunwald I, von Byern J, Power AM (2015) The chemistry of stalked barnacle adhesive (*Lepas anatifera*). Interface Focus 5(1):20140062. doi:10.1098/rsfs.2014.0062
Lane DD, Kaur S, Weerasakare GM, Stewart RJ (2015) Toughened hydrogels inspired by aquatic caddisworm silk. Soft Matter 11:6981–6990
Lee EH, Hsin J, Mayans O, Schulten K (2007) Secondary and tertiary structure elasticity of titin Z1Z2 and a titin chain model. Biophys J 93:1719–1735
Lucas F, Shaw JTB, Smith SG (1960) Comparative studies of fibroins: I. The amino acid composition of various fibroins and its significance in relation to their crystal structure and taxonomy. J Mol Biol 52:339–349
Malm T, Johanson KA, Wahlberg N (2013) The evolutionary history of Trichoptera (Insecta): a case of successful adaptation to life in freshwater. Syst Entomol 38:459–473
Merritt RW, Cummins KW (2008) An introduction to the aquatic insects of North America. Kendall Hunt Pub Co, Dubuque
Mitchinson NA (1974) Tanned silks. Proc R Soc Lond B 187:133–170
Mori K et al (1995) Production of a chimeric fibroin light-chain polypeptide in a fibroin secretion-deficient naked pupa mutant of the silkworm Bombyx mori. J Mol Biol 251:217–228
Nagy A et al (2005) Hierarchical extensibility in the PEVK domain of skeletal-muscle titin. Biophys J 89:329–336
Naskar D, Barua RR, Ghosh AK, Kundu SC (2014) In: Kundu SC (ed) Silk biomaterials for tissue engineering and regenerative medicine. Woodhead Publishing, Cambridge, p 15
Ohkawa K, Miura Y, Nomura T, Arai R (2012) Isolation of silk proteins from a Caddisfly Larva Stenopsyche Marmorata. J Fiber Bioeng Inform 5:125–137
Ohkawa K et al (2013) Long-range periodic sequence of the cement/silk protein of Stenopsyche marmorata : purification and biochemical characterisation. J Bioadhes Biofilm Res 29:357–367
Packard AS (1898) A text-book of entomology. The MacMillan Company, New York
Patra S, Singh RN, Raziuddin M (2012) Morphology and Histology of Lyonet' s Gland of the Tropical Tasar Silkworm, Antheraea mylitta Morphology and histology of Lyonet' s gland of the tropical tasar silkworm Antheraea mylitta. J Insect Sci 12:1–7
Pollock CM, Shadwick RE (1994) Relationship between body mass and biomechanical properties of limb tendons in adult mammals. Am J Physiol 266:1016–1021
Rapoport HS, Shadwick RE (2007) Reversibly labile, sclerotization-induced elastic properties in a keratin analog from marine snails: whelk egg capsule biopolymer. J Exp Biol 210:12–26
Riek E (1976) The marine caddisfly family Chathamiidae (Trichoptera). Aust J Entomol 15 (4):405–419
Rudall K, Kenchington W (1971) Arthropod silks: the problem of fibrous proteins in animal tissues. Annu Rev Entomol 16:73–96
Soudi M, Zamocky M, Jakopitsch C, Furtmüller PG, Obinger C (2012) Molecular evolution, structure, and function of peroxidasins. Chem Biodivers 9:1776–1793

Spanhoff B, Schulte U, Alecke C, Kaschek N, Meyer EI (2003) Mouthparts, gut contents, and retreat-construction by the wood-dwelling larvae of Lype phaeopa. Eur J Entomol 100:563–570
Stewart RJ, Wang CS (2010) Adaptation of caddisfly larval silks to aquatic habitats by phosphorylation of H-fibroin serines. Biomacromolecules 11:969–974
Stewart RJ, Ransom TC, Hlady V (2011) Natural underwater adhesives. J Polym Sci B Polym Phys 49:757–771
Strzelecki JW et al (2011) Nanomechanics of new materials-AFM and computer modelling studies of trichoptera silk. Cent Eur J Phys 9:482–491
Sun J-Y et al (2012) Highly stretchable and tough hydrogels. Nature 489:133–136
Tanaka K et al (1999) Determination of the site of disulfide linkage between heavy and light chains of silk fibroin produced by Bombyx mori. Biochim Biophys Acta 1432:92–103
Tszydel M et al (2009) Structure and physical and chemical properties of fibres from the fifth larval instar of caddis-flies of the species Hydropsyche angustipennis. Fibres Text East Eur 17:7–12
Tulachan B et al (2014) Electricity from the silk cocoon membrane. Sci Rep 4:5434
Vollrath F, Knight DP (2001) Liquid crystalline spinning of spider silk. Nature 410:541–548
Wang CS, Stewart RJ (2013) Multipart copolyelectrolyte adhesive of the sandcastle worm, Phragmatopoma californica (Fewkes): catechol oxidase catalyzed curing through peptidyl-DOPA. Biomacromolecules 14:1607–1617
Wang CS, Ashton NN, Weiss RB, Stewart RJ (2014) Peroxinectin catalyzed dityrosine cross-linking in the adhesive underwater silk of a casemaker caddisfly larvae Hysperophylax occidentalis. Insect Biochem Mol Biol 54C:69–79
Wang CS, Pan H, Weerasekare GM, Stewart RJ (2015) Peroxidase-catalysed interfacial adhesion of aquatic caddisworm silk. J R Soc Interface 12(112). doi:10.1098/rsif.2015.0710
Warwicker JO (1960) Comparative studies of fibroins II. The crystal structure of various fibroins. J Mol Biol 2(6):350
Wiggins GB (2004) Caddisflies: the underwater architects. University of Toronto Press, Toronto
Yonemura N, Sehnal F, Mita K, Tamura T (2006) Protein composition of silk filaments spun under water by caddisfly larvae. Biomacromolecules 7:3370–3378
Yonemura N, Mita K, Tamura T, Sehnal F (2009) Conservation of silk genes in Trichoptera and Lepidoptera. J Mol Evol 68:641–653
Zimmerman SS, Scheraga HA (1977) Influence of local interactions on protein structure. I. Conformational energy studies of N-acetyl-N′-methylamides of Pro-X and X-Pro dipeptides. Biopolymers 16:811–843

# Chapter 6
# Interfacial Phenomena in Marine and Freshwater Mussel Adhesion

**Eli D. Sone**

**Abstract** The attachment of mussels to underwater surfaces is one of the best-studied examples of biological adhesion, yet it continues to provide new insights into the phenomenon of wet adhesion. Both marine and freshwater mussels secrete a proteinaceous byssus, which attaches the mussel to a wide variety of surfaces with considerable strength. In marine mussels, Dopa (3,4-dihydroxyphenylalanine)-containing proteins have been shown to play key roles in this process, while less is known about the mechanism of adhesion in freshwater mussels, which contain much less Dopa. This chapter will highlight some of the recent developments in the mechanistic understanding of byssal adhesion, with a focus on the adhesive interface itself. In marine mussels, the extensive foundational work that preceded the last edition of this book has allowed a more recent emphasis on protein function, leading to important new insights regarding redox chemistry at the adhesive interface, the effect of surface chemistry on adhesion, and the role of amino acids other than Dopa in adhesion. Prior to this, adhesion of freshwater mussels will be reviewed. Here, the state of knowledge is much more nascent, with current emphasis still on protein sequencing and distribution. Nevertheless, careful structural characterization of the freshwater mussel adhesion has led to new understandings on the nature of the byssal adhesive interface. Furthermore, the comparison of the separately evolved adhesion strategies of freshwater and marine mussels may lead to new insights into the essential requirements of underwater adhesion.

---

E.D. Sone (✉)
Institute of Biomaterials and Biomedical Engineering, Department of Materials Science & Engineering, and Faculty of Dentistry, University of Toronto, 164 College Street, Toronto, ON, Canada, M5S 3G9
e-mail: eli.sone@utoronto.ca

© Springer International Publishing Switzerland 2016

A.M. Smith (ed.), *Biological Adhesives*, DOI 10.1007/978-3-319-46082-6_6

## 6.1 Introduction

Like other sedentary aquatic organisms, mussels have evolved the means to attach themselves to surfaces under conditions that are, from an engineering perspective, quite impressive: they are able to rapidly adhere underwater, at ambient temperature, to substrata with widely varied surface chemistry and with enough strength to withstand the action of currents, waves, and predators. In mussels, attachment is achieved by means of the proteinaceous byssus that is secreted by a muscular gland (the foot). The byssus consists of a series of threads, bundled at the stem, each of which connects the animal to an adhesive plaque deposited onto the substratum (Fig. 6.1). In both marine and freshwater mussels, the byssus is covered by a highly cross-linked protective coating (Bairati and Zuccarello 1976; Bonner and Rockhill 1994b). Byssal adhesion, in marine mussels especially, is one of the best-studied biological adhesive systems to date. Dopa (3,4-dihydroxyphenylalanine)-containing proteins in particular have been convincingly shown to play a key role in mediating both adhesion and cohesion (Sagert et al. 2006; Nicklisch and Waite 2012). Several key principles of underwater adhesion and techniques to understand its mechanism have emerged from the study of marine mussel byssal proteins, along with numerous biomimetic and bio-inspired materials based on those principles, including wet adhesives, coatings, tissue engineering scaffolds, and drug delivery vehicles, as reviewed in Chap. 15 of this book.

The most thoroughly characterized byssus is that of marine mussels of the *Mytilus* genus. In the thread of its byssus, collagen-like proteins of varying stiffness create a mechanical gradient along the length of the thread to minimize stresses at the substratum and on the mussel's internal tissue (Waite et al. 2004). In the plaque, several Dopa-containing proteins play various roles, including in the attachment of the mussel's adhesive plaque to the substratum, as well as in the formation of a protective cross-linked sheath over the threads (Lee et al. 2011). Five of these proteins are unique to the plaque: *Mytilus* foot proteins (Mfp) 2–6. Dopa, produced by posttranslational hydroxylation of tyrosine, can bind to metal oxide surfaces (Lee et al. 2006) and form metal cation-mediated complexation cross-links (Sever et al. 2004). It can also be oxidized to dopaquinone (Fig. 6.2), which may react via a variety of mechanisms to form covalent cross-links with Dopa and other residues (Rzepecki and Waite 1993a; Burzio and Waite 2000; Yang et al. 2014), thereby providing cohesive strength to the byssus.

In the separately evolved freshwater dreissenids (Morton 1993), which also attach to substrata through byssal threads and adhesive plaque, the mechanisms of adhesion are not as well understood. The freshwater zebra mussel, *Dreissena polymorpha*, and its cousin the quagga mussel, *Dreissena bugensis*, are both of particular interest as a biofouling species. The byssus of dreissenids is superficially similar to that of marine mussels (Eckroat 1993), but until recently, little was known about the composition and distribution of proteins which comprise the byssus or of the nanoscale structure of the adhesive interface. While Dopa-containing proteins have been identified in freshwater mussels, the levels of Dopa

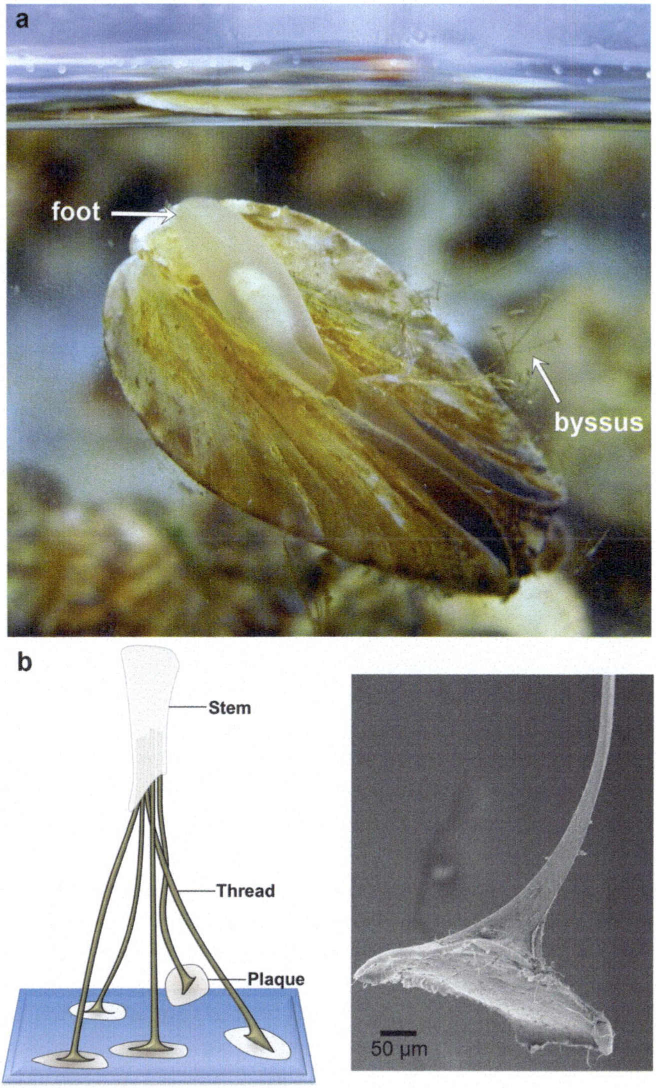


**Fig. 6.1** Byssal attachment in mussels. (**a**) A freshwater quagga mussel is shown attaching to the side of a glass aquarium. The byssus is visible, as is the foot of the mussel, which produces the byssus. (**b**) The byssus, consisting of several threads and plaques attached to a substratum, is shown schematically (*left*) and as a SEM micrograph of a detached plaque and thread (*right*)

are much lower than for most marine mussels (Rzepecki and Waite 1993b). The key question of whether the mechanism of attachment in freshwater mussels is fundamentally the same as for marine mussels remains unanswered.

The byssal proteins of marine mussels have been the subject of numerous reviews and book chapters since the last edition of this book (Silverman and

Fig. 6.2 Oxidation of protein-bound Dopa to dopaquinone

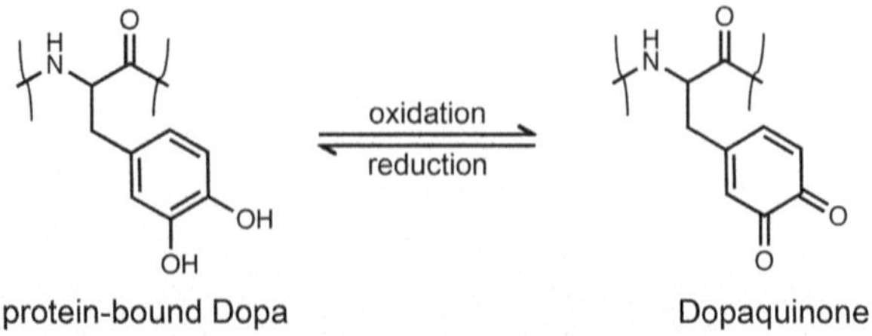


Roberto 2007, 2010; Lee et al. 2011; Stewart et al. 2011; Hwang et al. 2013; Bandara et al. 2013; Li and Zeng 2016; Zeng et al. 2015). This chapter will, therefore, focus on the adhesive interface itself, i.e., the plaque-substratum interface, and on the proteins and molecular processes that mediate adhesion, rather than on the entire byssus. This is not meant to imply that adhesion is the only interesting phenomenon occurring in byssal adhesion; complex coacervation of secreted proteins (Wei et al. 2014), the graded composition and mechanical properties of the thread (Harrington and Waite 2009), the hard yet compliant protective cuticle (Harrington et al. 2010), and the protein-protein, protein-metal, and covalent cross-linking chemistries that mediate cohesion in the byssus (Hwang et al. 2010; Wilker 2010) are all worthy topics in their own right. But the last 10 years have seen tremendous advances in unveiling the details of the adhesion phenomenon itself. A mechanistic understanding of adhesion is critical for the design of novel biomedical adhesives that need to function under similarly challenging conditions and will also enable the development of rationally designed antifouling surfaces.

This chapter will highlight some of the recent developments in the mechanistic understanding of byssal adhesion. In marine mussels, the extensive foundational work on protein discovery and localization as well as Dopa chemistry that preceded the last edition of this book (Sagert et al. 2006) has allowed a more recent focus on protein function, leading to important new insights, three of which will be covered here: (i) the importance of redox chemistry at the adhesive interface, (ii) the effect of surface chemistry and mode of Dopa interaction, and (iii) the role of amino acids other than Dopa in adhesion. Prior to this, adhesion of freshwater mussels will be reviewed. Here, the state of knowledge is much more nascent, with current emphasis still on protein sequencing and distribution. Nevertheless, careful structural characterization of the freshwater mussel adhesion has led to new understandings on the nature of the byssal adhesive interface. Furthermore, the comparison of the separately evolved adhesion strategies of freshwater and marine mussels may lead to new insights into the essential requirements of underwater adhesion.

## 6.2 Recent Advances in Understanding Freshwater Mussel Adhesion

Zebra mussels (*Dreissena polymorpha*) and their close relatives the quagga mussels (*Dreissena bugensis*) are freshwater mussels native to the Black and Caspian Seas that have spread rapidly as an invasive species (Strayer 2009). Their ability to attach to substrata such as ship hulls and water intake pipes has enabled them to spread widely, creating expensive biofouling problems and damaging native ecosystems (Sousa et al. 2014; Connely et al. 2007). Although Dopa-containing proteins are present in the byssus in both species, overall levels are low: 0.6 % in zebra mussels and 0.1 % in quagga mussels (Rzepecki and Waite 1993a; Anderson and Waite 2002). Unlike in marine mussels, Dopa levels are nearly identical in plaques and threads, which suggest Dopa may be playing more of a role in cross-linking of the byssus itself, rather than necessarily an adhesive role. Such compositional differences from the marine mussel byssus, despite the superficial structural similarity, raise the intriguing possibility that there may be differences in the molecular mechanism of adhesion. Yet much less is known about the adhesive proteins of freshwater mussels, and questions remain with respect to the structure of the plaque-substratum interface where the critical adhesive link occurs.

### *6.2.1 Structure and Composition of the Adhesive Interface*

The adhesive bond formed between the byssal plaque and the substratum is one of the most fascinating aspects of mussel adhesion; in most synthetic adhesives, wet conditions drastically reduce the adhesive strength due to the formation of a weak boundary layer and due to the high-dielectric constant of water (Waite et al. 2005; Lee et al. 2011). Characterization of the structure and composition of the adhesive interface is thus critical for understanding adhesion. Surprisingly, relatively few studies have focused on high-resolution structural characterization of the plaque-substratum interface. In zebra mussels, the adhesive interface was first imaged at high resolution by Bonner and Rockhill (1994b). Using transmission electron microscopy (TEM) of ultrathin sections taken from plaques deposited onto polyethylene substrata, they described an "electron-dense granular structure" on the plaque underside, which connects to the surface at discrete "attachment spots," separated by regions in which gaps of 75 nm separate the plaque and substratum. More recent studies from my lab used an epoxy substratum, to which plaques attach more robustly, leading to a reduced risk of detachment during sample processing (Farsad and Sone 2012). In heavy metal-stained sections, we showed that in fact the adhesive plaque makes direct and continuous contact with the substratum; no gaps were observed to within the nanometer level resolution of the images (Fig. 6.3a). This observation shows that plaque proteins interact directly with the surface, rather than bonding to any preexisting biofilm that may coat the surface, with the caveat

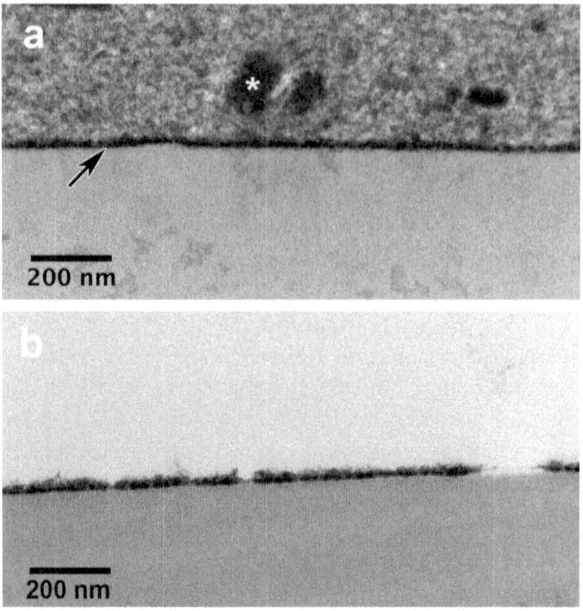


**Fig. 6.3** The plaque-substratum interface in zebra mussel. (**a**) TEM micrograph of an ultrathin section of a plaque deposited on an epoxy substratum. An *arrow* indicates the location of the interface, where a more darkly stained layer is apparent, which forms direct and continuous contact with the substratum. Adjacent to this layer in the plaque, some similarly stained granules are also visible (*asterisk*). (**b**) Upon removal of the plaque from the substratum, the thin adhesive layer is often left behind. Adapted from Farsad and Sone (2012), with permission from Elsevier

that the natural biofilm would be different from the one formed in the aquarium. Presumably, the mussel physically removes any existing material when it probes the surface with its foot, prior to the deposition of the plaques. We did sometimes observe bacteria within the bulk of the plaque, as did Bonner and Rockhill, suggesting that some parts of the biofilm may be engulfed by the forming plaque when they are loosened but not swept aside.

A striking feature of the adhesive interface in zebra mussel plaques is the presence of a rather uniform, thin (10–20 nm) layer directly apposed to the substratum, which stains more darkly than the rest of the plaque in TEM. When plaques are peeled off of the substratum, this layer is often left behind (Fig. 6.3b), highlighting its adhesive properties. We verified that the adhesive layer is part of the plaque itself, and not deposited separately (e.g., a mucous layer from the foot), or part of a preexisting biofilm, by confirming that areas of the substratum adjacent to the plaque do not contain this layer. The presence of a more darkly stained adhesive layer appears to be unique to zebra mussels; in various marine mussels, a somewhat thicker (~50 nm) continuous layer was also observed in direct contact with the interface, but it stained similarly to the bulk of the plaque and was thus originally assumed to be composed of the same proteins (Tamarin et al. 1976;

Waite 1986; Benedict and Waite 1986), until later biochemical studies showed that there is a distinct set of proteins, mostly very Dopa-rich, that is enriched at the interface (Zhao et al. 2006; Zhao and Waite 2006). The preferential staining of the adhesive layer in zebra mussels may also be indicative of the presence of Dopa and/or charged amino acids that bind the heavy metal stains (uranyl acetate and lead citrate). However, even unstained micrographs show a darker layer at the interface, indicating that a higher packing density may also play a role in this layer's increased contrast (Farsad and Sone 2012).

While structural characterization of the plaque-substratum interface emphasizes the importance of direct protein-surface interactions in adhesion, it also raises several questions. How does such a thin, uniform adhesive layer form, given that proteins are secreted into the distal depression of the foot prior to contact with the surface? It is notable in this regard that granules that stained similarly to this layer were observed in regions of the plaque adjacent to the surface (Fig. 6.3a), conceivably as a delivery vehicle driven by phase segregation. Does the nature or thickness of the adhesive layer change over time and/or as a function of surface chemistry of the substratum? Answers to these questions will advance our understanding of the mechanism of zebra mussel adhesion but of primary importance is to determine the composition of this layer.

Characterization of the composition of the adhesive layer poses several challenges: it is only 10–20 nm thick, it is "buried" under the adhesive plaque, and its proteins are cross-linked following secretion, all of which make extraction difficult. To overcome these challenges, we used matrix-assisted laser desorption/ionization (MALDI) mass spectrometry to directly probe the proteins of the adhesive interface in zebra mussels in two ways (Gilbert and Sone 2010) (Fig. 6.4a): (i) we examined the plaque "footprint," the residue left behind on the substratum after shaving off the plaque, as per Zhao et al. (2006), and (ii) we examined the intact plaque underside by first dissolving the substratum to which it has been attached. The latter method results in more reproducible spectra, as removing the plaque by shaving leaves behind variable amounts of interfacial and bulk plaque proteins. Since the MALDI laser penetrates ~200–300 nm into porous proteinaceous materials (Strupat et al. 1994) and the adhesive layer itself is only 10–20 nm thick, it is clear that when examining plaque undersides, both interfacial and bulk proteins are being sampled. Therefore, in order to determine which proteins are unique or enriched at the interface, it is necessary to compare these spectra to those obtained from the bulk of the plaque.

Comparing MALDI spectra from the interface with the bulk of the plaque, it is apparent that there is a set of proteins ranging in mass from 5.8–7 kDa that are present at only the interface (Fig. 6.4). This result supports the notion that the darkly stained adhesive layer observed in TEM is compositionally different from the bulk of the plaque. Potential indications of hydroxylation (shoulder peaks separated by ~16 Da) were found for some of the proteins at the interface, which could be indicative of partial conversion of tyrosine to Dopa (Gilbert and Sone 2010). In quagga mussels, similar experiments showed that although no proteins were uniquely present at the interface, some peaks were relatively much more intense,

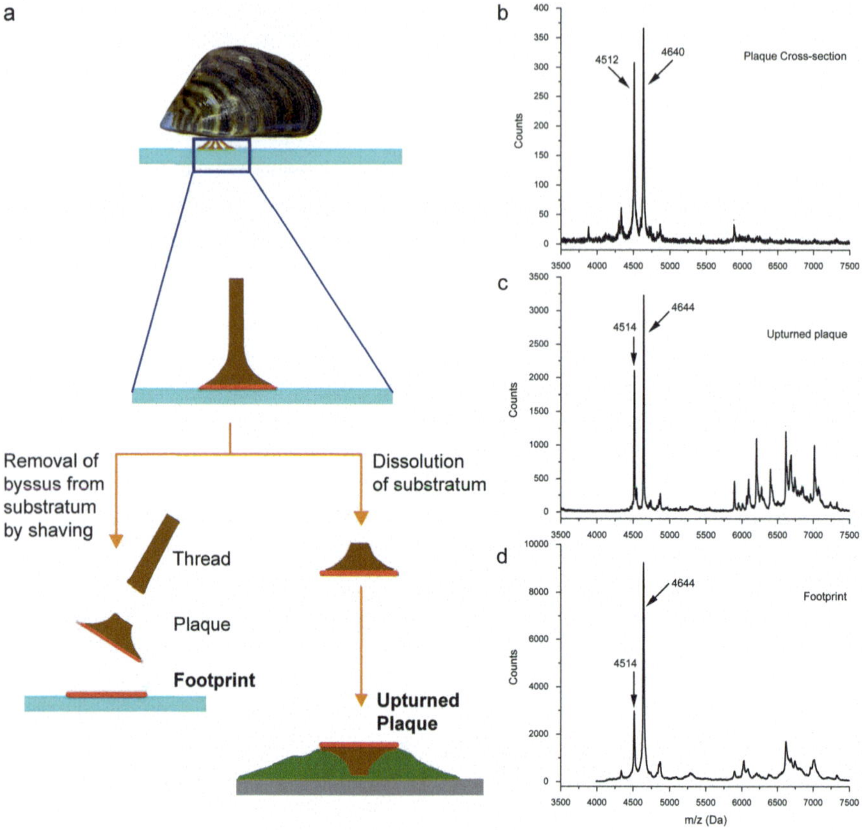


**Fig. 6.4** MALDI-TOF MS characterization of the zebra mussel plaque-substratum interface. (**a**) The interface was characterized by examining the "footprint" or upturned plaques and compared to spectra from cross-sections through the bulk of the plaque. (**b–d**) Both upturned plaque and footprint spectra show a set of peaks between 5.8–7 kDa that is not present in the plaque cross-section, indicating that there is a distinct set of proteins at the interface. Adapted from Gilbert and Sone (2010), with permission from Taylor & Francis (www.tandfonline.com)

suggesting preferential localization to the interface (Rees et al. 2016). Overall these data show that, similar to marine mussels, a distinct set of proteins is responsible for adhesion in zebra and quagga mussels. Unlike marine mussels, however, the sequences of these proteins are not yet known. Efforts in this direction are described in the following section.

### *6.2.2 Characterization of Zebra and Quagga Mussel Byssal Proteins*

The byssus in both marine and freshwater mussels is secreted by a muscular organ called the foot, containing a ventral groove into which thread proteins are deposited and a distal depression from which the plaque is formed (Bonner and Rockhill 1994a) (Figs. 6.5a, b). Because byssal proteins are cross-linked following deposition onto the substratum (Rzepecki and Waite 1993a), they were initially identified as precursor proteins isolated from the secretory granules of the foot, similar to marine mussels. Dopa-specific staining was used to distinguish byssal proteins from other foot proteins, as this unusual amino acid is not found in other mussel tissues. Three Dopa-containing proteins were identified from the zebra mussel foot, and thus named as *Dreissena polymorpha* foot proteins (Dpfp1–3), while four Dopa-containing proteins were identified from the quagga mussel foot (Dbfp0–4) (Rzepecki and Waite 1993b). Dpfp1, which contains 3–7 mol% Dopa, was isolated and sequenced, revealing a primary structure that consists mainly of two tandemly repeated consensus sequences (Anderson and Waite 1998). It was shown to be incorporated into threads based on immunolabeling of thread extracts and is considered most likely to be a component of the cross-linked cuticle that surrounds the byssus, but attempts to localize it in the mature byssus were not successful, presumably due to cross-linking (Anderson and Waite 2000). In quagga mussels, Dbfp1 is the only protein that has been characterized (Anderson and Waite 2002). It contains 0.5–3 mol% Dopa and also has a repetitive sequence, similar to Dpfp1 and also to Mytilus foot protein 1 (Mfp-1). However, the repetitive sequences are not homologous to Mfp-1 or to other marine mussel foot proteins. In addition, Dpfp1 and Dbfp1 are both acidic and glycosylated, in contrast with Mfps, which are basic and generally lack glycosylation.

While identification of byssal proteins from the foot of the mussel circumvents cross-linking, only Dopa-containing proteins can be unambiguously assigned as byssal precursors. Given that the overall levels of Dopa are quite low in zebra and quagga mussels, it would seem likely that there are non-Dopa-containing byssal components that are overlooked by this method. In my lab, we used the technique of KCl injection into the mussel foot, originally developed for marine mussels (Tamarin et al. 1976), to induce secretion of byssal proteins "on demand." The nascent byssal thread and plaque (Fig. 6.5b) can be removed from the foot of the mussel, and proteins can be directly extracted, such that all proteins recovered are byssal components. In both zebra (Gantayet et al. 2013) and quagga mussels (Rees et al. 2016), we showed that the previously identified Dopa-containing proteins represent only a fraction of the extractable byssal proteome (Fig. 6.5c).

Protein sequences for zebra mussels were determined by tryptic digestion of gel bands (Gantayet et al. 2013) or of the insoluble fraction (Gantayet et al. 2014), followed by peptide fragment fingerprinting against a cDNA library of foot (Xu and Faisal 2008). Although no genetic information is available for quagga mussels, my lab is currently following a similar bottom-up proteomic approach, using RNA-seq

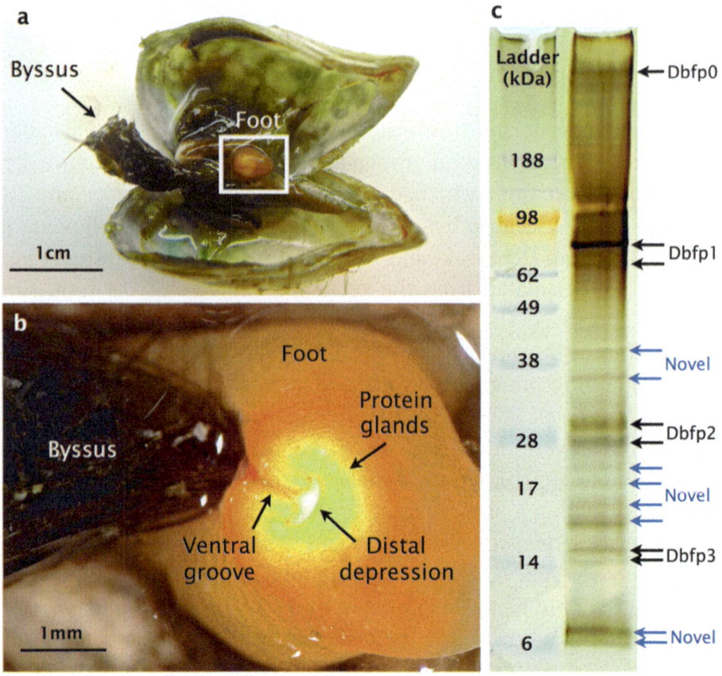


**Fig. 6.5** Protein extraction from freshly secreted byssal material following KCl injection. (**a**) An opened quagga mussel, showing the foot, into which KCl is injected to induce secretion, as well as mature byssal threads. (**b**) Magnification of the foot, showing the ventral groove and distal depression containing freshly secreted thread and plaque proteins, respectively. (**c**) Gel electrophoresis of proteins extracted directly from freshly secreted byssal material (silver-stained), showing both previously known Dopa-containing proteins extracted from the foot (Dbfp0–4) as well as several novel proteins. Reprinted from Rees et al. (2016) with permission from Taylor & Francis (www.tandfonline.com)

to produce a foot transcriptome, as was recently done for the green mussel, *Perna viridis* (Guerette et al. 2013) and other bioadhesive organisms (Hennebert et al. 2015). Our proteomic analysis of the zebra mussel byssus brings the total number of proteins sequenced to ten. The set of Dpfps shows little homology overall to Mfps and more generally tend to have fairly unusual sequences without strong homology to known proteins. However, they do have some characteristics in common with Mfps and adhesive proteins from other species, including multiple variants and low sequence complexity due to the presence of repeated sequences (Lee et al. 2011). Unlike marine mussel proteins that are generally basic, both acidic and basic proteins are found in the zebra mussel. Several proteins display block copolymer-like structure, with regions having distinct sequence repeats and isoelectric points, a feature found also in a sandcastle worm cement protein, Pc3A, which may be related to complex coacervation (Wang and Stewart 2013). Interestingly, many of the repeated sequence motifs are found in multiple zebra mussel byssal proteins: YP and PY diads, GGX triads and other glycine-rich repeats, and cysteine-rich sequences. Although present in different configurations, the presence of a few specific motifs interspersed among the ten proteins suggests that these may play key functional roles.

Although in some cases it is possible to make inferences about protein function from sequence data, bulk proteomic analysis, as was performed on zebra mussels, does not provide any direct functional or location information. Thus, it is not yet known which of these ten byssal proteins function as structural elements of the thread and/or plaque and which form part of the adhesive layer directly on the substratum surface. In order to obtain this information, it will be necessary to extract proteins from the adhesive interface itself, as has been done for marine mussels from plaque "footprints" (Guerette et al. 2013). Identification of the adhesive proteins will enable a mechanistic understanding of zebra mussel adhesion, similar to the detailed analysis that has been done for marine mussels, as described in the following sections. Until then, the key comparative question remains unresolved: is the drastically lower Dopa content of freshwater mussel byssal proteins indicative of a fundamentally different mechanism of adhesion or is it simply that less Dopa is required for effective adhesion in the freshwater environment, where both ionic strength of the water and the forces of waves are generally lower?

## 6.3 Redox Chemistry at the Adhesive Interface

The importance of redox balance for tuning adhesive vs. cohesive interactions of Dopa in marine mussel proteins has been recognized for some time (Waite 2002), due to the difference in behavior of Dopa vs. dopaquinone (Fig. 6.2). Single-molecule AFM experiments showed that while Dopa binds via coordination strongly and reversibly to $TiO_2$, the strength of this interaction is greatly reduced for the oxidized quinone form (Lee et al. 2006). On the other hand, oxidation increases the strength of interaction with amine-functionalized surfaces, presumably through similar oxidative cross-linking mechanisms that occur in the bulk of the byssus. Similarly, the catechol functionality in Dopa strongly coordinates soluble metal ions such as $Fe^{3+}$, leading directly to coordination cross-linking or to oxidative cross-linking in the bulk and possibly with organic surfaces (Wilker 2010). Clearly control over Dopa oxidation state is of utmost importance in balancing its many possible adhesive and cohesive interactions.

At the pH of seawater (8.2) and in the presence of dissolved $O_2$, Dopa in solution is readily oxidized to dopaquinone, which would compromise the ability of Dopa-containing proteins to attach to inorganic substrata. Mfp-3 and Mfp-5, for instance, are both Dopa-rich proteins that have been detected by MALDI mass spectrometry at the adhesive interface (Zhao et al. 2006; Zhao and Waite 2006). They have been shown by surface forces apparatus measurements to be highly adhesive between two mica sheets at acidic pH in vitro, but this adhesion is lost when the pH is raised, due to oxidation of Dopa (Yu et al. 2011b; Danner et al. 2012). How, then, could

Dopa exist in a reduced form in these proteins in seawater in order to maintain their adhesive capacity on inorganic surfaces? This is an important enough problem for mussels that there is recent evidence of at least two major strategies to keep Dopa in reduced form at the adhesive interface: lowered pH during initial secretion and the presence of antioxidant proteins at the interface. Furthermore, it is now apparent that oxidation of Dopa does not necessarily mean loss of the catechol functionality. Other suggested strategies for protection of Dopa from oxidation via shielding by hydrophobic amino acids (Wei et al. 2013) or by temporary $Fe^{3+}$ complexation (Hwang et al. 2010) also bear mentioning but are not covered here.

### *6.3.1 Control of pH During Protein Secretion*

Although the pH of seawater is 8.2, recent measurements have shown that that the mussel foot creates a much more acidic environment into which adhesive proteins are initially secreted while in contact with the substratum. The first evidence for this came from the work of Yu et al. (2011a), who inserted a microelectrode into the distal depression of a *Mytilus californianus* mussel that has been induced to secrete byssal material by KCl injection. They showed that within 2 min of injection, the pH dropped from 7.3 to 5.8. More recently, Rodriguez et al. (2015) were able to measure pH during natural plaque secretion, which eliminates potential artifacts caused by KCl injection. Their elegant approach to measuring interfacial pH involved attaching a pH-sensitive fluorochrome to the substratum onto which the plaque was deposited, such that the measured fluorescence during plaque deposition could be correlated to pH (Fig. 6.6). Predicted pH drops as low as 2.1 following protein secretion and rises to 3.5 just before the foot lifts off from the surface, at which point the plaque pH can equilibrate with seawater. The initial acidic pH allows Dopa to form H-bonds and coordination complexes with surfaces, which protects it from oxidation (Fig. 6.7). Furthermore, the acidic environment leads to

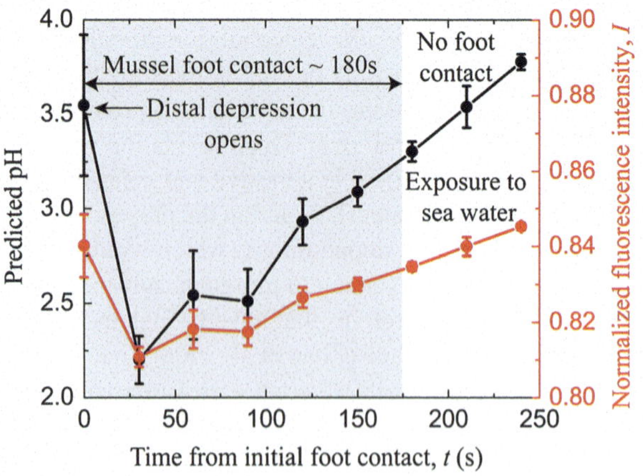


**Fig. 6.6** Correlation of pH (*black*) to normalized fluorescence intensity (*red*), measured at the plaque-substratum interface during plaque deposition by juvenile marine mussels (*Mytilus californianus*). Reprinted from Rodriguez et al. (2015) with permission from Taylor & Francis (www.tandfonline.com)

positive charges on basic amino acids (e.g., lysine) that can mediate electrostatic interactions with negatively charged surfaces. As the plaque pH equilibrates with seawater, uncoordinated Dopa residues would become oxidized and participate in cross-linking. The first direct measurement of interfacial pH during protein

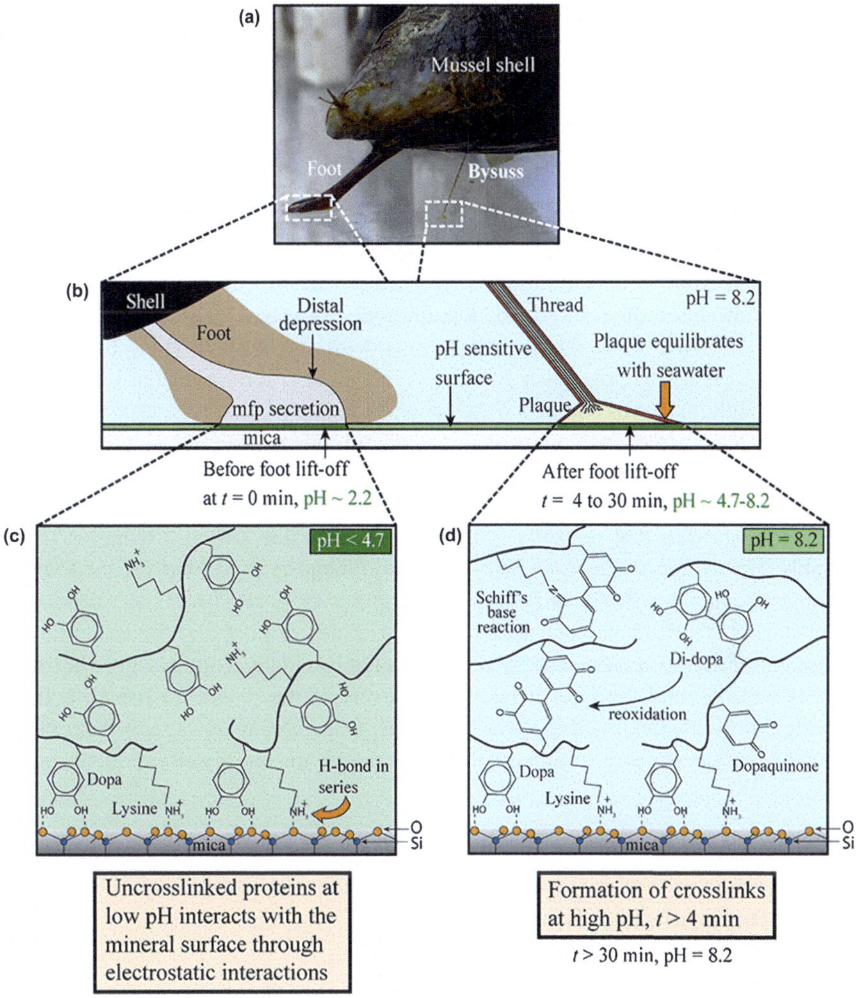


**Fig. 6.7** Control of pH during plaque deposition in marine mussels. During plaque deposition, the pH under the foot in the distal depression where proteins are being secreted is quite acidic (**a**, **b**). This maintains the catechol group of Dopa in reduced form, allowing it to form bidentate hydrogen bonds with the mica surface, while positively charged lysines interact electrostatically (**c**). Following lift-off of the foot, the pH of the plaque equilibrates with seawater (pH = 8.2), leading to oxidation of uncoordinated Dopa to dopaquinone and subsequent cross-linking. Reprinted from Rodriguez et al. (2015) with permission from Taylor & Francis (www.tandfonline.com)

secretion provides strong support for these important mechanistic concepts, many of which were proposed much earlier (Waite 1987).

### 6.3.2 Antioxidant Proteins at the Plaque-Substratum Interface

While the first proteins secreted onto the surface, Mfp-3 and Mfp-5, have very high Dopa content (20 and 30 mol%), Mfp-6, which is secreted shortly after, has much lower Dopa content (4 mol%) and much higher cysteine content (11 mol%) (Zhao and Waite 2006). Of the 11 cysteine residues, between 6 and 11 are present in reduced form (Yu et al. 2011a). Mfp-6 adheres to mica weakly compared to Mfp-3 and Mfp-5 but has recently been shown to enhance the adhesion of Mfp-3 at near neutral conditions, where adhesion is otherwise reduced due to oxidation of non-adsorbed Dopa to dopaquinone. Mfp-6 effectively functions as an antioxidant, whereby thiolates reduce oxidized dopaquinone back to Dopa (and in turn are oxidized), allowing more Dopa interactions with the surface (Fig. 6.8). Over time, Mfp-6 chains having fewer thiolate pairs can also form covalent bonds with Mfp-3 through the formation of 5-*S*-cysteinyldopa cross-links, increasing cohesive strength of the plaque. In support of this appealing idea, Miller et al. (2015) were able to directly measure antioxidant activity at the plaque underside, by placing plaques onto surfaces infused with a free radical redox sensor (DPPH). This clever experiment showed that not only is the plaque underside strongly reducing (presumably due to the presence of Mfp-6, as antioxidant activity was reduced when cysteine residues were blocked), this reducing power is persistent: 30 % remains after 20 days. Thus, Dopa is protected from oxidation not only by reduction of pH during initial deposition but also by secretion of antioxidant proteins whose activity persists well beyond the time it takes for the pH of the plaque to rise to that of seawater. Notwithstanding these two strategies, Dopa chemistry is such that oxidation will not necessarily eliminate the catechol group, as discussed in the next section.

### 6.3.3 Dopaquinone Tautomers

The redox chemistry of Dopa is in fact much more complex than the simplified schematic shown in Fig. 6.2. In the presence of $O_2$ and at neutral pH, the quinone form is not very stable and can undergo tautomerization, i.e., formation of structural isomers in equilibrium (Yu et al. 2011b) (Fig. 6.9). A potential role for tautomers of dopaquinone in cross-linking of byssal proteins was originally explored by Rzepecki et al. (1991) using peptide analogues. Twenty years later, Yu et al. (2011b) provided evidence for α,β-dehydrodopa (Δ-Dopa) formation in

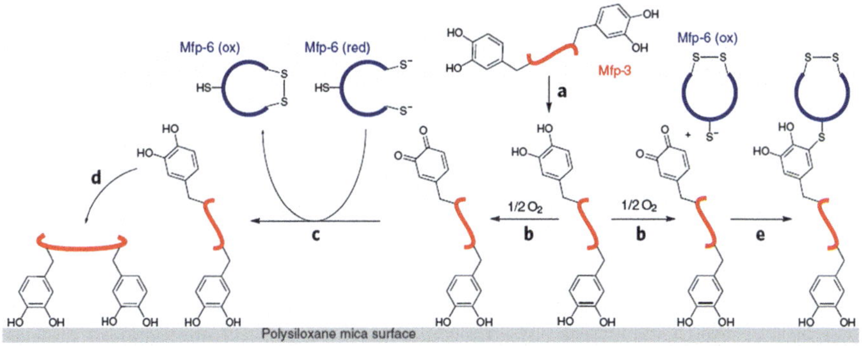


**Fig. 6.8** Schematic illustration of the antioxidant function of Mfp-6 at the plaque-substratum interface. Thiolates in Mfp-6 reduce dopaquinone back to Dopa in Mfp-3 (and in turn are oxidized), allowing more catechol interactions with the surface for Mfp-3. Over time, Mfp-6 proteins with fewer thiolate pairs can participate in covalent cross-linking. Reprinted from Yu et al. (2011a) by permission from Macmillan Publishers Ltd

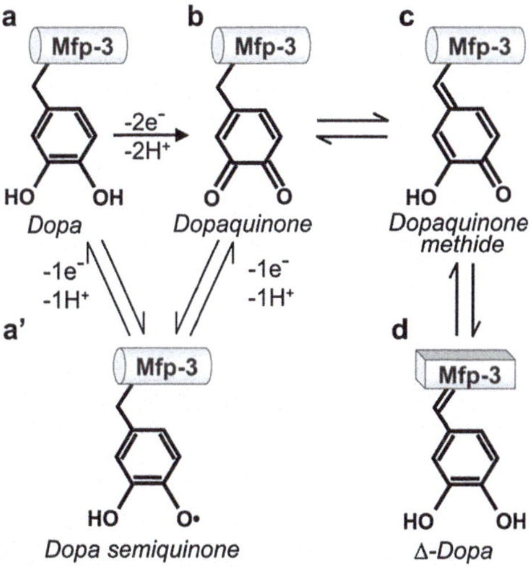


**Fig. 6.9** Potential redox products of protein-bound Dopa in Mfp-3. Two electron oxidation of Dopa (**a**) yield dopaquinone (**b**), while one electron oxidation leads to the semiquinone radical (**a**). Dopaquinone exists in equilibrium its tautomers, dopaquinone methide (**c**) and for α, β-dehydrodopa (Δ-Dopa) (**d**). Reprinted from Yu et al. (2011b) by permission from John Wiley and Sons

Mfp-3 on mica surfaces as pH is increased. The tautomerization results in a reversible swelling of the adsorbed film, as determined by hard wall expansion in surface forces apparatus measurements, due to loss of conformational flexibility. This year, Mirshafian et al. (2016) isolated different products from oxidation of a peptidyl-Dopa analogue and showed that Δ-Dopa is the major oxidation product at pH 7. Both cyclic voltammetry and UV spectroscopy of the peptidyl-Dopa analogue showed remarkable similarity to the same measurements of Mfp-3, providing strong evidence that it is an important species in the byssal proteins. Furthermore, ATR-FTIR spectroscopy showed an increase in proportion of β-sheet structure at pH 7 as compared to pH 4 (where it is primarily random coil), presumably due to formation of Δ-Dopa and its effect on the protein carbon backbone.

It is significant that when pH is increased at the adhesive interface (following release of the mussel's foot from the surface), the catechol moiety of Dopa can be retained through Δ-Dopa formation. The surface interactions of Δ-Dopa have only begun to be explored. Initial quartz crystal microgravimetry experiments show a 20-fold increase in adsorption capacity of the Δ-Dopa peptidyl-analogue on $TiO_2$ as compared to the Dopa analogue (Mirshafian et al. 2016). However, given that the film thickness was also much higher (12 nm vs. 0.6 nm), the increased adsorption capacity may be more reflective of increased intermolecular interactions than direct surface binding. Further investigation of Δ-Dopa interactions is clearly warranted. Moreover, the reported change in secondary structure accompanying the pH change is intriguing, raising the possibility of a role for protein conformation in adhesion. Recent work on the green mussel, *Perna viridis*, combing ATR-FTIR with molecular dynamics simulation, showed that while the plaque proteins are generally disordered, the β-sheet content of Pvfp-5 increased when adsorbed on $TiO_2$ (Petrone et al. 2015). A molecular level picture of structural changes that occur during adsorption, due to both oxidation and surface interactions, will ultimately be needed to fully understand mussel adhesive mechanisms.

## 6.4 The Effect of Surface Chemistry and Mode of Dopa Interaction

Given that Dopa forms strong coordination complexes with a variety of multivalent metal cations (Waite et al. 2005), the potential of Dopa coordination to mediate adhesion to metal oxide surfaces is clear. And indeed, on $TiO_2$ the spacing between Ti centers matches closely with the OH spacing of the catechol group, such that bidentate complexes can form with catechols and to a lesser extent with quinones (Fig. 6.10) (Anderson et al. 2010). On mica, an aluminosilicate mineral that can be easily cleaved to form atomically flat surfaces for surface forces apparatus measurements, catechols interact through H-bonds to O atoms on the surface, meaning oxidation to the quinone obviates this interaction. The spacing between O atoms in mica matches closely with the distance between OH groups in catechols, enabling bidentate binding. On other surfaces where the spacing is not well matched, the strength of this interaction is significantly reduced (Yu et al. 2013a). More recently, it was shown using Raman spectroscopy that the nature of Dopa binding to $TiO_2$ surfaces depends on pH, with coordination complexes forming above pH 7, while

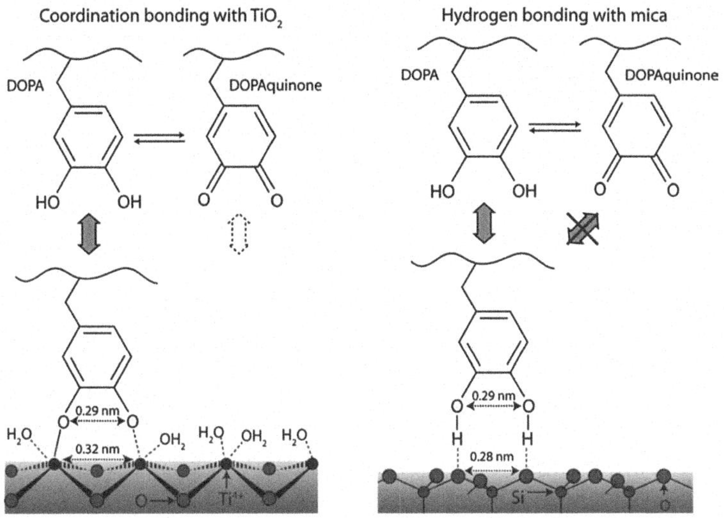


**Fig. 6.10** The spacing between $Ti^{4+}$ centers in $TiO_2$ (*left*) and O on mica (*right*) both match well with the spacing between catechol OH groups. This allows Dopa to form bidentate coordination bonds on $TiO_2$, and H-bonds on mica. Oxidation to dopaquinone results in weaker coordination on $TiO_2$ and eliminates H-bonding on mica. Reprinted from Anderson et al. (2010) by permission from John Wiley and Sons

H-bonded adsorption occurs below pH 5.5 (Yu et al. 2013b). On protein-containing surfaces (or those with similar functionality) on the other hand, oxidative cross-linking reactions similar to those involved in bulk cohesion could link adhesive proteins to the surface (Wilker 2010), as has been exploited in synthetic adhesives (Lee et al. 2011). While these schemes form a reasonable model for mussel adhesion to mineral and natural organic surfaces, mussels can also attach to a wide variety of synthetic surfaces, often where they are not wanted. Clearly, there is more to adhesion than Dopa coordination and H-bonding, which would not be relevant on hydrophobic surfaces such as polystyrene. In recent years, a much wider variety of engineered surfaces have been used to test surface interactions of mussel adhesive proteins, revealing a range of Dopa-surface interactions (described below), as well as the importance of other amino acids (described in Sect. 6.5).

A 2013 study (Lu et al.) was the first to systematically investigate the adhesion of the interfacial proteins Mfp-3 and Mfp-5 (as well as Mfp-1, a cuticle protein) to surfaces with different chemistries: mica, $SiO_2$, polymethyl methacrylate (PMMA), and polystyrene, having water contact angles ranging from 5° to 92°. These were all fabricated onto a mica support, producing very smooth surfaces (root mean square surface roughness of 1 nm or less), which eliminates confounding effects of surface topography. Mfp-3 and Mfp-5 adhere strongly to both hydrophilic and hydrophobic surfaces. The authors suggest that while bidentate H-bonding is the main contributor to adhesion on mica and silica surfaces, on polystyrene cation-π and π-π stacking interactions are most important. Cation-π interactions have been recently implicated in cohesive interactions in Dopa-deficient byssal proteins (Hwang et al. 2012; Kim et al. 2015) and in this case, would occur between benzene rings in polystyrene and charged amino acids in the protein, while π-π interactions would take place between Dopa and surface aromatics. On PMMA, which lacks aromatic residues, it appears that hydrophobic interactions are most important. Clearly, several hydrophobic amino acids in the protein can contribute in this regard (see below for further discussion). However, recent studies on hydrophobic and hydrophilic self-assembled monolayers suggest that Dopa may participate directly in hydrophobic interactions to methyl-terminated surfaces (Yu et al. 2013a). Together, these studies suggest that Dopa may contribute to adhesion to hydrophobic surfaces through π-π, cation-π, and hydrophobic interactions, to an extent that was not previously appreciated.

## 6.5 The Role of Amino Acids Other than Dopa in Adhesion

Surface forces apparatus (SFA) measurements (Israelachvili et al. 2010), in which the forces are measured as two surfaces are brought together at the nanometer scale can be precisely measured, have been pivotal in measuring adhesive (and cohesive) interactions of mussel proteins, while AFM has been used to make single-molecule measurements of Dopa binding to surfaces. These techniques, however, are both limited in that they apply a compressive force and therefore do not accurately reflect the ability of a protein to adhere in the absence of external forces. This is particularly relevant in the context of the ability of a protein to displace surface-bound water, a critical challenge in wet adhesion. A recent study by Akdogan et al. (2014) applied a spectroscopic approach termed Overhauser dynamic nuclear polarization (ODNP) relaxometry to address this question. By measuring the change in diffusion of water localized to different surfaces (spin-labeled silica and polystyrene nanobeads) upon protein adsorption, they were able to show that Mfp-5 did not achieve intimate contact with the polystyrene surface in acidic conditions, regardless of whether it had Dopa residues. Only adsorption of Mfp-3 slow (Mfp-3S), a hydrophobic variant of the Mfp-3 family, resulted in significant retardation of surface water diffusion, indicating it was able to displace waters of hydration from the surface. Apparently, hydrophobic residues are more important in this process than Dopa, and the hydrophobic Mfp-3S may play a critical role in drying the surface to allow adhesion to take place. A similar surface-priming role has been suggested for Pvfp-5, which was shown by ATR-IR to displace surface-bound water from $TiO_2$ (Petrone et al. 2015).

Along with an increased appreciation for the role of hydrophobic interactions in adhesion, the importance of electrostatic forces has also been recently brought to the fore. Wei et al. (2015) synthesized short Dopa-containing peptides, each corresponding to a different part of Mfp-5 and systematically tested their interactions with mica by SFA measurements in conditions where different peptides carried either neutral, positive, or negative charge. While increase in pH or injection of an oxidizing agent reduced adhesion of all peptides due to oxidation of Dopa, the effect was much less pronounced for positively charged peptides, which could interact electrostatically with the negatively charged mica surface. Furthermore, analogous peptides without Dopa followed a similar trend in adhesion energy to mica as a function of pH, albeit at lower values than their Dopa-containing counterparts. The authors suggest that such nonspecific, long-range electrostatic interactions of mussel proteins with biological surfaces may initiate interaction with the surface in adhesion, followed by more specific Dopa-mediated interactions.

While electrostatic interactions are important on their own, Maier et al. (2015) showed that a synergistic effect between lysine, a positively charged amino acid, and the catechol functionality of Dopa improves adhesion by displacing hydrated salt ions from the surface. Mfp-3 and Mfp-5 both contain abundant Lys residues, often in close proximity to Dopa. Although the mechanistic importance of this pairing was not yet understood, nearly 35 years ago, mussel-mimetic polymers incorporating this motif were first synthesized (Yamamoto and Hayakawa 1982). Recently, Maier et al. used SFA measurements to test the adhesion to mica of siderophores, bacterial iron chelators that also contain lysine and catechol, along with a series of synthetic analogues. They showed that siderophores displace surface-bound hydrated $K^+$ ions from the mica surface, greatly increasing the adhesion between two mica surfaces (~30 fold). By systematically varying the chemical functionality on a synthetic siderophore analogue, they showed that both Lys and catechol functionality are required for this effect, providing a compelling insight into the function of co-localization of Lys and Dopa in mussel adhesive proteins.

## 6.6 Concluding Remarks

It is remarkable that such a well-studied example of biological adhesion as byssal adhesion in marine mussels continues to provide new insights into the mechanism of underwater adhesion. This is both a testament to the beauty and complexity of the biological system and to the confluence of characterization techniques that have enabled new kinds of measurements. The use of the highly sensitive techniques to measure protein-surface interactions such as SFA and AFM has played a critical role in advancing our understanding of mussel adhesion over the last decade. More recently, the ability to make direct interfacial measurements (e.g., pH, antioxidant activity, protein secondary structure), along with the use of more varied surfaces and increasingly sophisticated protein-mimics, have accelerated the pace of discovery. In freshwater mussels, the field is much more nascent, but the careful characterization of the adhesive interface and the identification of new byssal proteins reveal several similarities but also intriguing differences from marine mussels. With the now ready availability of proteomic and transcriptomic techniques, new organisms and proteins are more accessible than ever before, promising new discoveries and insights in the years to come.

**Acknowledgments** I am extremely grateful to a number of talented postdoctoral, graduate, and undergraduate students in my lab, past and present, who have contributed to research efforts in this area. Our work is supported by the National Sciences and Engineering Research Council (NSERC) of Canada and the Canadian Foundation for Innovation (CFI).

## References

Akdogan Y, Wei W, Huang KY, Kageyama Y, Danner EW, Miller DR, Rodriguez NRM, Waite JH, Han S (2014) Intrinsic surface-drying properties of bioadhesive proteins. Angew Chem Int Ed 53(42):11253–11256

Anderson KE, Waite JH (1998) A major protein precursor of zebra mussel (*Dreissena polymorpha*) byssus: deduced sequence and significance. Biol Bull 194(2):150–160

Anderson KE, Waite JH (2000) Immunolocalization of Dpfp1, a byssal protein of the zebra mussel *Dreissena polymorpha*. J Exp Biol 203(20):3065–3076

Anderson KE, Waite JH (2002) Biochemical characterization of a byssal protein from *Dreissena bugensis* (Andrusov). Biofouling 18(1):37–45

Anderson TH, Yu J, Estrada A, Hammer MU, Waite JH, Israelachvili JN (2010) The contribution of DOPA to substrate-peptide adhesion and internal cohesion of mussel-inspired synthetic peptide films. Adv Funct Mater 20(23):4196–4205

Bairati A, Zuccarello LV (1976) Ultrastructure of the byssal apparatus of *Mytilus galloprovincialis* 4. Observations by transmission electron-microscopy. Cell Tissue Res 166(2):219–234

Bandara N, Zeng H, Wu J (2013) Marine mussel adhesion: biochemistry, mechanisms, and biomimetics. J Adhes Sci Technol 27(18–19):2139–2162

Benedict CV, Waite JH (1986) Composition and ultrastructure of the byssus of *Mytilus edulis*. J Morphol 189(3):261–270

Bonner TP, Rockhill RL (1994a) Functional morphology of the Zebra Mussel Byssus Dreissena polymorpha. Inform Rev 5:3

Bonner TP, Rockhill RL (1994b) Ultrastructure of the Byssus of the Zebra Mussel (Dreissena-Polymorpha, Mollusca, Bivalvia). Trans Am Microsc Soc 113(3):302–315

Burzio LA, Waite JH (2000) Crosslinking in adhesive quinoproteins: studies with model decapeptides. Biochemistry 39(36):11147–11153

Connely NA, O'Neill CR Jr, Knuth BA, Brown TL (2007) Economic impacts of zebra mussels on drinking water treatment and electric power generation facilities. Environ Manag 40:150–112

Danner EW, Kan YJ, Hammer MU, Israelachvili JN, Waite JH (2012) Adhesion of mussel foot protein Mefp-5 to mica: an underwater superglue. Biochemistry 51(33):6511–6518

Eckroat L (1993) The Byssus of the Zebra Mussel (Dreissena Polymorpha): morphology, byssal thread formation, and detachment. In: Nalepa TF, Schloesser DW (eds) Zebra Mussels: biology, impacts, and control. CRC Press, Boca Raton, pp 239–263

Farsad N, Sone ED (2012) Zebra mussel adhesion: structure of the byssal adhesive apparatus in the freshwater mussel, *Dreissena polymorpha*. J Struct Biol 177(3):613–620

Gantayet A, Ohana L, Sone ED (2013) Byssal proteins of the freshwater zebra mussel, *Dreissena polymorpha*. Biofouling 29(1):77–85

Gantayet A, Rees DJ, Sone ED (2014) Novel proteins identified in the insoluble byssal matrix of the freshwater Zebra Mussel. Mar Biotechnol 16(2):144–155

Gilbert T, Sone E (2010) The byssus of the zebra mussel (*Dreissena polymorpha*): spatial variations in protein composition. Biofouling 26:829–836

Guerette PA, Hoon S, Seow Y, Raida M, Masic A, Wong FT, Ho VHB, Kong KW, Demirel MC, Pena-Francesch A, Amini S, Tay GZ, Ding D, Miserez A (2013) Accelerating the design of biomimetic materials by integrating RNA-seq with proteomics and materials science. Nat Biotechnol 31(10):908–915

Harrington MJ, Waite JH (2009) How nature modulates a fiber's mechanical properties: mechanically distinct fibers drawn from natural mesogenic block copolymer variants. Adv Mater 21 (4):440–444

Harrington MJ, Masic A, Holten-Andersen N, Waite JH, Fratzl P (2010) Iron-clad fibers: a metal-based biological strategy for hard flexible coatings. Science 328(5975):216–220

Hennebert E, Maldonado B, Ladurner P, Flammang P, Santos R (2015) Experimental strategies for the identification and characterization of adhesive proteins in animals: a review. Interface Focus 5:20140064, http://dx.doi.org/20140010.20141098/rsfs.20142014.20140064

Hwang DS, Zeng HB, Masic A, Harrington MJ, Israelachvili JN, Waite JH (2010) Protein- and metal-dependent Interactions of a prominent protein in mussel adhesive plaques. J Biol Chem 285(33):25850–25858

Hwang DS, Zeng H, Lu Q, Israelachvili J, Waite JH (2012) Adhesion mechanism in a DOPA-deficient foot protein from green mussels. Soft Matter 8(20):5640–5648

Hwang DS, Wei W, Rodriguez-Martinez NR, Danner E, Waite JH (2013) A microcosm of wet adhesion: dissection protein interactions in mussel attachment plaques. In: Zeng H (ed) Polymer adhesion, friction, and lubrication. Wiley, Hoboken, pp 319–349

Israelachvili J, Min Y, Akbulut M, Alig A, Carver G, Greene W, Kristiansen K, Meyer E, Pesika N, Rosenberg K, Zeng H (2010) Recent advances in the surface forces apparatus (SFA) technique. Rep Prog Phys 73(3)

Kim S, Faghihnejad A, Lee Y, Jho Y, Zeng H, Hwang DS (2015) Cation-pi interaction in DOPA-deficient mussel adhesive protein mfp-1. J Mater Chem B 3(5):738–743

Lee H, Scherer NF, Messersmith PB (2006) Single-molecule mechanics of mussel adhesion. Proc Natl Acad Sci USA 103(35):12999–13003

Lee BP, Messersmith PM, Israelachvili JN, Waite JH (2011) Mussel-inspired adhesives and coatings. Annu Rev Mater Res 41:99–132

Li L, Zeng H (2016) Marine mussel adhesion and bio-inspired wet adhesives. Biotribology 5:44–51. doi:10.1016/j.biotri.2015.1009.1004

Lu QY, Danner E, Waite JH, Israelachvili JN, Zeng HB, Hwang DS (2013) Adhesion of mussel foot proteins to different substrate surfaces. J R Soc Interface 10:20120759, http://dx.doi.org/20120710.20121098/rsif.20122012.20120759

Maier GP, Rapp MV, Waite JH, Israelachvili JN, Butler A (2015) Adaptive synergy between catechol and lysine promotes wet adhesion by surface salt displacement. Science 349 (6248):628–632

Miller DR, Spahn JE, Waite JH (2015) The staying power of adhesion-associated antioxidant activity in Mytilus californianus. J R Soc Interface 12(111):10.1098/rsif.2015.0614

Mirshafian R, Wei W, Israelachvili JN, Waite JH (2016) α, β-Dehydro-dopa. A hidden participant in mussel adhesion. Biochemistry 55(5):743–750

Morton B (1993) The anatomy of Dreissena polymorpha and the evolution and success of the heteromyarian form in the Dreissenoidea. In: Nalepa TF, Schloesser D (eds) Zebra mussels—biology, impacts, and control. Lewis Publishers, Boca Raton, pp 185–215

Nicklisch SC, Waite JH (2012) Mini-review: the role of redox in Dopa-mediated marine adhesion. Biofouling 28(8):865–877

Petrone L, Kumar A, Sutanto CN, Patil NJ, Kannan S, Palaniappan A, Amini S, Zappone B, Verma C, Miserez A (2015) Mussel adhesion is dictated by time-regulated secretion and molecular conformation of mussel adhesive proteins. Nat Commun 6:8737. doi:10.1038/ncomms9737

Rees DJ, Hanifi A, Manion J, Gantayet A, Sone ED (2016) Spatial distribution of proteins in the quagga mussel adhesive apparatus. Biofouling 32(2):205–213

Rodriguez NRM, Das S, Kaufman Y, Israelachvili JN, Waite JH (2015) Interfacial pH during mussel adhesive plaque formation. Biofouling 31(2):221–227

Rzepecki LM, Waite JH (1993a) The byssus of the zebra mussel, *Dreissena polymorpha*. I: Morphology and in situ protein processing during maturation. Mol Mar Biol Biotechnol 2 (5):255–266

Rzepecki LM, Waite JH (1993b) The byssus of the zebra mussel, *Dreissena polymorpha*. II: Structure and polymorphism of byssal polyphenolic protein families. Mol Mar Biol Biotechnol 2(5):267–279

Rzepecki LM, Nagafuchi T, Waite JH (1991) α, β-Dehydro-3.4-dihydroxyphenylalanine derivates: potential sclerotization intermediates in natural composite materials. Arch Biochem Biophys 285:27–36

Sagert J, Sun C, Waite JH (2006) Chemical subtleties of the mussel and polychaete holdfasts. In: Smith AM, Callow JA (eds) Biological adhesives. Springer International Publishing, Germany, pp 125–143

Sever MJ, Weisser JT, Monahan J, Srinivasan S, Wilker JJ (2004) Metal-mediated crosslinking in the generation of a marine-mussel adhesive. Angew Chem Int Ed 43(4):448–450

Silverman HG, Roberto FF (2007) Understanding marine mussel adhesion. Mar Biotechnol 9 (6):661–681

Silverman HG, Roberto FF (2010) Byssus formation in *Mytilus*. In: von Byern J, Grunwald I (eds) Biological adhesive systems. Springer, Austria, pp 273–282

Sousa R, Novais A, Costa R, Strayer D (2014) Invasive bivalves in fresh waters: impacts from individuals to ecosystems and possible control strategies. Hydrobiologia 735:233–251

Stewart RJ, Ransom TC, Hlady V (2011) Natural Underwater Adhesives. J Polym Sci B Polym Phys 49(11):757–771

Strayer DL (2009) Twenty years of zebra mussels: lessons from the mollusk that made headlines. Front Ecol Environ 7(3):135–141

Strupat K, Karas M, Hillenkamp F, Eckerskorn C, Lottspeich F (1994) Matrix-assisted laser-desorption ionization mass-spectrometry of proteins electroblotted after polyacrylamide-gel electrophoresis. Anal Chem 66(4):464–470

Tamarin A, Lewis P, Askey J (1976) Structure and formation of the byssus attachment plaque in Mytilus. J Morph 149:199–221

Waite JH (1986) Mussel glue from Mytilus-Californianus conrad—a comparative-study. J Comp Physiol B Biochem Syst Environm Phys 156(4):491–496

Waite JH (1987) Nature's underwater adhesive specialist. Int J Adhes Adhes 7(1):9–14

Waite JH (2002) Adhesion a la moule. Integr Comp Biol 42(6):1172–1180

Waite JH, Lichtenegger HC, Stucky GD, Hansma P (2004) Exploring molecular and mechanical gradients in structural bioscaffolds. Biochemistry 43(24):7653–7662

Waite JH, Andersen NH, Jewhurst S, Sun CJ (2005) Mussel adhesion: finding the tricks worth mimicking. J Adhes 81(3-4):297–317

Wang CS, Stewart RJ (2013) Multipart copolyelectrolyte adhesive of the sandcastle worm, *Phragmatopoma californica* (Fewkes): catechol oxidase catalyzed curing through peptidyl-dOPA. Biomacromolecules 14:1607–1617

Wei W, Yu J, Broomell C, Israelachvili JN, Waite JH (2013) Hydrophobic enhancement of Dopa-mediated adhesion in a mussel foot protein. J Am Chem Soc 135(1):377–383

Wei W, Tan YP, Rodriguez NRM, Yu J, Israelachvili JN, Waite JH (2014) A mussel-derived one component adhesive coacervate. Acta Biomater 10(4):1663–1670

Wei W, Yu J, Gebbie MA, Tan YP, Rodriguez NRM, Israelachvili JN, Waite JH (2015) Bridging adhesion of mussel-inspired peptides: role of charge, chain length, and surface type. Langmuir 31(3):1105–1112

Wilker JJ (2010) Marine bioinorganic materials: mussels pumping iron. Curr Opin Chem Biol 14 (2):276–283

Xu W, Faisal M (2008) Putative identification of expressed genes associated with attachment of the zebra mussel (*Dreissena polymorpha*). Biofouling 24(3):157–161

Yamamoto H, Hayakawa T (1982) Synthesis of sequential polypeptides containing L-β-3, 4-dihydroxyphenyl-α-alanine (DOPA) and L-lysine. Biopolymers 21:1137–1151

Yang J, Stuart MAC, Kamperman M (2014) Jack of all trades: versatile catechol crosslinking mechanisms. Chem Soc Rev 43(24):8271–8298

Yu J, Wei W, Danner E, Ashley RK, Israelachvili JN, Waite JH (2011a) Mussel protein adhesion depends on interprotein thiol-mediated redox modulation. Nat Chem Biol 7(9):588–590

Yu J, Wei W, Danner E, Israelachvili JN, Waite JH (2011b) Effects of interfacial redox in mussel adhesive protein films on mica. Adv Mater 23(20):2362–2366

Yu J, Kan YJ, Rapp M, Danner E, Wei W, Das S, Miller DR, Chen YF, Waite JH, Israelachvili JN (2013a) Adaptive hydrophobic and hydrophilic interactions of mussel foot proteins with organic thin films. Proc Natl Acad Sci USA 110(39):15680–15685

Yu J, Wei W, Menyo MS, Masic A, Waite JH, Israelachvili JN (2013b) Adhesion of mussel foot protein-3 to TiO2 Surfaces: the effect of pH. Biomacromolecules 14(4):1072–1077

Zeng H, Lu Q, Yan B, Huang J, Li L, Liao Z (2015) Mussel Adhesives. In: Bianco-Peled H, Davidovich-Pinhas M (eds) Bioadhesion and biomimetics: from nature to applications. Pan Stanford Publishing, Singapore, pp 49–84

Zhao H, Waite JH (2006) Linking adhesive and structural proteins in the attachment plaque of Mytilus californianus. J Biol Chem 281(36):26150–26158

Zhao H, Robertson NB, Jewhurst SA, Waite JH (2006) Probing the adhesive footprints of *Mytilus californianus* byssus. J Biol Chem 281(16):11090–11096

# Chapter 7
# Barnacle Underwater Attachment

**Kei Kamino**

**Abstract** The barnacle, infraclass Cirripedia, is the only sessile crustacean. The adult firmly attaches its base to a foreign surface in the water via an underwater adhesive called cement. The multi-protein complex handles the multifunctionality of this underwater attachment, which is based on a different design from those of man-made adhesives in chemistry, structures, processing, and physics. The chemical structures and chemistry are actually substantially different from those of two other models, mussel byssus and tube-dwelling worm cement. In particular, barnacle adhesion is a physiological complex of events involved with molting, epicuticular membrane development, calcification of the shell, and secretion of the underwater adhesive. Thus, the molecular mechanism of the adhesion should be a result balanced on the complex physiology of the animal. This chapter summarizes barnacle underwater attachment and the adhesive. Perspectives in material science are also discussed.

## 7.1 Introduction

Biological localization at an interface is ubiquitous. An organism at an interface could benefit from access to resources, protection from predators, and improved gene transfer. Many aquatic organisms, therefore, specialize in localization at a liquid–solid interface and to do so utilize a specific molecular system for underwater attachment. Among them, the mussel, a mollusk; the sandcastle worm, an annelid; and the barnacle, a crustacean, are models for molecular research on underwater fixation by a permanent adhesive (Kamino 2010).

These model organisms employ a proteinaceous substance for underwater attachment, which forms a multi-protein complex. The biochemical structure of each is quite different though. Work on barnacle underwater cement has

K. Kamino (✉)
National Institute of Technology and Evaluation, and Research Institute for Science and Technology, Tokyo University of Science, Kisarazu, Chiba 292-0818, Japan
e-mail: kamino-kei@nite.go.jp

© Springer International Publishing Switzerland 2016

A.M. Smith (ed.), *Biological Adhesives*, DOI 10.1007/978-3-319-46082-6_7

characterized its multifunctionality and enabled a model for the molecular events to be proposed (Kamino 2013). The chemical structures and chemistry are actually substantially different from those of the other two models. Modes of attachment are also different among three models. In particular, barnacle adhesion is a physiological complex of events involved with molting, epicuticular membrane development, calcification of the shell, and secretion of the underwater adhesive. Thus, the view that is emerging in the last decade is that the molecular mechanism of the adhesion should be a result balanced on the complex physiology of the animal.

Identification of the multifunctionality in underwater attachment might be useful for protein-/peptide-based material research (Sarikaya et al. 2003; Zhang 2003) and may also allow unknown functions. Barnacle studies have especially indicated the significance of intermolecular non-covalent interaction (Kamino et al. 2012; Nakano and Kamino 2015), which is a likely trend in material science, i.e., supramolecular chemistry (Whitesides and Boncheva 2002). This chapter summarizes barnacle underwater attachment and the adhesive. Perspectives in material science are also discussed.

## 7.2 Barnacle Attachment

### *7.2.1 A Unique Sessile Crustacean*

The barnacle, infraclass Cirripedia, is the only sessile crustacean (Fig. 7.1). In the adult barnacle with acorn morphology, the soft body is armored with calcareous shell plates which are generally composed of four to six peripheral plates, four opercular valves, and a base plate (Fig. 7.2). All the shell plates are tightly pulled by a retractor muscle; thereby, they are intimately connected to each other, making the barnacle look like a "pot." The adult firmly attaches its base to a foreign surface in the water via an underwater adhesive called "cement" (Walker 1972). Once settled/attached in the last larval stage, neither the metamorphosed juvenile nor adult moves or self-detaches. This adult cement permanently bonds together two

**Fig. 7.1** Acorn barnacles attached on a synthetic polymer in water

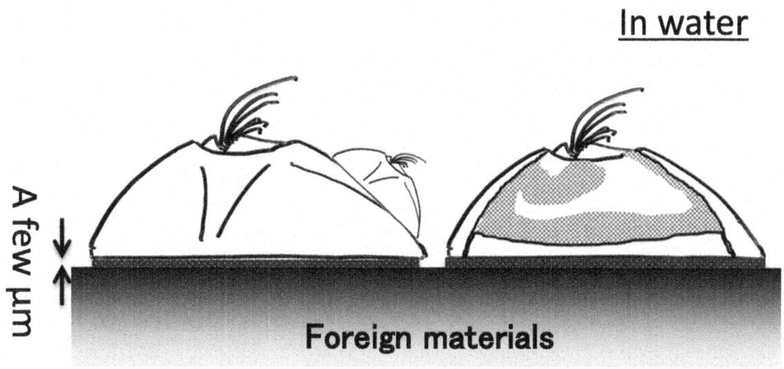


**Fig. 7.2** Schematic side view illustration of an acorn barnacle on a substratum. Barnacles are attached on foreign surfaces with the thin adhesive layer of a thickness of a few microns. The animal firmly fixes its base shell to a foreign material surface in water. The adlayer is noncellular proteinaceous material, which is biosynthesized in the cement gland and is secreted via the duct system. The *right* hand illustration indicates the cross section

different extraorganismic materials, i.e., the base and the foreign material, in water. Cirripedia are, in general, grouped into balanomorph (or acorn type) and pedunculate (or stalk type) by their morphology, the former including animals with calcareous bases and those with membranous bases. In the case of the acorn morphology with calcareous base, the cement attaches their base shell to foreign materials, while both acorn barnacles with membranous base and stalked barnacles attach their chitinous membranous base to foreign materials via the cement. The attachment strength of the cement is remarkably high, although an accurate survey of the strength on a naturally occurring substratum is usually difficult, because the body structure is destroyed by any attempt to dislodge it. The attachment strength of a small adult with a membranous base has been determined to be $9.252 \times 10^5$ ($Nm^{-2}$) (Yule and Walker 1984a), though this might involve tearing the soft body structure off. The adhesive strength of the adult cement between a calcareous base and a natural surface is much more than those values. This present review is concerned with the calcareous base acorn barnacle unless otherwise noted, because most information on the adhesive and adhesion has come from this species at present. This might be because they are a major fouling species on man-made surfaces in the marine environment.

### 7.2.2 *Attachment in the Life Cycle and Biosynthesis/ Secretion of the Cement*

The barnacle's free-swimming nauplius, which is the typical larval stage in crustaceans, metamorphoses into the cypris, which is the specific larval stage of the Cirripedia. The cypris is specialized to explore the settlement site to inhabit for their life-span (Aldred and Clare 2008). Because the selection of the settlement site is crucial for sessile animals, the selection criteria, including foreign surface properties, aeration, nutritional supply, and feeding in the environment, are important aspects that decide their fate (Zardus et al. 2008; Aldred et al. 2011; Pagett et al. 2012).

In exploring for a settlement site (Maleschlijski et al. 2015; Aldred et al. 2013), the larvae appear as if walking on stilts on underwater surfaces of natural, man-made, and living materials; a pair of antennules corresponds to the stilts. This behavior results in the cypris larva secreting a temporary adhesive, from the antennules, leaving footprints on the surface (Walker and Yule 1984), although its exact contribution to adhesion is currently unknown (Aldred and Clare 2008). Once the larva has settled, it undergoes metamorphosis and initiates the calcification process to form the shell plates (Maruzzo et al. 2012). The juvenile develops into an adult with repetitive molting and calcification and starts cementing their base.

There are, therefore, three attachment scenes in the barnacle life cycle: temporary attachment of the cyprid to allow it to explore underwater foreign surfaces, fixation or embedment of the tip of the antennules on the surface with cyprid cement secreted only once in the settlement, and permanent attachment of the adult by the adult cement (Walker 1981; Kamino 2013). Cyprid cement is secreted from a pair of larval cement glands (Walker 1971; Walker and Yule 1984; Yule and Walker 1984b; Okano et al. 1996; Ödling et al. 2006). It passes through the collecting canal, to the duct, and on to the adhesive disk, whereby the adhesive disk and fourth segment of the antennules are embedded by the cypris cement onto the foreign substratum. Histological study has suggested that adult cement glands start to secrete adult cement approximately 40 days later (Yule and Walker 1987), although recent in situ observation indicated an earlier start of the secretion (Burden et al. 2014). Cypris cement and adult cement have been thought to be different substances. However, occurrence of homologous genes in the cyprid transcriptome (Chen et al. 2011) may indicate that a portion of proteins are shared, though it might need further confirmation. This is discussed in a later section (Sect. 3.5.3). In this chapter, "barnacle" and "cement" refer to "adult" and "adult cement," respectively, unless otherwise stated.

Histological studies of the adults (Lacombe 1970; Saroyan et al. 1970; Walker 1970; Fyhn and Costlow 1976) have identified possible cement glands to be giant cells (200–300 μm in diameter) which are laid on connective tissue in close association with ovarian tissue. These are joined together by ducts. In the stalked barnacle, similar large single-cell glands (70–180 μm) have been identified (Jonker et al. 2012). The hypothesis that the histologically identified gland biosynthesizes the cement proteins is supported by northern blotting data. This has indicated that the cement protein genes (Sect. 3.5) are only expressed in the basal portion, where the cement gland is located, and not in the upper portion, where almost all the viscera are located (Kamino et al. 2000, 2012; Urushida et al. 2007; Mori et al. 2007). Both portions include body fluids, so expression in the body fluid cells is negligible. Immunohistochemical study with polyclonal antibodies of some cement proteins conclusively indicated that the gland biosynthesizes the cement proteins (Jonker et al. 2014).

The ducts from the adult cement glands eventually lead to the attachment site in the base of the animal. A physical pushing movement to secrete the cement out has been suggested by the occurrence of a node-like structure at the joint of the duct (Saroyan et al. 1970). It has occasionally been observed that double or triple "dumpling"-like secondary cement secretion (Sect. 3.2) was piled up on the barnacle calcareous base. This observation probably indicates that the first secretion is puffed out by the second secretion after rapid hardening of the first secretion, supporting the occurrence of an external pressure to push the cement secretion out.

While the inward soft tissue of the animal grows with repetitive molting, the peripheral and base plates grow by epicuticle membrane development and calcification. The enlarged marginal area of the calcareous base must be further fixed to the substratum, so it is assumed that the timing of cement secretion and/or gene expression is tuned to the process. The growing edge of the base of the animal on a glass substratum was visualized from the reverse side by using autofluorescence of biological materials with confocal fluorescent microscopy (Burden et al. 2014). This allowed in situ observation of biological events in the growing area, thus allowing a glimpse of precisely controlled processes. These data suggested that the secretion of the cement from the duct is not a single event, but is a complicated process. The study would be a future signpost for unraveling the physiology and molecular events, while it should be remembered that it was unclear what was visualized, because the contrast was given by unspecified autofluorescence. Specific in-depth characterization of each molecular event would open the path to novel findings at the molecular level to micro-/mesoscopic level.

A histological study (Fyhn and Costlow 1976) has confirmed that a positive region of nucleic acid staining in the cement gland cell, i.e., a possible biosynthesis region, became enlarged immediately after molting and subsequently reduced in size. It was also confirmed by northern blotting analysis that the expression level of cement protein genes (Sect. 3.5) in a barnacle that has naturally attached to the substratum was higher after molting, before subsequently dropping to the base level (Urushida Y, Yoshida M, Kado R, Kamino K, unpubl.). These results gave an assumption that the biosynthesis of cement proteins and molting are under the same regulation, though a transcriptome based on RNA-seq technology did not follow the speculation (Wang et al. 2015). Facts of individual differences in the secondary cement secretion (Saroyan et al. 1970), as well as temporal interruption in their growth by environmental factors (Burden et al. 2014), may suggest that the contradiction comes from perturbation of their physiology by environmental factors. Multangular investigation would need to continue to unravel the physiology in growth, molting, cementing, and calcification.

## 7.3 Barnacle Underwater Cement

### 7.3.1 Adhesive Layer of the Cement

The cement between the calcareous base and foreign substratum is a transparent rigid layer approximately a few μm in thickness (Saroyan et al. 1970; Raman and Kumar 2011). The cement layer, however, formed on a man-made foul-release coating is much thicker than that mentioned above. Foul-release coatings, typically poly(dimethylsiloxane) [PDMS], have been developed to reduce the biological adhesion substantially (Griffith and Bultman 1980; Baier and Meyer 1992; Swain and Schultz 1996; Watermann et al. 1997). The adlayer on a foul-release coating is a white rubbery mass, reaching up to 1 mm thickness. The formation of this thick cement layer has been explained by Weigemann and Watermann (2003). Briefly, at the time of cement secretion, the retractor muscle pulls the peripheral shell plate downward. Since the fixation of the base shell plate to the antifouling coating is much weaker than to the natural substratum, this animal's pulling action partially separates the base plate from the coating. The cement permeates to either the newly formed interspace or to the separated old cement layer, with repeated cementing making the cement layer thicker. In the case of attachment to a foul-release coating, the animal has been dislodged from the substratum with a force of $0.1–0.3 \times 10^5$ ($Nm^{-2}$) (Watermann et al. 1997; Berglin et al. 2001; Kavanagh et al. 2003).

Stalked barnacles attach to foreign surfaces by cementing their base of organic tegument. The natural cement adlayer is a rubbery, thicker mass (Jonker et al. 2014), and the appearance of the normal cement of stalked barnacle might be similar to those of the cement of acorn barnacles on a foul-release coating mentioned above and of the secondary cement mentioned in the next section.

The cement layer seems not to have a macroscopic type of segment structure like the mussel holdfast (i.e., the byssus) (Sect. 4), but instead has a microscopic structure. Several studies have tried to identify the microscopic structure of the cement (Kamino 2013). In the investigation, there are two opposite aims, i.e., to improve the performance of foul-release coatings by seeing how barnacle adhesion fails and to learn more about the natural system by observing how adhesion succeeds. For the former, studies usually observe the cement on a "nonstick" surface such as silicone, while for the latter, studies need to observe the cement on "easy-to-attach-to" surface such as natural minerals/metals. The swollen cement adlayer of a barnacle attached to a "nonstick" surface is easy to observe after dislodgement from the substratum and is formed by a loose network of hydrated adhesive threads in the microscopic structure (Burden et al. 2012). Identification of the microscopic structure of the cement on "easy-to-attach-to" surfaces is, however, much more difficult in the access of the intact adlayer for observation with higher magnification/resolution (Kamino 2013). In the stalked barnacle, the intact microscopic structure of the cement was a solid foam structure (Zheden et al. 2014). The solid foam microscopic structure might be similar to those found in the mussel byssal disk and the cement of tube-dwelling sandcastle worms (Sect. 4). It may be also that barnacles control the mechanics and/or microscopic structure of the cement adlayer to adapt to foreign materials/surfaces, i.e., changing stiffness of the cement and/or their base (Hui et al. 2011), though this has not been shown yet. In the respect of learning the natural system, identification of the microscopic structure of the natural cement on "easy-to-attach-to" surfaces needs to be continuously studied, such study would include the specific localization of the proteins.

A stalked species, *Dosima fascicularis*, is particularly interesting. This species clusters on the cement of conspecifics in addition to foreign substrata (Zheden et al. 2014). They secrete layer by layer an excess amount of solid foam-like cement, wherein gas is contained, and can drift on the surface of water by the buoyancy of the cement (Zheden et al. 2015). Thus, the cement of the species has a dual use of adhesion and buoyancy.

### 7.3.2 Authentic Sample of the Cement

How to take an authentic sample is crucial to analyzing the nature of the cement. It is straightforward if the adlayer of the animal is directly analyzed. The access to the natural cement is, however, difficult due to its almost hidden localization and the limited amount of the thin adlayer. Contamination of biological materials of other functions such as those originated from the membranous base and from the organic matrix of the base shell is a serious problem. How to take the authentic sample needs to respectively consider for the aim of the study, i.e., chemical analysis of the component, functional study, identification of the intact micro-structure, or the processing. Researchers have, so far, developed special procedures for the collection (Kamino 2013). The cement has historically been characterized into two types, primary and secondary, based on the mode of formation (Saroyan et al. 1970). Primary cement is a material found in the narrow gap at the joint between the calcareous base and substratum when the animal is naturally attached in the water (Fig. 7.2). Secondary cement is an underwater secretion by the intact barnacle through the calcareous base after it has been dislodged from the substratum and is usually hardened, resulting in a white opaque rubbery material. Both types of cement are proteinaceous substances (Walker 1972; Kamino et al. 1996). Primary cement is difficult to collect for detailed analysis, due to the reasons mentioned above. Secondary cement, however, is easier to collect. The reattachment of a dislodged intact barnacle is accomplished by the secondary cement (Saroyan et al. 1970; Dougherty 1990), and the similarity in amino acid composition (Naldrett 1993), the similarity of peptide maps by cyanogen bromide (CNBr)-protein fragmentation (Kamino et al. 1996), and cross-reactivity of an antibody have enabled the secondary cement to be identified as almost the same as the primary cement in its protein composition, thus simplifying further biochemical analyses. Similarity of in situ FTIR spectra between the primary cement on $CaF_2$ substrate and the secondary cement of reattached barnacles from silicone (Barlow et al. 2010) further supported it. Barnacles attached on a foul-release coating might be easily dislodged without damage to the animal; thus, this has been used to collect the secondary cement or for reattachment studies on new surfaces. Although the sample would be useful for analyses of the chemical composition, it is noticed that the cement layer on a foul-release coating might not be exactly the same as the primary cement on a natural surface in terms of microscopic structure and mechanical properties. Many papers have analyzed either or both of the cements. General readers need to pay attention as to what sample was analyzed to avoid misinterpretation (Kamino 2013).

The natural cement is difficult to collect in the case of acorn barnacles with a membranous base, because separation of the cement from the organic membranous base is difficult. The situation is also similar in the cyprid cement.

### *7.3.3 Cement Nature*

The cement is primarily a proteinaceous substance. Quantification by amino acid analysis has revealed the cement to consist of more than 90 % protein (Walker 1972; Yan and Tang 1981; Kamino et al. 1996; Zheden et al. 2014). Crude cement from several barnacle species has shown a similar whole amino acid composition (Walker 1972; Naldrett 1993; Kamino et al. 1996) including stalked barnacles (Jonker et al. 2015; Zheden et al. 2014).

3,4-Dihydroxyphenylalanine (DOPA), which is a common component in the byssal proteins of the mussel (Sect. 4), has not been found in barnacle cement (Naldrett 1993; Kamino et al. 1996; Zheden et al. 2014; Jonker et al. 2015). Careful ATR-FTIR investigation of the natural cement on Germanium wafers gave the whole chemical signature and confirmed the proteinaceous nature, which is rich in beta-sheet secondary structure (Barlow et al. 2009). The beta-sheet-rich spectra were also found in stalked barnacles (Zheden et al. 2014; Jonker et al. 2015). The whole spectra were actually the sum of information from several cement proteins mentioned in the latter section. Future studies may assign the contribution of each protein to the whole chemical signature. The overall chemical nature of the cement seemed to be similar among acorn types and stalked types.

Because the biological permanent adhesives are, in general, hardened irreversibly to join two surfaces together, the insoluble nature is an essential obstacle for research. Approaches to render soluble the insoluble proteinaceous cement involved partial proteolytic (Kamino et al. 1996) and non-proteolytic methods. A development of the non-proteolytic method, whereby more than 90 % of *M. rosa* cement could be rendered soluble without any hydrolysis of the proteins (Kamino et al. 2000), has revealed the cement to be a multi-protein complex. The method depends on heat denaturation with higher concentrations of dithiothreitol and guanidine hydrochloride, both being indispensable to rendering soluble the cement proteins. The treatment has so far enabled three major cement proteins (CPs) and three minor CPs to be identified (Kamino 2013). The six CPs each has characteristics distinct from each other; five of them being novel in their primary structure (see Sect. 3.5). These results have confirmed the cement to be a unique protein complex.

The nature of the cyprid cement has been long unclear due to the size of the adhesive plaque and, thus, the quantity being very small. Recent confocal microscopy observation of the cyprid adhesive plaques on materials with different surface chemistries (Aldred et al. 2013) restarted the debate long after several histological studies. The combination of transmission electron microscopy (TEM), light microscopy, and confocal microscopy with several staining techniques indicated that the adhesive plaque is a three layered material, and the inner two layers are proteinaceous, while the outermost layer is lipidaceous, the latter probably being lipid protein and/or lipopolysaccharide (Gohad et al. 2014). A specific staining suggested that the proteinaceous layers may be composed of phosphoprotein, though direct evidence would be required to confirm it. The lipidaceous nature of the outermost layer was also supported by several facts including the lipidaceous nature of the content of granules in the cyprid cement gland, and it also agreed with previous indirect evidence. The lipidaceous component has not been found in the biological underwater adhesives. Further study is awaited to characterize the intriguing chemistry.

### *7.3.4 Multifunctionality in Underwater Attachment*

Underwater attachment is actually an instantaneous multistep event (Waite 1987) and needs to fulfill all the requirements with the defined order and/or proper timing. That is to say, an underwater cement would need to prevent random aggregation during accumulation and transport via the duct, to displace bound water and cation species (Maier et al. 2015; Akdogan et al. 2014), to spread on foreign surfaces, to couple strongly with a variety of foreign materials and the mixed patch, and to self-assemble itself to join two interfaces together. After the initial instantaneous event, it is then necessary to cure the cement so that the holdfast remains stiff and tough, to protect it from continuous water erosion and microbial degradation, and/or to repair it. The underwater cement can orchestrate all these requirements, thus allowing the barnacle to inhabit the liquid–solid interface for more than a year. Since underwater attachment is a multifunctional event and the cement is a multi-protein complex, this, from a different standpoint, poses obstacles to measuring the activity of each protein in the research. Several microanalytical methods for each function, thus, need to be developed. These obstacles are common to all biotic underwater adhesive studies. Each cement protein has unique characteristics as described in the next section. The individual cement proteins are presumed to cooperate for underwater attachment, each protein specializing in at least one of the functions required to achieve underwater attachment. The fact that biological adhesives including the cement are self-organizing, multi-protein, and multifunctional complexes, it necessitates a multidisciplinary approach.

## 7.3.5 Cement Proteins and Possible Functions

The cement is composed of at least six cement proteins (CPs). Five of the six CPs are novel in respect to their primary structure, and the six CPs can be classified into four categories. The cement proteins are identified by their apparent molecular weight, as estimated by SDS-PAGE, and the unique characteristics of these six CPs are summarized next.

### 7.3.5.1 CP-100k and CP-52k

CP-100k and CP-52k are two major proteins in the *M. rosa* cement with similar abundances (Kamino et al. 2000). Both proteins also have similar behavior during solubilization, as the DTT-denaturing treatment is the only non-hydrolytic method known to render these proteins in *M. rosa* cement soluble. Both cement proteins were first isolated from *M. rosa* cement, and further isolation of CP-100k in *A. amphitrite* cement allowed universal PCR primers to be designed, whereby CP-100k homologous genes were detected in other barnacle species. cDNA cloning indicated both CP-100k and CP-52k to be novel from database searches and to have low homology to each other (Kamino et al. 2000, 2012). The primary structure of CP-52k contains an approximately 120-amino acid-long repetitive sequence, while there is no defined repetitive sequence in CP-100k. CP-100k and CP-52k are both highly hydrophobic and composed of a relatively low number of Cys residues (1.4 and 1.1 %, respectively). Although the exact reason why an excess amount of a reducing agent is required in the solubilization is unclear yet, the insoluble nature would not depend on simple polymerization via disulfide bonding. The disulfide bonding might be buried inside the self-assembled proteins, and the hydrophobic barrier may make reducing the disulfide bonding difficult. A denaturing treatment by using GdnHCl and heating would loosen the intermolecular and intramolecular hydrophobic barrier, whereby an excess amount of a reductant could cleave the disulfide bonds inside the proteins. Cleavage of the disulfide bonds may make the molecular conformation flexible, thus further weakening the intermolecular interaction and finally rendering these proteins soluble. The disulfide bonds in CP-52k are intramolecular, while there is no data to indicate whether the disulfide bonds in CP-100k are intramolecular or intermolecular. This author leans toward the hypothesis that non-covalent intermolecular interaction is the major contributor in initial self-assembly of the cement and that intramolecular disulfide bonds in the bulk proteins contribute to the stability of each molecular conformation, thereby optimizing the intermolecular interaction. Non-covalent intermolecular interaction is unusually strong when the molecular conformation is optimized for mutual interaction. This has particularly been shown in the assembly of beta-sheet protein (Otzen et al. 2000). The primary structure of CP-52k contained amyloid-like short sequences (Nakano and Kamino 2015), and this fact might indicate that non-covalent interaction of the regions, hydrophobic and pi interactions, are involved with the intermolecular self-assembly of the proteins. The occurrence of the sequence in the protein may also be involved with a trigger mechanism of the self-assembly when the protein is secreted to the attachment site via the duct (Nakano and Kamino 2015). This hypothesis does not exclude further intermolecular cross-linking, including other types, such as coordination bonding, and exchange of disulfide bonds in the curing process.

CP-100k in *A. amphitrite* has lower homology in the primary structure but very similar amino acid composition with that in *M. rosa*; it is higher in hydrophobic amino acids, has a rather basic pI, and is lower in Cys residues (0.9 % in *A. amphitrite*). The high contents of Arg resides among the basic amino acids in both species may be also involved with the effectiveness of GdnHCl in the solubilization from insoluble natural cement due to the similarity in the guanidino functional group.

A similar 114 kDa protein was further found in the transcriptome of adult *A. amphitrite* (Wang et al. 2015). Although LC-MS detected the tryptic peptides of the protein, the material, from which the paper detected the peptides, may also include the chitinous materials above the cement layer, originated from the base shell after the decalcification, in the current author's experience. The protein may need further careful confirmation of its localization.

The similarities in insolubility, contents, and high hydrophobicity have suggested that both CP-100k and CP-52k play a similar role in underwater attachment. They actually provide an insoluble framework for other proteins, for example, while the CP-68k and CP-19k cement proteins are partially soluble in a GdnHCl solution, they are completely soluble in DTT/GdnHCl, possibly due to destruction of the insoluble framework formed by CP-100k and CP-52k. Thus, CP-100k and CP-52k are the bulk materials in the cement and are thought to tie up other cement proteins and give the framework that provides adhesive bulk mechanical strength. In the current author's experience, the heat-denaturing treatment with both DTT and GdnHCl may be yet insufficient for the complete solubilization of CP-100k in *M. rosa*. This could be confirmed by the CNBr treatment of remained material after the DTT/GdnHCl treatment. Investigation of the curing process, including unknown types of cross-linking and the contribution of inorganic ions, needs to be continued.

#### 7.3.5.2 CP-68k

Another protein, CP-68k, is partially soluble in a GdnHCl solution and rendered completely soluble by the DTT-GdnHCl treatment which rendered CP-100k and CP-52k soluble for the first time (Kamino et al. 2000). CP-68k is, thus, not covalently connected to CP-100k and CP-52k, but is non-covalently embedded in their framework. CP-68k is characterized by an amino acid composition of Ser, Thr, Ala, and Gly evenly comprising 60 % (Kamino et al. 1996). CP-68k has been isolated from both *M. rosa* and *A. amphitrite*, each with a similar amino acid composition, although the sequence homology was not high (47 identical and

65 % similar). An attempt to design a universal PCR primer failed, probably due to lower constraint of the amino acid substitution in the protein. The reason why no fragment peptide of the protein was derived from CNBr treatment of the *M. rosa* cement (Kamino et al. 1996) was due to the absence of Met residues in the protein.

The alignment of the two CP-68ks indicates that the primary structure is divided into two regions: a long Ser-, Thr-, Ala-, and Gly-rich N-terminal regions, and a short C-terminal region composed of much less of these four amino acids and more of such amino acids as Lys, Pro, Trp, Cys, and hydrophobic amino acids. The prevalence of hydroxyl groups in the N-terminal region of CP-68k could be an essential element for its function. The C-terminal region may have a compact conformation, because Pro, Trp, and the hydrophobic amino acids are generally advantageous for packing into this structure by their backbone fixation, pi-pi stacking, and hydrophobic core formation. The function of the C-terminal region is thought to owe to amino acids displayed onto the molecular surface of the region in solution. An underwater glue first approaches and spreads over the liquid–solid interface without being dispersed in the surrounding seawater. Functions of priming and/or spreading are the most elusive functions in underwater attachment. Waite (1987) has suggested that a macromolecule with many hydroxyl groups may be a good primer to displace the water layer bound onto a foreign material surface. In this context, the prevalence of hydroxyl groups in the N-terminal region of the protein would be useful for the priming activity. In an antifreeze protein, which binds to ice nuclei to inhibit crystal growth in the cytosolic space of several organisms (Fletcher et al. 2001), the Ala and/or methyl group of Thr on the molecular surface are known to be essential in the process of binding to the ice nucleus (Zhang and Laursen 1998; Jia and Davies 2002), although the role of the amino acids is not yet clearly understood. Analogously, the hydrophobicity of the side chains on abundant Ala and part of Thr in the N-terminal region of CP-68k may play a pivotal role in interacting with the aqueous layer bound to the material surface. Though what amino acids are displayed onto the molecular surface of the C-terminal region of CP-68k are unknown yet, charged amino acids and a portion of abundant hydrophobic/aromatic amino acids may be surface displayed and involved with the interaction with foreign material.

#### 7.3.5.3 CP-20k

Several other proteins are also present in small amounts in the underwater cement. The relative amounts of proteins, however, may not reflect their importance in the function. For example, surface functional components might be found in more limited quantities than bulk adhesive components. CP-20k has been identified as a simple polypeptide form of protein in *M. rosa* cement with low abundance (Kamino 2001). This protein is characterized by the abundance of charged amino acids and Cys residues. Of these charged amino acids, Asp, Glu, and His comprise 32 and Cys comprises 17 % of the protein. Time-of-flight mass spectrometry has shown the protein recovered in the GdnHCl-soluble fraction designated GSF1 to have the mass of a monomer, thus precluding any intermolecular disulfide bond formation among CP-20k in the cement. The fact that this protein has not been recovered in the DTT/GdnHCl-treated soluble fraction, designated GSF2 also indicates no formation of intermolecular disulfide bonds among the other cement proteins. If anything, the lack of reactivity with alkylating agents suggests that all Cys in the protein has formed intramolecular disulfide bonds. NMR analysis of the recombinant form in *E. coli* has indicated that intramolecular disulfide bonds in the protein are essential for the conformation (Suzuki et al. 2005). Thus, the abundant Cys in CP-20k would be more essential for the structure than for the function. Since the protein was composed of a much lower number of hydrophobic amino acids than usual, many intramolecular disulfide bonds may have been introduced to compensate for the poor hydrophobic core formation in the protein structure. Although polymerization by CP-20k based on intermolecular disulfide bonds has been speculated for the cement curing elsewhere (Weigemann and Watermann 2003), experimental data do not support this hypothesis for the function of this protein (Mori et al. 2007).

The primary structure of CP-20k was found to comprise six repetitions when the Cys residues were appropriately aligned. However, except for Cys and Pro, the amino acids were not at all conserved among the repetitive modules. The results of an NMR study indicated the notion of three units being the minimum structural unit (Suzuki et al. 2005).

Such abundant charged amino acids as Glu, Asp, His, Lys, and Arg positioned on the molecular surface, even at the cost of poor hydrophobic core for the molecular conformation, would thus be essential for the function in underwater attachment. Analysis of the bacterial MrCP-20k recombinant (Mori et al. 2007) has indicated that the protein had no adsorption activity to a glass surface in seawater, but had adsorption activity to metals and minerals including calcite in seawater. Since the base and periphery of a barnacle shell are composed of calcite, CP-20k may be a specific coupling agent to the barnacle's own calcareous base and to another barnacle shell in the gregarious nature.

Recent omics data of *A. amphitrite* detected similar sequences to CP-20k, which are homologous genes to *B. albicostatus* CP20k (Chen Z-F et al. 2011). The *B. albicostatus* CP20k was originally found by PCR of the adult cDNA with oligo-DNA primers designed from *M. rosa* CP20k (Mori et al. 2007). The repetitive units have homology to *M. rosa* CP20k especially in the spacing of Cys residues, and both are rich in charged amino acids, though the repetitive numbers were different. *M. rosa* CP20k protein has been isolated from the cement (Kamino 2001), though it is unclear whether *B. albicostatus* CP20k is a functional orthologue to *M. rosa* CP20k, because no data on the *B. albicostatus* protein has been reported yet. In *A. amphitrite,* two homologous but different genes were identified, being different in expression pattern and localization (He et al. 2013). Further careful confirmation might be needed.

#### 7.3.5.4 CP-19k

CP-19k (Urushida et al. 2007) in *M. rosa* is a simple polypeptide form of protein and a minor component in the amount of the cement. Like CP-68k, this protein has been recovered in both GSF1 and GSF2 in the solubilization process from the cement. CP-19k is characterized by a similar amino acid bias to that of CP-68k, with the six amino acids, Ser, Thr, Gly, Ala, Lys, and Val, comprising more than 67 % of the total. The development of a universal PCR primer has allowed us to amplify homologous cDNA from other species, *B. albicostatus* and *B. improvisus*. The homology among the three CP-19ks is as follows: 51 % for the *Mr* vs *Bal* proteins, 54 % for the *Mr* vs *Bi* proteins, and 60 % for the *Bal* vs *Bi* proteins. Careful observation and a bioinformatics analysis of the primary structures have suggested four alternating repetitions of two short segments, these being a [Ser, Thr, Gly, and Ala]-rich segment and [charged and hydrophobic amino acid]-rich one. Although there is no exact homology and different polypeptide lengths between CP-19k and CP-68k, segments/domains might be similar in the amino acid compositions. This may indicate that the biased amino acids are probably of functional significance, and these proteins probably contribute to a similar function.

A recombinant form of CP-19k has been successfully expressed in the soluble fraction of *E. coli* (Urushida et al. 2007). Purification of the recombinant form allowed it to be characterized, and adsorption to an Au surface, and a $TiO_2$ surface in seawater has been demonstrated. The possible function of CP-19k is coupling to a foreign material surface, although other surface activities, including priming and spreading, also need to be carefully investigated. CP-68k and CP-19k should play similar but distinct roles in underwater attachment. Finding the difference between these proteins awaits a detailed analysis.

#### 7.3.5.5 CP-16k

CP-16k is a minor cement protein and can be extracted in GSF1 from the cement. The primary structure shares 47 % homology with lysozyme P in *Drosophila melanogaster*, with all catalytic amino acids and Cys being conserved. The crude cement shows lytic activity to the *Micrococcus luteus* membrane fraction, so CP-16k is probably a lysozyme (Kamino and Shizuri 1998). Its possible functions include removing biofilm from the substratum and/or protecting the cement from microbial degradation.

Both the primary and secondary cement of *M. rosa* are usually collected by very gently scraping the surface of the calcareous base. This raises the possibility of contamination by calcification proteins in the base shell. If this was so, this protein would also be contained in the peripheral shell. In contrast, if the protein cannot be found in the peripheral shell, this would confirm it to be a cement protein. The peripheral shell and base shell were each decalcified and subjected to western blotting. All CPs, except for CP-16k (the antibody for CP-16k was not then available), were conclusively confirmed to be cement proteins (Urushida et al. 2007; Mori et al. 2007; Kamino et al. 2012).

### 7.3.6 *Possible Molecular Model for Barnacle Underwater Attachment*

Adhesion would begin with cement being extruded from the duct to the compartmentalized narrow interspace between the calcareous base and the substratum (Burden et al. 2014), and several molecular events in underwater attachment would occur. In the event, each cement protein interacts with some of the targets for the contribution to underwater attachment (Fig. 7.3). The target molecules for cement proteins can be placed in these categories: the foreign material surface, barnacle base material, cation species on the material, water molecules, and homogenous/heterogeneous cement protein molecules. The amino acid bias in the composition seems to be ubiquitous in cement proteins, and the biased amino acids are presumed to be essential for interaction with the targets. Taken together with characterization of the recombinant forms and indirect evidence, the model is suggested for the molecular events in barnacle underwater attachment (Fig. 7.4).

The instantaneous function would include the displacement of water bound to the material surface, spreading and coupling, after the extrusion of cement from the duct. CP-19k and CP-68k are candidates for these functions. The characterization of the bacterial CP-20k recombinant also suggests a specific coupling agent to its own base or to the peripheral shell of another barnacle.

Almost simultaneously with successful coupling to both of the substratum and its own calcareous base, cement proteins must be self-assembled together to join the two materials. CP-100k and CP-52k, by the stimuli responsive conformation

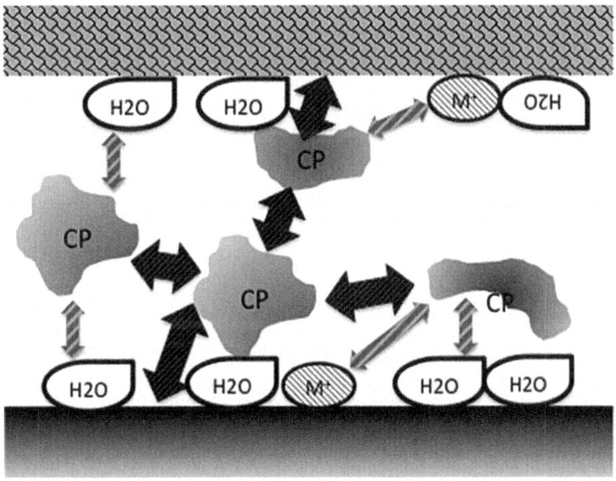


**Fig. 7.3** Interaction with several molecules as crucial function of underwater adhesive proteins. Each cement protein interacts with some of the targets for the contribution to underwater attachment. The target molecules include the foreign material surface, barnacle base material, cation species on the materials, water molecules, and homogenous/heterogeneous cement protein molecules

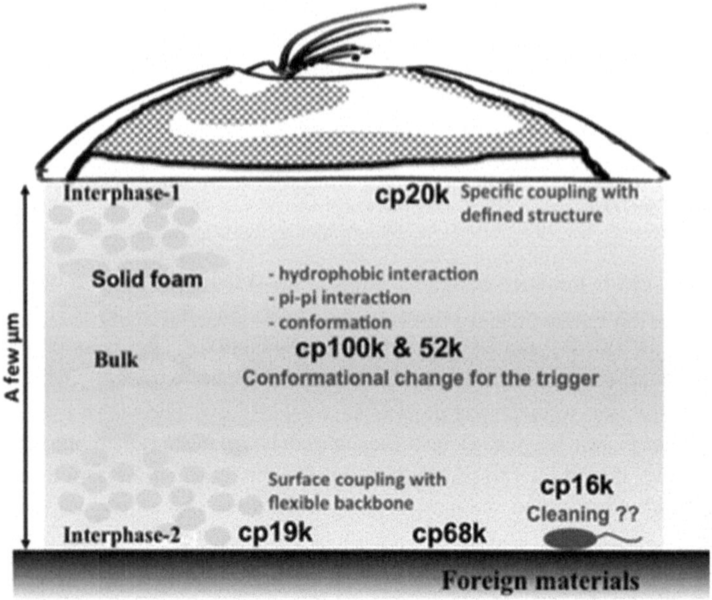


**Fig. 7.4** Barnacle cement as a protein complex. Possible functions of each protein are shown. Each protein does not necessarily have the responsibility of a single function

change, would quickly form the adhesive bulk by intermolecular hydrophobic interaction or by another mechanism, so that other CPs are associated together. It is unclear yet if a further curing mechanism of the adhesive bulk occurred to make the holdfast stiff and tough. Daily protection from bacterial degradation might also be significant for forming a durable holdfast, and CP-16k is a likely candidate for this function. Furthermore, the adhesive joint including both of the adhesive interface and the bulk would be exposed to continuous water erosion. A man-made synthetic adhesive usually relies on water being absent because of the use in air. If we use the design criteria for a man-made synthetic adhesive, protection against water erosion would be vital. However, barnacle underwater cement is protein made, and the constitutive water molecules would be essential to stabilize the molecular conformation. It is therefore likely that a biotic underwater adhesive works in concert with water. This may be an element common to the structure of all cement proteins.

## 7.4 Comparison with Other Holdfasts and Proteins

The best counterpart biotic underwater adhesives to consider are mussel byssus (Rzepecki and Waite 1995; Taylor and Waite 1997; Waite 1999; Holten-Andersen et al. 2007; Lee et al. 2011; Waite and Broomell 2012; Lu et al 2013) and sandcastle worm cement (Stewart 2011). The barnacle cement, the mussel byssus, and the tubeworm cement, all are protein materials specialized for underwater adhesion, yet they have substantial differences ranging from the macroscopic level to the molecular level. The comparison of differences and common traits would lead to insights for underwater adhesives and adhesion (Kamino 2010). The mussel holdfast, the byssus, consists of macroscopically and microscopically separate modular structures. The mussel adhesive plaque is distally coupled with the foreign surface. The bulk of the plaque consists of a solid foam-like structure composed primarily of Cys-rich DOPA-containing byssal proteins. The boundary between the plaque and substratum consists of small DOPA-containing byssal proteins, while the thread consists of collagenous proteins forming a stiff distal zone and an elastic proximal zone. Specific proteins in two junctions, i.e., the adhesive interface-plaque bulk and plaque bulk-distal end of the thread, might occur. The surface of the whole byssus is coated with a thin layer (5–10 μm) that is rich in a DOPA-containing byssal protein. A mussel holds itself onto a foreign substratum with the modular composite material, the byssus, as a whole against strong and repeated shock by waves and currents. On the other hand, a barnacle attaches directly to the substratum by its whole base via cement; this cement seeming not to have a macroscopically modular structure. The tubeworm cement, in contrast, attaches environmental particulates that are several hundred microns in diameter to construct the tube where the worm inhabits inside. The tubeworm cement joins particulates and is uniform at the macroscopic level, while the bulk has a similar solid foam microscopic structure with that of the mussel byssal disk. No specific localization of the components at the molecular level probably occurs in the tubeworm cement. The difference mentioned above is probably involved with at least the distribution of the secretory glands. Several glands specializing in the biosynthesis of individual byssal proteins are appropriately aligned along the longitudinal axis of the ventral groove in the mussel foot (Waite 1992), making possible a macroscopically modular structure of the byssus. In the barnacle and tubeworm (Wang and Stewart 2012), however, only the unicellular cement gland is involved with the biosynthesis, and all cement proteins may be secreted together via the duct system. The difference of secretory glands seemed to be reflected in the molecular mechanisms.

The biological materials for underwater adhesion in these organisms consist of a multi-protein complex. A substantial distinction between them at the molecular level would be a posttranslational modification, peptidyl DOPA, found in almost all mussel byssal proteins and tubeworm cement proteins, but not found in barnacle cement. The functional catechol group in the DOPA residue is versatile and crucial in the underwater adhesion. The functional group is involved in such functions as surface coupling, self-assembly, curing, and fiber elasticity. The catechol group of DOPA is essential for both surface coupling to foreign materials such as metal/metal oxide and the di- and tri-coordination bonding via $Fe^{3+}$, while the quinone form possibly undergoes reactions with sulfhydryl, catechol, and amine groups, thereby forming irreversible cross-links via the other residues for curing. The coordination bonding might make the adhesive bulk durable with reversible cross-linking. In the "DOPA-system," the balance between the catechol group and the quinone form is essential in adhesion, and it may be redox controlled with SH-SS exchange of Cys residues in Cys-rich FP6 (Yu et al. 2011).

Besides DOPA, the primary structures in the barnacle cement proteins and the mussel byssal proteins or the tubeworm cement proteins have no homology at all. The mussel adhesive proteins found at the adhesive interface (FP-3 and FP-5) are typically small (6–10 kD) and hydrophilic with substantial posttranslational modification to DOPA, *O*-phosphoserine, and/or 4-hydroxyarginine and are involved in surface coupling to foreign surface. A variant of FP-3, FP-3 slow, may also be involved with the displacement of the water layer on the material surface (Wei et al. 2013). The barnacle CP-20k, which is not posttranslationally modified, has been suggested to be a calcite-specific coupling protein. The function seems to be provided by the side chains in the common amino acids, optimized by the protein-molecular conformation to perform the same functions as the versatile modified amino acid residues in the mussel. The characterization of the bacterial CP-19k recombinant and the CP-68k has suggested that these proteins are possible coupling and/or priming agents of underwater surface with their common amino acids on flexible backbone structures. The molecular mechanisms to meet the requirement in underwater surface functions are actually diversified.

CP-100k and CP-52k of the barnacle cement proteins are hydrophobic, resulting in the formation of an insoluble adhesive bulk. The bulk of the adhesive plaque in the mussel is formed by FP-2. No hydrophobic protein has so far been found among mussel foot proteins and tubeworm cement, so cross-linking has been suggested as the major contributor to the organization of FPs or the tubeworm cement proteins. A single thread is formed at a time in producing the mussel byssus. Thus, all functional processes including curing must be completed at once, very rapid processing being required to achieve this. The condition in the adhesion would be similar to that of the tubeworm. In contrast, most barnacle cement is added to the enlarged marginal area as the animal grows or repairs the cement layer that has already been formed. Curing may not be urgent in barnacle cement, because the cement layer which has already been formed would assist the holdfast. The curing process for barnacle cement may therefore be relatively slow. The distinct conditions for the adhesion might be reflected by the molecular mechanisms in the bulk curing.

Protection from environmental bacterial degradation is a possible function required by an extracellular proteinaceous structure. The mussel byssus is evidently protected by being varnished with FP-1 which is cross-linked via DOPA residues. Barnacle cement provides lytic activity originating from CP-16k, a lysozyme homologue, and this is probably useful for protecting the cement. These typical sessile organisms have evidently evolved individually in respect to their molecular systems for cementing. Further discussion of comparisons among biological adhesives and adhesion should refer to Kamino (2012).

## 7.5 Impacts to Material Science

How material science learns from biological adhesives is challenging yet. There would be several viewpoints for this, including chemical structures; chemistry; shape of the molecules; micro-, meso-, and macroscopic structures; processing; and physical nature. In terms of chemical structure and chemistry, the DOPA or catechol moiety found in mussel byssus and sandcastle worm cement has a great impact on polymer science (Lee et al. 2007; Holten-Andersen et al. 2011; Ahn et al. 2015). DOPA's notorious susceptibility to oxidation, however, poses significant challenges to the practical translation of mussel adhesion (Wei et al. 2013), thus substantial efforts were continued, including utilization of the physicochemical property of complex coacervation, found in the sandcastle worm cement and the mussel byssus (Stewart et al. 2011). In contrast to DOPA, barnacle cement studies have had less impact on material science yet, probably due to the absence of a simple and versatile chemical structure such as DOPA (Kamino 2012). The biological underwater adhesives are a multi-protein complex, and each respective protein has responsibility in multifunctional underwater attachment. The characterization of each bacterial recombinant form of the cement protein has indicated that the multifunctionality in the cement can be dissected and might be applied to each single function. The functions involved in underwater attachment are attractive to material science, making each cement protein a possible model for developing a protein-/peptide-based material. For instance, molecular self-assembly is one of the most attractive molecular events in supramolecular chemistry (Whitesides and Boncheva 2002), nanoscale technology (Zhao and Zhang 2004), and tissue engineering. Several peptide-based materials have so far been designed, and an amphiphilicity and a beta-sheet structure have been preferentially studied (Zhang 2003). Peptides offer great advantages in interdisciplinary biotechnological material science due to their huge molecular diversity, simple design from protein-based biomolecular materials, easy insertion of a biological motif, and freedom from contamination of viral and prion origin in their production. Peptides whose primary structures have been designed from the repetitive sequences in barnacle CP-20k were chemically synthesized, and their ability for self-assembly under several salt concentrations was evaluated (Nakano et al. 2007). The peptides were found to self-assemble in an ionic strength-responsive manner, typically at more than 0.4 M of NaCl concentration, while the peptide was soluble in pure water. The self-assembly had a consistent rod shape of 100 nm in width. A higher peptide concentration resulted in the formation of a macroscopically observable and handleable membranous material. An analysis of the self-assembly process indicated that self-assembly upon the addition of salt was based on non-covalent molecular interactions. The peptides examined in the study contained numerous charged amino acids, few hydrophobic amino acids, had no simple amphiphilicity, and had no beta-sheet structure; therefore, self-assembly of these peptides may lead to novel principles for peptide self-assembly.

Peptides designed from the primary structure of CP-52k had amyloid-like properties (Nakano and Kamino 2015). The self-assembly was pH or ionic strength responsive and formed nano-spicules or extended long fibers with defined width. The self-assembly occurred by the conformation change of the peptide to a beta-sheet-rich form. Improving the peptide sequences by molecular engineering and different self-assembly conditions might lead to the development of various peptide-based materials. Peptide-based design of materials from underwater cement proteins would be a promising future application.

Recombinant forms of whole CP-20k, CP-19k, and CP-68k would be available for anchoring any functionality by fusion or conjugation with a functional structure. Because barnacle cement proteins have limited or no posttranslational modifications, their recombinant forms are ready to use as is and are free from inconvenient enzymatic post-conversion. The genetic fusion of any functional protein with a cement protein would be practical to anchor and/or arrange the functionality onto material surfaces (Kamino 2008). If any self-assembly motif is selected for the functionality, the combination may be an approach for an underwater mimetic adhesive.

On a foul-release coating, the cement's microscopic structure had a fibrous nature. The microscopic structure was inspiration for the artificial design of a fibrous protein-based adhesive (Zhong et al. 2014). Overall, the path from barnacle adhesive and adhesion to material science applications is slowly but certainly becoming clear.

## 7.6 Concluding Remarks

Barnacle underwater attachment is the combination of molecular events with the cellular process of the secretion. The multi-protein complex handles the multifunctionality of this underwater attachment, which is based on a different design from those of man-made adhesives in chemistry, structures, processing, and physics. A detailed analysis of the multi-protein complex and challenge of the reconstitution would help to elucidate the specific functions involved in underwater attachment and might lead to the future design of novel materials. The study of barnacle cement has indicated the significance of non-covalent molecular interactions and highlighted the importance of the orientation of the functional group, thus solving the defined protein structure-function relationships are awaited. Recombinant production of several cement proteins and fusion forms with other proteins was promising in addition to the self-assembly of the peptides. Further trial and error of molecular design would lead to a reliable application. In contrast, finding a crucial functional group, e.g., DOPA in a mussel system, has had a direct impact on material science. Efforts to find such a simple structural unit in barnacle cement, i.e., posttranslational modification in the curing process, are to be encouraged. Recent barnacle studies have led to a preliminary molecular model for underwater attachment, although verification and repeated refining of this model are essential.

Further research objectives also include characterizing the molecular mechanism of the adaptable coupling to various material surfaces, the cell biology, the physiology, and antifouling technology supported by the chemistry of the cement. In the biological attachment/adhesive study, the cypris is also intriguing in its size range. Several instrumentation approaches might fit to this scale range (Maleschlijski et al. 2015), thus cypris adhesion might be a good model for such research. In the view of biological adhesive and adhesion, the barnacle is doubly interesting.

## References


Ahn BK, Das S, Linstadt R, Kaufman Y, Martinez-Rodriguez NR, Mirshafian R, Kesselman E, Talmon Y, Lipshutz BH, Israelachvili JN, Waite JH (2015) High-performance mussel-inspired adhesives of reduced complexity. Nat Commun 6:8663

Akdogan Y, Wei W, Huang KY, Kageyama Y, Danner EW, Miller DR, Martinez Rodriguez NR, Waite JH, Han S (2014) Intrinsic surface-drying properties of bioadhesive proteins. Angew Chem Int Ed Engl 53:11253–11256

Aldred N, Clare AS (2008) The adhesive strategies of cyprids and development of barnacle-resistant marine coatings. Biofouling 24:351–363

Aldred N, Ekblad T, Andersson O, Liedberg B, Clare AS (2011) Real-time quantification of microscale bioadhesion events in situ using imaging surface plasmon resonance (iSPR). ACS Appl Mater Interfaces 3:2085–2091

Aldred N, Høeg JT, Maruzzo D, Clare AS (2013) Analysis of the behaviours mediating barnacle cyprid reversible adhesion. PLoS One 8, e68085

Baier RE, Meyer AE (1992) Surface analysis of fouling-resistant marine coatings. Biofouling 6:165–180

Barlow DE, Dickinson GH, Orihuela B, Rittschof D, Wahl KJ (2009) In situ ATR-FTIR characterization of primary cement interfaces of the barnacle *Balanus amphitrite*. Biofouling 25:359–366

Barlow DE, Dickinson GH, Orihuela B, Kulp JL 3rd, Rittschof D, Wahl KJ (2010) Characterization of the adhesive plaque of the barnacle Balanus amphitrite: amyloid-like nanofibrils are a major component. Langmuir 26:6549–6556

Berglin M, Larsson A, Jonsson PR, Gatenholm P (2001) The adhesion of the barnacle, *Balanus improvisus*, to poly(dimethylsiloxane) fouling-release coatings and poly(methyl methacrylate) panels: the effect of barnacle size on strength and failure mode. J Adhesion Sci Technol 15:1485–1502

Burden DK, Barlow DE, Spillmann CM, Orihuela B, Rittschof D, Everett RK, Wahl KJ (2012) Barnacle Balanus amphitrite adheres by a stepwise cementing process. Langmuir 28:13364–13372

Burden DK, Spillmann CM, Everett RK, Barlow DE, Orihuela B, Deschamps JR, Fears KP, Rittschof D, Wahl KJ (2014) Growth and development of the barnacle *Amphibalanus amphitrite*: time and spatially resolved structure and chemistry of the base plate. Biofouling 30:799–812

Chen ZF, Matsumura K, Wang H, Arellano SM, Yan X, Alam I, Archer JA, Bajic VB, Qian PY (2011) Toward an understanding of the molecular mechanisms of barnacle larval settlement: a comparative transcriptomic approach. PLoS One 6, e22913

Dougherty WJ (1990) Barnacle adhesion: reattachment of the adult barnacle *Chthamalus fragilis* Darwin to polystyrene surfaces followed by centrifugational shearing. J Crustac Biol 10:469–478

Fletcher GL, Hew CL, Davies PL (2001) Antifreeze proteins of teleost fishes. Annu Rev Physio 63:359–390
Fyhn UE, Costlow JD (1976) A histochemical study of cement secretion during the intermolt cycle in barnacles. Biol Bull 150:47–56
Gohad NV, Aldred N, Hartshorn CM, Jong Lee Y, Cicerone MT, Orihuela B, Clare AS, Rittschof D, Mount AS (2014) Synergistic roles for lipids and proteins in the permanent adhesive of barnacle larvae. Nat Commun 5:4414
Griffith JR, Bultman JD (1980) Fouling release coatings. Naval Eng J 92(2):129–132
He LS, Zhang G, Qian PY (2013) Characterization of two 20kDa-cement protein (cp20k) homologues in Amphibalanus amphitrite. PLoS One 8(5), e64130
Holten-Andersen N, Fantner GE, Hohlbauch S, Waite JH, Zok FW (2007) Protective coatings on extensible biofibers. Nat Mater 6:669–672
Holten-Andersen N, Harrington MJ, Birkedal H, Lee BP, Messersmith PB, Lee KY, Waite JH (2011) pH-induced metal-ligand cross-links inspired by mussel yield self-healing polymer networks with near-covalent elastic moduli. Proc Natl Acad Sci U S A 108:2651–2655
Hui CY, Long R, Wahl KJ, Everett RK (2011) Barnacles resist removal by crack trapping. J R Soc Interface 8:868–879
Jia Z, Davies PL (2002) Antifreeze proteins: an unusual receptor-ligand interaction. Trends Biochem Sci 27:101–106
Jonker JL, von Byern J, Flammang P, Klepal W, Power AM (2012) Unusual adhesive production system in the barnacle Lepas anatifera: an ultrastructural and histochemical investigation. J Morphol 273:1377–91
Jonker JL, Abram F, Pires E, Varela Coelho A, Grunwald I, Power AM (2014) Adhesive proteins of stalked and acorn barnacles display homology with low sequence similarities. PLoS One 9, e108902
Jonker JL, Morrison L, Lynch EP, Grunwald I, von Byern J, Power AM (2015) The chemistry of stalked barnacle adhesive (Lepas anatifera). Interface Focus 5:20140062
Kamino K (2001) Novel barnacle underwater adhesive protein is a charged amino acid-rich protein constituted by a Cys-rich repetitive sequence. Biochem J 356:503–507
Kamino K (2008) Underwater adhesive of marine organisms as the vital link between biological science and material science. Mar Biotechnol 10:111–112
Kamino K (2010) Molecular design of barnacle cement in comparison with those of mussel and tubeworm. J Adhes 86:96–110
Kamino K (2012) Chapter 12. Diversified molecular design in the biological underwater adhesives. In: Thomopoulos S, Birman V, Genin GM (eds) Structural interfaces and attachments in biology. Springer, Berlin, Heidelberg, pp 175–199
Kamino K (2013) Mini-review: barnacle adhesion and the adhesive. Biofouling 29:735–749
Kamino K, Shizuri Y (1998) Structure and function of barnacle cement proteins. In: Le Gal Y, Halvorson H (eds) New developments in marine biotechnology. Plenum Press, New York, pp 77–80
Kamino K, Odo S, Maruyama T (1996) Cement proteins of the acorn barnacle, *Megabalanus rosa*. Biol Bull 190:403–409
Kamino K, Inoue K, Maruyama T, Takamatsu N, Harayama S, Shizuri Y (2000) Barnacle cement proteins. J Biol Chem 275:27360–27365
Kamino K, Nakano M, Kanai S (2012) Significance of the conformation of building blocks in curing of barnacle underwater adhesive. FEBS J 279:1750–1760
Kavanagh CJ, Swain GW, Kovach BS, Stein J, Darkangelo-Wood C, Truby K, Holm E, Montemarano J, Meyer A, Wiebe D (2003) The effects of silicone fluid additives and silicone elastomer matrices on barnacle adhesion strength. Biofouling 19:381–390
Lacombe D (1970) A comparative study of the cement glands in some balanid barnacles (cirripedia, balanidae). Biol Bull 139:164–179
Lee H, Lee BP, Messersmith PB (2007) A reversible wet/dry adhesive inspired by mussels and geckos. Nature 448:338–341

Lee BP, Messersmith PB, Israelachvili JN, Waite JH (2011) Mussel-inspired adhesives and coatings. Annu Rev Mater Res 41:99–132
Lu Q, Danner E, Waite JH, Israelachvili JN, Zeng H, Hwang DS (2013) Adhesion of mussel foot proteins to different substrate surfaces. J R Soc Interface 10:20120759
Maier GP, Rapp MV, Waite JH, Israelachvili JN, Butler A (2015) Biological adhesives. Adaptive synergy between catechol and lysine promotes wet adhesion by surface salt displacement. Science 349:628–632
Maleschlijski S, Bauer S, Aldred N, Clare AS, Rosenhahn A (2015) Classification of the pre-settlement behaviour of barnacle cyprids. J R Soc Interface 12:20141104
Maruzzo D, Aldred N, Clare AS, Høeg JT (2012) Metamorphosis in the cirripede crustacean Balanus amphitrite. PLoS One 7, e37408
Mori Y, Urushida Y, Nakano M, Uchiyama S, Kamino K (2007) Calcite-specific coupling protein in barnacle underwater cement. FEBS J 274:6436–6446
Nakano M, Kamino K (2015) Amyloid-like conformation and interaction for the self-assembly in barnacle underwater cement. Biochemistry 54:826–835
Nakano M, Shen JR, Kamino K (2007) Self-assembling peptide inspired by a barnacle adhesive protein. Biomacromolecules 8:1830–1835
Naldrett MJ (1993) The importance of sulphur cross-links and hydrophobic interactions in the polymerization of barnacle cement. J Mar Bio Assoc UK 73:689–702
Ödling K, Albertsson C, Russell JT, Mårtensson LG (2006) An in vivo study of exocytosis of cement proteins from barnacle Balanus improvisus (D.) cyprid larva. J Exp Biol 209:956–964
Okano K, Shimizu K, Satuito C, Fusetani N (1996) Visualization of cement exocytosis in the cypris cement gland of the barnacle Megabalanus rosa. J Exp Biol 199:2131–2137
Otzen DE, Kristensen O, Oliveberg M (2000) Designed protein tetramer zipped together with a hydrophobic Alzheimer homology: a structural clue to amyloid assembly. Proc Natl Acad Sci U S A 97:9907–9912
Pagett HE, Abrahams JL, Bones J, O'Donoghue N, Marles-Wright J, Lewis RJ, Harris JR, Caldwell GS, Rudd PM, Clare AS (2012) Structural characterisation of the N-glycan moiety of the barnacle settlement-inducing protein complex (SIPC). J Exp Biol 215:1192–1198
Raman S, Kumar R (2011) Interfacial morphology and nanomechanics of cement of the barnacle, *Amphibalanus reticulatus*, on metallic and non-metallic substrata. Biofouling 27:569–577
Rzepecki LM, Waite JH (1995) Wresting the muscle from mussel beards: research and applications. Mol Mar Biol Biotech 4:313–322
Sarikaya M, Tamerler C, Jen AK, Schulten K, Baneyx F (2003) Molecular biomimetics: nanotechnology through biology. Nat Biotech 2:577–585
Saroyan JR, Lindner E, Dooley CA (1970) Repair and reattachment in the balanidae as related to their cementing mechanism. Biol Bull 139:333–350
Stewart RJ (2011) Protein-based underwater adhesives and the prospects for their biotechnological production. Appl Microbiol Biotechnol 89:27–33
Stewart RJ, Wang CS, Shao H (2011) Complex coacervates as a foundation for synthetic underwater adhesives. Adv Colloid Interf Sci 167:85–93
Suzuki R, Mori Y, Kamino K, Yamazaki T (2005) NMR assignment of the barnacle cement protein Mrcp-20k. J Biomol NMR 32:257
Swain G, Schultz MP (1996) The testing and evaluation of non-toxic antifouling coatings. Biofouling 10:187–197
Taylor SW, Waite JH (1997) Marine adhesives: from molecular dissection to application. In: McGrath K, Kaplan D (eds) Protein-based materials. Birkhauser, Boston, pp 217–248
Urushida Y, Nakano M, Matsuda S, Inoue N, Kanai S, Kitamura N, Nishino T, Kamino K (2007) Identification and functional characterization of a novel barnacle cement protein. FEBS J 274:4336–4346
Waite JH (1987) Nature's underwater adhesive specialist. Int J Adhes Adhes 7:9–14
Waite JH (1992) Results and problems in cell differentiation. In: Case ST (ed) Biopolymers, vol 19. Springer, Berlin Heidelberg New York, pp 27–53

Waite JH (1999) Reverse engineering of bioadhesion in marine mussels. Ann NY Acad Sci 875:301–309
Waite JH, Broomell CC (2012) Changing environments and structure-property relationships in marine biomaterials. J Exp Biol 215:873–883
Walker G (1970) The histology, histochemistry and ultrastructure of the cement apparatus of three adult sessile barnacles, *Elminius modestus, Balanus balanoides* and *Balanus haemri*. Mar Biol 7:239–248
Walker G (1971) A study of the cement apparatus of the cypris larva of the barnacle *Balanus balanoides*. Mar Biol 9:205–212
Walker G (1972) The biochemical composition of the cement of two barnacle species, *Balanus hameri* and *Balanus crenatus*. J Mar Biol Assoc UK 52:429–435
Walker G (1981) The adhesion of barnacles. J Adhes 12:51–58
Walker G, Yule AB (1984) Temporary adhesion of the barnacle cyprid: the existence of an antennular adhesive secretion. J Mar Biol Assoc UK 64:679–686
Wang CS, Stewart RJ (2012) Localization of the bioadhesive precursors of the sandcastle worm, Phragmatopoma californica (Fewkes). J Exp Biol 215:351–361
Wang Z, Leary DH, Liu J, Settlage RE, Fears KP, North SH, Mostaghim A, Essock-Burns T, Haynes SE, Wahl KJ, Spillmann CM (2015) Molt-dependent transcriptomic analysis of cement proteins in the barnacle Amphibalanus amphitrite. BMC Genomics 16:859
Watermann B, Berger H-D, Sönnichsen H, Willemsen P (1997) Performance and effectiveness of non-stick coatings in seawater. Biofouling 11:101–118
Wei W, Yu J, Broomell C, Israelachvili JN, Waite JH (2013) Hydrophobic enhancement of dopa-mediated adhesion in a mussel foot protein. J Am Chem Soc 135:377–383
Weigemann M, Watermann B (2003) Peculiarities of barnacle adhesive cured on non-stick surfaces. J Adhes Sci Technol 17:1957–1977
Whitesides GM, Boncheva M (2002) Beyond molecules: self-assembly of mesoscopic and macroscopic components. Proc Natl Acad Sci U S A 99:4769–4774
Yan W, Tang Y (1981) The biochemical composition of the secondary cement of *Balanus reticulates* Utinomi and *Balanus amaryllis* Darwin. Nanhai Studia Marina Sinica 2:145–152
Yu J, Wei W, Danner E, Ashley RK, Israelachvili JN, Waite JH (2011) Mussel protein adhesion depends on interprotein thiol-mediated redox modulation. Nat Chem Biol 7:588–590
Yule AB, Walker G (1984a) The adhesion of the barnacle, *Balanus balanoides,* to slate surfaces. J Mar Biol Assoc UK 64:147–156
Yule AB, Walker G (1984b) The temporary adhesion of barnacle cyprids: Effects of some differing surface characteristics. J Mar Biol Assoc UK 64:429–439
Yule AB, Walker G (1987) Adhesion in barnacles. In: Southward AJ (ed) Barnacle biology. Balkema, Rotterdam, pp 389–402
Zardus JD, Nedved BT, Huang Y, Tran C, Hadfield MG (2008) Microbial biofilms facilitate adhesion in biofouling invertebrates. Biol Bull 214:91–98
Zhang S (2003) Fabrication of novel biomaterials through molecular self-assembly. Nat Biotech 21:1171–1178
Zhang W, Laursen RA (1998) Structure-function relationships in a type I antifreeze polypeptide. The role of threonine methyl and hydroxyl groups in antifreeze activity. J Biol Chem 273:34806–34812
Zhao X, Zhang S (2004) Fabrication of molecular materials using peptide construction motifs. Trends Biotech 22:470–476
Zheden V, Klepal W, von Byern J, Bogner FR, Thiel K, Kowalik T, Grunwald I (2014) Biochemical analyses of the cement float of the goose barnacle Dosima fascicularis-a preliminary study. Biofouling 30:949–963
Zheden V, Kovalev A, Gorb SN, Klepal W (2015) Characterization of cement float buoyancy in the stalked barnacle Dosima fascicularis (Crustacea, Cirripedia). Interface Focus 5:20140060
Zhong C, Gurry T, Cheng A, Downey J, Deng ZT, Stultz M, Lu TK (2014) Strong underwater adhesives made by self-assembling multi-protein nanofibers. Nat Nanotechnol 9:858–866

# Chapter 8
# The Biochemistry and Mechanics of Gastropod Adhesive Gels

**Andrew M. Smith**

**Abstract** Several different types of gastropods produce tough adhesive gels. These gels consist of more than 95 % water, yet they are far stiffer and tougher than the mucus used for other activities. This chapter reviews the structure and properties of adhesive gels from limpets, marsh periwinkles, terrestrial snails, and terrestrial slugs. While some of these gels contain large, entangled, carbohydrate-rich polymers such as those found in the mucus that is secreted for locomotion and lubrication, what sets the adhesive gels apart is the additional presence of substantial quantities of gel-stiffening proteins. Adhesive gels have been studied in depth in the slug *Arion subfuscus*. In this glue, the gel-stiffening proteins bind iron, and metals such as iron and calcium stiffen the gel. Metals bind directly to polymers in the glue and can cross-link them through coordinate covalent bonds. Additionally, key proteins in the glue are rich in carbonyls, presumably due to metal-catalyzed oxidation. These carbonyls appear to serve as sites for reversible cross-links between proteins. Furthermore, the action of the proteins does not merely stiffen the gel; it creates a second network that interpenetrates the tangled carbohydrate network creating a double network. In a double network gel, the two networks act together to achieve toughness values that are orders of magnitude greater than what they can achieve separately.

## 8.1 Introduction

A wide variety of organisms attach to surfaces using glues that are gels. These are unusual biomaterials. Their structure and properties are strikingly different from common commercial glues or gels. Commercial glues are generally solids; they may be applied in liquid form and then solidify, or they may be deformable, tacky solids (Wake 1982). In either case, their final form consists entirely of polymers or cross-linked material. In contrast, adhesive gels typically consist of dilute polymer

A.M. Smith (✉)
Department of Biology, Ithaca College, Ithaca, NY 14850, USA
e-mail: asmith@ithaca.edu

© Springer International Publishing Switzerland 2016 
A.M. Smith (ed.), *Biological Adhesives*, DOI 10.1007/978-3-319-46082-6_8

networks that contain more than 95 % water. These gels are highly deformable. One would not expect such a dilute hydrogel to be suited for adhesion. In fact, dilute polymer gels are often excellent lubricants. Nevertheless, many animals use such gels as powerful glues (Smith 2002).

Because they are gels, these glues have a variety of interesting and useful properties. Foremost among these are their great flexibility and their ability to bond to wet, untreated surfaces. Furthermore, despite being dilute and easily deformable, these adhesive gels provide surprisingly strong attachments. Snails such as limpets can be extraordinarily difficult to detach by hand. They use gels to create tenacities (attachment force per unit area) ranging from 100 to 500 kPa (Branch and Marsh 1978; Grenon and Walker 1981; Smith 1992; Smith and Morin 2002). This approaches the adhesive strength of the solid cements of mussels and barnacles, which is typically 500–1000 kPa (Waite 1983; Yule and Walker 1987). In addition, because these gels are so deformable, it can require a surprisingly large amount of energy to break the bond (Wilks et al. 2015).

What features of these gels make them strong adhesives instead of lubricants? The goal of this chapter is to describe the structure and mechanics of adhesive gels focusing on those of gastropod mollusks. Gastropods are particularly interesting because of the diversity and impressive performance of their adhesive gels. The chapter will look at the overall structure of different gastropod adhesive gels, and then it will focus on terrestrial slug glue. Terrestrial slugs provide a useful model system to analyze the biochemical structure and function of such glues in greater depth. The results of these analyses have shown that the performance of adhesive gels from gastropods depends on a number of interesting structural features. These include the polymer composition of the gel, the presence of specific proteins that differentiate adhesive from nonadhesive gels, the manner in which the polymers cross-link, and synergistic actions of different types of polymers. All of these factors work together to create unusually tough adhesives.

## 8.2 Adhesive Gels Used by Different Animals

There are probably many animals that use adhesive gels. Echinoderms adhere using highly hydrated secretions that contain either mucopolysaccharides or protein (Flammang 1996; Chap. 9). A wide variety of worm-like invertebrates also adhere using such viscous secretions (Hermans 1983). Any time a polymeric adhesive secretion contains a high percentage of water and is easily deformable but strikingly viscous and even elastic, it is likely to be a gel. Some species of frogs can produce a sticky gel that is markedly different from common mucous secretions (Chap. 10). Many microscopic organisms also adhere with gels (Callow and Callow 2006).

Adhesive gels have been studied in depth for four gastropod mollusks. These represent a range of habitats and functions. Of these, the attachment of limpets has been studied for the greatest length of time. Most limpets live on rocky intertidal coasts. Limpets in the genus *Lottia* produce a nonadhesive gel for locomotion, and they can use suction to create strong attachments without restricting mobility (Smith 1991b, 1992). They can also produce an adhesive gel to glue down when they are exposed and inactive during low tide (Smith 1992). The adhesive strength protects them from dislodgement by predators such as shorebirds (Hahn and Denny 1989). When the tide returns, the limpets typically become active and at this point rely on suction for adhesion (Smith 1992). It is likely that other limpets also alternate between attachment mechanisms, though the cues may be different.

The marsh periwinkle, *Littoraria irrorata*, can also produce adhesive and nonadhesive gels. These snails forage along mud flats, gliding along a relatively nonadhesive mucus, but when the tide returns, they climb marsh grass stems and glue the lip of their shell down. In this way, they avoid aquatic predators such as crabs and fish (Warren 1985; Vaughn and Fisher 1988). When the tide recedes, they break their adhesion and return to the mud flats. The shear tenacity created by their adhesive gel can exceed 100 kPa. This is an order of magnitude greater than the tenacity these snails create using suction and any viscous contributions from the mucus they crawl upon (Smith and Morin 2002). As with limpets, the adhesive gel is surprisingly elastic and significantly firmer than the mucus the animal crawls upon (Fig. 8.1a, b). One difference from limpets is that the glue forms a thin strip along the edge of the shell, while limpets secrete the glue under the sole of the foot. This means that periwinkle glue, unlike limpet glue, is exposed to the elements. Thus, it may dry into a solid sheet in warm, dry weather. In some species of periwinkle, such as *L. aspera*, the glue may always dry (Denny 1984), while in

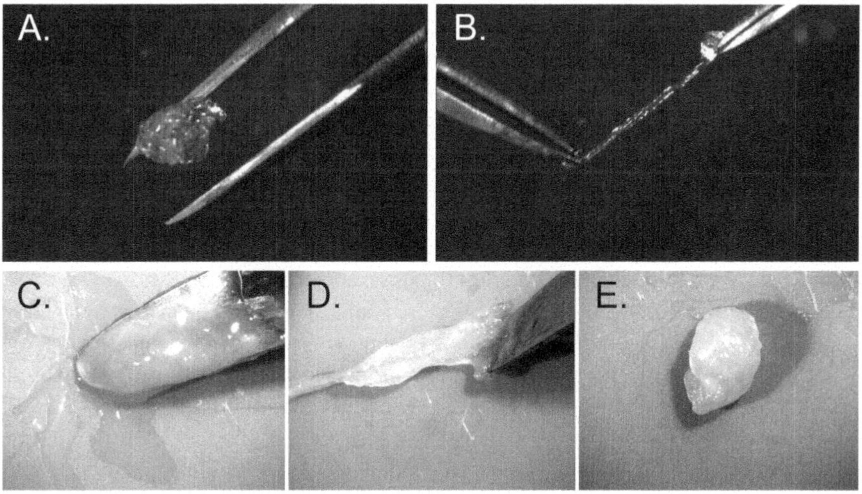


**Fig. 8.1** (**a–e**) The physical characteristics of two gastropod adhesive gels: (**a**, **b**) the glue from the marsh periwinkle *L. irrorata* forms an irregular mass (**a**) that can be stretched extensively (**b**). This deformation is reversible. The glue in (**a**) and (**b**) is held by fine-tip forceps; (**c–e**) the glue from the slug *A. subfuscus* is often secreted in a form that appears fairly fluid (**c**), but which sets into an elastic gel that sticks strongly to most surfaces and can also be stretched extensively (**d**). This deformation is also reversible (**e**). In (**c**) and (**d**) the glue is attached to a metal spatula

others it may stay gelled. The peak force required to detach marsh periwinkles using the gelled glue, though, is not significantly different from that of the dried film (Smith, unpublished). If anything, the flexibility of the gel would provide better adhesive performance by absorbing energy during detachment rather than failing as a brittle solid. While *L. irrorata* has been studied in depth, many other periwinkles also use glues to attach to rocks or vegetation, often switching between active and inactive states.

Some land snails such as *Helix aspersa* also glue the lip of their shell onto the substrate during periods of inactivity. In this case they stay glued for longer periods, sometimes months, estivating until conditions are sufficiently moist (Wells 1944; Campion 1961; Barnhart 1983). Unlike marsh periwinkles, the glue always seems to dry into a tough film. It also forms a seal around the entire circumference of the opening instead of just the anterior end. As with limpets and marsh periwinkles, adhesion is not immediate. Marsh periwinkles take roughly 10 min to form the glue (Bingham 1972). The limpets and land snails that have been studied seem to take at least that long to form the fully functional adhesive bond (personal observation), though limpet feet can become tacky to the touch within seconds (Smith 1991b).

Finally, some terrestrial slugs use an adhesive gel as a defensive secretion (Mair and Port 2002). When disturbed, *A. subfuscus* secretes a markedly sticky, elastic, orange gel from its dorsal surface. This typically appears as a liquid on the dorsal surface but stiffens into a rubbery mass when removed (Fig. 8.1c–e). Like the glues of limpets and marsh periwinkles, it is a gel composed of roughly 95 % water. This glue can sustain stresses over 100 kPa (Wilks et al. 2015). As with the other gastropods, the mechanical properties of the glue are quite different from those of the normal mucus used in locomotion and lubrication.

All of the gastropod adhesive gels are based on mucus. The use of the term mucus, however, is probably misleading in that it implies a unity of structure that does not exist. A key step in understanding how gastropods can create strong adhesion with gels was the recognition that the glue was a different form of gel than the normal mucus. As anyone who has handled an invertebrate can appreciate, the normal mucus covering their outer surface is slippery rather than sticky. Unless they are producing strong suction attachments, gastropods that are not glued down are relatively easy to slide across smooth surfaces (Smith 1991b, 2002). Thus, the mucus used in locomotion provides minimal adhesive strength on its own. The mucus that is used in adhesion, though, is different. When limpets glue down in an aquarium, they are easily distinguished from limpets that are not glued down (Smith 1992). In addition to having a high shear tenacity, detachment of limpets that are glued down occurs abruptly and usually leaves a thin film of gel stuck firmly to the glass. One can remove this gel with a razor blade to get an elastic mass that is unlike the loose slime that many snails produce across their general body surface. This functional difference corresponds with important structural differences between gels. In all four types of gastropods studied, a primary difference between adhesive and nonadhesive mucus is the presence of specific proteins, and these proteins have substantial mechanical effects (Smith et al. 1999; Pawlicki et al. 2004). To understand these effects, it is necessary to understand gel mechanics and the detailed structure of the glue.

## 8.3 Principles of Gel Mechanics

A gel is a dilute polymer network within a liquid (Tanaka 1981). The liquid, which is water in biological systems, keeps the polymer network from collapsing. The polymers trap the liquid so that it cannot easily flow. Thus, a gel has solid properties, despite its high water content. Gels are typically viscoelastic; when a stress is applied to them, they have a viscous and an elastic resistance to deformation. For a full description of viscosity and elasticity, see Wainwright et al. (1976) or Denny and Gosline (1980). In brief, a purely viscous material will flow in response to an applied stress, resulting in a permanent deformation. Higher viscosities give more resistance to flow. A purely elastic material will not flow. Instead, it will deform in proportion to the applied stress and will maintain that stable deformation as long as the force is applied. When the force is removed, the material will spring back to its original configuration. A viscoelastic material may resist deformation elastically at first, but over time the molecules rearrange and flow in response to the stress. Note that these properties depend on the rate of deformation (Wainwright et al. 1976). Gels range from highly viscous secretions that have little elasticity to elastic materials that barely flow (Tanaka 1981).

The mechanics of gels depend largely on interactions between polymers (Williams and Phillips 2000). Polymers can entangle or cross-link to form a network that will not dissolve in water. Only certain kinds of polymers are likely to interact in this way. If each polymer occupies a relatively small volume because of its size and configuration, it is unlikely to interact with its neighbors, and the solution will not have elasticity (Fig. 8.2a). In contrast, if a polymer occupies a large volume, it may begin to overlap its neighbors at low concentrations (Fig. 8.2b). There will be a critical concentration at which this overlap begins. Once there is overlap, the viscosity of the solution increases markedly and begins to depend on the rate of shear (Williams and Phillips 2000). Very large, extended molecules may begin to overlap near a concentration of 1 %, while similar-sized molecules that fold into a compact shape may not overlap until the concentration reaches 20 % or higher. Smaller compact molecules may not interact at all even though they become noticeably more viscous (Williams and Phillips 2000). Thus, most gels contain molecules that take on an extended configuration in order to occupy a large volume and achieve overlap. In many cases, these molecules will be unusually large. Because overlap is so important, increasing the concentration of gels will increase their overlap and thus their viscosity. It is notable that limpet glue and marsh periwinkle glue may be twice as concentrated as the nonadhesive mucus, though still less than 3 % organic material by weight (Smith et al. 1999; Smith and Morin 2002). The increased concentration is unlikely, however, to play a large role in glue mechanics. Data from Smith (2002) and Ben-Zion and Nussinovitch (1997) for a

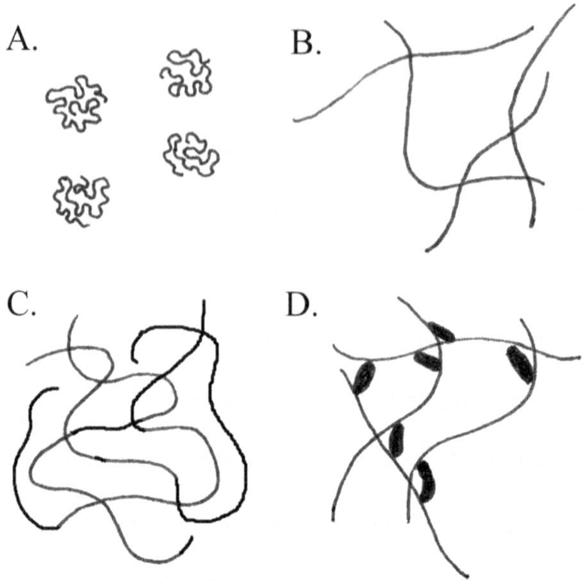


**Fig. 8.2** (**a**–**d**) The effect of size, configuration, and interactions of polymers on gel mechanics: (**a**) compact polymers do not interact at low concentrations; (**b**) if the polymers take on an extended configuration, they are more likely to entangle, increasing viscosity and stiffness; (**c**) longer polymers entangle to a greater extent, creating a more viscous material; (**d**) cross-links between polymers can dramatically increase the stiffness of the material

wide variety of polymers show that increased concentration can improve adhesive strength somewhat, but even a much greater change in concentration would not come close to accounting for the great adhesive strength of molluskan adhesive gels.

If the only interactions among polymers are through physical entangling due to overlap, the polymer solution will be viscoelastic but may not solidify to form a classic gel. The behavior is dominated by reptation, which is the untangling process that occurs in response to stress (Doi and Edwards 1988; deGennes and Leger 1982). Initially, the stress deforms the polymers elastically as each polymer is stretched against the resistance of its neighbors. The polymers can flow, though, creeping through the boundaries imposed by their neighbors. The ability of the polymers to move in this way determines the mechanics of the material.

To create a more elastic gel, polymers generally form more stable cross-links (Williams and Phillips 2000). When the material is strained, the polymers are deformed, but the cross-links prevent them from flowing appreciably to alleviate the stress. In this case, the elastic contribution to the mechanics is greater than the viscous contribution, and neither depends much on shear rate (Williams and Phillips 2000). Thus, the material may behave more like a solid, even though it may still consist of over 95 % water.

Thus, two major characteristics of polymers affect gel mechanics. These are (1) the size and configuration of the polymers and (2) the ability of the polymers to cross-link. Large, extended molecules entangle to a greater extent and thus have greater difficulty working their way through the twisted path imposed by the network (Fig. 8.2c) (Doi and Edwards 1988). Branching also impedes this flow. Increased concentration also increases the extent of entangling, restricting the movements of the polymers further (Doi and Edwards 1988). On the other hand, if a gel is chemically cross-linked (Fig. 8.2d), the size of the individual polymers does not matter as much. For example, with gelatin, the strength of the gel increases as the gelatin fragments get bigger, up to 100 kDa. Increasing the size of the polymers beyond this does not strengthen the gel further (Williams and Phillips 2000). Instead, the elasticity of the gel would depend on the number and strength of the cross-links (Denny 1983). Biological adhesive gels are typically cross-linked to produce the higher stiffness needed to be an effective adhesive.

In addition to having a high stiffness, a good adhesive must also be tough. Toughness is determined by the energy required to fracture the material (Wainwright et al. 1976). Some gels can be cross-linked to achieve high stiffness, but they are brittle and lack toughness; they fail by simple crack propagation at small strains. Commercial gels such as agar and gelatin behave this way. In contrast, complex biological gels deform extensively, leading to energy dissipation and high toughness (Wainwright et al. 1976). This energy dissipation can occur through reptation of large polymers and failure of weaker cross-links as the material is strained. It is particularly effective to combine two networks together. The combination of a deformable network of large polymers and a stiff, highly cross-linked network can increase the toughness of a gel by several orders of magnitude relative to each network separately (Gong 2010). This is called a double-network mechanism (see Sect. 8.5), and this seems to account for the high toughness of slug glue (Wilks et al. 2015).

The way the system behaves as it is stressed also affects the glue's performance. During detachment, the adhesive may be stretched uniformly, but in most biological systems, this is probably the exception. Attachment organs are often flexible, and the detaching forces are not perfectly centered and normal to the surface. Thus peeling is likely to occur, which markedly changes the performance (Gay 2002). If the adherends are more rigid, failure will depend on the relative viscosity and elasticity of the gel. If it is more viscous, regions of instability can trigger localized flow, resulting in "fingers" of air or water being sucked between the surfaces (Gay 2002). The glue would deform dramatically during this process, dissipating energy. Since these are hydrogels, bubble formation by cavitation can also occur (Gay 2002). In cavitation, the detaching force creates a reduced pressure in the water that makes up the gel. This may be sufficient to trigger the expansion of microscopic air pockets (Smith 1991a). As with fingering, cavitation would initiate failure, but in doing so would create scattered regions of high deformation and thus energy dissipation (Gay 2002). Finally, the failure of a thin adhesive film may occur adhesively or cohesively.

## 8.4 Adhesive Gel Structure

Adhesive gels are different from typical mucous secretions. The structure and function of a typical mucus is dominated by giant, space-filling carbohydrate-rich macromolecules such as proteoglycans and mucopolysaccharides (glycosaminoglycans) (Denny 1983; Davies and Hawkins 1998). These entangle to give a viscoelastic material that can often serve as a lubricant. In contrast, molluskan adhesive gels also contain a substantial fraction of proteins, and the proteins play a central role in stiffening and toughening the gels. In particular, in all four species studied, the primary characteristic that distinguishes the adhesive gel from the nonadhesive mucus is the presence of specific proteins. These were named glue proteins (Pawlicki et al. 2004). In marsh periwinkles (*L. irrorata*) and land snails (*H. aspersa*), the lubricating mucus consists of giant, usually carbohydrate-rich polymers ($\geq$1000 kDa), while the glue has those but also has a roughly equal amount of two or three smaller glue proteins (Figs. 8.3 and 8.4) (Smith and Morin 2002; Pawlicki et al. 2004). Limpet glue is unusual in that it seems to be constructed primarily of 20–200 kDa proteins with no giant, carbohydrate-rich molecules (Figs. 8.3a and 8.4a) (Grenon and Walker 1980; Smith et al. 1999). The nonadhesive mucus from the limpet *L. limatula* is structurally similar to the glue,

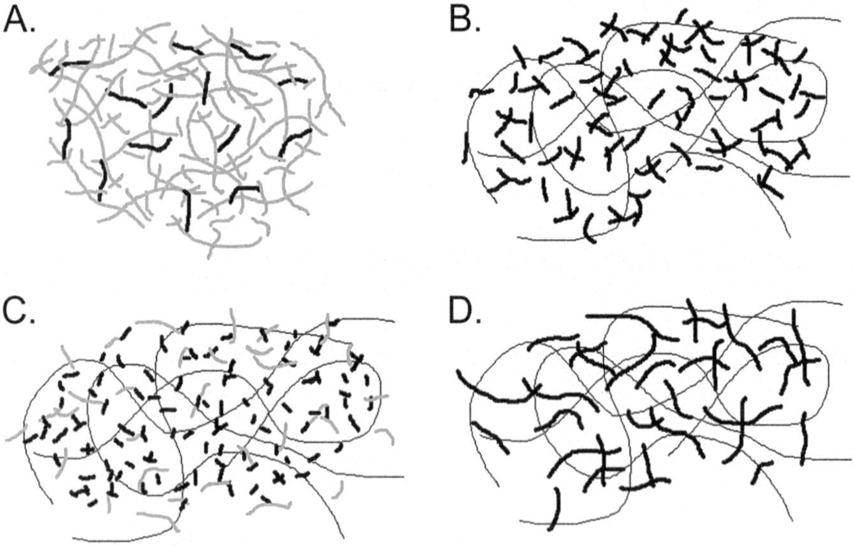


**Fig. 8.3** (**a–d**) Schematic illustrations of the relative size and abundance of the components of adhesive gels from: (**a**) the limpet *L. limatula*, (**b**) the periwinkle *L. irrorata*, (**c**) the slug *A. subfuscus*, (**d**) the land snail *H. aspersa*. Polymers of roughly 1000 kDa or larger are drawn as *thin black lines*, smaller proteins (10–200 kDa) are drawn as *thicker lines*, with glue proteins in *black* and other proteins in *gray*. The size of the polymers and relative amounts of each are depicted to scale using data from Smith et al. (1999), Smith and Morin (2002), and Pawlicki et al. (2004)

but the glue has an additional 118 kDa glue protein (Fig. 8.4a). In the terrestrial slug *A. subfuscus*, the nonadhesive mucus has giant, carbohydrate-rich molecules and some proteins, and the glue has two additional proteins. These two glue proteins constitute roughly half of the protein in the material (Pawlicki et al. 2004). The combination of giant polymers and smaller proteins is significant since the combination of two such networks can work synergistically to enhance toughness by a double-network mechanism (see Sect. 8.5; Wilks et al. 2015).

The glue proteins vary in size, but they are relatively polar or charged and lysine-rich. In limpet glue, 65 % of the amino acids would be polar or charged at neutral pH (Smith et al. 1999), whereas for marsh periwinkles, the two glue proteins have 49 and 52 % polar or charged amino acids (Smith and Morin 2002). While many of the proteins are acidic (Smith et al. 1999; Smith and Morin 2002), the 15 kDa glue protein from slugs is notably basic (Wilks et al. 2015). In all the species tested, the glue proteins contain 8–9 % lysine residues (Smith et al. 1999; Smith and Morin 2002; Bradshaw et al. 2011). The glue proteins differ in size; the masses estimated by sodium dodecyl sulfate polyacrylamide gel electrophoresis (SDS-PAGE) for the most abundant glue proteins in each species were 15 kDa (slugs), 41 kDa (periwinkles), 97 kDa (land snails), and 118 kDa (limpets) (Smith et al. 1999; Smith and Morin 2002; Pawlicki et al. 2004).

The carbohydrate-rich polymers in the glue are notable primarily for their size and negative charge. In Sephacryl S-400 gel filtration, they elute in the void volume, implying a mass greater than 1,000 kDa (Smith and Morin 2002; Pawlicki et al. 2004). In most invertebrate mucus, such polymers are carbohydrate-rich and

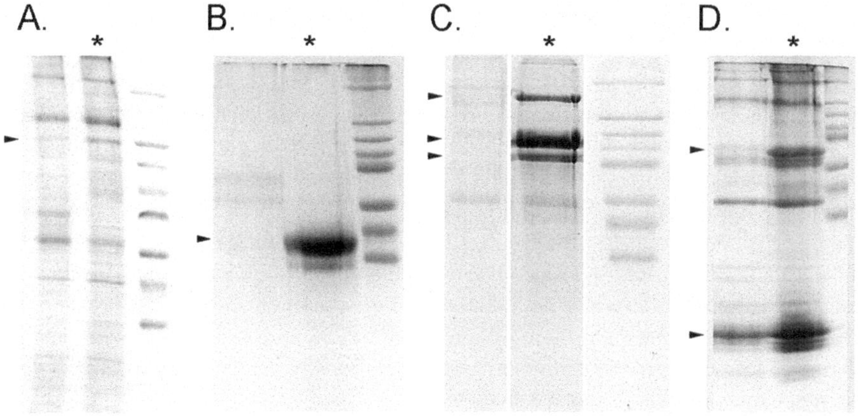


**Fig. 8.4** (**a–d**) Sodium dodecyl sulfate polyacrylamide gel electrophoresis comparison of nonadhesive (*left lanes*) and adhesive mucus (*asterisk*) from four mollusks: (**a**) the limpet *L. limatula*, (**b**) the marsh periwinkle *L. irrorata*, (**c**) the land snail *H. aspersa*, and (**d**) the slug *A. subfuscus*. Note the difference in specific proteins (*arrowheads*) between the adhesive and nonadhesive lanes. These are identified as glue proteins. MW markers (*right lanes*) are 205, 116, 97, 84, 66, 55, 45, and 36 kDa. For each species, the same amount of dried sample was present in the adhesive and nonadhesive lanes. From Smith et al. (1999), Smith and Morin (2002), and Pawlicki et al. (2004)

anionic. The charge typically comes from sulfated or carboxylated sugars, with the former more common in seawater (Denny 1983). A large amount of negative charge would generally cause the polymers to take on an extended configuration, which would assist gel formation. In the slug *A. subfuscus*, the giant polymers have been tentatively characterized as proteoglycans containing the glycosaminoglycan heparan sulfate (Wilks et al. 2015). In marsh periwinkle glue, the giant polymers consist primarily of carbohydrate (Smith and Morin 2002), while in land snail glue, the giant polymers appearing to consist mostly of protein with much less carbohydrate (Pawlicki et al. 2004). The low carbohydrate content in the latter case could be due to the use of assays that don't detect all sugars found in glycosaminoglycans. It is possible that carbohydrate concentrations are significantly greater than measured by these assays (Smith and Morin 2002).

Another essential feature of the structure of adhesive gels is their metal content. This has only been determined in depth for the slug *A. subfuscus*. The glue is notably rich in calcium (40 $mMol^{-1}$) and magnesium, but has unusually high concentrations of zinc (1–3 $mMol^{-1}$), iron, and manganese (0.1 $mMol^{-1}$) (Werneke et al. 2007; Braun et al. 2013). Metals are effective cross-linkers and contribute in multiple ways to the performance of this glue (Smith 2013).

## 8.5 Cross-Linking and the Mechanics of Adhesive Gels: Stiffness and Double Networks

The cross-linking and mechanics of glue from the slug *A. subfuscus* have been studied in depth, and multiple mechanisms have been shown to contribute to the glue's performance (Smith 2013). These include physical entangling, electrostatic interactions, coordinate covalent bonds involving metals, and reversible covalent bonds. The giant, carbohydrate-rich polymers seem to depend most on physical tangling because of their size, and electrostatic interactions would likely play a role because of their high negative charge density and the presence of divalent metals and basic glue proteins (Smith 2013). Electrostatic cross-links would be weakened substantially, however, by the presence of water. Water has a high dielectric constant; it interacts strongly with charges on the polymers, effectively weakening the electrical field between them (Waite et al. 2005). In essence, water masks the charges. Nevertheless, the charge on multivalent ions typically has a large effect on the mechanics of gels (Tanaka 1981).

In the case of the proteins, the cross-links depend on metals. Metals can directly cross-link proteins via coordinate covalent bonds (Sagert et al. 2006; Lichtenegger et al. 2008). They can also catalyze oxidation reactions that initiate cross-linking (Bradshaw et al. 2011). The 15 kDa glue protein binds iron tightly (Werneke et al. 2007) and would presumably bind calcium and magnesium as well, since they have similar ligand specificities (Smith 2013). Other proteins in the glue may bind metals, leading to a cross-linked network. Removing metals breaks up the glue and causes it to solubilize (Werneke et al. 2007). The effect is much greater if calcium and magnesium are removed, whereas zinc removal does not weaken the glue (Braun et al. 2013). Iron also probably plays a role, since it is bound so tightly that strong chelators could not remove it (Braun et al. 2013). These direct metal-based cross-links provide significant strength, and they are reversible.

The proteins are also linked indirectly by metals. Metal-catalyzed oxidation involving iron or copper can cleave the amines off of lysine side chains, creating aldehydes in their place. Many of the proteins in slug glue are heavily oxidized, giving them a high aldehyde content (Bradshaw et al. 2011). These aldehydes are tethered on the end of flexible chains and can react easily with the primary amines of other lysine residues. This results in the formation of imine bonds. Such bonds appear to play a significant role in strengthening slug glue (Bradshaw et al. 2011). Disrupting these imine bonds with nucleophiles markedly weakens the glue (Braun et al. 2013; Wilks et al. 2015). Thus, the protein network appears to be held together by a combination of direct coordinate covalent cross-links and oxidatively derived imine bonds.

All of these cross-links would stiffen the glue, and this stiffening is necessary to convert a material that would be a lubricating gel into a strong adhesive (Eagland 1990; Smith 2002). Nevertheless, stiffening alone would not necessarily lead to increased toughness. Many stiff gels, such as gelatin or agar gels, are brittle and fail easily through crack formation (Wilks et al. 2015). There must also be a mechanism to ensure toughness. Toughness can be gained through energy dissipation, often by viscous flow or the breaking of weak, sacrificial bonds. One specific, powerful way to toughen a gel is through a double-network mechanism (Gong 2010; Haque et al. 2012). In this mechanism, a loose, deformable network interpenetrates a stiffer, highly cross-linked network. The synergistic effect of the two networks can increase the toughness of a gel by a factor of a thousandfold over the separate networks. The rigid network provides stiffness, but after it begins to fracture, the deformable network must still be ruptured. The deformable network requires extensive strain before it can fracture, and this strain will continually break up more bonds in the rigid network. Thus, the rigid network must be disrupted throughout a large volume before failure occurs, rather than fracture occurring in a simple crack plane. The large volume of damage that is required greatly increases the energy required to fracture the material (Smith 2013; Wilks et al. 2015).

Slug glue appears to gain its toughness from a double-network mechanism. The glue contains a deformable network of tangled giant polymers and a network of proteins linked by reversible, metal-dependent covalent bonds. These two networks interpenetrate, but aren't tightly linked to each other (Wilks et al. 2015). Targeted disruption of either network eliminates the toughness of the gel as predicted by the double-network mechanism (Wilks et al. 2015). Thus, both networks must be present to have any measurable toughness. Either network alone is not sufficient, and they are not merely additive. The result of these two networks working together is that slug glue has a stiffness comparable to gels like gelatin or agar, yet it can be stretched to over ten times its original length before failure. This leads to a high fracture energy.

## 8.6 Protein Functions in the Glue

A full understanding of molluskan adhesive gels depends on more precise characterization of the structure and function of individual proteins. Because a major difference between the adhesive gels and nonadhesive forms of mucus that have been studied is the presence of specific glue proteins, the structure and function of these proteins is of great interest. The glue proteins from slugs, periwinkles, and land snails have been shown to have a non-specific stiffening action on polyanionic gels (Pawlicki et al. 2004) (Fig. 8.5). This may be due to direct or indirect (metal-catalyzed) cross-linking. The glue proteins from *A. subfuscus* glue bind iron strongly, suggesting a central role in cross-linking (Werneke et al. 2007), yet these proteins are not tightly linked into the protein network (Smith et al. 2009; Wilks et al. 2015). Instead, the protein network consists primarily of the higher molecular weight proteins, which are heavily oxidized and tend to form strong aggregations (Bradshaw et al. 2011; Braun et al. 2013; Wilks et al. 2015). The 40 kDa protein in particular appears to be joined by imine bonds into giant complexes (Smith et al. 2009; Wilks et al. 2015). Thus, the glue proteins may act by catalyzing oxidation to trigger the other proteins to cross-link via imine bonds. It is also notable that the 15 kDa glue protein in slug glue is associated with a number of other protein bands in that size range on SDS-PAGE, and these may be related. One of these proteins is strongly oxidized (Bradshaw et al. 2011). All of them may be different size variants of the same protein, or they may have slightly different structures and functions.

Some of the proteins are likely to play different roles. Recent work with barnacles (Kamino, Chap. 7) and mussels (Sagert et al. 2006) has made it clear that attachment in animals often depends on the action of a variety of proteins, each with somewhat different functions. The 165 kDa protein from slug glue is interesting as it is the only one that is strongly linked to the giant, carbohydrate-rich polymers (Wilks et al. 2015). Of the two proteins unique to the glue, the 61 kDa glue protein is different from the 15 kDa protein in that it appears to be negatively

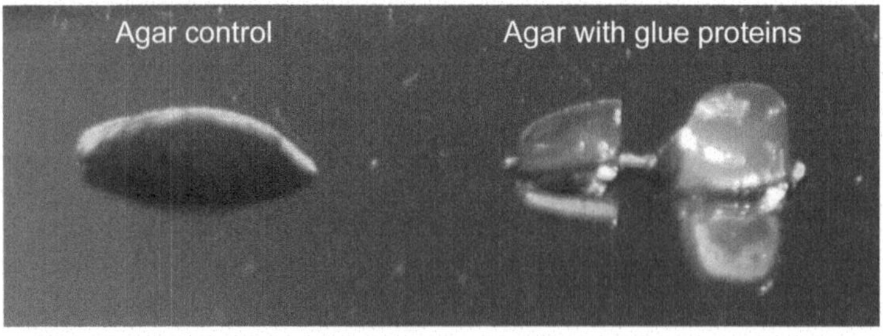


**Fig. 8.5** The gel-stiffening effect of gastropod glue proteins. The following shows 0.6 % agar with either bovine serum albumen (*left*) or glue proteins from *L. irrorata* (*right*) added. The control is a viscous liquid, while the glue proteins stiffen the agar into a firm gel. From Pawlicki et al. (2004)

charged rather than basic (Wilks et al. 2015), and it does not bind iron (Werneke et al. 2007). It may have another function. Some of the proteins may function in interfacial adhesion. Presumably there are specific proteins that are designed to interact non-specifically with any surfaces that the animal may encounter. This is relevant to the other species as well, since there are typically multiple glue proteins, or two apparently closely related glue proteins (Pawlicki et al. 2004).

Another important function is ensuring proper mixing of the glue after release. The glue needs to form rapidly after release rather than inside the secretory organ. This must depend on how the material is secreted, and how the components mix. In all the species studied, multiple glands are involved, and there are often separate glands for mucopolysaccharides and proteins (Smith 2010). Identifying which proteins are present in the different glands would be useful. In the terrestrial slug *Ariolimax*, the mucus is secreted in packets that rupture after secretion (Deyrup-Olsen et al. 1983). Similar packets have been seen in land snails and limpets (Smith 2010). When the packets rupture, the components must mix to interact. Some proteins may be essential for this mixing and initial interaction to occur, and others may be involved in cross-linking. Electrostatic interactions are likely to play a prominent role in this mixing, possibly by a coacervation mechanism as described by Stewart et al. (2004). The carbohydrates are polyanionic as are many of the proteins, but there are positively charged metals and some basic proteins. Electrostatic forces would draw oppositely charged components together, and under the right conditions, a coacervate could form, thus creating a structured material that could cross-link further. Alternatively, the mucopolysaccharides may provide a uniform structure, and the mixing could occur within that network.

Another important question is how these animals detach. In periwinkles it has been reported that the animal simply eats the glue or tears it down with its radula (Bingham 1972). In other snails, the detachment mechanism has not been tested. In *H. aspersa*, it has been suggested that an unidentified protease present in the mucus breaks down the glue (Campion 1961). In limpets, the glue forms a thin layer between the foot and substratum, so one possibility is that they secrete a layer of nonadhesive mucus over the top of the glue. This mucus could include molecules that compete for binding sites and block them, or it could include an enzyme that breaks bonds in the glue. There is a 68 kDa protein that is unique to the nonadhesive mucus of the limpet *L. limatula* (Smith et al. 1999). This may play a role in detachment, though it is found in the pedal mucus used during locomotion, not solely during detachment.

## 8.7 Comparison of Gel Structure Among Gastropods

Most of the work on adhesive gel structure and function has been performed on the slug *A. subfuscus*. It remains to be seen if the same principles apply to limpets, periwinkles, and land snails. In this context, it is notable that there are differences among glue proteins and differences among gastropod species in the overall composition of the adhesive gel. It is possible that the differences merely reflect evolutionary heritage, where animals evolved glue proteins to work with whatever polymers were present in the trail mucus. In this case, the differences may be unrelated to adhesive performance. Alternatively, they may reflect adaptations for different performance requirements. Limpets are well known among gastropods for adhesive strength, producing the highest recorded adhesive strengths for gels. Slug glue is a defensive secretion and must adhere and set much more quickly than the other adhesive gels and that may necessitate differences in structure. At 15 kDa, the glue protein from the slug *A. subfuscus* is markedly smaller than the other glue proteins, which may promote more rapid mixing, adhesion, and setting. It is often advantageous to make glues out of smaller polymers that can flow more readily to interact with the adhering surfaces before cross-linking (Bikerman 1958; Wake 1982; Waite 1983). In the case of the land snail and marsh periwinkle, both glues can be exposed to air for a longer period, and thus may dry into a tough film. The structural requirements for a glue in this case may be different as well.

## 8.8 Conclusion

The adhesive gels produced by mollusks have unusual and potentially useful properties. They are highly flexible, strong, and adhere well underwater. Analyzing how they achieve this performance could lead to the development of new biomimetic adhesives. These gels are different from typical mucus gels, which derive their viscosity primarily from tangling interactions between giant, carbohydrate-rich polymers. In contrast, adhesive gels may have similar giant polymers, but also contain a large fraction of proteins. These proteins play an essential role in stiffening the glue. In the glue from the slug *A. subfuscus*, the proteins are cross-linked by several metal-dependent mechanisms. They can use metals to form direct, coordinate covalent cross-links, and they can use metal-catalyzed oxidation to create aldehyde groups that serve as cross-linking sites. This cross-linking stiffens the glue to help it resist deformation, but to function well, the glue must also be tough. The combination of a deformable network of proteoglycans and a stiff, cross-linked network of proteins functions as a double network to dramatically increase the toughness of the gel. The glue is stiff, but it can sustain an unusually large deformation before failure, and thus requires a great deal of energy to fracture. Using a double network is a promising way to create tough gels, and it is likely that many other animals create tough gels using a double-network mechanism.

**Acknowledgments** I would like to thank R. Shadwick for comments on the manuscript and J. H. Waite and S. Werneke for helpful discussions. This work was supported by an Ithaca College summer grant.

## References

Barnhart MC (1983) Gas permeability of the epiphragm of a terrestrial snail, *Otala lactea*. Physiol Zool 56:436–444

Ben-Zion O, Nussinovitch A (1997) Hydrocolloid wet glues. Food Hydrocoll 11:429–442

Bikerman JJ (1958) The rheology of adhesion. In: Eirich FR (ed) Rheology: theory and applications, vol III. Academic, New York, pp 479–503

Bingham FO (1972) The mucus holdfast of *Littorina irrorata* and its relationship to relative humidity and salinity. Veliger 15:48–50

Bradshaw A, Salt M, Bell A, Zeitler M, Litra N, Smith AM (2011) Cross-linking by protein oxidation in the rapidly setting gel-based glues of slugs. J Exp Biol 214:1699–1706

Branch GM, Marsh AC (1978) Tenacity and shell shape in six *Patella* species: adaptive features. J Exp Mar Biol Ecol 34:111–130

Braun M, Menges M, Opoku F, Smith AM (2013) The relative contribution of calcium, zinc and oxidation-based cross-links to the stiffness of *Arion subfuscus* glue. J Exp Biol 216:1475–1483

Callow JA, Callow ME (2006) The *Ulva* spore adhesive system. In: Smith AM, Callow JA (eds) Biological adhesives. Springer, Berlin, pp 63–78

Campion M (1961) The structure and function of the cutaneous glands in *Helix aspersa*. Quart J Microscop Sci 102:195–216

Davies MS, Hawkins SJ (1998) Mucus from marine molluscs. Adv Mar Biol 34:1–71

DeGennes PG, Leger L (1982) Dynamics of entangled polymer gels. Annu Rev Phys Chem 33:49–61

Denny MW (1983) Molecular biomechanics of molluscan mucous secretions. In: Wilbur K, Simkiss K, Hochachka PW (eds) The mollusca, vol I. Academic, New York, pp 431–465

Denny MW (1984) Mechanical properties of pedal mucus and their consequences for gastropod structure and performance. Am Zool 24:23–36

Denny MW, Gosline JM (1980) The physical properties of the pedal mucus of the terrestrial slug *Ariolimax columbianus*. J Exp Biol 88:375–393

Deyrup-Olsen I, Luchtel DL, Martin AW (1983) Components of mucus of terrestrial slugs (Gastropoda). Am J Physiol 245:R448–R452

Doi M, Edwards SF (1988) The theory of polymer dynamics. Clarendon, Oxford

Eagland D (1990) What makes stuff stick? Chemtech (April):248–254

Flammang P (1996) Adhesion in echinoderms. In: Jangoux M, Lawrence JM (eds) Echinoderm studies, vol 5. Balkema, Rotterdam, pp 1–60

Gay C (2002) Stickiness—some fundamentals of adhesion. Integr Comp Biol 42:1123–1126

Gong JP (2010) Why are double network hydrogels so tough? Soft Matter 6:2583–2590

Grenon JF, Walker G (1980) Biomechanical and rheological properties of the pedal mucus of the limpet, *Patella vulgata* L. Comp Biochem Physiol B 66:451–458

Grenon JF, Walker G (1981) The tenacity of the limpet, *Patella vulgata* L.: an experimental approach. J Exp Mar Biol Ecol 54:277–308

Hahn T, Denny M (1989) Tenacity-mediated selective predation by oystercatchers on intertidal limpets and its role in maintaining habitat partitioning by *'Collisella' scabra* and *Lottia digitalis*. Mar Ecol Prog Ser 53:1–10

Haque MA, Kurokawa T, Gong JP (2012) Super tough double network hydrogels and their application as biomaterials. Polymer 53:1805–1822

Hermans CO (1983) The duo-gland adhesive system. Oceanogr Mar Biol Annu Rev 21:283–339

Lichtenegger HC, Birkedal H, Waite JH (2008) Heavy metals in the jaws of invertebrates. Met Ions Life Sci 4:295–325

Mair J, Port GR (2002) The influence of mucus production by the slug, *Deroceras reticulatum*, on predation by *Pterostichus madidus* and *Nebria brevicollis* (Coleoptera: Carabidae). Biocontrol Sci Technol 12:325–335

Pawlicki JM, Pease LB, Pierce CM, Startz TP, Zhang Y, Smith AM (2004) The effect of molluscan glue proteins on gel mechanics. J Exp Biol 207:1127–1135

Sagert J, Sun C, Waite JH (2006) Chemical subtleties of mussel and polychaete holdfasts. In: Smith AM, Callow JA (eds) Biological adhesives, 1st edn. Springer, Berlin, pp 125–143
Smith AM (1991a) Negative pressure generated by octopus suckers: a study of the tensile strength of water in nature. J Exp Biol 157:257–271
Smith AM (1991b) The role of suction in the adhesion of limpets. J Exp Biol 161:151–169
Smith AM (1992) Alternation between attachment mechanisms by limpets in the field. J Exp Mar Biol Ecol 160:205–220
Smith AM (2002) The structure and function of adhesive gels from invertebrates. Integr Comp Biol 42:1164–1171
Smith AM (2010) Gastropod secretory glands and adhesive gels. In: von Byern J, Grunwald I (eds) Biological adhesive systems: from nature to technical and medical application. Springer, Berlin, pp 41–51
Smith AM (2013) Multiple metal-based cross-links: protein oxidation and metal coordination in a biological glue. In: Santos R, Aldred N, Gorb S, Flammang P (eds) Biological and biomimetic adhesives: challenges and opportunities. Royal Society of Chemistry, Cambridge, pp 3–15
Smith AM, Morin MC (2002) Biochemical differences between trail mucus and adhesive mucus from marsh periwinkles. Biol Bull 203:338–346
Smith AM, Quick TJ, St. Peter RL (1999) Differences in the composition of adhesive and non-adhesive mucus from the limpet *Lottia limatula*. Biol Bull 196:34–44
Smith AM, Robinson TM, Salt MD, Hamilton KS, Silvia BE, Blasiak R (2009) Robust cross-links in molluscan adhesive gels: testing for contributions from hydrophobic and electrostatic interactions. Comp Biochem Physiol B Biochem Mol Biol 152:110–117
Stewart RJ, Weaver JC, Morse DE, Waite JH (2004) The tube cement of *Phragmatopoma californica*: a solid foam. J Exp Biol 207:4727–4734
Tanaka T (1981) Gels. Sci Am 244:124–138
Vaughn CC, Fisher FM (1988) Vertical migration as a refuge from predation in intertidal marsh snails: a field test. J Exp Mar Biol Ecol 123:163–176
Wainwright SA, Biggs WD, Currey JD, Gosline JM (1976) Mechanical design in organisms. Princeton University Press, Princeton
Waite JH (1983) Adhesion in bysally attached bivalves. Biol Rev Camb Philos Soc 58:209–231
Waite JH, Andersen NH, Jewhurst S, Sun C (2005) Mussel adhesion: finding the tricks worth mimicking. J Adhes 81:1–21
Wake WC (1982) Adhesion and the formulation of adhesives. Applied Science Publishers, London
Warren JH (1985) Climbing as an avoidance behaviour in the salt marsh periwinkle, *Littorina irrorata* (Say). J Exp Mar Biol Ecol 89:11–28
Wells GP (1944) The water relations of snails and slugs III. Factors determining activity in Helix pomatia L. J Exp Biol 20:79–87
Werneke SW, Swann C, Farquharson LA, Hamilton KS, Smith AM (2007) The role of metals in molluscan adhesive gels. J Exp Biol 210:2137–2145
Wilks AM, Rabice SR, Garbacz HS, Harro CC, Smith AM (2015) Double-network gels and the toughness of terrestrial slug glue. J Exp Biol 218:3128–3137
Williams PA, Phillips GO (2000) Introduction to food hydrocolloids. In: Phillips GO, Williams PA (eds) Handbook of hydrocolloids. Woodhead Publishers, Cambridge, pp 1–20
Yule AB, Walker G (1987) Adhesion in barnacles. In: Southward AJ (ed) Crustacean issues: barnacle biology, vol 5. Balkema, Rotterdam, pp 389–402

# Chapter 9
# Adhesive Secretions in Echinoderms: A Review

**Patrick Flammang, Mélanie Demeuldre, Elise Hennebert, and Romana Santos**

**Abstract** Echinoderms are quite exceptional in the sense that most species belonging to this group use adhesive secretions extensively. Two different adhesive systems may be recognised in these animals: the tube feet, organs involved in attachment to the substratum or food capture, and the Cuvierian tubules, organs involved in defence. These two systems rely on different types of adhesion and therefore differ in the way they operate, in their structure and in the composition of their adhesive. Although tube feet are present in every extant echinoderm species, only those of asteroids and regular echinoids have been studied in detail in terms of adhesion. These organs are involved in temporary adhesion, functioning as duo-gland adhesive systems in which adhesive cells release a proteinaceous secretion, while de-adhesive cells allow detachment. To date, only two adhesive proteins have been characterized in echinoderm tube feet, i.e., Sfp1 in sea stars and Nectin in sea urchins. These two proteins do not appear to be related, but they share similar protein–carbohydrate interaction domains. Cuvierian tubules occur only in some holothuroid species. These single-use organs rely on instantaneous adhesion, their contact with a surface triggering the release of the protein-based adhesive from a single cell type. Some proteins have been identified in the adhesive, but no confirmation of their adhesive function has been provided so far.

P. Flammang (✉) • M. Demeuldre
Research Institute for Biosciences, Biology of Marine Organisms and Biomimetics, University of Mons—UMONS, Mons, Belgium
e-mail: Patrick.Flammang@umons.ac.be

E. Hennebert
Research Institute for Biosciences, Biology of Marine Organisms and Biomimetics, University of Mons—UMONS, Mons, Belgium

Research Institute for Biosciences, Cell Biology, University of Mons—UMONS, Mons, Belgium

R. Santos
Science Faculty, Marine and Environmental Sciences Centre, University of Lisbon, Lisbon, Portugal

Science Faculty, Chemistry and Biochemistry Centre, University of Lisbon, Lisbon, Portugal

© Springer International Publishing Switzerland 2016 
A.M. Smith (ed.), *Biological Adhesives*, DOI 10.1007/978-3-319-46082-6_9

## 9.1 Introduction

Members of the phylum Echinodermata are among the most familiar sea creatures, and representatives, such as the sea stars, have become virtually a symbol of sea life. The phylum contains about 7250 living species of relatively large invertebrates, all being exclusively marine and largely bottom-dwellers (Ruppert et al. 2003). There are five extant classes of echinoderms: the crinoids (sea lilies and feather stars), the asteroids (sea stars), the ophiuroids (brittle stars), the echinoids (sea urchins and sand dollars) and the holothuroids (sea cucumbers). The most striking characteristics of the group are the pentamerous radial symmetry, the endodermal calcareous skeleton, the mutable collagenous tissues and the water–vascular system, a unique system of coelomic canals and surface appendages.

Echinoderms are also quite exceptional in the sense that most species belonging to this group use adhesive secretions extensively. Two different adhesive systems may be recognised in post-metamorphic individuals: the tube feet or podia, organs involved in attachment to the substratum or food capture, and the Cuvierian tubules, organs involved in defence. The former are present in every extant echinoderm species, whereas the latter occur only in some holothuroid species. These two systems rely on different types of adhesion and therefore differ in the way they operate, in their structure and in the composition of their adhesive.

## 9.2 Tube Feet

Being almost exclusively benthic animals, echinoderms have activities and adaptations that are correlated with a relationship with the sea bottom. Most of these activities, such as attachment to the substratum, locomotion, handling of food and burrow-building, rely on adhesive secretions allowing the animal to stick to or to manipulate a substratum. In post-metamorphic echinoderms, these adhesive secretions are always produced by specialised organs, the podia or tube feet. These are the external appendages of the water–vascular system and are also probably the most advanced hydraulic organs in the animal kingdom (Nichols 1966). Tube foot attachment is typically temporary adhesion. Indeed, although tube feet can adhere very strongly to the substratum, they are also able to detach easily and voluntarily from the substratum before reinitiating another attachment–detachment cycle (Thomas and Hermans 1985; Flammang 1996).

### *9.2.1 Morphology and Adhesion Strength*

From their presumed origin as simple respiratory evaginations of the ambulacral system (Nichols 1966), tube feet have diversified into the wide range of specialised structures found in extant echinoderms. This morphological diversity reflects the variety of functions of tube feet (Lawrence 1987). In some groups, a single type of tube foot fulfils different functions; in others, different types of tube feet are specialised, each in one particular function. Based on their external morphology only, tube feet can be classified into six broad types: disc-ending (Fig. 9.1a), penicillate, knob-ending, lamellate, digitate (Fig. 9.1b) and ramified (Flammang 1996). Adhesive areas are organised differently according to the morphotype, and this organisation represents the first stage of specialisation of the tube feet. Tube feet that capture or manipulate small particles present an adhesive area fragmented into small, discrete zones (Flammang 1996). This is the case, for example, in the adhesive papillae scattered on the tube feet of filter-feeding ophiuroids (Fig. 9.1b). Discrete adhesive zones are presumably more efficient in the handling of small particles; conversely, a large adhesive area provides a strong attachment site for tube feet involved in locomotion or in maintaining position (Flammang 1996). Such large adhesive areas occur on the distal surface of the disc in disc-ending podia (Fig. 9.1a).

For practical reasons (relatively large size and high adhesion force of the tube feet), only disc-ending tube feet of asteroids and regular echinoids have been studied in detail in terms of adhesion. These tube feet consist of a basal hollow cylinder, the stem and an enlarged and flattened apical extremity, the disc (Figs 9.1a and 9.2a, b). The different constituents making up these two parts act cooperatively to make tube feet an efficient holdfast, allowing sea stars and sea urchins to resist hydrodynamically generated forces, but also to perform rather elaborate tasks such as climbing, righting, covering or shell opening (Lawrence 1987). The stem acts as a tough tether connecting the disc to the animal's body. It is also mobile and flexible and thus gives the tube foot the capacity to perform various movements. The disc, on the other hand, makes contact with the substratum (Fig. 9.2a). It adapts to the surface profile, produces the adhesive secretion that fastens the tube foot to the substratum and encloses support structures that bear the tensions associated with adhesion. It also produces the de-adhesive secretion that allows detachment of the tube foot.

Tube foot adhesive strength has been evaluated by measuring their tenacity, which is the adhesion force per unit area and is expressed in Pascals (Pa). The normal tenacity of single disc-ending tube feet has been quantified in several species of asteroids and echinoids under different conditions. Mean tenacity ranges from 0.17 to 0.43 MPa in asteroids and from 0.09 to 0.54 MPa in echinoids (Table 9.1). Tenacity was shown to be dependent on the chemical and physical characteristics of the surface to which the tube foot adheres (Santos et al. 2005a; Santos and Flammang 2006). In the sea urchin *Paracentrotus lividus*, the tenacity of single tube foot discs on four different smooth substrata was compared and showed that both the total surface energy and the ratio of polar to nonpolar forces at the surface influence tube foot attachment strength. In both asteroids and echinoids, it was demonstrated that tube feet show increased adhesion on a rough substratum in comparison to its smooth counterpart (Santos et al. 2005a). This is because the disc adhesive surface is highly compliant, replicating the substratum profile. The

**Fig. 9.1** Morphological diversity in echinoderm tube feet (for comparison, tube feet have been oriented distal end up). (**a**) Disc-ending tube foot of the echinoid *Heterocentrotus trigonarius*. (**b**) Digitate tube foot of the ophiuroid *Ophiothrix fragilis*. *Arrows* indicate large adhesive areas and *arrow heads* small adhesive zones (see text for details). Modified from Santos et al. (2009a)

increase in contact area between the disc and the substratum leads to a higher adhesion force (Santos et al. 2005a). Tube foot discs and their adhesive secretions therefore appear to be well-tailored to provide an efficient attachment to natural rocky substrata, allowing echinoderms to resist hydrodynamically generated forces, but also to a large range of artificial substrata.

Suction has long been regarded as the primary functional mean for attachment in sea star and sea urchin tube feet (Nichols 1966; Lawrence 1987). However, detailed morphological and biomechanical observations clearly showed that echinoderm tube feet rely on adhesive secretions and not on suction (Thomas and Hermans 1985; Flammang et al. 1994; Hennebert et al. 2012a). Indeed, microscopy observations of tube feet rapidly fixed while they were attached to a smooth substratum showed that their distal surfaces are totally flat and lack a suction cavity (Thomas and Hermans 1985; Hennebert et al. 2012a). Moreover, detachment force and tenacity of single tube feet do not vary with pulling angle or surface perforation, as would be expected for a sucker (Santos et al. 2005a; Hennebert et al. 2012a).

### 9.2.2 Histology and Ultrastructure

The histological structure of the tube feet is remarkably constant for all echinoderm species. Their tissue stratification consists of four layers: an inner myomesothelium surrounding the water–vascular lumen, a connective tissue layer, a nerve plexus and

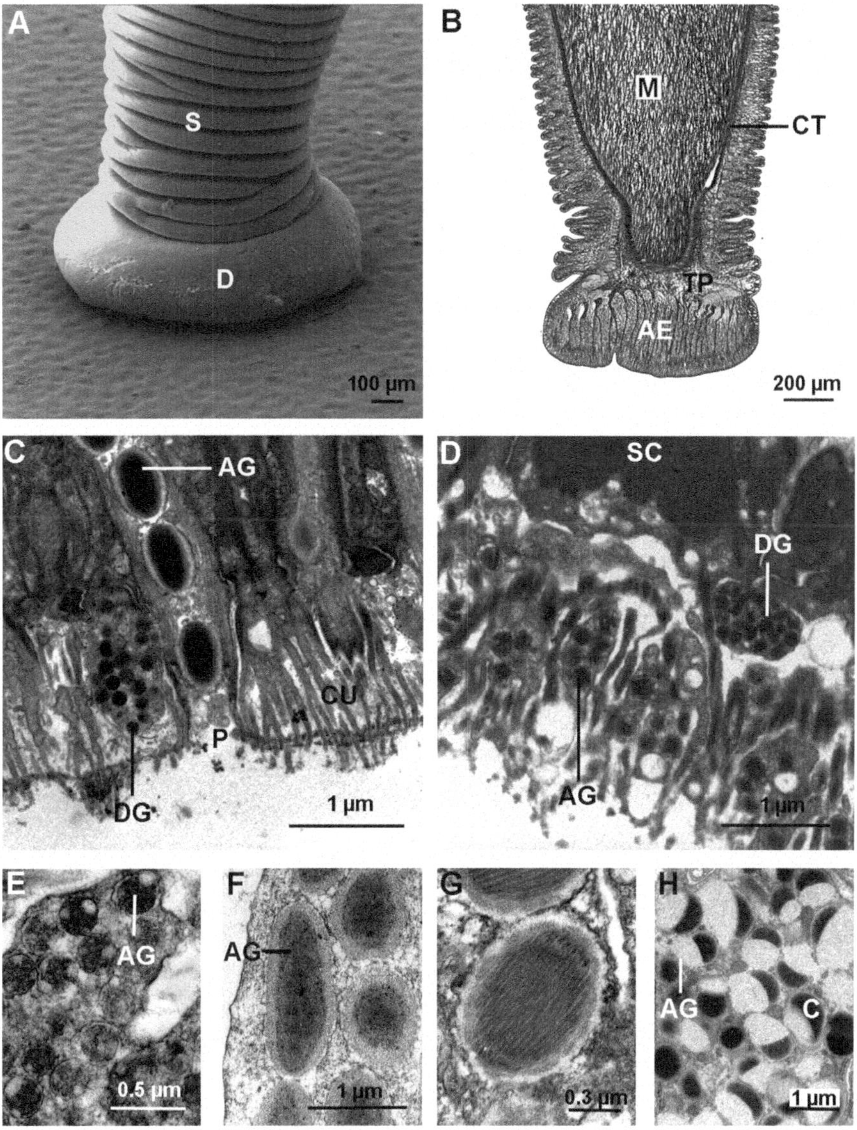


**Fig. 9.2** Fine structure of echinoderm tube feet (modified from Flammang 1996 and Santos et al. 2009a). (**a**) SEM photograph of a disc-ending tube foot of the sea star *Asterias rubens* attached to a textured polymer substratum. (**b**) Longitudinal LM section through a tube foot of *A. rubens*. (**c**, **d**) Longitudinal TEM sections through the adhesive epidermis of tube foot discs of the asteroid *Marthasterias glacialis* and of the echinoid *Sphaerechinus granularis*, respectively. (**e–h**) Ultrastructure of the secretory granules of the adhesive cells from echinoderm tube feet. Heterogeneous granules in the echinoid *S. granularis* (**e**), dense-cored granules in the ophiuroid *Asteroxyx loveni* (**f**), granules with a central filamentous bundle in the asteroid *A. rubens* (**g**), capped granules in the holothuroid *Holothuria forskali* (**h**). *AE* adhesive epidermis, *AG* adhesive secretory granule, *C* cap, *CT* connective tissue layer, *CU* cuticle, *D* disc, *DG* de-adhesive secretory granule, *M* myomesothelium, *P* secretory pore, *S* stem, *SC* support cell, *TP* terminal plate

**Table 9.1** Adhesion strength measured for single tube feet of sea stars and sea urchins on various smooth substrata

| Species | Substratum | Tenacity (MPa) | Reference |
|---|---|---|---|
| Asteroids | | | |
| *Asterias rubens* | Glass | 0.20–0.24 | Flammang and Walker (1997) and Hennebert et al. (2010) |
| | PMMA | 0.18 | Santos et al. (2005a) |
| *Asterias vulgaris* | Glass | 0.17 | Paine (1926) |
| *Marthasterias glacialis* | Glass | 0.43 | Hennebert et al. (2010) |
| Echinoids | | | |
| *Arbacia lixula* | Glass | 0.09 | Santos and Flammang (2006) |
| *Colobocentrotus atratus* | PMMA | 0.54 | Santos and Flammang (2008) |
| *Echinometra mathaei* | PMMA | 0.22 | Santos and Flammang (2008) |
| *Heterocentrotus trigonarius* | PMMA | 0.25 | Santos and Flammang (2008) |
| *Paracentrotus lividus* | Glass | 0.29–031 | Santos and Flammang (2006) |
| | PMMA | 0.34 | Santos et al. (2005a) and Santos and Flammang (2006) |
| | PP | 0.14–0.17 | Santos et al. (2005a) and Santos and Flammang (2006) |
| | PS | 0.29 | Santos and Flammang (2006) |
| *Sphaerechinus granularis* | Glass | 0.20 | Santos and Flammang (2006) |
| *Stomopneustes variolaris* | PMMA | 0.21 | Santos and Flammang (2008) |

*PMMA* poly(methylmetacrylate), *PP* polypropylene, *PS* polystyrene

an outer epidermis covered externally by a cuticle (Flammang 1996). At the level of the tube foot disc, these tissue layers are specialised for strong attachment. The discs of both asteroid and echinoid tube feet consist of two superposed layers of approximately equal thickness: a proximal supporting structure bearing the tensions associated with adhesion and a distal adhesive pad making contact with the substratum and producing the adhesive secretion that fastens the tube foot to this substratum (Fig. 9.2b; Santos et al. 2005a, 2009a). There are, however, differences in the organisation of these layers between sea star and sea urchin discs.

The supporting structure consists mostly of a circular plate of connective tissue, the so-called terminal plate that is composed of densely packed collagen fibres (Fig. 9.2b). In both sea stars and sea urchins, numerous branching connective tissue septa (made up mostly of collagen fibres) emerge from the distal surface of the terminal plate, manoeuvring themselves between the epidermal cells of the adhesive pad. The thinnest, distal branches of these septa attach apically to the support cells of the epidermis. In sea stars, these septa are arranged as well-defined radial lamellae (Fig. 9.2b), whereas in sea urchins they form a more irregular meshwork (Santos et al. 2005a). On its proximal side, the terminal plate is continuous with the connective tissue sheath of the stem.

The adhesive pad is composed of a thick adhesive epidermis reinforced by the branching connective tissue septa (Fig. 9.2b). This epidermis is much thicker than the stem epidermis. Externally, it is covered by a well-developed, multilayered glycocalyx, the so-called cuticle (Ameye et al. 2000). As a general rule, epidermal adhesive areas of echinoderm tube feet always consist of four cell categories: support cells, sensory cells, adhesive cells of one (in echinoid tube feet) or two types (in asteroid tube feet) and de-adhesive cells (see Flammang 1996, and Santos et al. 2009a, for review). The study of the ultrastructure of the adhesive and de-adhesive cells during a complete cycle of attachment–detachment of the tube foot in *Asterias rubens* demonstrated that they function as a duo-gland adhesive system as originally proposed by Hermans (1983), and in which adhesive cells (types 1 and 2) release an adhesive secretion and de-adhesive cells a de-adhesive secretion (Flammang et al. 1994; Flammang 1996; Flammang et al. 1998; Hennebert et al. 2008). In this species, polyclonal antibodies have been raised against footprint material and were used to locate the origin of footprint constituents in the tube feet (Flammang et al. 1998). Extensive immunoreactivity was detected in the secretory granules of both types of adhesive cells, confirming that their secretions make up together the bulk of the adhesive material.

Two modes of granule secretion can be recognised according to the morphology of the apex of the adhesive cell (McKenzie 1988; Flammang and Jangoux 1992; Flammang 1996; Santos et al. 2009a). In 'apical duct' cells, secretory granules are extruded through a duct delimited by a ring of microvilli and opening onto the tube foot surface as a cuticular pore (Fig. 9.2c). This kind of adhesive cell occurs in asteroid, ophiuroid and crinoid tube feet, as well as in holothuroid locomotory tube feet (Flammang 1996). In 'apical tuft' cells, secretory granules are released at the tip of microvillar-like cell projections which are arranged in a tuft at the cell apex (Fig. 9.2d). This second kind of adhesive cell has been observed only in echinoid tube feet and holothuroid locomotory tube feet and buccal tentacles (Flammang 1996).

Although the ultrastructure of the de-adhesive cell granules is remarkably constant from one echinoderm taxon to another (Fig. 9.2c,d), the one of the adhesive cell granules varies extensively. These secretory granules are usually made up of at least two materials of different electron density, which gives them a complex ultrastructure. Five broad categories can be recognised (Flammang 1996):

1. Homogeneous granules apparently made up of only one material;
2. Heterogeneous granules in which two different materials are mixed in an irregular pattern (Fig. 9.2e);
3. Dense-cored granules consisting of an electron-denser core surrounded by less dense material (Fig. 9.2f);

4. Granules with a central filamentous bundle resembling granules of the previous group but in which the core is made up of a parallel arrangement of fibrils or rods (Fig. 9.2g);
5. Capped granules in which an electron-lucent material is covered, on one side, by a cap of electron-dense material (Fig. 9.2h).

The significance of these ultrastructural differences between different echinoderm taxa is unknown at present. However, in asteroids, Engster and Brown (1972) pointed out a relationship between the internal organisation of adhesive cell secretory granules and species habitat: asteroids confined to hard rocky substratum have complex granules enclosing a highly organised core, whereas soft substratum dwelling species have granules of considerably simpler ultrastructure. They suggested that the different substructure of the adhesive cell granules would depend on the nature and composition of their contents that, in turn, could be related to the possible adhesive strength of the tube feet.

### *9.2.3 Fine Structure and Composition of the Adhesive Material*

In all echinoderm species investigated so far, after detachment of the tube foot, the adhesive secretion usually remains firmly bound to the substratum as a footprint (Fig. 9.3a). The material constituting these footprints can be stained, allowing the observation of their morphology under the light microscope (Chaet 1965; Thomas and Hermans 1985; Flammang 1996; Santos and Flammang 2006; Hennebert et al. 2008). In both sea stars and sea urchins, the footprints have the same shape and the same diameter as the distal surface of the tube foot discs (Fig. 9.3a). Various techniques have been used to study the fine structure of the material constituting the footprints, and whatever the method used, the adhesive material always appears as a foam-like or sponge-like material made up of a fibrillar matrix with numerous holes in it (Flammang et al. 1994; Flammang et al. 1998; Flammang 2006; Hennebert et al. 2008; Santos et al. 2009a). This aspect has been observed in LM, SEM (Fig. 9.3b) and AFM (Fig. 9.3c); and it does not differ according to whether the footprint has been fixed or not (Flammang et al. 1998; Flammang 2006; Hennebert et al. 2008; Higgins and Mostaert 2013). In both asteroid and echinoid footprints, one can distinguish a very thin and homogeneous priming film covering the substratum on which the fibrillar matrix is deposited (Fig. 9.3b,c). In sea stars, the fibrils tend to form a loose meshwork with relatively large meshes, about 2–5 μm in diameter (Fig. 9.3b, c). The walls delimiting the meshes may be quite thick (up to 1 μm), and under the AFM, they appear as strings of little beads (Fig. 9.3c; Hennebert et al. 2008). In sea urchin and sea cucumber footprints, the meshwork appears denser, with smaller meshes ($<1$ μm) delimited by very fine fibrils (about 50 nm in diameter) (Santos et al. 2009a). These differences in ultrastructure could be linked to the way the adhesive secretions are delivered to

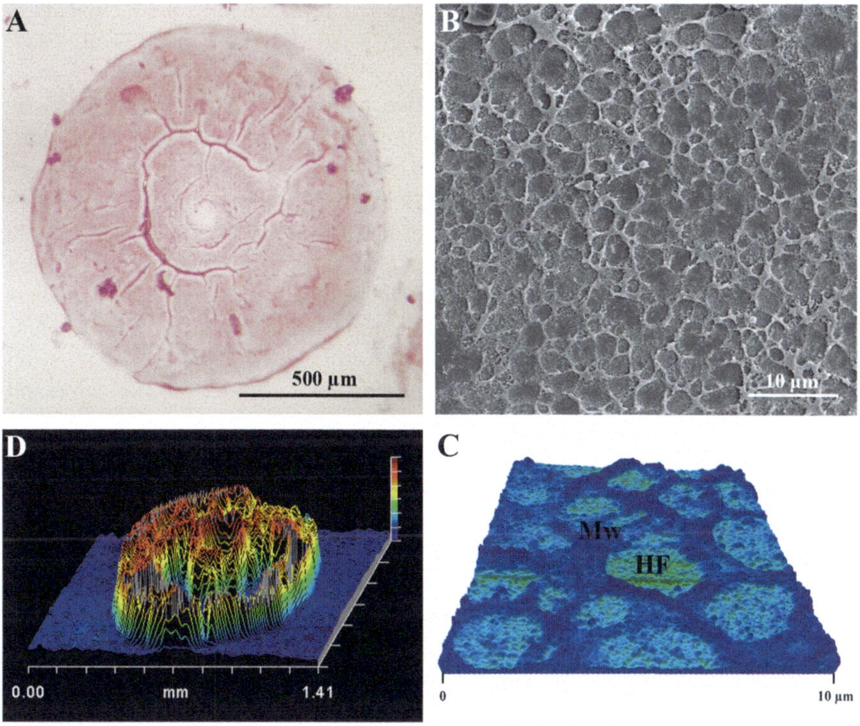


**Fig. 9.3** Micro- and nanostructure of the adhesive footprints left by echinoderm tube feet (modified from Flammang et al. 2005; Hennebert et al. 2008, 2012a). (**a**) LM photograph of a footprint of *Paracentrotus lividus* stained with a 0.05 % aqueous solution of the cationic *dye crystal violet*. (**b–c**) Details of footprints deposited by the tube feet of *Asterias rubens* on pieces of glass and observed with SEM and AFM, respectively. Both views show areas where the adhesive material forms a meshwork deposited on a thin homogeneous film. (**d**) 3-D topographical view of a footprint from P. lividus deposited on a glass substratum and air-dried (vertical scale: 0–80 nm). *HF* homogeneous film, *Mw* meshwork

the substratum, viz. through secretory pores in asteroids and at the apex of microvillar-like cell projections in echinoids. Indeed, the loose meshwork of sea star footprints reflects approximately the distribution of the secretory pores on the tube foot disc surface (Hennebert et al. 2008), while the denser meshwork of sea urchin footprints is more reminiscent of the dense array of cell projections covering their disc surface.

The thickness of the fibrillar matrix may vary from one footprint to another but also between different areas of the same footprint (Flammang et al. 1994; Hennebert et al. 2008; Santos et al. 2009a). However, footprint thickness is difficult to estimate. Using an interference–optical profilometer, which generated three-dimensional images of the footprint surface, the mean maximum thickness of dry footprints was found to be of 100 nm in the echinoid *P. lividus* (Fig. 9.3d) and of 230 nm in the asteroid *A. rubens* (Flammang et al. 2005). On the other hand, based on TEM observations, the thickness of the adhesive layer ranges from 0.2 to 9 μm in *A. rubens* and, at least, from 0.3 to 2 μm in *P. lividus* (Flammang et al. 1994; Hennebert et al. 2008; Santos et al. 2009a).

The composition of echinoderm adhesive material was first investigated by histochemical tests performed on tube foot longitudinal sections and on footprints. In tube feet, acidic glycosaminoglycans (GAGs) were detected in the secretory granules of adhesive cells (Defretin 1952; Chaet and Philpott 1964; Chaet 1965; Souza Santos and Silva Sasso 1968; Engster and Brown 1972). Later, Perpeet and Jangoux (1973) also demonstrated the presence of proteins associated to these GAGs in the sea star *A. rubens*. The footprints of the sea urchin *Sphaerechinus granularis* contain GAGs but no proteins (Flammang and Jangoux 1993), whereas the footprints of the sea stars *Asterias forbesi*, *A. rubens*, *Leptasterias hexactis* and *Marthasterias glacialis* stain for both proteins and GAGs (Chaet 1965; Thomas and Hermans 1985; Flammang et al. 1994).

At present, data on the biochemical composition of the adhesive footprints is only available for the sea star *A. rubens* and the sea urchin *P. lividus*. The water content of the adhesive material has never been measured, but in terms of dry weight, footprints are made up mainly of proteins (20.6 % in sea stars and 6.4 % in sea urchins), carbohydrates (8 % in sea stars and 1.2 % in sea urchins) and a large inorganic fraction (40 % in sea stars and 45.5 % in sea urchins) (Fig. 9.7a; Flammang et al. 1998; Santos et al. 2009b). This composition is in accordance with the previously mentioned histochemical tests. In addition, lipids were also detected in footprints (5.6 % in sea stars and 2.5 % in sea urchins), although they have not been detected in the secretory granules of adhesive cells by histochemistry (Perpeet and Jangoux 1973; Flammang et al. 1998; Santos et al. 2009b).

#### 9.2.3.1 Protein Fraction

Sea star and sea urchin adhesive footprints have been analysed in terms of amino acid composition, and due to their insolubility, all the analyses had to be performed under hydrolytic conditions (Table 9.2). The footprint material of *A. rubens* contains slightly more polar (55 %) than nonpolar (45 %) residues and, among the former, more charged (34 %) than uncharged residues (21 %). Sea star adhesive presents a strong bias towards asparagine/aspartic acid (11.8 %), glutamine/glutamate (10.2 %) and glycine (9.7 %), followed by threonine (7.8 %) and serine (7.6 %) (Fig. 9.7b; Flammang et al. 1998). As for sea urchins, the footprints of *P. lividus* present more nonpolar (57.4 %) than polar amino acids (42.6 %), and equivalent amounts of both charged (20.2 %) and uncharged residues (22.4 %). The adhesive material presents a significant bias towards glycine (14.7 %), followed by alanine (9.8 %), valine (8.9 %), serine (8.6 %) and threonine (7.4 %) (Fig. 9.7b; Santos et al. 2009b). In addition, both sea star and sea urchin footprints present higher levels of half-cystine (3.2 and 2.6 %, respectively) than the average eukaryotic proteins.

**Table 9.2** Amino acid composition of the adhesive secretions from echinoderm tube feet (values in residues per thousand)

| Amino acid | *Asterias rubens*[a] | *Paracentrotus lividus*[b] |
|---|---|---|
| ASX | 118 | 48 |
| THR | 78 | 74 |
| SER | 76 | 86 |
| GLX | 102 | 74 |
| PRO | 61 | 68 |
| GLY | 97 | 147 |
| ALA | 62 | 98 |
| CYS/2 | 32 | 26 |
| VAL | 67 | 89 |
| MET | 17 | 19 |
| ILE | 45 | 50 |
| LEU | 61 | 72 |
| TYR | 27 | 38 |
| PHE | 38 | 31 |
| HIS | 56 | 13 |
| LYS | 21 | 27 |
| ARG | 41 | 40 |

[a]Flammang et al. (1998)
[b]Santos et al. (2009b)

### Adhesive Proteins

The use of strong denaturing and reducing extraction conditions allowed the solubilisation of sea star and sea urchin footprint proteins (Santos et al. 2009b; Hennebert et al. 2012b). The need for these harsh solubilisation conditions plus the above-mentioned biased amino acid composition provides evidence for the importance of non-covalent interactions and disulphide bonds between the adhesive proteins in footprint cohesion. Indeed, hydrophobic and electrostatic interactions involve nonpolar and charged polar amino acids, respectively, both present in significant amounts in sea star and sea urchin adhesives (Santos et al. 2009b; Hennebert et al. 2012b). As for disulphide bonds, they probably reinforce the cohesive strength and insolubility of the adhesive footprints, either by intermolecular bonds that covalently cross-link the proteins or by intramolecular bonds that hold proteins in the specific shape required for interaction with their neighbours (Flammang et al. 1998; Hennebert et al. 2012b). In sea urchins, the presence of disulphide bonds was further corroborated by the observation of a mobility shift for three proteins in 2D nonreducing/reducing diagonal gels, attributed to the presence of intra- or intermolecular disulphide bonds in these proteins (Santos et al. 2009b).

Upon gel electrophoresis, sea star footprint protein extracts separated into 11 major and several minor protein bands. Using mass spectrometry (MS) analyses and homology-database search, it was shown that most of the minor protein bands correspond to known intracellular proteins, presumably contaminants from cellular epidermal material, while no homolog proteins were identified for the major protein bands. The later were further analysed by tandem MS (MS/MS), yielding 43 de novo-generated peptide sequences (Hennebert et al. 2012b). The same approach was applied to sea urchin footprint protein extracts, highlighting 13 major protein bands, 6 of which were known intracellular proteins such as actins, tubulins and histones. The remaining unidentified protein bands were further processed by automated de novo peptide sequencing, but as for sea stars, no homologies were found for the deduced peptide sequences, suggesting that these adhesive proteins might be either novel or highly modified (Santos et al. 2009b).

More recently, the proteome of sea star adhesive footprints was established using high-throughput sequencing of expressed tube foot mRNAs (transcriptome analysis) combined to MS-based identification of footprint proteins. The tube foot transcripts coding for proteins identified in the adhesive footprints were then functionally annotated by similarity searches against the NCBI nr database (Hennebert et al. 2015a). The results showed that the adhesive secretion is made up of 34 proteins. Most of these proteins were not annotated in public databases and probably correspond to novel adhesive proteins (Hennebert et al. 2015a). Regarding the annotated proteins, some present a strong potential to play a role in sea star adhesion. One is similar to tachylectin-like proteins, lectins able to bind to various carbohydrates. Such a protein would be a good candidate as a component of the footprint homogeneous priming film, where it would promote adhesion to the biofilm present on the surface of the substratum. Two proteins are similar to the IgGFc binding protein, a mucin-like protein forming structural networks through oligomerisation. In sea star adhesive footprints, the two mucin-like proteins could be involved in the formation of structural networks through their potential ability to oligomerise and/or cross-link to other adhesive molecules. Finally, some footprint proteins were annotated on the basis of the presence of functional domains such as hyalin, EGF and discoidin domains, known from other studies to mediate protein–protein, protein–carbohydrate or protein–metal interactions.

To date, only one of the unannotated proteins, the sea star footprint protein 1 (Sfp1, UniProt X2KZ73), has been fully characterised (Hennebert et al. 2014). This large protein of 3,853 amino acids is the second most abundant constituent of the secreted adhesive. It contains 23 of the 43 de novo-generated peptides obtained from the MS/MS analysis of the major protein bands (Hennebert et al. 2012b). MS and Western blot analyses showed that Sfp1 is translated from a single mRNA and then cleaved into four subunits linked together by disulphide bridges in sea star tube foot adhesive cells (Fig. 9.4a). The four subunits display specific protein-, carbohydrate- and metal-binding domains that mediate interactions with other proteins present in the adhesive material and on the tube foot surface (Hennebert et al. 2014). In situ hybridisation located the mRNA coding for Sfp1 in the tube foot adhesive epidermis (Fig. 9.4b) (Hennebert and Flammang unpubl. obs.). Using immunohisto- and immunocytochemistry, the precise location of Sfp1 was revealed at the level of the rods of the secretory granules enclosed in type 1 adhesive cells (Fig. 9.4c). Within the adhesive footprints, Sfp1 was located in the fibrillar

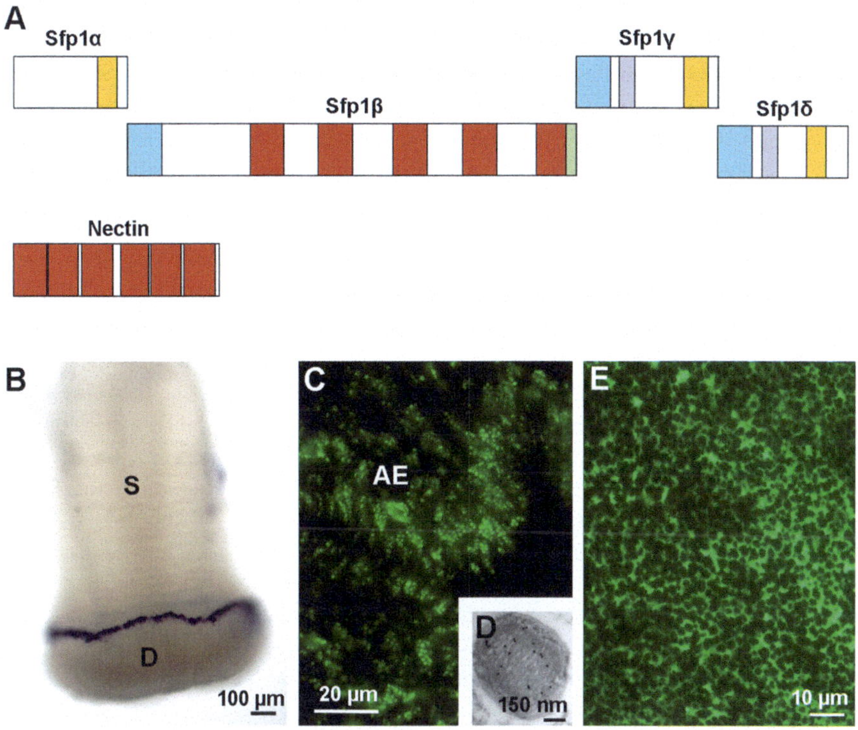


**Fig. 9.4** Echinoderm adhesive proteins (adapted in part from Hennebert et al. 2014). (**a**) Subunits and predicted structural domains of Sfp1 (*top*) and Nectin (*bottom*). The two proteins are drawn at the same scale. *Green*, calcium-binding EGF-like domain; *yellow*, galactose-binding lectin domain; *red*, discoidin domain (also known as coagulation factor 5/8 C-terminal domain); *blue*, von Willebrand Factor; *purple*, C8 domain. (**b**) In situ hybridisation on a tube foot of *Asterias rubens* using probes designed on the basis of the cDNA coding for Sfp1 (whole mount). (**c**–**e**) Immunolabelling of tube feet and footprints in *A. rubens* with antibodies directed against Sfp1β. Longitudinal section through the adhesive epidermis in which secretory granules of adhesive cells are clearly labelled (**c**, immunofluorescence). Immunocytochemical localisation of Sfp1 in the secretory granules from type 1 adhesive cells (**d**, immunogold labelling in TEM). In the adhesive footprints left on the substratum after tube foot detachment, immunoreactivity demonstrates that Sfp1 is localised at the level of the fibrillar meshwork (**e**, immunofluorescence). *AE* adhesive epidermis, *D* disc, *S* stem

meshwork (Fig. 9.4d) and therefore seems to provide cohesion to the adhesive layer, rather than adhesive properties (Hennebert et al. 2014).

As for sea urchins, a first attempt to obtain the tube foot disc proteome successfully identified 328 nonredundant proteins, but since the disc presents a complex histological structure, only 2 % were categorised as putative adhesive proteins (Santos et al. 2013). More recently, high-resolution quantitative mass spectrometry was used to perform the first study combining the analysis of sea urchin tube foot differential proteome with the proteome of its secreted adhesive (Lebesgue et al. 2016). The differential tube foot proteome of *P. lividus* allowed comparing protein expression in the adhesive part (the disc) versus the nonadhesive part (the stem), resulting in the identification of 163 highly over-expressed disc proteins. In addition, the analysis of the footprint proteome yielded 611 proteins among which more than 70 % fall within five protein groups: actins (27.9 %), histones (24.4 %), tubulins (11.9 %), ribosomal proteins (7.9 %) and myosins (1.4 %). In all these proteomic studies, one protein was repeatedly pinpointed as a putative adhesive protein, Nectin (Santos et al. 2013; Lebesgue et al. 2016).

Nectin (Uniprot Q70JA0), a cell adhesion protein secreted by the eggs and embryos of *P. lividus*, was shown to significantly increase the binding of dissociated embryonic cells to the substratum (Matranga et al. 1992). Its identification in tube feet led to the hypothesis that it could also be involved in substratum attachment in adult sea urchin (Santos et al. 2013). Indeed, nectin contains six galactose-binding discoidin-like domains (Fig. 9.4a; Costa et al. 2010) and can therefore bind molecules bearing galactose and *N*-acetylglucosamine carbohydrate moieties on the substratum, on the cuticle or within the adhesive material. Several variants differing only by a few amino acids were identified in the adhesive material, totalising 1.2 % of the footprint proteins in terms of relative abundance (Lebesgue et al. 2016). All these nectin variants are also highly over-expressed (5.4- to 13-fold) in the tube foot disc relatively to the stem (Lebesgue et al. 2016). Nectin cDNA was amplified from tube foot mRNAs showing that, in addition to the known embryonic Nectin mRNA called Nectin-1, a new mRNA sequence called Nectin-2 (GenBank KT351732), differing by 15 nucleotide substitutions, is also expressed in the tube feet of adult sea urchins. These Nectin variants most likely derive from nucleotide substitutions (SNPs) during DNA replication due, for example, to high gene expression (Toubarro et al. 2016). The two Nectin mRNAs were found to be highly over-expressed in tube foot discs comparatively to stems (Toubarro et al. 2016). Finally, Nectin was successfully immunolocalised in the adhesive cells of the tube foot disc as well as in the footprints, confirming the adhesive function of the protein (Lebesgue et al. 2016). This adhesive function is further corroborated by the fact that Nectin expression might be regulated according to the hydrodynamic conditions. Indeed, its expression is significantly higher in tube feet from freshly collected sea urchins than in tube feet from sea urchins maintained in an aquarium (Toubarro et al. 2016).

### De-adhesive Proteins

When tube foot detachment occurs, it always takes place at the level of the outermost layer of the disc cuticle, the fuzzy coat, leaving the adhesive material strongly attached to the substratum as a footprint (Fig. 9.3a) (Flammang 1996). In *A. rubens*, the polyclonal antibodies raised against footprint material strongly label the fuzzy coat, but no immunoreactivity is detected in the secretory granules of de-adhesive cells (Flammang et al. 1998). This pattern of immunoreactivity suggests that secretions of de-adhesive cells are not incorporated into the footprints, but instead might function enzymatically to jettison the fuzzy coat thereby allowing the tube foot to detach (Flammang 1996; Flammang et al. 1998). In this connection, enzymes (both proteases and glycosylases) were searched in the recently obtained proteomes as they could represent putative de-adhesive proteins. Two proteases were identified in the sea star footprint proteome, which are similar to enzymes presenting a metalloendopeptidase activity (Metalloproteinase SpAN and Tolloid-like protein 2) (Hennebert et al. 2015a). Similarly, several proteases and glycosylases are over-expressed in the sea urchin tube foot disc compared to the stem, indicating that they might be potential components of the de-adhesive secretions. These comprise proteases, such as aminopeptidases, dipeptidases, bleomycin hydrolase-like and cathepsin *z*, and glycosylases such as *N*-(beta-*n*-acetylglucosaminyl)-L-asparaginase, carbohydrate binding module 9-containing protein and sialidases (Lebesgue et al. 2016). Although the carbohydrate fraction of the adhesive of *P. lividus* is still poorly characterised, the identified glycosylases might be an indication that it could contain sialylated oligosaccharides presumably conjugated to proteins through asparagine residues, similarly to sea star adhesive (Hennebert et al. 2011).

#### 9.2.3.2 Carbohydrate Fraction

Based on colorimetric assays, Flammang et al. (1998) showed that the carbohydrate fraction (in dry weight) of sea star footprints is made up of neutral sugars (3 %), amino sugars (1.5 %) and uronic acids (3.5 %) (Fig. 9.7a). In sea urchin footprints, the presence of neutral carbohydrates (1.2 % of the footprint dry weight) was also demonstrated (Santos et al. 2009b), but no quantification was yet performed for amino sugars and uronic acids (Fig. 9.7a).

In sea stars, the composition of the carbohydrate moiety was further investigated using lectins, molecules that specifically recognise carbohydrate residues. These lectins were used on tube foot histological sections, on footprints and on adhesive protein extracts upon separation on polyacrylamide gels. The results indicate that at least two glycoproteins, as well as larger molecules such as proteoglycans, compose the carbohydrate fraction of sea star footprints. The sugar chains of both glycoproteins and proteoglycans appear to enclose mannose, galactose and sialic acid residues, and to a lesser extent *N*-acetylgalactosamine and fucose residues (Hennebert et al. 2011).

## 9.3 Cuvierian Tubules

Cuvierian tubules are peculiar defence organs found in about 60 species of sea cucumbers all belonging to the family Holothuriidae. Two main types of tubules can be differentiated on the basis of their gross external morphology, lobulated and smooth (Lawrence 2001). Lobulated tubules occur exclusively in the genus *Actinopyga*; they are never expelled and are not sticky (VandenSpiegel and

Jangoux 1993). These tubules are used as a toxic decoy to deter predators (Van Dyck et al. 2010). On the other side, smooth tubules are present in the genera *Bohadschia*, *Holothuria* and *Pearsonothuria*, in which they generally appear as sticky white threads that function as a defence mechanism (Hamel and Mercier 2000; Flammang et al. 2002). Indeed, once ejected, they can elongate and release a glue allowing instantaneous adhesion on any object and can therefore entangle a predator in a matter of seconds (Zahn et al. 1973; VandenSpiegel and Jangoux 1987).

Smooth Cuvierian tubules occur in great numbers (from 50 to 600 according to the species considered) in the posterior part of the body cavity of the holothuroid (Becker and Flammang 2010). Proximally they are attached to the basal part of the left respiratory tree, and their distal, blind ends float freely in the coelomic fluid. The mechanism leading to their discharge is as follows: when the animal is disturbed, it directs its posterior end toward the stimulating source and undergoes a general body contraction. Consequently, the wall of the cloaca breaks and the free ends of a few tens of tubules are expelled through the tear and the cloacal orifice. The water of the respiratory tree is then forcefully injected into the lumen of the tubules causing their elongation, up to 20 times their initial length. Upon contact with any surface (e.g., a predator integument), the lengthened tubules become instantly sticky. Finally, the elongated tubules autotomise at the attachment point on the respiratory tree and are left behind as the sea cucumber crawls away (VandenSpiegel and Jangoux 1987; Becker and Flammang 2010). Lost tubules are then regenerated in a few weeks (VandenSpiegel et al. 2000; Hamel and Mercier 2000). As only a portion of the tubules are emitted at one time, the total number may suffice for several responses (Hamel and Mercier 2000).

Four characteristics concur to make Cuvierian tubules very efficient as a defence system: (1) their large number, (2) their adhesiveness, (3) their mechanical design and (4) their regeneration capacities. Indeed, the adhesiveness of Cuvierian tubules combines with their tensile properties to entangle and immobilise potential predators (Zahn et al. 1973; Hamel and Mercier 2000; Flammang et al. 2002). On the other side, their large number, sparing use and regeneration dynamics make them almost inexhaustible line of defence (VandenSpiegel and Jangoux 1987; VandenSpiegel et al. 2000; Hamel and Mercier 2000).

### *9.3.1 Fine Structure and Adhesion Strength*

Cuvierian tubules consist of, from the inside to the outside, an epithelium surrounding the narrow lumen, a thick connective tissue layer and a mesothelium lining the surface of the tubule that is exposed to the coelomic cavity (Fig. 9.5a, b). The mesothelium is the layer responsible for adhesion. In quiescent tubules (i.e., non-expelled and non-elongated tubules), it is a pseudostratified epithelium made up of two superposed cell layers, an outer layer of peritoneocytes and an inner layer of granular cells which is highly folded along the long axis of the tubule (Fig. 9.5b).

Granular cells are flattened cells filled with electron-dense membrane-bound granules enclosing a proteinaceous material (Endean 1957; VandenSpiegel and Jangoux 1987; Delmeudre et al. 2014). They are organised in V-shaped structures (Fig. 9.5c). Peritoneocytes are T-shaped, displaying a flattened apical part lining the coelomic cavity and a thin elongated basal part running between the granular cell folds (Fig. 9.5c). They bear a single cilium and a few short microvilli, and their apical cytoplasm contains mucous vesicles (VandenSpiegel and Jangoux 1987; Delmeudre et al. 2014). During elongation, the structure of the mesothelium is modified: the protective outer layer of peritoneocytes disintegrates, and the granular cell layer, now unfolded, thus becomes outermost on the tubule (Fig 9.5d; VandenSpiegel and Jangoux 1987; Delmeudre et al. 2014). Granular cells empty the contents of their granules when the elongated tubule comes into contact with a surface, resulting in adhesion (VandenSpiegel and Jangoux 1987; Delmeudre et al. 2014). Once released, this material changes in aspect, swells and spreads readily on any type of substrate where it forms a thin homogeneous adhesive layer (Fig. 9.5e; Becker and Flammang 2010; Delmeudre et al. 2014).

The interspecific diversity of Cuvierian tubule histology among the three genera possessing smooth Cuvierian tubules is relatively low. However, some differences were observed in their fine structure, especially at the level of the mesothelium, with the highest variability within the genus *Holothuria* (Becker and Flammang 2010). In *H. hilla* and *H. leucospilota*, the granular cell layer is less folded than in *H. forskali*. In *H. maculosa*, the mesothelium presents an unusual morphology, being thicker with seemingly over-sized granular cells. In *H. nobilis* and in all the species from the genus *Bohadschia*, peritoneocytes lack mucous vesicles (Becker and Flammang 2010).

Cuvierian tubule adhesive strength on glass has been measured in seven species of sea cucumbers belonging to the genera *Bohadschia*, *Holothuria* and *Pearsonothuria* (Flammang et al. 2002). The mean normal tenacity observed varied from about 0.03 to 0.14 MPa. In the species *H. forskali*, tubule tenacity is influenced by the nature of the substratum: tubules adhere more strongly to polar than to nonpolar substrata, indicating the importance of polar interactions in adhesion (Flammang et al. 2002). A similar trend was observed for the tubules of *H. dofleinii* using peel tests (Peng et al. 2011). Moreover, in several species of the genus *Holothuria*, adhesive forces have been shown to vary with the temperature, salinity and pH of seawater (Zahn et al. 1973; Flammang et al. 2002; Peng et al. 2011).

### *9.3.2 Ultrastructure and Composition of the Adhesive Material*

After detachment of Cuvierian tubules from a substrate, the material left on the surface is called the tubule print material (TPM). The quantity of TPM deposited on

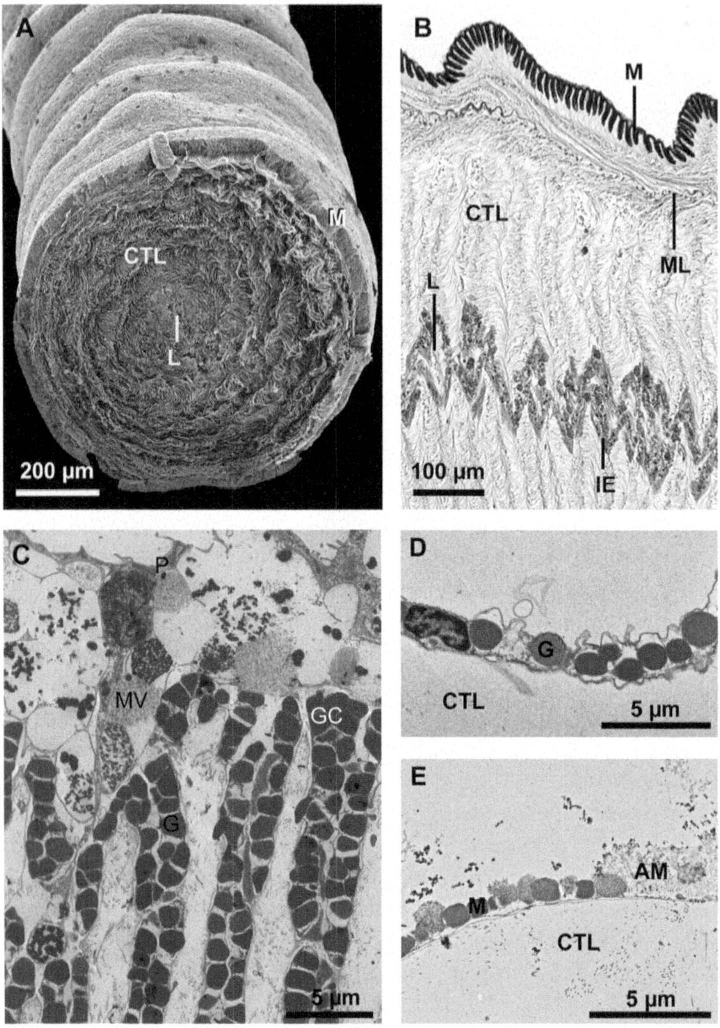


**Fig. 9.5** Morphology and ultrastructure of the Cuvierian tubules of *Holothuria impatiens* (**a**, **b**) and *Holothuria forskali* (**c**–**e**) (modified from Flammang et al. 2005, and Delmeudre et al. 2014). SEM photograph of a transversally sectioned tubule (**a**), and longitudinal histological section showing the arrangement of the tissue layers (**b**). (**c**) TEM of the apical part of the mesothelium of tubules before the elongation process. (**d**, **e**) TEM of the mesothelium of an elongated tubule before and after contact with a surface, respectively. *AM* adhesive material, *CTL* connective tissue layer, *IE* inner epithelium, *G* granule, *GC* granular cell, *L* lumen, *ML* muscle layer, *M* mesothelium, *MV* mucus vesicle, *P* peritoneocyte

a surface varies from one tubule print to another or even within a single print. The adhesive material appears as a thin homogeneous film which clearly derives from the secretory granules of granular cells (Fig. 9.6a). Different structures such as intact granules from granular cells or contaminating collagen fibres from the

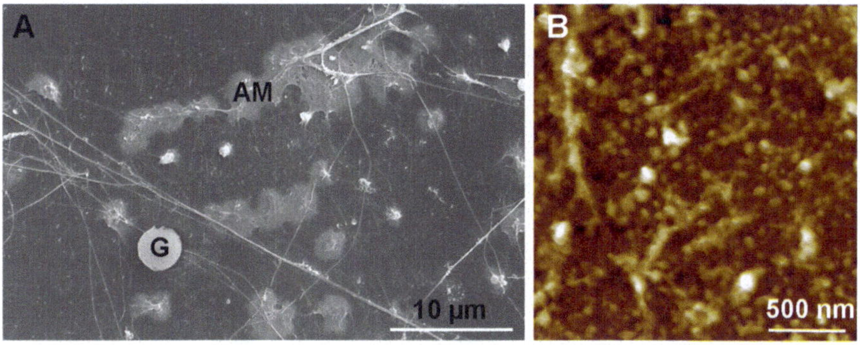


**Fig. 9.6** Tubule print material from *Holothuria forskali* (modified from Delmeudre et al. 2014). (**a**) SEM picture showing that the adhesive material derives from the secretory granules of granular cells. (**b**) AFM image of the adhesive material. *AM* adhesive material, *G* granule

connective tissue layer can be distinguished on this adhesive film (Fig. 9.6a). AFM observations demonstrated that the adhesive film is composed of nano-globular structures of about 70 nm in diameter (Fig. 9.6b) (Delmeudre et al. 2014).

In *H. forskali*, the TPM is composed of an organic fraction consisting of 54 % protein and 36 % carbohydrate, and of an inorganic fraction accounting for 10 % of the dry weight (Fig. 9.7a; De Moor et al. 2003). The proteinaceous nature of the adhesive material is confirmed by the observation that proteolytic enzymes and protein denaturation agents reduce the adhesive strength of Cuvierian tubules (Zahn et al. 1973; Peng et al. 2011). The amino acid compositions of the protein fraction in *H. forskali*, *H. leucospilota*, *B. subrubra* and *P. graeffei* indicate that their adhesives are closely related (Table 9.3). All are rich in small side-chain amino acids, especially glycine, and in charged and polar amino acids. The amino acid composition of the TPM from *H. maculosa* stands apart from all other Cuvierian tubule adhesives with only half their content in glycine but a much higher proportion of glutamate/glutamine and serine (Table 9.3).

Protein extractions using strong denaturing buffers containing both chaotropic and reducing agents were performed on TPM of *H. forskali* and *H. dofleinii*. In both species, the extracts contain about ten major protein bands with apparent molecular masses ranging from 17 to 220 kDa (De Moor et al. 2003; Flammang et al. 2009; Peng et al. 2011, 2014). In *H. forskali*, these proteins possessed closely related amino acid compositions, rich in glycine and in glutamine/glutamic acid residues (De Moor et al. 2003). In the same species, Baranowska et al. (2011) identified an 18 kDa protein cross reacting with antibodies raised against precollagen D, a byssal protein from the mussel *Mytilus galloprovincialis*. These authors claimed that this 18 kDa protein would be a major adhesive protein, but no sequence was reported. More recently, Peng et al. (2014) identified some proteins extracted from the adhesive material of *H. dofleinii*. Among the nine protein bands detected by gel electrophoresis, tandem mass spectrometry-based sequencing of tryptic peptides allowed the authors to identify two novel proteins, one C-type lectin and three

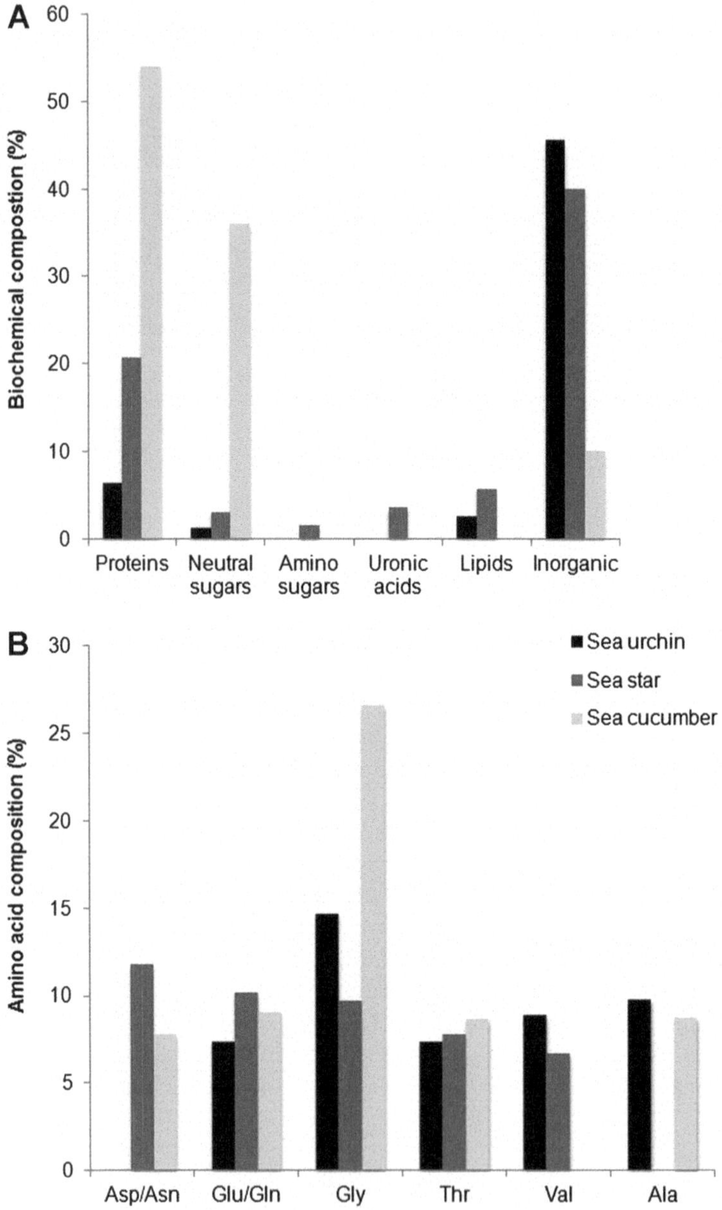


**Fig. 9.7** Comparison of the biochemical compositions (**a**) and of proportions of the five most abundant amino acids (**b**) between the adhesives of the sea star *Asterias rubens*, of the sea urchin *Paracentrotus lividus* and of the sea cucumber *Holothuria forskali*

**Table 9.3** Amino acid compositions of adhesive secretions from the Cuvierian tubules of several species of holothuroids (values in residues per thousand)

| Amino acid | *Bohadschia subrubra*[a] | *Holothuria forskali*[b] | *Holothuria leucospilota*[a] | *Holothuria maculosa*[c] | *Pearsonothuria graeffei*[a] |
|---|---|---|---|---|---|
| HYP | 8 | 0 | 24 | 0 | 8 |
| ASX | 64 | 78 | 74 | 97 | 62 |
| THR | 65 | 87 | 69 | 96 | 80 |
| SER | 58 | 60 | 42 | 99 | 58 |
| GLX | 106 | 91 | 122 | 162 | 124 |
| PRO | 69 | 55 | 74 | 41 | 63 |
| GLY | 298 | 266 | 267 | 125 | 254 |
| ALA | 91 | 88 | 115 | 63 | 85 |
| CYS/2 | 9 | 14 | 3 | 16 | 4 |
| VAL | 35 | 38 | 29 | 53 | 37 |
| MET | 1 | 10 | 9 | 12 | 9 |
| ILE | 25 | 28 | 24 | 23 | 32 |
| LEU | 37 | 37 | 31 | 50 | 38 |
| TYR | 17 | 20 | 14 | 29 | 17 |
| PHE | 20 | 20 | 16 | 37 | 20 |
| HIS | 8 | 26 | 13 | 17 | 20 |
| HLYS | 12 | 0 | 5 | 0 | 3 |
| LYS | 29 | 31 | 12 | 45 | 22 |
| ARG | 46 | 50 | 57 | 36 | 63 |

[a]Flammang et al. (2005)
[b]De Moor et al. (2003)
[c]Demeuldre and Flammang (unpublished obs)

enzymes associated to the pentose phosphate cycle and glycolysis. Partial cDNA sequences of three of these proteins (one novel protein, one C-type lectin and one transketolase) were retrieved by RT–PCR experiments. No confirmation of their adhesive function was provided, however.

The second most abundant fraction composing the Cuvierian tubule adhesive, the carbohydrate fraction, was investigated by histochemical experiments. Lectin labelling was used to detect the presence of oligosaccharidic structures on tubule sections from *H. forskali*. No labelling was found in the granular cells using seven lectins specific for neutral sugar containing oligosaccharides (Becker and Flammang 2010). However, lectin blots performed on TPM extracts with another set of six lectins suggest that at least three glycoproteins, containing galactose, *N*-acetylgalactosamine, *N*-acetylglucosamine and sialic acid residues, would be present in the adhesive material (Demeuldre and Flammang, unpublished obs.). In addition to glycosylation, another protein post-translational modification, phosphorylation, has been highlighted in the adhesive of Cuvierian tubules (Flammang et al. 2009). Using specific antibodies, phosphoserine residues were detected in the granular cells from the tubules of three different species (*B. subrubra*, *H. forskali* and *P. graeffei*) (Flammang et al. 2009). Immunoblots and amino acid analyses confirmed the presence of polyphosphoproteins in the adhesive secretion of Cuvierian tubules from these species (Flammang et al. 2009).

## 9.4 Comparisons Between Echinoderm Adhesives and with Other Marine Bioadhesives

Adhesion (attachment with adhesive secretions) is a way of life in the sea (Waite 1983). Indeed, representatives of bacteria, protoctists (including macroalgae) and all animal phyla, living in the sea, attach to natural or artificial surfaces. Adhesion ability is particularly developed and diversified in invertebrates, which adhere during their larval and adult life (see other chapters in this book). It is involved in various functions such as the handling of food, the building of tubes or burrows and, especially, the attachment to the substratum (Walker 1987; Tyler 1988; Whittington and Cribb 2001; Flammang et al. 2005). Indeed, seawater, being a dense medium, denies gravity to hold organisms to the bottom. Thus, to withstand the hydrodynamic forces, marine organisms rely on specialised adhesive mechanisms. Adhesion to the substratum may be permanent, transitory, temporary or instantaneous (Tyler 1988; Whittington and Cribb 2001; Flammang et al. 2005). Permanent adhesion involves the secretion of a cement and is characteristic of sessile organisms staying at the same place throughout their adult life (e.g., the attachment of barnacles on rocks). Transitory adhesion allows simultaneous adhesion and locomotion: the animals attach by a viscous film they lay down between their body and the substratum and creep on this film, which they leave behind as they move (e.g., the ventral secretions of turbellarian platyhelminths). Temporary adhesion allows organisms to attach firmly but momentarily to a substratum (e.g., the adhesion of echinoderm tube feet). The boundary between transitory and temporary adhesion is not always clear, however. Indeed, gastropod molluscs may use either transitory adhesion (in conjunction with suction) when they are moving, or temporary adhesion when stationary for a long period of time; the latter giving by far the greatest adhesive strength to the animal (see, e.g., Smith et al. 1999a). Instantaneous adhesion relies on single-use organs or cells and is used in functions other than attachment to the substratum requiring a very fast formation of adhesive bonds. Prey capture by collocyte-bearing tentacles of ctenophorans and defence reaction involving Cuvierian tubules in holothuroids are typical examples of this type of adhesion (Flammang et al. 2005).

The evaluation of the adhesive strength in marine invertebrates is usually done by measuring their tenacity, which is the adhesion force per unit area. According to the taxonomic group considered, tenacities of marine organisms range from about 0.001 to 2 MPa (see, e.g., Walker 1987, for review). Many studies have shown that several factors may profoundly influence the tenacity of invertebrates (see, e.g., Grenon and Walker 1981). For example, the physical (e.g., roughness) as well as chemical characteristics (e.g., hydrophobicity, surface charges) of the substratum are known to change the tenacity of organisms by up to one order of magnitude (Young and Crisp 1982; Yule and Walker 1987). As a consequence, great care should be exercised when comparing values of tenacity extracted from different studies. The mean tenacity of echinoderm tube feet on polymer and glass substrata ranges from 0.09 to 0.54 MPa. These values are in the same range as those observed in other marine invertebrates known to adhere strongly to such substrata (e.g., 0.17–0.23 MPa in limpets, Grenon and Walker 1981; 0.08–0.52 MPa in barnacles, Yule and Walker 1987; 0.12–0.75 MPa in mussels, Waite 2002). On the other hand, the mean normal tenacity measured for sea cucumber Cuvierian tubules, i.e., 0.03–0.14 MPa, falls within the lower range of adhesive strengths described for marine organisms (Flammang et al. 2002).

In marine invertebrates, adhesive secretions are always predominantly made up of proteins. Yet, their biochemical composition varies from one taxonomic group to another (Flammang 2006). As a general rule, permanent adhesives consist almost exclusively of proteins. On the other hand, nonpermanent adhesives are made up of an association of proteins and carbohydrates, the latter being mostly in the form of acid and sulphated sugars (see Whittington and Cribb 2001, for review). The ratio of proteins to carbohydrates is usually about 2:1, but there may be substantial variation on this figure though there is typically more protein than carbohydrate (Grenon and Walker 1980; Davies et al. 1990; Smith et al. 1999a; Smith and Morin 2002). These adhesives usually also comprise a large inorganic fraction. The echinoderm tube foot adhesive composition corresponds to a typical nonpermanent adhesive (Fig. 9.7a). Lipids were also detected in tube foot adhesive secretions (Fig. 9.7a). This lipid fraction might come from the membranes of the adhesive granules or could be a contaminant in the footprint material (Flammang et al. 1998). However, an actual role of lipids in marine adhesion cannot be discarded since, recently, the permanent adhesive of barnacle cyprid larvae was shown to be a biphasic system containing both lipids and phosphoproteins, working synergistically to maximise adhesion to diverse surfaces under hostile conditions. Lipids were shown to be secreted first, possibly to displace water from the surface interface creating a conducive environment for introduction of phosphoproteins while simultaneously modulating the spreading of the protein phase and protecting the nascent adhesive from bacterial biodegradation (Gohad et al. 2014). The composition of the instantaneous adhesive of the Cuvierian tubule adhesive is reminiscent of nonpermanent adhesives by its association of proteins and carbohydrate in a 3:2 ratio (De Moor et al. 2003). However, it differs from them by the fact that the carbohydrate fraction is in the form of neutral sugars and not acidic sugars, and by its lower inorganic content (Fig. 9.7a).

As far as the amino acid composition of the protein fraction is concerned, all marine bioadhesives characterised so far have in common their richness in small side-chain amino acids as well as in charged and polar amino acids (Flammang 1996). These traits are common to many marine adhesives and are pointed out as key factors for their high cohesion and adhesive strength. Small side-chain amino acids are often found in large quantities in elastomeric proteins (Tatham and Shewry 2000). These proteins are able to withstand significant deformations

**Table 9.4** SΔQ for comparison of amino acid compositions among adhesives in echinoderms

| | Tube feet | | Cuvierian tubules | | | | |
|---|---|---|---|---|---|---|---|
| | Ar | Pl | Hm | Hf | Hl | Bs | Pg |
| Ar | **0** | – | – | – | – | – | – |
| Pl | 114 | **0** | – | – | – | – | – |
| Hm | **73** | 164 | **0** | – | – | – | – |
| Hf | 345 | 217 | 290 | **0** | – | – | – |
| Hl | 422 | 280 | 337 | **40** | **0** | – | – |
| Bs | 484 | 312 | 397 | **28** | **34** | **0** | – |
| Pg | 335 | 211 | 245 | **20** | **23** | **34** | **0** |

Marchalonis and Weltman (1971) reported that values of SΔQ ≤ 100 (in bold) indicate relatedness, SΔQ being calculated by pairwise comparison of the percentages of each amino acid constituting the proteins. Here, this method has been extended to whole adhesives, which are usually blends of different proteins, based on the assumption that if they include closely related proteins, their whole amino acid compositions will be similar too. Ar *Asterias rubens*, Bs *Bohadschia subrubra*, Hf *Holothuria forskali*, Hl *Holothuria leucospilota*, Hm *Holothuria maculosa*, Pg *Pearsonothuria graeffei*, Pl *Paracentrotus lividus*

without rupture before returning to their original state when the stress is removed (Smith et al. 1999b). Charged and polar amino acids, on the other hand, are probably involved in adhesive interactions with the substratum through hydrogen and ionic bonding (Waite 1987). Some adhesives, like those of barnacles are also rich in nonpolar, hydrophobic residues which could be involved in hydrophobic interactions with the substratum or within the adhesive material (Naldrett 1993). Adhesives from the three model echinoderm species present all these characteristics of marine adhesives. Their amino acid compositions resemble each other by the fact that they share some of their most abundant amino acids (Fig. 9.7b). However, the sea star adhesive possesses more charged residues, the sea urchin adhesive more hydrophobic residues and the sea cucumber adhesive more glycine residues (Fig. 9.7b). When these differences are quantified by the method of Marchalonis and Weltman (1971), no relatedness is found between the adhesive secretions of *A. rubens*, *P. lividus* and *H. forskali* (Table 9.4). The values of SΔQ for comparisons between the Cuvierian tubule adhesives of five sea cucumber species show that all these adhesives are closely related except the one of *H. maculosa* which stands apart and shows relatedness with the adhesive from sea star footprint (Table 9.4).

Although the detailed composition of echinoderm adhesives is only known for four species (i.e., for the tube feet of the sea star *A. rubens* and of the sea urchin *P. lividus* as well as for the Cuvierian tubules of the sea cucumbers *H. forskali and H. dofleinii*), the variability of the adhesive secretions within the phylum has been investigated by immunohistochemistry (Santos et al. 2005b, 2009a; Becker and Flammang 2010; Santos and Flammang 2012). This was done using polyclonal antibodies raised against the adhesive material from model species to evaluate the differences in the composition of the contents of the tube foot or Cuvierian tubule adhesive cells by looking for antibody cross reactivity on histological sections made from different species.

Polyclonal antibodies raised against the adhesive material of *H. forskali* were used to locate the origin of TPM constituents in the tubule. Granular cells showed extensive immunoreactivity, suggesting that their secretions make up the bulk of the adhesive material (De Moor et al. 2003). These antibodies were tested on ten other species from the genera *Holothuria*, *Bohadschia* and *Pearsonothuria* (Becker and Flammang 2010). Granular cells are strongly labelled in all species of the genera *Bohadschia* and *Holothuria* possessing sticky tubules, except in *H. maculosa* in which only the very basal part of granular cells is labelled, but not the apex. In *P. graeffei*, the contents of granular cells are immunoreactive, but the labelling is weaker than in *Bohadschia* and *Holothuria*. These results indicate that Cuvierian tubules adhesives are closely related, probably sharing many identical molecules or, at least, many identical epitopes on their constituents, but also that a certain variability occurs in the composition of these adhesive materials (Becker and Flammang 2010). In particular, *H. maculosa* is confirmed as possessing peculiar Cuvierian tubules (Demeuldre and Flammang, unpublished obs.).

A similar comparative immunohistochemical study was conducted with antibodies raised against the footprint material of the echinoid *Sphaerechinus granularis* on seven other sea urchin species belonging to three orders and five families. It showed that the adhesive secretions of sea urchins do not share any or little common epitopes on their constituents and thus seem to be more or less 'species-specific' (Santos and Flammang 2012). In sea urchins, variations in the composition of adhesive secretions could therefore explain the observed interspecific differences in disc tenacity and in adhesive granule ultrastructure reported by Santos and Flammang (2006, 2008). On the contrary, when the variability of the adhesive secretions from 14 sea star species representing five orders and families was investigated using polyclonal antibodies raised against the footprint material of *A. rubens*, a very strong immunolabelling was always observed at the level of the tube foot adhesive cells in every species investigated, irrespective of the taxon considered, of the tube foot morphotype or function or of the species habitat (Santos et al. 2005b). This immunoreactivity indicates that, contrary to sea urchin adhesives, sea star adhesives share many identical epitopes. However, differences in the adhesive secretion composition may exist, that are not detected by the antibodies used and that could account for the differences observed in the structure and function of asteroid tube feet (Santos et al. 2005b). For instance, Sfp1 was detected in the transcriptome of the species *Pisaster ochraceus* (EchinoDB; Janies et al. 2016), which belongs to the same order as *A. rubens*, but not in the genome of *Patiria miniata* (EchinoBase; Cameron et al. 2009) which belongs to another order (Hennebert and Flammang, unpublished obs.).

When the antibodies directed against sea star footprints were assayed in the other echinoderm classes, a phylogeny-related immunoreactivity pattern emerged. Crinoid and ophiuroid digitate tube feet were strongly immunoreactive, their labelling being restricted to the adhesive epidermal areas (Santos et al. 2009a). As for echinoids, of the five orders investigated, only members of the order Echinoida (the order comprising *P. lividus*) presented clusters of immunolabelled adhesive cells, whereas members of the other four orders did not present any labelling in the tube foot adhesive areas (Santos et al. 2009a). Finally, in holothuroids, the antibodies did not recognise the tube foot adhesive epidermis (Santos et al. 2009a). These results suggest that both crinoids and ophiuroids possess adhesive secretions sharing many similarities with the adhesive material of asteroids. On the other hand, in echinoids and holothuroids, the immunoreactivity was clearly weak or even absent indicating that there are no common epitopes between their adhesive secretions and those of *A. rubens* (Santos et al. 2009a). These observations are congruent with the phylogenetic hypothesis on the evolution of echinoderm adhesive systems (McKenzie 1988) according to which asteroids, crinoids and ophiuroids would share a common ancestral adhesive system in which the adhesive is extruded through apical duct cells, while a common echinoid/holothuroid adhesive system would have arisen later in the evolution in which the adhesive is released through apical tuft cells. This model fits well with the most commonly accepted echinoderm phylogeny in which echinoids and holothuroids form a derived clade, the Echinozoa (Littlewood et al. 1998, David and Mooi 1998). A drawback to this model is the moderate immunoreactivity observed in the tube feet of *P. lividus*, meaning that its adhesive shares some common epitopes with that of *A. rubens*. There is a possibility that these two species convergently acquired their similarity because of common selective pressures. Indeed, although they are clearly not homologous, Sfp1 and Nectin share the presence in their sequence of several discoidin-like domains. More studies are needed to address this question and to understand the evolution and functioning of echinoderm adhesive systems.

Further comparisons between echinoderm adhesives and those of other marine invertebrates will require a detailed knowledge of their protein composition, of the sequences of these proteins and of their post-translational modifications. So far, none of these information are available for the Cuvierian tubule adhesive, whereas only two tube foot adhesive proteins have been characterised. In recent years, the combined use of transcriptomics and proteomics has emerged as the best way leading to the identification of novel adhesive proteins and retrieval of their complete sequences (see Hennebert et al. 2015b for review). In addition to novel proteins, attention should also be paid to proteins such as actins or histones, which have been detected in the footprints of both sea stars and sea urchin. Although these proteins have been considered as contaminants, their abundance raises the possibility that they might be specific components of the adhesive material involved in exocytosis or in protection against microbes (Lebesgue et al. 2016).

**Acknowledgements** This work was supported in part by the Fund for Scientific Research of Belgium (F.R.S.–FNRS), by the 'Service Public de Wallonie—Programme Winnomat 2', by the 'Communauté française de Belgique—Actions de Recherche Concertées' and by COST Action TD0906. P.F. is Research Director of the F.R.S.-FNRS. RS is supported by Fundação para a Ciência e Tecnologia through a post-doctoral grant (SFRH/BPD/109081/2015). This study is a contribution from the 'Centre Interuniversitaire de Biologie Marine'.

# References


Ameye L, Hermann R, Dubois P, Flammang P (2000) Ultrastructure of the echinoderm cuticle after fast freezing/freeze substitution and conventional chemical fixation. Microsc Res Tech 48:385–393

Baranowska M, Schloßmacher U, McKenzie JD, Muller WEG, Schroder HC (2011) Isolation and characterization of adhesive secretion from cuvierian tubules of sea cucumber *Holothuria forskali* (Echinodermata: Holothuroidea). Evid Based Complement Alternat Med 2011:486845

Becker P, Flammang P (2010) Unravelling the sticky threads of sea cucumbers. A comparative study on Cuvierian tubule morphology and histochemistry. In: von Byern J, Grunwald I (eds) Biological adhesive systems—from nature to technical and medical application. Springer, Wien, pp 87–98

Cameron RA, Samanta M, Yuan A, He D, Davidson E (2009) SpBase: the sea urchin genome database and web site. Nucl Acids Res 37(Suppl 1):D750–D754

Chaet AB (1965) Invertebrate adhering surfaces: secretions of the starfish, *Asterias forbesi*, and the coelenterate, *Hydra pirardi*. Ann N Y Acad Sci 118:921–992

Chaet AB, Philpott DE (1964) A new subcellular particle secreted by the starfish. J Ultrastruct Res 11:354–362

Costa C, Cavalcante C, Zito F, Yokota Y, Matranga V (2010) Phylogenetic analysis and homology modelling of *Paracentrotus lividus* nectin. Mol Divers 14:653–665

David B, Mooi R (1998) Major events in the evolution of echinoderms viewed by the light of embryology. In: Mooi R, Telford M (eds) Echinoderms: San Francisco. Balkema, Rotterdam, pp 21–28

Davies MS, Jones HD, Hawkins SJ (1990) Seasonal variation in the composition of pedal mucus from *Patella vulgata* L. J Exp Mar Biol Ecol 144:101–112

De Moor S, Waite JH, Jangoux M, Flammang P (2003) Characterization of the adhesive from the Cuvierian tubules of the sea cucumber *Holothuria forskali* (Echinodermata, Holothuroidea). Mar Biotechnol 5:37–44

Defretin R (1952) Etude histochimique des mucocytes des pieds ambulacraires de quelques échinodermes. Recueil des travaux de la Station Marine d'Endoume 6:31–33

Delmeudre M, Chinh Ngo T, Hennebert E, Wattiez R, Leclère P, Flammang P (2014) Instantaneous adhesion of Cuvierian tubules in the sea cucumber *Holothuria forskali*. Biointerphases 9 (2):029016

Endean R (1957) The Cuvierian tubules of *Holothuria leucospilota*. Q J Micros Sci 98:455–472

Engster MS, Brown SC (1972) Histology and ultrastructure of the tube foot epithelium in the phanerozonian starfish, *Astropecten*. Tissue Cell 4:503–518

Flammang P (1996) Adhesion in echinoderms. In: Jangoux M, Lawrence JM (eds) Echinoderm studies, vol 5. Balkema, Rotterdam, pp 1–60

Flammang P (2006) Adhesive secretions in echinoderms: an overview. In: Smith AM, Callow JA (eds) Biological adhesives. Springer, Berlin, Heidelberg, pp 183–206

Flammang P, Jangoux M (1992) Functional morphology of the locomotory podia of *Holothuria forskali* (Echinodermata, Holothuroidea). Zoomorphology 11:167–178

Flammang P, Jangoux M (1993) Functional morphology of coronal and peristomeal podia in *Sphaerechinus granularis* (Echinodermata, Echinoida). Zoomorphology 113:47–60

Flammang P, Walker G (1997) Measurement of the adhesion of the podia in the asteroid *Asterias rubens* (Echinodermata). J Mar Biol Ass UK 77:1251–1254

Flammang P, Demeuleneare S, Jangoux M (1994) The role of podial secretions in adhesion in two species of sea stars (Echinodermata). Biol Bull 187:35–47

Flammang P, Michel A, Van Cauwenberge A, Alexandre H, Jangoux M (1998) A study of the temporary adhesion of the podia in the sea star *Asterias rubens* (Echinodermata, Asteroidea) through their footprints. J Exp Biol 201:2383–2395

Flammang P, Ribesse J, Jangoux M (2002) Biomechanics of adhesion in sea cucumber Cuvierian tubules (Echinodermata, Holothuroidea). Integr Comp Biol 42:1107–1115

Flammang P, Santos R, Haesaerts D (2005) Echinoderm adhesive secretions: from experimental characterization to biotechnological applications. In: Matranga V (ed) Marine molecular biotechnology: echinodermata. Springer, Berlin, pp 201–220

Flammang P, Lambert A, Bailly P, Hennebert E (2009) Polyphosphoprotein-containing marine adhesives. J Adhes 85:447–464

Gohad NV, Aldred N, Hartshorn CM, Jong Lee Y, Cicerone MT, Orihuela B, Clare AS, Rittschof D, Mount AS (2014) Synergistic roles for lipids and proteins in the permanent adhesive of barnacle larvae. Nat Commun 5:4414

Grenon JF, Walker G (1980) Biochemical and rheological properties of the pedal mucus of the limpet, Patella vulgata L. Comp Biochem Physiol B 66:451–458

Grenon JF, Walker G (1981) The tenacity of the limpet, *Patella vulgata* L.: An experimental approach. J Exp Mar Biol Ecol 54:277–308

Hamel J-F, Mercier A (2000) Cuvierian tubules in tropical holothurians: usefulness and efficiency as a defence mechanism. Mar Fresh Behav Physiol 33:115–139

Hennebert E, Viville P, Lazzaroni R, Flammang P (2008) Micro- and nanostructure of the adhesive material secreted by the tube feet of the sea star *Asterias rubens*. J Struct Biol 164:108–118

Hennebert E, Haesaerts D, Dubois P, Flammang P (2010) Evaluation of the different forces brought into play during tube foot activities in sea stars. J Exp Biol 213:1162–1174

Hennebert E, Wattiez R, Flammang P (2011) Characterisation of the carbohydrate fraction of the temporary adhesive secreted by the tube feet of the sea star *Asterias rubens*. Mar Biotechnol 13:484–495

Hennebert E, Santos R, Flammang P (2012a) Echinoderms don't suck: evidence against the involvement of suction in tube foot attachment. In: Kroh A, Reich M (eds) Echinoderms 2010: proceedings of the 7th European Conference on echinoderms, Zoosymposia, vol 7., pp 25–32

Hennebert E, Wattiez R, Waite JH, Flammang P (2012b) Characterization of the protein fraction of the temporary adhesive secreted by the tube feet of the sea star *Asterias rubens*. Biofouling 28:289–303

Hennebert E, Wattiez R, Demeuldre M, Ladurner P, Hwang DS, Waite JH, Flammang P (2014) Sea star tenacity mediated by a protein that fragments, then aggregates. Proc Natl Acad Sci USA 111:6317–6322

Hennebert E, Leroy B, Wattiez R, Ladurner P (2015a) An integrated transcriptomic and proteomic analysis of sea star epidermal secretions identifies proteins involved in defense and adhesion. J Proteomics 128:83–91

Hennebert E, Maldonado B, Ladurner P, Flammang P, Santos R (2015b) Experimental strategies for the identification and characterization of adhesive proteins in animals: a review. Interface Focus 5:20140064

Hermans CO (1983) The duo-gland adhesive system. Oceanogr Mar Biol Ann Rev 21:281–339

Higgins LJ, Mostaert AS (2013) Qualitative and quantitative study of spiny starfish (*Marthasterias glacialis*) footprints using atomic force microscopy. In: Santos R, Aldred N, Gorb S, Flammang P (eds) Biological and biomimetic adhesives: challenges and opportunities. RSC Publishing, Cambridge, pp 26–37

Janies DA, Witter Z, Linchangco GV, Foltz DW, Miller AK, Kerr AM, Jay J, Reid RW, Wray GA (2016) EchinoDB, an application for comparative transcriptomics of deeply-sampled clades of echinoderms. BMC Bioinf. doi:10.1186/s12859-016-0883-2

Lawrence JM (1987) A functional biology of echinoderms. Croom Helm, London

Lawrence JM (2001) Function of eponymous structures in echinoderms: a review. Can J Zool 79:1251–1264

Lebesgue N, da Costa G, Ribeiro RM, Ribeiro-Silva C, Martins GG, Matranga V, Scholten A, Cordeiro C, Heck AJR, Santos R (2016) Deciphering the molecular mechanisms underlying sea urchin reversible adhesion: a quantitative proteomics approach. J Proteomics 138:61–71

Marchalonis JJ, Weltman JK (1971) Relatedness among proteins: a new method of estimation and its application to immunoglobins. Comp Biochem Physiol B 38:609–625

Matranga V, Di Ferro D, Zito F, Cervello M, Nakano E (1992) A new extracellular matrix protein of the sea urchin embryo with properties of a substrate adhesion molecule. Roux's Arch Dev Biol 201:173–178

McKenzie JD (1988) The ultrastructure of tube foot epidermal cells and secretions: Their relationship to the duo-glandular hypothesis and the phylogeny of the echinoderm classes. In: Paul CRC, Smith AB (eds) Echinoderm phylogeny and evolutionary biology. Clarendon, Oxford, pp 287–298

Naldrett MJ (1993) The importance of sulphur cross-links and hydrophobic interactions in the polymerization of barnacle cement. J Mar Biol Assoc UK 73:689–702

Nichols D (1966) Functional morphology of the water vascular system. In: Boolootian RA (ed) Physiology of echinodermata. Interscience Publishers, New York, pp 219–244

Paine VL (1926) Adhesion of the tube feet in starfishes. J Exp Zool 45:361–366

Peng YY, Glattauer V, Skewes TD, White JF, Nairn KM, McDevitt AN, Elvin CM, Werkmeister JA, Graham LD, Ramshaw JAM (2011) Biomimetic materials as potential medical adhesives—Composition and adhesive properties of the material coating the Cuvierian tubules expelled by *Holothuria dofleinii*. In: Pignatello R (ed) Biomaterials- physics and chemistry. InTech Press, Rijeka, pp 245–258

Peng YY, Glattauer V, Skewes TD, McDevitt A, Elvin CM, Werkmeister JA, Graham LD, Ramshaw JAM (2014) Identification of proteins associated with adhesive prints from *Holothuria dofleinii* Cuvierian tubules. Mar Biotechnol 16:695–706

Perpeet C, Jangoux M (1973) Contribution á l'etude des pieds et des ampoules ambulacraires d'Asterias rubens (Echinodermata, Asteroides). Forma et functio 6:191–209

Ruppert EE, Fox RS, Barnes RD (2003) Invertebrate zoology: a functional evolutionary approach. Saunders College Publishers, Fort Worth

Santos R, Flammang P (2006) Morphology and tenacity of tube foot disc of three common European sea urchin species: a comparative study. Biofouling 22:187–200

Santos R, Flammang P (2008) Estimation of the attachment strength of the shingle sea urchin, *Colobocentrotus atratus*, and comparison with three sympatric echinoids. Mar Biol 154:37–49

Santos R, Flammang P (2012) Is the adhesive material secreted by sea urchin tube feet species-specific? J Morphol 273:40–48

Santos R, Gorb S, Jamar V, Flammang P (2005a) Adhesion of echinoderm tube feet to rough surfaces. J Exp Biol 208:2555–2567

Santos R, Haesaerts D, Jangoux M, Flammang P (2005b) Comparative histological and immunohistochemical study of sea star tube feet (Echinodermata, Asteroidea). J Morphol 263:259–269

Santos R, Hennebert E, Varela Coelho A, Flammang P (2009a) The echinoderm tube foot and its involvement in temporary underwater adhesion. In: Gorb S (ed) Functional surfaces in biology, vol 2. Springer, Netherlands, pp 9–41

Santos R, da Costa G, Franco C, Gomes-Alves P, Flammang P, Coelho AV (2009b) First insights into the biochemistry of tube foot adhesive from the sea urchin Paracentrotus lividus (Echinoidea, Echinodermata). Mar Biotechnol 11:686–698

Santos R, Barreto A, Franco C, Coelho AV (2013) Mapping sea urchins tube feet proteome—a unique hydraulic mechano-sensory adhesive organ. J Proteomics 79:100–113

Smith AM, Morin MC (2002) Biochemical differences between trail mucus and adhesive mucus from marsh periwinkle snail. Biol Bull 203:338–346

Smith AM, Quick TJ, St Peter RL (1999a) Differences in the composition of adhesive and non-adhesive mucus from the limpet *Lottia limatula*. Biol Bull 196:34–44

Smith BL, Schäffer TE, Viani M, Thompson JB, Frederick NA, Kindt J, Belcher A, Stucky GD, Morse DE, Hansma PK (1999b) Molecular mechanistic origin of the toughness of natural adhesives, fibres and composites. Nature 399:761–763

Souza Santos H, Silva Sasso W (1968) Morphological and histochemical studies on the secretory glands of starfish tube feet. Acta Anat 69:41–51

Tatham AS, Shewry PR (2000) Elastomeric proteins: biological roles, structures and mechanisms. Trends Biochem Sci 25:567–571

Thomas LA, Hermans CO (1985) Adhesive interactions between the tube feet of a starfish, *Leptasterias hexactis*, and substrata. Biol Bull 169:675–688

Toubarro D, Gouveia A, Ribeiro RM, Simões N, da Costa G, Cordeiro C, Santos R (2016) Cloning, characterization and expression levels of the Nectin gene from the tube feet of the sea urchin *Paracentrotus lividus*. Mar Biotechnol 18:372–383

Tyler S (1988) The role of function in determination of homology and convergence—examples from invertebrates adhesive organs. Fortsch Zool 36:331–347

Van Dyck S, Gerbaux P, Flammang P (2010) Qualitative and quantitative saponin contents in five sea cucumbers from the Indian Ocean. Mar Drugs 8:173–189

VandenSpiegel D, Jangoux M (1987) Cuvierian tubules of the holothuroid *Holothuria forskali* (Echinodermata): a morphofunctional study. Mar Biol 96:263–275

Vandenspiegel D, Jangoux M (1993) Fine structure and behaviour of the so-called Cuvierian organs in the holothuroid genus *Actinopyga* (Echinodermata). Acta Zool 74:43–50

VandenSpiegel D, Jangoux M, Flammang P (2000) Maintaining the line of defense: Regeneration of Cuvierian tubules in the holothuroid *Holothuria forskali* (Echinodermata). Biol Bull 198:34–49

Waite JH (1983) Adhesion in byssally attached bivalves. Biol Rev 58:209–231

Waite JH (1987) Nature's underwater adhesive specialist. Int J Adhes Adhes 7:9–14

Waite JH (2002) Adhesion à la moule. Integr Comp Biol 42:1172–1180

Walker G (1987) Marine organisms and their adhesion. In: Wake WC (ed) Synthetic adhesives and sealants. John Wiley & Sons, Chichester, pp 112–135

Whittington ID, Cribb BW (2001) Adhesive secretions in the Platyhelminthes. Adv Parasitol 48:101–224

Young GA, Crisp DJ (1982) Marine animals and adhesion. In: Allen KW (ed) Adhesion, vol 6. Applied Sciences, London, pp 19–39

Yule AB, Walker G (1987) Adhesion in barnacles. In: Southward AJ (ed) Crustacean issues, vol 5, Biology of Barnacles. Balkema, Rotterdam, pp 389–402

Zahn RK, Müller WEG, Michaelis M (1973) Sticking mechanisms in adhesive organs from a *Holothuria*. Res Mol Biol 2:47–88

# Chapter 10
# An Adhesive Secreted by Australian Frogs of the Genus *Notaden*

**Lloyd D. Graham, Veronica Glattauer, Yong Y. Peng, Paul R. Vaughan, Jerome A. Werkmeister, Michael J. Tyler, and John A.M. Ramshaw**

**Abstract** When provoked, the Australian fossorial frog *Notaden bennettii* secretes from its dorsal skin an exudate which rapidly forms a tacky elastic hydrogel. This protein-based material acts as a promiscuous pressure-sensitive adhesive which works even in wet conditions. The largest protein, Nb-1R, is rich in Gly, Pro/Hyp and Glx and appears to be the key structural component; it probably contains extensive segments of intrinsic disorder along with some well-folded domains. Indeed, the material properties of adhesive secretions from both amphibians (vertebrates) and onychophorans (invertebrates) may rely upon large proteins containing long and intrinsically unstructured regions composed of imperfect tandem repeats. Although the *N. bennettii* secretion contains sterols, carotenoids and other undesirable metabolites, in vitro, ex vivo and in vivo studies suggest that the structural matrix of the set glue is highly biocompatible. Its open porous structure is likely to encourage cell infiltration, and glue pellets implanted subcutaneously in mice were fully resorbed by the surrounding tissues within two months. Ex vivo studies in sheep showed that the frog glue bonded meniscal tears more strongly than traditional protein-based medical glues, while conventional tendon attachment repairs were approximately doubled in strength when augmented with the frog adhesive. Overall, the properties of the frog glue suggest that a recombinant mimic would have great potential for medical applications.

---

L.D. Graham
CSIRO Alumnus, 52 Eastcote Road, North Epping, NSW 2121, Australia
e-mail: Lloyd.Graham@cantab.net

V. Glattauer • Y.Y. Peng • P.R. Vaughan • J.A. Werkmeister • J.A.M. Ramshaw (✉)
Ian Wark Laboratory, CSIRO Manufacturing, Bag 10, Clayton South, VIC 3169, Australia
e-mail: Veronica.Glattauer@csiro.au; Yong.Peng@csiro.au; Paul.Vaughan@csiro.au; Jerome.Werkmeister@csiro.au; john.ramshaw@csiro.au

M.J. Tyler
School of Earth and Environmental Sciences, University of Adelaide, Adelaide, SA 5005, Australia
e-mail: Michael.Tyler@adelaide.edu.au

© Springer International Publishing Switzerland 2016

A.M. Smith (ed.), *Biological Adhesives*, DOI 10.1007/978-3-319-46082-6_10

## 10.1 Introduction

The main body of research into skin secretions from amphibians has been directed towards biologically active compounds responsible for the various toxic, pharmacological, antimicrobial and poisonous nature of various amphibian granular secretions. A very important difference between the amphibian skin and that of other vertebrates is the diversity of glandular structures that they contain. Significant bioactivity has been found within the secretions of amphibious skins. For example, pharmacological activity testing of Australian frog skins over a period of more than 30 years has shown the presence of an extensive array of simple and complex aliphatic, aromatic and heterocyclic molecules, as well as a range of small and large peptides (Doyle et al. 2003; Apponyi et al. 2004).

There has been little research into adhesive exudates. The scarcity of research in this area has been attributed to the difficulty of getting the adhesive gland products into a solution for analysis (Williams and Anthony 1994), even though the adhesives of these amphibians generally do not seem to entail a covalent curing step to give tenacity of adhesion. Brodie and Gibson (1969) were unable to analyse the granular gland secretions of a salamander, *Ambystoma gracile*, due to difficulties in dissolving the adhesive secretions. Williams and Larsen (1986) used histological methods to show that the adhesive secretions from *A. macrodactylum* were proteinaceous. Subsequently, Williams and Anthony (1994) examined the secretions of three species of salamander known to secrete an adhesive. Stimulating the animals both above and under water revealed that the skin secretions only became adhesive when exposed to air. Also, Evans and Brodie (1994) examined the tensile strength of the adhesive secretions from 12 species of amphibian, including 5 frog species. Frogs that produce secretions with tenacious adhesive properties include those from a South African species, *Breviceps poweri* (Channing 2001; Tyler 2010).

In 2002, we began a detailed study of the adhesives produced by members of the genus of Australian fossorial frogs, *Notaden*. These produce an exudate, secreted by glands on their backs (Tyler 2010), that quickly sets into an adhesive and elastic material. Three Australian frogs have so far been identified as producing adhesive, these being *N. melanoscaphus*, *N. bennettii* and *N. nichollsi*. Initial observations on adhesive efficacy were made using the exudates of *N. melanoscaphus*. Subsequently, however, our studies focused the exudates from *N. bennettii*. Commonly known as the 'holy cross frog' (Fig. 10.1), *N. bennettii* is found on plains near larger rivers, in savannah woodland and in mallee areas across central New South Wales and the interior of southern Queensland (Barker et al. 1995). There appears to be little, if any, difference between the properties of the adhesive from these two species.

*Notaden* frogs secrete the sticky material from their dorsal skin when they are provoked, probably in an attempt to deter potential predators such as snakes (Evans and Brodie 1994). *Limnodynastes convexiusculus*, another Australian limnodynastid frog, also secretes a glue-like material when threatened; the material loses stickiness as it dries, so snakes of the death adder family wait approximately

**Fig. 10.1** The 'holy cross frog', *Notaden bennettii*

10 min after their strike before consuming frogs of this species (Phillips and Shine 2007). Although interest in amphibian and other specialty bioadhesives is growing (Gross 2011), the *Notaden* glue has yet to be compared with the glue produced by frogs of the Australian genera *Heleioporus* and *Neobatrachus* (Williams et al. 2000).

*Notaden* frogs typically live up to 1 m underground in dried mud for nine months of the year, emerging only during and after torrential rain. On these occasions, they are vulnerable to insect attacks, and so another possible use of the exudate may be to jam the jaws of biting insects like ants, sticking them to the frog's skin, which it later sheds and eats (Tyler 2010). The exudate sets rapidly as a yellow-coloured tacky elastic solid and sticks well even in the frog's moist habitat.

## 10.2 Preliminary Field and Laboratory Data

The potential of the *Notaden* exudate as a novel adhesive system was first identified around the year 2000 during field studies (reviewed by Tyler 2010). These data indicated that the liquid exudate had a fast setting time, ~10 s to ~3 min, giving a strong adhesive bond. The adhesion worked on moist surfaces, for example where moisture had condensed onto a cool surface, and remained adhesive in cool conditions. It adhered to a wide range of both hydrophilic and hydrophobic surfaces, including metal, wood, paper, plastics and glass. Subsequent preliminary laboratory trials confirmed the field observations that the *Notaden* exudate adheres tightly and with comparable affinity to a wide variety of surfaces such as glass, metal, wood and plastic including polypropylene, polystyrene and even the 'non-stick' polymer Teflon® (Graham et al. 2005; Tyler 2010), and also to biological tissues such as soft tissue, cartilage and bone. Further data indicated that the adhesive remained flexible in an indoor environment for over 6 months and that it was resistant to many solvents and to mild abrasion (Tyler and Ramshaw 2002; Tyler 2010).

## 10.3 Adhesive Collection

Skin secretions are elicited once the frog is provoked. In the field, simply approaching the frogs is generally sufficient to elicit secretion. However, for laboratory studies, a small number of frogs were held in captivity, where they become accustomed to handling and the frog no longer secretes the adhesive exudate when approached. In these cases, mild stimulation is required. A method for achieving this is to wash the animals briefly in water to remove debris and then stimulate the back of the frog with a mild electrical discharge to contract the dermal muscles. The magnitude and time of the pulse is designed to lead to contraction of only the dermal muscles, including glandular muscles, and thereby empty the glands of their contents onto the surface of the frog. This can be achieved by using a two-pole electrode and a physiological stimulator that provides a square wave output. At low voltage (<20 V) and frequencies of up to 60 Hz (Tyler et al. 1992), the secretion is readily induced.

The skin secretion can be harvested by taking the liquid from the back of the frog by dipping an implement such as a spatula into the liquid. The secretions also accumulate on the electrodes and can be removed by scraping from the metal, for example, with a spatula. The quantity of skin secretion that is harvested varies with the size of the frog, the length in captivity and the species, but a quantity in the range of 20–500 mg can be collected. This process needs to be carried out with speed as the exudate rapidly starts to become more viscous and harder to handle. The time period between effective collections of reasonable amounts of material has not been established, but our experience suggests that generally the frog exudate can be harvested about once a month in captivity.

The skin secretion can also be collected by washing the material from the skin into a collection vessel using an aqueous washing solution that is not harmful to the frog. For example, the washing solution could be phosphate-buffered saline (PBS) or PBS containing other solutes. Thus, washing the backs of exudate-producing frogs under argon in PBS/10 mM Cys.HCl, pH 4.5, emulsified the solids which, on settling, consolidated into a sticky and elastic plug of translucent yellow solid (Fig. 10.2a) (Graham et al. 2005). When the frogs were irrigated using 20 mM phosphate buffer, pH 6.2, the solids were not emulsified but rather formed aggregates of translucent yellow elastic solid. Irrigation with 50 mM ascorbic acid resulted in clear viscous solutions (pH ~2.5) rather than the formation of any solid material. For samples collected without irrigation, the material typically bonded avidly to the metal electrode poles or other sample collecting devices, and the solid that formed was typically much more cohesive (i.e., firmer or more rubber-like) and much less tacky than the adhesive that was collected by emulsification followed by settling (Graham et al. 2005).

The frog adhesive binds most avidly to the surface on which it first solidifies. It is likely that the adhesive determinants in the newly secreted material seek interaction partners, and, in the absence of suitable adherends, they simply interact with each other and/or the buffer components, resulting in a solid that is still sticky but not

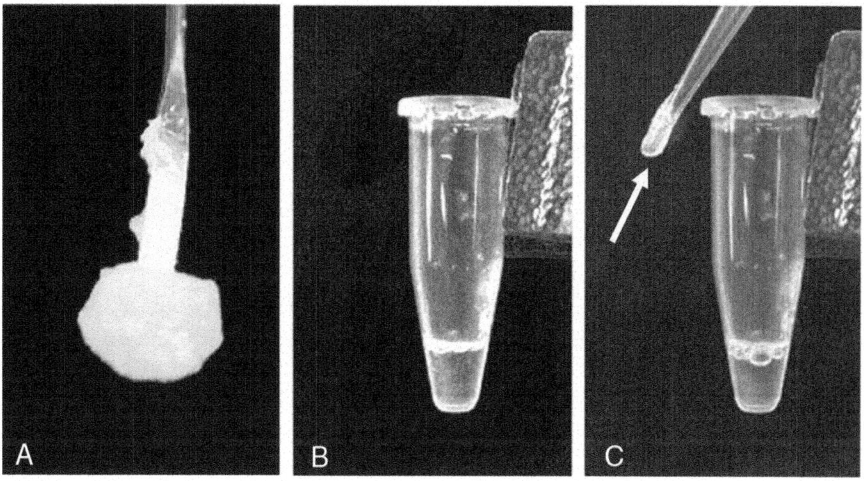


**Fig. 10.2** (**a**) *Notaden* adhesive solids, after collection by washing the backs of exudate-producing frogs under argon in PBS/10 mM Cys.HCl, pH 4.5, have consolidated into a sticky yellow solid. (**b**) A clear yellow solution obtained by dissolving *Notaden* adhesive in 5 % (v/v) acetic acid. (**c**) Reconstituted adhesive (harvested onto a disposable pipette tip, as indicated by the *arrow*) recovered from solution by adding 5 M NaCl solution to a final concentration of about 0.8–1.0 M NaCl

nearly as adhesive as the nascent exudate. In all cases, the collected material is a yellow-orange colour and is initially pliable and compressible, for example, using a finger, and is elastic in that it can stretch when adhered between two surfaces. Samples of solid that were stored at −70 °C in their cognate washings (if any) appeared to retain all the functionality of the unfrozen solid (Graham et al. 2005).

## 10.4 Solubilisation and Solidification

Initial studies had suggested that the frog exudate, once it had formed a firm, dehydrated product, was not readily dissolved by a number of solvents, including water, hexane, xylene, hydrochloric acid and sulphuric acid. However, it was subsequently found that the solid material obtained both from undiluted exudates and those collected with irrigation could be fully dissolved at room temperature in 5 % (v/v) acetic acid (Fig. 10.2b). This could be achieved in a matter of hours for samples that had not become fully dehydrated and was slower (but still possible) for material that had been dried. Sometimes a small proportion of the sample would remain insoluble, in which case the trace solids could be removed by brief centrifugation. Other solvent systems that have similar effects include 10 % (w/v) SDS, 5 M guanidinium hydrochloride (pH 5) and 10 mM $H_3PO_4$ (Graham et al. 2005). Thus, the solubility of frog material appears to be greater than that of marine mussel plaque (Burzio et al. 1997; Deming 1999) and much greater than that of barnacle cement (Burzio et al. 1997; Naldrett and Kaplan 1997; Kamino et al. 2000). As a routine solvent, 5 % (v/v) acetic acid is preferred, and this permits concentrations of up to 10 mg protein/ml to be obtained. Although 0.05 % (v/v) acetic acid was not effective at dissolving the adhesive, concentrated solutions in this solvent could be achieved by dissolving the adhesive in 5 % (v/v) acetic acid and then dialysing the solution against 0.05 % (v/v) acetic acid at 4 °C (Graham et al. 2005).

It was particularly interesting to find that lumps of frog adhesive could be dissolved and then resolidified in a functionally adhesive form (Graham et al. 2005). For example, raising the ionic strength of solutions of dissolved material caused the components to self-assemble spontaneously into a tacky and elastic solid which retained most of the functionality of the original adhesive. Thus, if a small volume of 5 M NaCl solution was gradually dispensed to a final concentration of 0.8–1.0 M NaCl into a solution of ≥4 mg protein/ml in 5 % or 0.05 % (v/v) acetic acid, then adhesive formed as fibrils or particles that could be grown or harvested as a translucent yellow sticky solid (Fig. 10.2c). Removing trapped acetic acid and salt from reconstituted adhesive by washing in water increased the translucence and adhesiveness of the solid, although it seemed to remain softer and less adhesive than the adhesive collected directly from the frogs (Graham et al. 2005).

## 10.5 Mechanical Properties

To provide an initial assessment on the strength of the adhesion provided by the *Notaden* adhesive, a simple test involving wooden craft stick pairs that were bonded in an in-line lap-joint configuration was used. Moist adhesive pieces were sandwiched between the overlapped ends of the craft sticks, and the joint was allowed to dry completely while being held in mild compression. The test samples, along with joints formed using readily available commercial glues, were then examined using an Instron testing machine. These bond strength data (Table 10.1) showed that the *Notaden* adhesive was stronger than PVA and UHU® Stic glues and comparable in strength to cyanoacrylate adhesive (Graham et al. 2005). Overall, in these strength tests, there seemed to be no difference between adhesives sourced from different *N. bennettii* frogs. Also, adhesive that had been dissolved and reconstituted before use provided dry bond strengths similar to adhesive that had not (L.D. Graham, unpublished results). Possibly the most striking property of the *Notaden* adhesive is its capability to adhere to many diverse materials. Indeed, when the adhesive was used to bond a nickel spatula to a hard surface bound with Teflon® tape and allowed to dry, the spatula could not be removed without tearing a hole in the tape (Graham et al. 2005).

The strength tests were expanded to include data for moist frog adhesive as a bonding agent for polypropylene (Graham et al. 2005), tests that were designed to be comparable to those used in assessment of other amphibian secretions (Evans

**Table 10.1** Shear strength of wood bonded using selected adhesives (Graham et al. 2005)

| Adhesive | Shear strength mean ± SD (MPa) ($n=6$) |
|---|---|
| *Notaden* | 1.7 ± 0.3 |
| Cyanoacrylate glue | 1.7 ± 0.7 |
| PVA glue | 1.3 ± 0.2 |
| UHU® Stic | 0.9 ± 0.4 |

and Brodie 1994). Freshly secreted neat exudate was observed to bond 10 mm diameter polypropylene discs in an elastic manner with a prompt tensile strength of 57 ± 6 kPa ($n=8$). On testing, each separated joint showed a mixture of adhesive and cohesive failure. When separated discs were rejointed and stored humidified for 24 h at 4 °C, they were found to have a tensile strength of 78 ± 8 kPa ($n=8$). When the joints were again reformed and stored humidified for a further 1 h at 21 °C, they were found to have a tensile strength of 64 ± 5 kPa ($n=8$). These three sets of tensile strength data were statistically indistinguishable (Graham et al. 2005). It is clear from these data that reforming separated joints by hand rapidly restores most of their strength, with more than 80 % of the original bond strength returning within 1 h. The elastic modulus was also unaffected by breaking and reforming the joint (Graham et al. 2005).

Overall, it appears that moist *Notaden* adhesive functions as a pressure-sensitive adhesive whose bond strength is largely unaffected by breaking and reforming of the joint. Like other pressure-sensitive adhesives, the frog adhesive is most effective at bonding rigid surfaces which can be placed under pressure. In tests similar to our studies with polypropylene discs, other researchers determined tensile strengths of ~100 kPa for freshly secreted tomato frog exudate and 20–63 kPa for freshly secreted adhesives from other frogs and salamanders (Evans and Brodie 1994). With a tensile strength of 57–78 kPa, freshly secreted *N. bennettii* exudate sets with a bond strength comparable to those for adhesive secretions from other amphibians (Graham et al. 2005). The *N. bennettii* values are intermediate between the tensile strengths reported for the adhesion of marine mussel plaque to hydrophobic substrates (~13 kPa) and to hydrophilic ones (320–860 kPa) (Crisp et al. 1985).

Mechanical testing of the glue in biomedical contexts is described below under the heading Applications (Sect. 10.9).

## 10.6 Biocompatibility

Another feature of this adhesive is that it appears to be biocompatible, in that it does not, in a general sense, evoke an adverse reaction and that the structural components appear to be non-toxic.

Water extracts of the glue were applied at approximately 1 μg/ml to toad *rectus abdominis* muscle preparation (striated muscle), guinea pig ileum, vas deferens from rats and isolated rabbit ear artery preparations. No activity was observed with any of these tissue preparations, not even when the secretions were tested on the very sensitive rabbit ear artery model (Tyler and Ramshaw 2002; Tyler 2010).

The adhesive was also assessed in cell and organ culture experiments. Before these experiments could be undertaken, it was essential to find an effective means of sterilising the adhesive. Many standard treatments were ineffective. For example, dissolved adhesive could not be filter-sterilised because the material blocked 0.2 μm filters. Chemical sanitisation (e.g., acid treatment plus germicidal UV irradiation) did not kill all of the bacteria present. High-pressure treatments of solid samples were also unsuccessful: subjecting the material to 400 MPa for 2 min did not kill fungal contaminants, and although treatment at 600 MPa for 5 min appeared to sterilise the sample, it adversely affected the functionality of the adhesive. Likewise, autoclave treatments sterilised the adhesive but rendered it non-functional. However, gamma irradiation at 25 kGy sterilised the adhesive without causing structural or functional damage (Fig. 10.3), although higher levels of irradiation (35 kGy) did impair the stickiness and elasticity of the adhesive (L.D. Graham, unpublished results).

In organ culture experiments, it was found that collagen-coated plastic lenticules could be adhered effectively to debrided bovine corneas using material dissolved in 0.05 % (w/v) acetic acid. Epithelial regrowth was not inhibited by exposure to the solution, and regrowth proceeded smoothly over the lenticule-cornea junction even where the corneal edge was uneven. Paraffin histology of corneal sections stained with H&E stain confirmed that epithelium had successfully mounted the edge of the lenticule in all cases (Graham et al. 2010).

Tissue culture plates were coated with the adhesive by allowing the material to adsorb from solution and then rinsing the plates with sterile PBS. Bovine corneal epithelial cells attached and proliferated well on these coated plates, with no impairment relative to uncoated surfaces, while the migration of these cells from

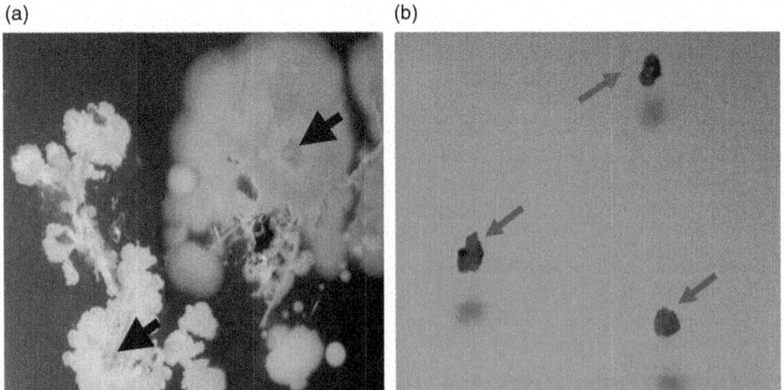


**Fig. 10.3** (**a**) Unsterilised *Notaden* adhesive pieces (shown by *arrows*) on nutrient agar, showing growth of microorganisms after 3 days at 37 °C. (**b**) *Notaden* adhesive pieces (shown by *arrows*) that had been sterilised with 25 kGy γ-irradiation, showing no growth of microorganisms under similar culture conditions

a tissue explant was ~70 % of that observed on uncoated surfaces (Graham et al. 2010). The edge of the migrating cell sheets appeared thickened (Fig. 10.4), in the manner typical of how these cells respond to hydrophobic surfaces. No cytotoxicity was detected from dissolved frog adhesive at concentrations up to ~8 μg protein/ml, beyond which the acetic acid present in the stock solution began to adversely affect cell viability (Graham et al. 2010).

To test biocompatibility in vivo, small pellets of glue were implanted subcutaneously into mice (Graham et al. 2010). The pellets were resorbed by surrounding tissues, and all of the animals made a full recovery. Within 7 days, the skin around the wound became necrotic in all glue-implanted animals, but these lesions had largely healed after a further 7 days. One month after implantation, the implants had mostly been resorbed, and the skin around the implant sites had regrown its fur. By now, the glue-implanted mice were visually indistinguishable from the sham surgery control mice, although histology revealed that the implanted areas were still undergoing repair. At 2 months, normal dermal architecture had returned. The initial but transient skin necrosis at the implant site was probably caused by toxic metabolites present in the frog secretion, which has been shown to contain sterols and carotenoids, as well as fatty alcohols, aldehydes, ketones, acids and aromatic compounds (Graham et al. 2010).

Despite these encouraging findings on the biocompatibility of the *Notaden* adhesive, other issues such as immunological reactivity need to be ascertained.

## 10.7 Biochemical Studies

General biochemical analyses have shown the *Notaden* adhesive to be a protein-based hydrogel. Vacuum drying of fully hydrated adhesive pellets suggested that the fully hydrated adhesive contained around 85–90 % (w/w) water, while a

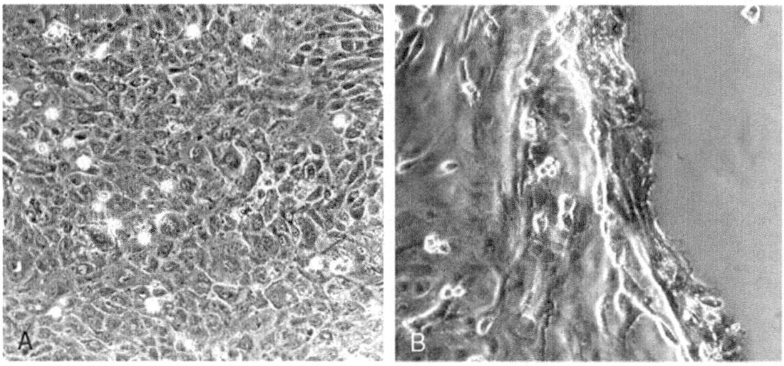


**Fig. 10.4** Culture of bovine corneal epithelial cells on tissue culture plates that had been coated with *Notaden* adhesive (by adsorption from solution, followed by rinsing with sterile PBS). (**a**) Confluent cells attached to the surface. (**b**) Migrating cells on the surface, showing a thickened layer at the leading edge of migration

colorimetric protein analysis indicated that the hydrated adhesive contained ~10 % (w/w) protein. Despite the very high water content of fully hydrated adhesive, the material handled more like an elastic solid than a viscous dilute hydrogel (Graham et al. 2005). For simplicity, and to distinguish it from liquid and emulsion forms, we will refer to the set adhesive as a solid. A colorimetric assay for carbohydrate, using the phenol-sulphuric acid assay (Dubois et al. 1956) with D-glucose as standard, indicated that vacuum dried material contained only 0.75 % (w/w) glucose equivalents (Graham et al. 2005). This assay detects individual aldoses, ketoses, deoxy-sugars and sugar acids with somewhat different efficiencies but is still one of the best general-purpose biochemical assays for carbohydrate content. Gas chromatography–mass spectrometry of acetone extracts indicated the presence of a significant amount of cholesterol in the glue (Graham et al. 2010).

### *10.7.1 Colour*

Some of the yellow-orange colour can be removed from the *Notaden* adhesive by organic solvents, and the UV–visible absorbance spectrum of the resulting extracts is characteristic of carotenoid chromophores (Schmidt-Dannert et al. 2000). Such compounds are known to contribute to the colour of amphibian skin (Frost-Mason et al. 1994). Acetone extraction of frog glue removed much of its pigment; analysis of the component chromophores by thin-layer chromatography and reversed-phase HPLC confirmed the presence of carotenoid chromophores, mainly β-carotene (Graham et al. 2010). Depigmented *Notaden* adhesive is translucent and white but otherwise seems to be fully functional (Graham et al. 2010).

### *10.7.2 CD Spectra*

Since protein was the major non-water component of the *Notaden* adhesive, CD spectroscopy was used to examine the overall secondary structure of the exudate. The CD spectrum of the adhesive dissolved in 10 mM $H_3PO_4$ at 0.1 mg protein/ml is shown in Fig. 10.5. Deconvolution of the spectrum (Greenfield and Fasman 1969) indicated a poorly structured system, dominated by random coil and/or containing non-standard elements, with the presence of little or no alpha helix, about 36 % β-elements and about 64 % non-standard elements or random coil (Graham et al. 2005).

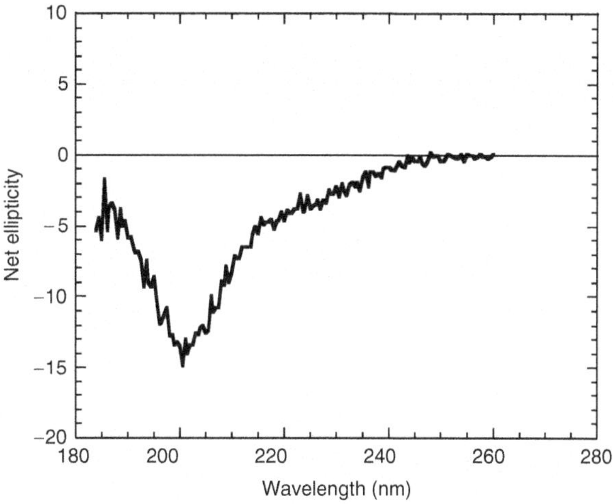


**Fig. 10.5** Circular dichroism (CD) spectrum of *Notaden* adhesive dissolved in 10 mM $H_3PO_4$ at 0.1 mg protein/ml. The spectrum indicates a relatively unstructured system dominated by random coil and/or containing non-standard elements

## *10.7.3 Amino Acid Analysis*

Amino acid analysis of the *Notaden* adhesive indicated that the material was rich in Gly, Pro and Glx compared to other vertebrate proteins (Doolittle 1986), but was low in Ala, Ser and Met. Structural proteins often have high levels of Gly and Pro, while elastomeric proteins show compositions that may be rich in Gly, in Gly and Pro, or in Gly, Pro and Gln (Tatham and Shewry 2000). It was of interest that there was a significant 4-hydroxyproline (Hyp) content (Table 10.2) (Graham et al. 2005). This imino acid, which is obtained by post-translational modification of proline, is characteristically found in collagens or other proteins that form triple-helical structures (Brodsky and Ramshaw 1997). It is also seen in adhesive plaque proteins from the mussel, *Mytilus edulis* (Waite 1993; Burzio et al. 1997; Deming 1999). Specific analyses for dihydroxyphenylalanine (L-dopa), which is obtained by post-translational modification of tyrosine, were done using modified hydrolysis conditions (Gieseg et al. 1993), but no L-dopa was detected in the *Notaden* adhesive (Table 10.2) (Graham et al. 2005). The presence of this amino acid is a key part of the adhesive mechanism of the glues from several organisms, including *M. edulis* (Waite 1993; Monahan and Wilker 2004).

**Table 10.2** Amino acid composition (mol%) of *Notaden* adhesive (Graham et al. 2005) and its major large protein, Nb-1R (Graham et al. 2013)

| Amino acid | Frog glue | Nb-1R |
|---|---|---|
| Asx[a] | 7.2 | 6.7 |
| Thr | 4.4 | 4.3 |
| Ser | 3.8 | 7.2 |
| Glx[a] | 14.1 | 16.9 |
| Gly | 15.8 | 17.2 |
| Ala | 2.8 | 4.0 |
| Cys/2 | 0.7 | 0.7 |
| Val | 6.2 | 6.0 |
| Met | 1.1 | 0.9 |
| Ile | 4.8 | 3.4 |
| Leu | 6.9 | 6.7 |
| Tyr | 2.2 | 2.2 |
| Phe | 3.8 | 2.7 |
| Lys | 5.8 | 4.1 |
| His | 3.1 | 1.2 |
| Arg | 3.8 | 4.8 |
| Pro | 8.8 | 6.9 |
| Hyp | 4.6 | 4.0 |
| L-Dopa | 0 | n.d. |
| Trp | n.d. | n.d. |

[a]Deamidation during acid hydrolysis means that Asn cannot be distinguished from Asp and Gln cannot be distinguished from Glu; the two groups are presented together as Asx and Glx, respectively. *n.d.* not determined

### *10.7.4 Protein Fractionation*

Various electrophoretic methods have been used to examine the protein components of the *Notaden* adhesive. It was found that sodium dodecyl sulphate–polyacrylamide gel electrophoresis (SDS-PAGE) using precast tricine or Tris/glycine 10–20 % gradient gels gave good resolution of the proteins found in *Notaden* exudates (Fig. 10.6). The gel pattern was essentially the same for reduced and unreduced samples except for one key difference: material that in non-reduced samples could not enter the gel migrated in reduced samples as the most abundant and slowest-moving component, Nb-1R (Fig. 10.6). The characteristic pattern of proteins has apparent molecular masses ranging from ~15 to ~400 kDa. Other electrophoretic separations were attempted, including native gels run in both acidic and basic conditions, acid urea gels, and isoelectric focusing. These separated some components but did not give the full resolution afforded by the SDS-PAGE gradient gels (V. Glattauer and L.D. Graham, unpublished results).

Various approaches for larger-scale fractionation of the protein components have also been examined. These included fractional precipitation with various agents, including ammonium sulphate, PEG, NaCl and EtOH, but none gave well-resolved components. Similarly, attempts to fractionate using column

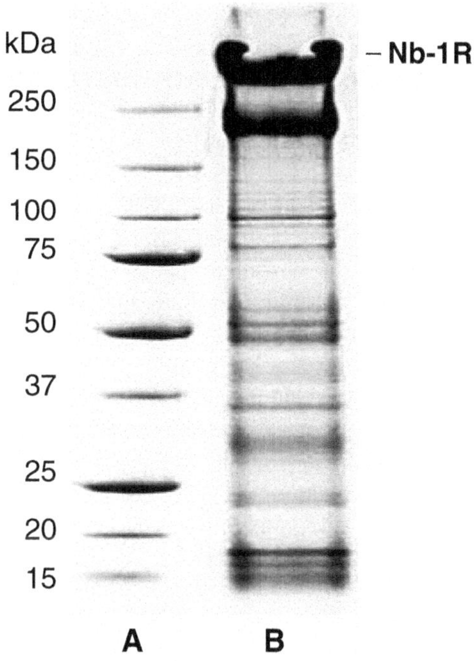


**Fig. 10.6** SDS-PAGE of reduced *Notaden* adhesive, showing a range of protein components. (**a**) Electrophoretic markers, with the molecular masses (kDa) shown at the *left*. (**b**) The characteristic pattern of proteins, with the largest and most abundant protein, Nb-1R, indicated

chromatography approaches, including gel permeation and ion-exchange chromatography, were also not as successful as had been hoped. In many cases, the adhesive components bound irreversibly to the chromatography matrices and rendered the columns inoperable, even when strong solvents containing high concentrations of denaturants were used (Graham et al. 2005; L.D. Graham and V. Glattauer, unpublished results).

Thus, although attempts at larger-scale fractionation were unsatisfactory and did not permit the immediate examination of the adhesive effects of individual components or specific mixtures, SDS-PAGE does provide a convenient approach to resolving the mixture. Separations of this type have allowed the amino acid composition of the Nb-1R protein to be determined (Table 10.2). Given the sensitivity of present analytical methods, including gas-phase Edman degradation and sequence analysis of tryptic peptides by mass spectrometry, it is anticipated that much structural information on the individual components will be obtained from material separated by SDS-PAGE.

## 10.8 Comparative Studies

Recently, we showed that the *Notaden bennettii* glue and the prey-capture 'slime' ejected by *Euperipatoides* sp. velvet worms (Benkendorff et al. 1999; Haritos et al. 2010) look similar and handle similarly. Both secretions consist largely of protein (55–60 % of dry weight), which provides the structural scaffold. The major protein of the onychophoran glue (Er_P1 for *E. rowelli*) and the dominant frog glue protein (Nb-1R) are both very large (260–500 kDa), and both give odd-looking wavy bands in SDS-PAGE. Both major proteins, which are rich in Gly (16–17 mol%) and Pro (7–12 mol%) and contain 4-hydroxyproline (Hyp, 4 mol%) (see Table 10.2 for Nb-1R), have a composition suggestive of intrinsically unstructured proteins. The observed Gly and Pro/Hyp content places them in an ambivalent zone where both elastomeric and amyloid structures are possible (Rauscher et al. 2006). We presume that the elastomeric state prevails under physiological conditions, but note that wide-angle X-ray scattering of a rapidly stretched and air-dried glue fibre of frog glue yielded the diffraction signature of a cross β-fold. The low carbohydrate content of both glues is consistent with conventional protein glycosylation; interestingly, the *N*-linked sugars of Nb-1R appear to prevent inappropriate self-aggregation (Graham et al. 2013).

Initial Edman N-terminal sequencing and de novo sequencing of tryptic peptides derived from the major SDS-PAGE bands of *N. bennettii* glue suggest that the majority of the bands are novel proteins, although some may be multidomain proteins that contain modules known from other contexts. Preliminary de novo sequencing of tryptic peptides from Nb-1R supports the idea that this protein contains imperfect repeats rich in Pro, Gly, acidic and hydrophobic residues (Graham et al. 2013). Some recurring motifs are reminiscent of the repeated sequences found in elastin and in the elastic PEVK domain of mammalian titin, two 'entropic spring' proteins that are believed to be largely unstructured (Tatham and Shewry 2000; Tompa 2010; Huber et al. 2012; Graham et al. 2013). From this and the amino acid composition data, we predicted that Nb-1R would probably contain extensive segments of low complexity that were intrinsically disordered, in addition to some well-folded globular domains (Graham et al. 2013). Regarding the latter, recent cDNA sequencing (Y.Y. Peng, P.R. Vaughan, J.A.M. Ramshaw and L.D. Graham, unpublished results) indicates that Nb-1R possesses two domains in its N-terminal half that are highly homologous to the H-domain of human IgG-Fc-binding protein (IgG-FcBP) (Harada et al. 1997). The role of the H-domain is unclear, but IgG-FcBP also contains mucin-like repeats, and the protein is thought to form a hydrogel (Harada et al. 1997). Interestingly, skin secretions from *Bombina* frogs contain an L- to D-amino acid isomerase that is highly homologous to the H-domain of IgG-FcBP (Jilek et al. 2005) and therefore to the Nb-1R domains as well (L.D. Graham, unpublished results). It is unclear whether the Nb-1R domains are catalytically competent or whether they represent another example of an enzyme scaffold being recruited for structural purposes. They may even fulfil both roles, as in the case of lens crystallins that retain lactate dehydrogenase activity (Atta et al. 2014).

Studies on the composition of salamander adhesive (Hamning et al. 2000; Largen and Woodley 2008) are still at an early stage compared to studies on the frog glue, but the two types of amphibian secretion seem to be broadly similar. For example, *Plethodon shermani* secretes from its dorsal skin a proteinaceous adhesive that is physically similar to the frog and velvet worm glues (Largen and Woodley 2008; von Byern et al. 2015). In addition to proteins of approximately 14–120 kDa

(Largen and Woodley 2008), SDS-PAGE of mercaptoethanol-reduced samples of this salamander glue shows a wide and dark protein band near the top of the gel (260-330 kDa) that potentially corresponds to the large Nb-1R and Er_P1 proteins of the frog and velvet worm secretions, respectively (Graham et al. 2013).

Overall, there are enough similarities between the frog, velvet worm and salamander glues to suspect that they employ related mechanisms for setting and adhesion; accordingly, we have proposed a common paradigm for amphibian and onychophoran adhesives. The model relies upon large proteins containing long regions that are rich in Pro/Hyp and charged residues; these regions of low complexity are organised into imperfect tandem repeats and form intrinsically unstructured polypeptides. The proteins need not have homologous sequences or be derived from a common ancestor. Under normal conditions, we envisage that the secretions begin as viscous protein solutions or emulsions that, upon release, set into elastomeric hydrogels. High shear probably favours a thread-like morphology (as in the velvet worm ejectate), whereas low shear probably favours a three-dimensional matrix (as in the frog and salamander secretions), but both forms are able to dehydrate rapidly to a glass-like state, which has the material properties of a strong and brittle solid (Graham et al. 2013). If proven correct, the identification of a molecular template utilised independently by vertebrates and invertebrates for fast-acting protein-based glues should prove helpful in the design of biomimetic adhesives for use in surgery and wound healing (Graham et al. 2013; Yang et al. 2014).

## 10.9 Applications

The adhesive from *Notaden* has several useful properties compared with other materials. These include its ability to bond to moist surfaces or those covered by water droplets. It is also very adaptable in the types of surfaces to which it can effectively bond, including metals, glass, plastics and biological tissues. Although for convenience we have been referring to the set adhesive as a solid, the moist material is actually a hydrogel containing a large proportion of trapped water. Examination of the set adhesive by scanning electron microscopy (Fig. 10.7) shows an open, porous structure formed by a meshwork of fibres and sheets. The pores in this structure can exceed 5–10 μm. The sponge-like network of pores and channels within the adhesive explains how the moist solid can accommodate a water content of 85–90 % (w/w). Along with the adhesive's ability to bond biological tissues, this porosity suggests that the adhesive could potentially be useful in biomedical applications, as the porosity will allow the diffusion of nutrients, gases and waste products and may also allow the migration and infiltration of cells.

A novel adhesive for medical applications would be particularly useful as there is a significant unmet clinical need for a strong and flexible surgical adhesive that is highly biocompatible (MedMarket 2002). Current biological adhesives (e.g., fibrin, albumin, gelatin–resorcinol–formaldehyde (GRF), etc.) typically suffer from low bond strength and in some cases are derived from blood products, with associated

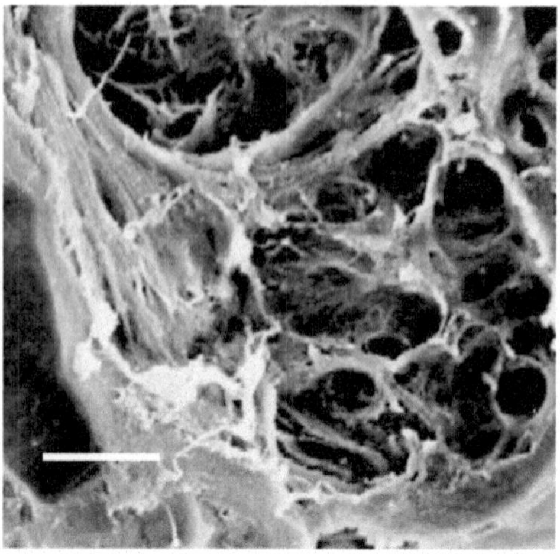

**Fig. 10.7** Scanning electron microscopy of the *Notaden* adhesive. Bar = 5 μm

risk of viral or prion contamination. On the other hand, synthetic glues (e.g., cyanoacrylate adhesives) are very strong, but they are also toxic to living tissues and form rigid, brittle, non-porous films that can hinder wound healing.

Since meniscal tears are a prevalent injury in humans (Roeddecker et al. 1994), preliminary studies on the frog glue were done using an in vivo meniscal defect model in sheep. These suggested the usefulness of the adhesive in meniscal repair. A study using 10 test and 6 control animals with artificial meniscal tears showed after 10 weeks that the adhesive could hold the torn fragments together, especially if the meniscal fragments were not completely separated, and that tissue repair had commenced with new fibrous tissue present (cited in Tyler and Ramshaw 2002). The limitations of current treatments of the meniscus are most evident in those parts of the tissue that are poorly vascularised and where it is difficult to keep the meniscus fragments in good contact for the healing procedure.

A complementary ex vivo investigation involved a tear propagation test on glued longitudinal tears in sheep menisci. Fresh medial menisci were dissected from intact sheep joints, and a longitudinal cut was created. Holding sutures were inserted into the free end of the meniscal fragments. Fresh *Notaden* adhesive (from *N. melanoscaphus*) was applied to one side of the cut and the fragments pressed firmly together. For comparison, gelatin–resorcinol–formaldehyde (GRF) glue, fibrin glue and cyanoacrylate glue were also tested. Glued samples were incubated in humid conditions at room temperature for 24 h, following which a tear propagation method was used to evaluate bond strength (Szomor et al. 2008). In all cases, tearing occurred along the tissue-adhesive interface; cyanoacrylate formed a hard, brittle film on the glued surfaces, while the other glues remained flexible. The peel strengths (Table 10.3) show that the cyanoacrylate glue was the strongest, followed by the *Notaden* adhesive, which was significantly stronger than both the GRF and fibrin adhesives (Szomor et al. 2008).

Noting that rotator cuff tears are a significant cause of morbidity in the adult human shoulder, a further ex vivo investigation explored whether the application of

**Table 10.3** The average peel strengths of selected adhesives (Szomor et al. 2008)

| Adhesive | Contact pressure duration (s) | Peel strength mean ± SD (N/m) ($n = 11$–12) |
|---|---|---|
| Cyanoacrylate | 30 | 149 ± 10 |
| *Notaden* | 90 | 97 ± 9 |
| Gelatin (GRF) | 90 | 39 ± 8 |
| Fibrin | 60 | 20 ± 3 |

*N. bennettii* glue could enhance conventional tendon attachment repairs (Millar et al. 2009). Three techniques were used to repair sheep infraspinatus tendons with a mattress suture configuration. In each group, seven shoulders were repaired with the addition of frog glue to the infraspinatus 'footprint', while another seven were used as nonadhesive controls. Failure occurred in all constructs at the tendon–bone–suture interface. For all three repair techniques, the presence of the frog glue significantly increased the load to failure, the total energy required for failure, and the maximum energy at failure ($P < 0.01$) (Millar et al. 2009). The load to failure of the two types of repair that utilised metallic anchors was doubled by the presence of the glue (failure at 143 ± 8 and 165 ± 20 N), while the load to failure for a transosseous repair (86 ± 8 N) was increased 1.7-fold over control values (Millar et al. 2009).

If the *Notaden* adhesive is to be developed into a medical adhesive, then we must be able to produce a sterile material that is reproducible in composition and available in sufficient quantity. This strongly suggests that the natural extract, which has collection difficulties (due to seasonal availability and environmental concerns) and which is difficult to harvest and sterilise, would not be the material of choice. Rather, if we can arrive at an understanding of which protein(s) represent the key structural agent(s), then production of a recombinant version of the adhesive may present an excellent option. Such material would be very well defined and highly reproducible. Further, the knowledge of the key component(s) and the mechanism of adhesion could lead not only to recombinant versions of the natural material but also to artificial mimics produced by synthetic organic chemistry. These could minimise immunogenicity issues or other adverse phenomena and prove easier to progress through to regulatory approval.

Nonadhesive biomedical applications could also emerge for a recombinant form of Nb-R1. We found that peroxidase treatment of dissolved frog glue introduced dityrosine cross-links between Nb-1R proteins, which in turn caused a soft transparent hydrogel to form (Graham et al. 2010). This artificial material was very different in appearance and material properties to the natural set glue, which was much denser, as indeed was every other known type of frog glue including NaCl-reconstituted material (Sect. 10.4). The rheological properties of the cross-linked hydrogel were comparable to those of mammalian mucus and mucin preparations (Graham et al. 2010). Artificial hydrogels like this might prove useful where mucosal hydration and/or drug delivery is required. In addition, hydrogels that transiently acquire mechanical strength could be particularly useful in emergency dressings for wounds and burns.

## 10.10 Conclusions

The yellow adhesive substance that frogs of the genus *Notaden* secrete on their dorsal skin is a fascinating material that may help to inspire the development of new medical glues. This pressure-sensitive adhesive bonds rapidly to a wide range of polar and non-polar materials, including moist and cool surfaces. Biochemical studies have shown that the adhesive is a highly hydrated protein matrix that is rich in Gly, Pro/Hyp and Glx residues. Nb-1R, the band of highest molecular mass in the *N. bennettii* adhesive, appears to be its key component. The similarities between frog, velvet worm and salamander glues have led us to suggest that the mechanism of all three may rely upon large proteins containing long regions of imperfect tandem repeats that form intrinsically unstructured polypeptides.

An ex vivo study of the reattachment of sheep infraspinatus tendons revealed that augmentation using frog glue approximately doubled the strength of the repairs. Ex vivo tests of peel strength in glued sheep menisci have indicated that, although not as strong as cyanoacrylate, the frog glue is stronger than other protein-based medical glues, and a preliminary in vivo study of meniscal repair in sheep has given promising results. In vitro experiments have shown that the glue is compatible with cell attachment and growth, while its open porous structure is likely to encourage cell infiltration. When small pellets of glue were implanted subcutaneously into mice, they caused an initial necrotic response in the skin, which was probably caused by the toxic metabolites (sterols, carotenoids, etc.) known to be present within the glue matrix. However, all of the animals made a full recovery; by 2-week post-implantation, the lesions had largely healed; by 2 months, the pellets had been fully resorbed by the surrounding tissues and normal dermal architecture had returned.

While small quantities of adhesive can be harvested from frogs in the laboratory, other strategies such as recombinant protein production will be needed for larger-scale trials. Recombinant production of the key structural protein(s) would afford much-needed certainty in regard to composition and scale, and reliability in regard to supply; it would also eliminate the complications caused by the bioactive metabolites that are present in the natural product. Overall, the properties of the frog adhesive suggest that a recombinant mimic would have great potential for medical applications.

**Acknowledgements** The authors wish to thank Joel Mackay (Sydney University) for the CD spectrum; Raju Adhikari, Lawry McCarthy, Helmut Thissen and Tony Cripps (CSIRO) for the help with the adhesive strength tests; and Graham Johnson and Meg Evans (CSIRO) for the bovine corneal organ culture and epithelial cell culture tests.

# References

Apponyi MA, Pukala TL, Brinkworth CS, Maselli VM, Bowie JH, Tyler MJ, Booker GW, Wallace JC, Carver JA, Separovic F, Doyle J, Llewellyn LE (2004) Host-defence peptides of Australian anurans: structure, mechanism of action and evolutionary significance. Peptides 25:1035–1054

Atta A, Ilyas A, Hashim Z, Ahmed A, Zarina S (2014) Lactate dehydrogenase like crystallin: a potentially protective shield for Indian spiny-tailed lizard (*Uromastyx hardwickii*) lens against environmental stress? Protein J 33:128–34

Barker J, Grigg GC, Tyler MJ (1995) A field guide to Australian frogs. Surrey Beatty, Chipping Norton

Benkendorff K, Beardmore K, Gooley AA, Packer NH, Tait NN (1999) Characterisation of the slime gland secretion from the peripatus, *Euperipatoides kanangrensis* (Onychophora : Peripatopsidae). Comp Biochem Physiol B 124:457–465

Brodie ED, Gibson LS (1969) Defensive behaviour and skin glands of the northwestern salamander, *Ambystoma gracile*. Herpetologica 25:187–194

Brodsky B, Ramshaw JAM (1997) The collagen triple-helix structure. Matrix Biol 15:545–554

Burzio LO, Burzio VA, Silva T, Burzio LA, Pardo J (1997) Environmental bioadhesion: themes and applications. Curr Opin Biotechnol 8:309–312

Channing AE (2001) Amphibians of central and southern Africa. Comstock, Ithica

Crisp DJ, Walker G, Young GA, Yule AB (1985) Adhesion and substrate choice in mussels and barnacles. J Colloid Interface Sci 104:40–50

Deming TJ (1999) Mussel byssus and biomolecular materials. Curr Opin Chem Biol 3:100–105

Doolittle RF (1986) Of URFs and ORFs: a primer on how to analyze derived amino acid sequences. University Science, California, p 55

Doyle J, Brinkworth CS, Wegener KL, Carver JA, Llewellyn LE, Olver IN, Bowie JH, Wabnitz PA, Tyler MJ (2003) nNOS inhibition, antimicrobial and anticancer activity of the amphibian skin peptide, citropin 1.1 and synthetic modifications. The solution structure of a modified citropin 1.1. Eur J Biochem 270:1141–1153

Dubois M, Gilles KA, Hamilton JK, Rebers PA, Smith F (1956) Colorimetric method for determination of sugars and related substances. Anal Chem 28:350–356

Evans CM, Brodie ED (1994) Adhesive strength of amphibian skin secretions. J Herpetol 28:502–504

Frost-Mason S, Morrison R, Mason K (1994) Pigmentation. In: Heatwole H, Barthalmus GT (eds) Amphibian biology, vol 1. Surrey Beatty, Chipping Norton, pp 64–97

Gieseg SP, Simpson JA, Charlton TS, Duncan MW, Dean RT (1993) Protein-bound 3,4-dihydroxyphenylalanine is a major reductant formed during hydroxyl radical damage to proteins. Biochemistry 32:4780–4786

Graham LD, Glattauer V, Huson MG, Maxwell JM, Knott RB, White JF, Vaughan PR, Peng Y, Tyler MJ, Werkmeister JA, Ramshaw JAM (2005) Characterization of a protein-based adhesive elastomer secreted by the Australian frog *Notaden bennettii*. Biomacromolecules 6:3300–3312

Graham LD, Danon SJ, Johnson G, Braybrook C, Hart NK, Varley RJ, Evans MDM, McFarland GA, Tyler MJ, Werkmeister JA, Ramshaw JAM (2010) Biocompatibility and modification of the protein-based adhesive secreted by the Australian frog *Notaden bennetti*. J Biomed Mater Res A 93:429–441

Graham LD, Glattauer V, Li D, Tyler MJ, Ramshaw JAM (2013) The adhesive skin exudate of *Notaden bennetti* frogs (Anura: Limnodynastidae) has similarities to the prey capture glue of *Euperipatoides* sp. velvet worms (Onychophora: Peripatopsidae). Comp Biochem Physiol B 165:250–259

Greenfield N, Fasman GD (1969) Computed circular dichroism spectra for the evaluation of protein conformation. Biochemistry 8:4108–4116

Gross M (2011) Getting stuck in. Chem World 8(12):52–55

Hamning VK, Yanites HL, Peterson NL (2000) Characterization of adhesive and neurotoxic components in skin granular gland secretions of *Ambystoma tigrinum*. Copeia 2000:856–859

Harada N, Iijima S, Kobayashi K, Yoshida T, Brown WR, Hibi T, Oshima A, Morikawa M (1997) Human IgGFc binding protein (FcgammaBP) in colonic epithelial cells exhibits mucin-like structure. J Biol Chem 272:15232–15241

Haritos VS, Niranjane A, Weisman S, Trueman HE, Sriskantha A, Sutherland TD (2010) Harnessing disorder: onychophorans use highly unstructured proteins, not silks, for prey capture. Proc R Soc B 277:3255–3263

Huber T, Grama L, Hetényi C, Schay G, Fülöp L, Penke B, Kellermayer MS (2012) Conformational dynamics of titin PEVK explored with FRET spectroscopy. Biophys J 103:1480–1489

Jilek A, Mollay C, Tippelt C, Grassi J, Mignogna G, Müllegger J, Sander V, Fehrer C, Barra D, Kreil G (2005) Biosynthesis of a D-amino acid in peptide linkage by an enzyme from frog skin secretions. Proc Natl Acad Sci U S A 102:4235–4239

Kamino K, Inoue K, Maruyama T, Takamatsu N, Harayama S, Shizuri Y (2000) Barnacle cement proteins. J Biol Chem 275:27360–27365

Largen W, Woodley SK (2008) Cutaneous tail glands, noxious skin secretions, and scent marking in a terrestrial salamander (*Plethodon shermani*). Herpetologica 64:270–280

MedMarket (2002) MedMarket Diligence Report #S120, Worldwide Wound Sealant Market, Medmarket Diligence, Foothill Ranch, CA

Millar NL, Bradley TA, Walsh NA, Appleyard RC, Tyler MJ, Murrell GA (2009) Frog glue enhances rotator cuff repair in a laboratory cadaveric model. J Shoulder Elbow Surg 18:639–645

Monahan J, Wilker JJ (2004) Cross-linking the protein precursor of marine mussel adhesives: bulk measurements and reagents for curing. Langmuir 20:3724–3729

Naldrett MJ, Kaplan DL (1997) Characterization of barnacle (*Balanus eburneus* and *B. cenatus*) adhesive proteins. Marine Biol 127:629–635

Phillips B, Shine R (2007) When dinner is dangerous: Toxic frogs elicit species-specific responses from a generalist snake predator. Am Nat 170:936–942

Rauscher S, Baud S, Miao M, Keeley FW, Pomès R (2006) Proline and glycine control protein self-organization into elastomeric or amyloid fibrils. Structure 14:1667–1676

Roeddecker K, Muennich U, Nagelschmidt M (1994) Meniscal healing: a biomechanical study. J Surg Res 56:20–27

Schmidt-Dannert C, Umeno D, Arnold FH (2000) Molecular breeding of carotenoid biosynthetic pathways. Nat Biotechnol 18:750–753

Szomor ZL, Murrell GAC, Appleyard RC, Tyler MJ (2008) Meniscal repair with a new biological glue: an ex vivo study. Tech Knee Surg 7:261–265

Tatham AS, Shewry PR (2000) Elastomeric proteins: biological roles, structures and mechanisms. Trends Biochem Sci 25:567–571

Tompa P (2010) Structure and function of intrinsically disordered proteins. CRC Press/Chapman and Hall, New York, p 166

Tyler MJ (2010) Adhesive dermal secretions of the amphibia, with particular reference to the Australian Limnodynastid Genus *Notaden*. In: Biological adhesive systems: from nature to technical and medical application. Springer, Vienna and New York, pp 181–186

Tyler MJ, Ramshaw JAM (2002) International Patent Application WO200222756-A1

Tyler MJ, Stone DJ, Bowie JH (1992) A novel method for the release and collection of dermal, glandular secretions from the skin of frogs. J Pharmacol Toxicol Methods 28:199–200

von Byern J, Dicke U, Heiss E, Grunwald I, Gorb S, Staedler Y, Cyran N (2015) Morphological characterization of the glandular system in the salamander *Plethodon shermani* (Caudata, Plethodontidae). Zoology 118:334–47

Waite JH (1993) Polyphenolic substance of *Mytilus edulis*: novel adhesive containing L-Dopa and hydroxyproline. Science 212:1038–1040

Williams TA, Anthony CD (1994) Technique to isolate salamander granular gland products with a comment on the evolution of adhesiveness. Copeia 1994:540–541

Williams TA, Larsen JH (1986) New function for the granular skin glands of the eastern long-toed salamander, *Ambystoma macrodactylum columbianum*. J Exp Zool 239:329–333
Williams CR, Brodie ED Jr, Tyler MJ, Walker SJ (2000) Antipredator mechanisms of Australian frogs. J Herpetol 34:431–443
Yang YJ, Jung D, Yang B, Hwang BH, Cha HJ (2014) Aquatic proteins with repetitive motifs provide insights to bioengineering of novel biomaterials. Biotechnol J 9:1493–1502

# Chapter 11
# Properties, Principles, and Parameters of the Gecko Adhesive System

**Kellar Autumn and Jonathan Puthoff**

> *The designers of the future will have smarter adhesives that do considerably more than just stick.* (Fakley 2001)

**Abstract** Current understanding of the adhesion system of geckos is the culmination of efforts by investigators throughout the biological and applied sciences. Continuing research in this area promises dividends in areas such as biomechanics, evolution, ecology, adhesive technology, and robotics. We describe here the key topics involved in the gecko adhesion system: the notable properties, the underlying physical principles, and the parameters that govern system performance. Additionally, we highlight the most important unresolved issues and propose productive directions related to gecko adhesion research and bioinspired engineering.

## 11.1 Introduction

Gecko toe pads are sticky because they feature an extraordinary hierarchy of structure that functions as a smart adhesive. Gecko toe pads (Russell 1975) operate under perhaps the most severe conditions of any adhesive application. Geckos are capable of attaching and detaching their adhesive toes in milliseconds (Autumn et al. 2006a) while running with seeming reckless abandon on vertical and inverted surfaces, a challenge no conventional adhesive is capable of meeting. Structurally, the adhesive on gecko toes differs dramatically from that of conventional adhesives. Conventional pressure-sensitive adhesives (PSAs) such as those used in adhesive tapes are fabricated from materials that are sufficiently soft and sticky to flow and

K. Autumn
Department of Biology, Lewis & Clark College, Portland, OR 97219-7899, USA

J. Puthoff (✉)
Department of Chemical and Materials Engineering, California State Polytechnic University Pomona, Pomona, CA 91767, USA
e-mail: jbputhoff@cpp.edu

© Springer International Publishing Switzerland 2016

A.M. Smith (ed.), *Biological Adhesives*, DOI 10.1007/978-3-319-46082-6_11

make intimate and continuous surface contact (Pocius 2012). Because they are soft and sticky, PSAs also tend to degrade, foul, self-adhere, and attach accidentally to inappropriate surfaces. Gecko toes typically bear a series of scansors covered with uniform microarrays of hairlike setae formed from β-keratin (Wainwright et al. 1982; Russell 1986; Alibardi 2003; Rizzo et al. 2006), a material order of magnitude stiffer than those used to fabricate PSAs. Each seta branches to form a nanoarray of hundreds of spatular structures; these structures make ultimate contact with the surface.

Functionally, the properties of gecko setae are as extraordinary as their structure: the gecko adhesive (1) is directional, (2) attaches strongly with minimal preload, (3) detaches quickly and easily, (4) sticks to nearly every material, (5) exhibits rate-dependent adhesion, (6) does not stay dirty or (7) self-adhere, and (8) is nonsticky by default. While some of the principles underlying these eight functional properties are now well understood, much more research will be necessary to fully map out the parameters of this complex system.

Over two millennia ago, Aristotle (1910) commented on the ability of the gecko to "run up and down a tree in any way, even with the head downwards." How geckos adhere has attracted substantial and sustained scientific scrutiny (Cartier 1872, 1874a; Gadow 1901; Weitlaner 1902; Schmidt 1904; Hora 1924; Dellit 1934; Mahendra 1941; Maderson 1964; Ruibal and Ernst 1965; Hiller 1968, 1969, 1976; Gennaro 1969; Russell 1975, 1986; Williams and Peterson 1982; Stork 1983; Schleich and Kästle 1986; Irschick et al. 1996; Autumn et al. 2000, 2002b; Autumn and Peattie 2002; Arzt et al. 2003; Huber et al. 2005a; Hansen and Autumn 2005). The unusual hairlike microstructure of gecko toe pads has been recognized for well over a century (Cartier 1872, 1874a, b; Braun 1878). Setal branches were discovered using light microscopy (Schmidt 1904), but the discovery of multiple split ends (Altevogt 1954) and spatular nanostructure (Ruibal and Ernst 1965) at the tip of each seta was made only after the development of electron microscopy.

A single seta of the tokay gecko (*Gekko gecko*) is approximately 110 μm in length and 4.2 μm in diameter (Ruibal and Ernst 1965; Russell 1975; Williams and Peterson 1982) (Fig. 11.1). Setae are similarly oriented and uniformly distributed on the scansors. Setae branch a number of times at the tips into 100–1000 terminal structures (Ruibal and Ernst 1965; Schleich and Kästle 1986; Rizzo et al. 2006) known as spatulae. A single spatula consists of a stalk with a thin, roughly triangular end whose apex connects the spatula to its stalk. Spatulae are approximately 0.2 μm long with a similar width at the tip (Ruibal and Ernst 1965; Williams and Peterson 1982).

While the tokay is currently the best studied of any adhesive gecko species, there are over a thousand species of gecko (Han et al. 2004), encompassing an impressive range of morphological variation at the spatula, seta, scansor, and toe levels (Maderson 1964; Ruibal and Ernst 1965; Russell 1975, 1981, 1986; Peterson and Williams 1981; Williams and Peterson 1982; Stork 1983; Schleich and Kästle 1986; Russell and Bauer 1988, 1990a, b; Röll 1995; Irschick et al. 1996; Autumn et al. 2002a; Arzt et al. 2003; Peattie and Full 2007). Setae have even evolved on the tails of some gecko species (Bauer 1998). Remarkably, setae have evolved convergently in iguanian lizards of the genus *Anolis* (Braun 1878; Ruibal and

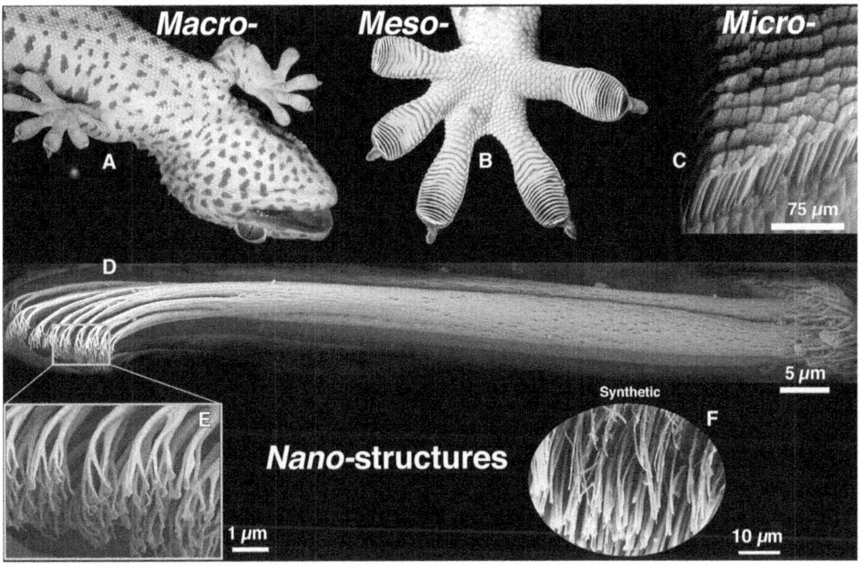


**Fig. 11.1** Structural hierarchy of the gecko adhesive system; (**a**) ventral view of a tokay gecko climbing a vertical glass surface; (**b**) ventral view of the foot of a tokay gecko, showing a mesoscale array of seta-bearing scansors (adhesive lamellae) (images **a** and **b** by 1 Mark Moffett); (**c**) microscale array of setae are arranged in a nearly grid-like pattern on the ventral surface of each scansor. In this scanning electron micrograph, each *diamond-shaped* structure is the branched end of a group of four setae clustered together in tetrad; (**d**) micrograph of a single gecko seta assembled from a montage of five Cryo-SEM images (image by S. Gorb and K. Autumn). Note individual keratin fibrils comprising the setal shaft; (**e**) nanoscale array of hundreds of spatular tips of a single gecko seta; (**f**) synthetic spatulae fabricated from polyimide using nanomolding (Campolo et al. 2003)

Ernst 1965; Peterson and Williams 1981) and in scincid lizards of the genus *Prasinohaema* (Williams and Peterson 1982; Irschick et al. 1996).

This chapter aims broadly at identifying the known properties of the gecko adhesive system, possible underlying principles, and quantitative parameters that affect system function. However, much of what is known is based on studies of a single species—the tokay gecko—and the degree of variation in function among species remains an open question that should be kept in mind.

## 11.2 Adhesive Properties of Gecko Setae

Two front feet of a tokay gecko can withstand 20.1 N of force parallel to the surface across 227 mm$^2$ of toe-pad area (Irschick et al. 1996). The foot of a tokay bears approximately 3600 tetrads of setae per mm$^2$; this gives a corresponding setal density $\rho = 14{,}400$ setae per mm$^2$ (Schleich and Kästle 1986). Consequently, a single seta should produce an average force of 6.2 μN and an average shear stress of 0.090 N/mm$^2$ (0.9 atm). However, single setae have proved both much less sticky and much more sticky than predicted by whole-animal measurements under varying experimental conditions, implying that attachment and detachment in gecko setae are mechanically controlled (Autumn et al. 2000).

A similar analysis can be performed using the clinging force and morphology data from other gecko species; these results do not indicate many universal scaling principles relating to setal density and animal-level clinging force across species. A phylogenetic analysis indicates that there is no apparent relationship between setal density and the body mass of the animal (Peattie and Full 2007). Other recent experiments (Hagey et al. 2014) indicate that, while the setal arrays of different species operate by similar mechanics, the exact connection between the morphological details and the performance is unclear.

### *11.2.1 Properties (1) and (2): Anisotropic Attachment and High Adhesion Coefficient ($\mu'$)*

Using a newly developed microelectromechanical system (MEMS) force sensor (Chui et al. 1998), Autumn et al. (2000) measured the adhesive and shear force of a single isolated gecko seta. The angle of the setal shaft was particularly important in achieving an adhesive bond. Strong attachment occurred when using proper orientation and a motion based on the dynamics of gecko legs during climbing (based on force-plate data; Fig. 11.2) (Autumn et al. 2006b). A small normal preload force yielded a shear force of ≈ 40 μN, six times the force predicted by whole-animal measurements (Irschick et al. 1996). The small normal preload force, combined with a 5-μm proximal lateral/shear displacement, yielded a very large shear force of 200 μN, 32 times the force predicted by whole-animal measurements of Irschick et al. and 100 times the frictional force measured with the seta oriented with spatulae facing away from the surface. The preload and drag steps were also necessary to initiate significant adhesion in isolated gecko setae, consistent with the load dependence and directionality of adhesion observed at the whole-animal scale by Haase (1900) and Dellit (1934). The ratio of preload to pull off force is the adhesion coefficient ($\mu'$), which represents the strength of adhesion as a function of the preload (Bhushan 2013). In isolated tokay gecko setae, a 2.5-μN preload yielded adhesion between 20 μN (Autumn et al. 2000) and 40 μN (Autumn et al. 2002b), and thus $\mu'$ lies between 8 and 16.

#### 11.2.1.1 Large Safety Factor for Adhesion and Friction?

All 6.5 million (Schleich and Kästle 1986; Irschick et al. 1996) setae of a 50-g tokay gecko attached maximally could theoretically generate 1300 N (133 kg force) of shear force—enough to support the weight of two humans. This suggests that a gecko need only attach 3 % of its setae to generate the greatest forces measured in the whole animal (≈20 N) (Irschick et al. 1996). Only less than 0.04 % of a gecko's

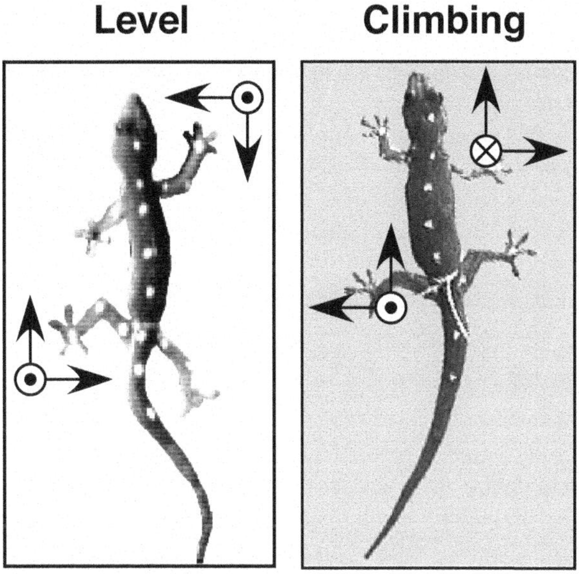


**Fig. 11.2** Single-leg ground reaction forces in running geckos (*Hemidactylus garnotii*): (**a**) during level running, geckos' front legs produce deceleratory ground reaction forces while their hind legs produce acceleratory forces (Autumn et al. 2006b). All legs push away from the body, producing ground reaction forces aimed through the joints toward the center of mass, minimizing joint moments, and possibly stabilizing the animal as it runs. *Circles with dots* represent positive ground reaction forces normal to the surface. During level running these represent the forces that support the animal's weight; (**b**) during vertical climbing, geckos have similar kinematics, but alter dramatically their kinetics in comparison to level running (Chen et al. 2006). While climbing, all legs accelerate the body up the wall, and all legs pull in toward the center of mass, engaging the adhesive setae and claws. Front limbs pull away from the surface and hind limbs push into the surface, producing a torque that tips the head toward the wall, counteracting the tendency of the head to pitch back as it climbs

setae attached maximally are needed to support its weight (0.49 N) on a wall. At first glance, gecko feet seem to be enormously overbuilt; they have a safety factor of at least (20 N/0.49 N) – 1 = 41 (or 4100 %). However, it is unlikely that all setae are able to achieve the optimal orientation simultaneously. The proportion of spatulae attached may be greatly reduced on rough surfaces, particularly those with roughness on the same scale as spatulae or setae (Persson and Gorb 2003). On dusty or exfoliating surfaces, attachment to a well-anchored substrate will not be possible for every seta, perhaps even rendering the scansor surfaces useless (Delaugerre et al. 2015). Geckos may use a significant portion of their safety margin while withstanding high winds during tropical storms, resisting predator attack, or recovering adhesion after a fall (Hecht 1952; Vinson and Vinson 1969; Russell 1976; Autumn and Peattie 2002; Pianka and Sweet 2005).

Geckos have been observed to recover from a fall by reattaching their toes to leaves or branches as they plummet (Vitt and Zani 1997; Pianka and Sweet 2005). A simple calculation suggests that recovery from a fall may require a large proportion of a gecko's safety margin in adhesion or friction. Consider a 50-g gecko falling from rest. If the gecko attaches a foot to a vertical surface after it has fallen 10 cm (neglecting air resistance), it will be moving at 1.4 m/s. If the foot is able to produce 5 N of friction, the gecko will be able to come to a stop in 15 ms after sliding 1.1 cm. In this theoretical example, recovering from a fall of the very modest distance of 10 cm would require 50 % of the shear capacity of one foot based on whole-animal measurements (Irschick et al. 1996) but still less than 4 % of the theoretical maximum shear stress calculated for single setae (Autumn and Peattie 2002).

### *11.2.2 Property (3): Low Detachment Force*

The surprisingly large forces generated by single setae raised the question of how geckos manage to detach their feet in just 15 ms with no measurable detachment forces (Autumn et al. 2006b). Autumn et al. (2000) discovered that simply increasing the angle ($\alpha$) that the setal shaft makes with the substrate to 30° causes detachment. As $\alpha$ increases, sliding stops and stress increases at the trailing edge of the seta, causing fracture of the seta-substrate bonds (Autumn et al. 2000) and returning the seta to the unloaded default state. This scenario is supported by models of setae as cantilever beams (Sitti and Fearing 2003; Gao et al. 2005; Spolenak et al. 2005a) and by finite element method (FEM) modeling of the seta (Gao et al. 2005). FEM simulation of the single-seta pull off experiment in Autumn et al. (2000) revealed more than an order of magnitude decrease in adhesive force as $\alpha$ increased from 30° to 90°. The FEM simulation also identified a transition from sliding to peeling that occurs at $\alpha = 30°$, consistent with cantilever beam-based models (Sitti and Fearing 2003) and empirical observations that setae slide at $\alpha < 30°$ but detach at $\alpha > 30°$ (Autumn et al. 2000).

Later measurements (Autumn et al. 2006a) indicate that setal arrays obey a tribological law called "frictional adhesion." In frictional adhesion, adhesive force depends on the shear/friction forces established; a decrease in shear loading decreases adhesion, and an increase in shear force increases adhesion. This relationship is consistent with a critical release angle in gecko toes, isolated setal arrays, and single setae and indicates why attachment is initiated proximally along the axis of the toe (Haase 1900) and why geckos only stick in the presence of lateral forces (Hora 1924).

This principle of directional or anisotropic adhesion is an important feature of the climbing system. The gecko adhesive can be thought of as the first known "programmable adhesive." Preload and drag steps turn on and modulate stickiness, while increasing the shaft/array/toe angle to 30° turns off stickiness. There are many physical explanations for the directional adhesion effect that involve a variety of geometrical and material parameters across a number of length scales (see Sect. 11.2.5.5) (Gao et al. 2005; Tian et al. 2006; Chen et al. 2009; Sauer 2009; Yao and Gao 2006; Hu and Greaney 2014).

### *11.2.3 Integration of Body and Leg Dynamics with Setal Attachment and Detachment*

How attachment and detachment of millions of setae during locomotion are integrated with the function of the scansor, toe, foot, leg, and body remains a topic of interest and ongoing research (Russell 1986; Autumn and Peattie 2002; Autumn et al. 2005; Gao et al. 2005). Since gecko setae require a preload in the normal axis for adhesion, large forces could potentially be associated with attachment of the foot. The tremendous adhesive capacity of gecko setae suggests that large forces could also occur during detachment. In fact, no measurable ground reaction forces were associated with either attachment or detachment during vertical climbing on a force plate of the house gecko *Hemidactylus garnotii* (Autumn et al. 2006b), indicating that these actions are either mechanically decoupled from the center of mass in this species or result in forces so small as to be undetectable. Russell (2002) suggested that in the tokay gecko, the perpendicular-preload and 5-μm drag requirements (Autumn et al. 2000; Autumn and Peattie 2002) are controlled by hydrostatic pressure in the highly derived blood sinuses and lateral digital tendon system, respectively. However, control of inflation and deflation of the sinuses remains to be demonstrated. This mechanism would not be available to those species that lack blood sinuses.

Setal preload and drag could also be a consequence of force development during the stride (Fig. 11.2). The force necessary to bend even thousands of setae into an adhesive orientation is probably quite small (at most 10 mN) (Autumn and Peattie 2002) and possibly below the sensitivity of the force plate used (Autumn et al. 2006b). Another possibility is that during attachment, which is a reversal of the peeling process of toe detachment, the required motion is decoupled from the center of mass. The gecko's foot approaches the substrate without pressing into it and reapplies the adhesive by unrolling its toes like a new year's party favor. This process is called digital hyperextension (Russell 1975, 2002; Xu et al. 2015). In this case, setal preload forces would be spread out over time and would likely be far below the 1-mN sensitivity of force plates used in measurements of whole-body and single-leg dynamics of small animals (Biewener and Full 1992). Peeling may reduce detachment and attachment forces, but may limit speed during vertical climbing. If toe peeling and uncurling in climbing geckos require some minimum time, then speed cannot be increased by reducing contact time as is typical in level running. *Hemidactylus garnotii* increased velocity by increasing stride length (Autumn et al. 2006b), and attachment and detachment occupied a constant value of approximately 20 ms.

The experiments of Gravish et al. (2008) indicate that certain detachment pathways are more energetically conservative than others. The release of stored elastic energy in the attached setal array helps drive the detachment process, and detachment at optimal foot-withdrawal angles apparently occurs spontaneously, meaning that little external force needs to be applied. This mechanical effect likely contributes significantly to the efficiency of locomotion. It is also worth noting that, after attachment, a gecko can maintain adhesion with little effort. In fact, deceased geckos continue to cling to the last surface that they attached themselves to (Stewart and Higham 2014).

## *11.2.4 Molecular Mechanism of Gecko Adhesion*

While setal structures of many gecko species are well documented, a complete understanding of what makes them adhere has been more elusive. At the turn of the twentieth century, Haase (1900) noted that attachment is load dependent and only occurs in one direction: proximally along the axis of the toe. Haase was also the first to suggest that geckos stick by intermolecular forces ("adhesion"), noting that under this hypothesis the attractive force should increase as the space between the feet and the substrate decreases. However, at least seven possible mechanisms for gecko adhesion have been discussed over the past 185 years.

### 11.2.4.1 Unsupported Mechanisms: Glue, Suction, Electrostatics, Microinterlocking, and Friction

Since geckos lack glandular tissue on their toes, sticky secretions similar to those used by insects (Dewitz 1882) were ruled out early in the study of gecko adhesion (Wagler 1830; Cartier 1872; Simmermacher 1884). The hypothesis that the toe pads acted as suction cups was first proposed by Wagler (1830) (who classified geckos as amphibians). The hypothesis that individual setae act as miniature suction cups was first under debate in the insect adhesion literature (Blackwall 1845; Hepworth 1854). Dewitz (1882) argued against suction as an explanation for gecko adhesion, but Simmermacher (1884) considered suction to be the most likely mechanism. However, there are no data to support suction as an adhesive mechanism, and the adhesion experiments carried out in a vacuum by Dellit (1934) suggest that suction is not involved. Furthermore, measurements of $\approx 9$ atm of adhesive stress (Autumn et al. 2002b) strongly contradict the suction hypothesis, since the adhesive stress produced by suction is limited by the ambient atmospheric pressure. Despite substantial evidence against it, the suction hypothesis has been surprisingly tenacious in the popular literature (e.g., Gennaro 1969). Electrostatic attraction (Schmidt 1904) was another hypothesized mechanism for adhesion in gecko setae. Experiments using X-ray bombardment (Dellit 1934) eliminated electrostatic attraction as a necessary mechanism for setal adhesion since geckos could still adhere to metal in ionized air. However, electrostatic effects could enhance adhesion even if another mechanism is operating (Maderson 1964).

Setae are recurved such that their tips point proximally, leading Dellit (1934) to hypothesize that setae act as micro- or nanoscale hooks, catching on surface irregularities (microinterlocking). Mahendra (1941) suggested that setae were analogous to the crampons of a climber's boot. The microinterlocking hypothesis was challenged by the ability of geckos to adhere while inverted on polished glass. A mechanism like this could play a secondary role under some conditions, but the presence of large adhesive forces on a molecularly smooth $SiO_2$ MEMS semiconductor surfaces (Autumn et al. 2000) demonstrates that surface irregularities are not necessary for adhesion. The friction hypothesis (Hora 1924) can be dismissed since, by definition, friction only acts in shear (Bhushan 2013) and therefore cannot in itself explain the adhesive capabilities of geckos on inverted surfaces.

#### 11.2.4.2 Potential Intermolecular Mechanisms: Van der Waals and Capillary Forces

Using the greatly enhanced resolution of electron microscopy, Ruibal and Ernst (1965) described the spatular structures at the setal tips. They concluded that spatulae were unlikely to function like the spikes on climbing boots and postulated that the spatulae lie flat against the substrate while the seta is engaged. It was clear to them that these flattened tips increased the realized contact area, and they concluded that gecko adhesion was the result of molecular interactions, not mechanical interlocking.

The turning point in the study of gecko adhesion came with a series of experiments (Hiller 1968, 1969) that suggested that the surface energy of the substrate, rather than its texture, determined the strength of attachment. By providing evidence that intermolecular forces were responsible, Hiller paved the way for the application of modern methods of surface science (Israelachvili 2011) in studies of gecko adhesion (Autumn et al. 2000, 2002b; Autumn and Peattie 2002; Huber et al. 2005a; Hansen and Autumn 2005). Hiller (1968, 1969, 1976) showed that shear force was correlated with the water droplet contact angle of the surface, and thus with the surface energy of the substrate, providing the first direct evidence that intermolecular forces are responsible for attachment in geckos.

Intermolecular capillary forces are the principal mechanism of adhesion in many insects (Gillett and Wigglesworth 1932; Edwards and Tarkanian 1970; Lee et al. 1986; Lees and Hardie 1988; Brainerd 1994), frogs (Emerson and Diehl 1980; Green 1981; Hanna et al. 1991), and even mammals (feathertail gliders) (Rosenberg and Rose 1999). Unlike many insects, geckos lack glands on the surface of their feet (Cartier 1874a; Bellairs 1970). However, this does not preclude the role of capillary adhesion (von Wittich 1854, quoted directly in Simmermacher 1884; Baier et al. 1968; Israelachvili 2011; Stork 1980) since layers of water molecules are commonly present on hydrophilic surfaces at ambient humidity and can cause strong attraction between surfaces. The observation (Hiller 1968, 1971, 1976) that geckos cannot adhere to polytetrafluoroethylene (PTFE or Teflon®) is consistent with the capillary hypothesis, since PTFE is strongly hydrophobic. Indeed, the apparent correlation between adhesive force and hydrophobicity (as indicated by the water contact angle $\theta$) (Hiller 1968, 1971, 1976) suggested that the polarity of the surface might be an important factor in the strength of adhesion (Autumn and Peattie 2002).

A non-mutually exclusive alternative mechanism is dispersive van der Waals (vdW) forces (Stork 1980; Autumn et al. 2000). These vdW forces are strongly dependent on the distance between surfaces, increase with the polarizability of the two surfaces, and are not related directly to surface polarity (Israelachvili 2011). The observation (Hiller 1968) that geckos cannot adhere to PTFE is consistent with both van der Waals and capillary hypotheses, since PTFE is weakly polarizable and hydrophobic.

#### 11.2.4.3 Contact Angle Estimates of Surface Energy

Hiller's experiments (Hiller 1968, 1969, 1976) were groundbreaking because they provided the first direct evidence for adhesion sensu stricto. The precise nature of the adhesion remained unknown until 2002 (Autumn and Peattie 2002; Autumn et al. 2002b). The intermolecular attraction between any fluid droplet and a surface is due to a combination of dispersive vdW and polar components (Israelachvili 2011; Pocius 2012). Water contact angle by itself cannot be used to determine the relative contributions of vdW and polar interactions. Complete liquid droplet contact angle analyses require a series of fluids ranging from primarily dispersive (e.g., methylene iodide) to primarily polar (e.g., water) in order to partition the relative contributions of the different intermolecular forces (Baier et al. 1968; Israelachvili 2011; Pocius 2012). However, it is possible to test the hypothesis that vdW forces are sufficient for gecko adhesion by reanalyzing Hiller's data to linearize the relationship between water contact angle and adhesion energy (Autumn and Peattie 2002). Hiller's data (1968, 1969, 1976) when linearized yield a strong correlation between force and adhesion energy for $\theta > 60°$, consistent with the van der Waals hypothesis (Autumn and Peattie 2002).

### *11.2.5 Property (4): Material-Independent Adhesion*

#### 11.2.5.1 Testing the van der Waals and Capillary Adhesion Hypotheses

To test directly whether capillary adhesion or the dispersive vdW force is a sufficient mechanism of adhesion in geckos, Autumn and colleagues (2002b) measured the hydrophobicity of the setal surface and measured adhesion and friction on two polarizable semiconductor surfaces that varied greatly in hydrophobicity. If capillary adhesive forces dominate, we expect a lack of adhesion on the strongly hydrophobic surfaces. In contrast, if the dispersive vdW forces are sufficient, large adhesive forces on the hydrophobic, but polarizable, GaAs and Si surfaces of the MEMS devices were predicted. In either case, strong adhesion to the hydrophilic $SiO_2$, control surfaces should be present. The tokay gecko setae are ultrahydrophobic ($\theta = 160.9°$) (Autumn et al. 2002b; Autumn and Hansen 2006), probably as a consequence of the hydrophobic side groups of β-keratin proteins

(Landmann 1986). The strongly hydrophobic nature of setae suggests that they interact primarily via dispersive vdW forces whether water is present or not.

Shear stress of live gecko toes on GaAs ($\theta = 110°$) and $SiO_2$ ($\theta = 0°$) semiconductors was not significantly different, and adhesion of a single gecko seta on the hydrophilic $SiO_2$ and hydrophobic (HF-etched) Si cantilevers differed by only 2 %. These results reject the hypothesis that the polarity (as indicated by $\theta$) of a surface predicts attachment forces in gecko setae, as suggested by Hiller (1968, 1969), and are consistent with the Hiller data following their reanalysis using adhesion energies (Autumn and Peattie 2002). Since vdW dispersion forces are the only mechanism that can cause two hydrophobic surfaces to adhere in air (Israelachvili 2011), the GaAs and hydrophobic semiconductor experiments provide direct evidence that these vdW forces are a sufficient mechanism of adhesion in gecko setae and that water-based capillary forces are not required. Setal adhesion is strong on polar and nonpolar surfaces, perhaps because of the strongly hydrophobic material they are made of and due to the very large contact areas made possible by the spatular nanoarray. Gecko setae thus have the property of material independence: they can adhere strongly to a wide range of materials, largely independently of surface chemistry.

#### 11.2.5.2 The Role of Water in Gecko Adhesion

Property (4), material-independent adhesion, does not preclude an effect of water on gecko adhesion under some conditions. Water is likely to alter contact geometry and adhesion energies when present between hydrophobic (e.g., spatula) and hydrophilic (e.g., glass) surfaces, but it is exceedingly difficult to predict what the effect will be in gecko setae because of the complexity of the system. Experimental evidence indicates that adhesion is enhanced in environments with high relative humidity (Sun et al. 2005; Huber et al. 2005b; Niewiarowski et al. 2008; Puthoff et al. 2010).

The humidity-dependent adhesion forces appear to contradict the fact that setae are strongly hydrophobic (Autumn et al. 2002b; Autumn and Hansen 2006). This raises the question of how to explain the apparent effect of humidity. Interpretations based on the presence of water bridges predict that the adhesion forces at fixed humidity should be proportional to the absolute temperature (Kim and Bhushan 2008), but these predictions do not agree with the experimental observations (Huber et al. 2005b; Niewiarowski et al. 2008). Water could reduce adhesion on rough surfaces by preventing spatular penetration into gaps, thus decreasing the contact fraction, but the humidity effects are present on atomically flat surfaces (Huber et al. 2005b; Puthoff et al. 2010). The data of Huber et al. rejected "true" capillary forces involving a water bridge since only a few monolayers of water were present at the spatula–substrate interface, even at high humidity. Instead, they concluded that humidity (a) modifies the contact geometry, increasing adhesion, and (b) decreases the Hamaker coefficient ($A$) that determines the strength of the vdW forces, reducing adhesion. Puthoff et al. (2010) confirmed that contact forces increase with humidity in isolated gecko setal arrays but not because of capillary effects. Contrary to the predictions of a capillary mechanism, contact forces and humidity effects were similar on hydrophobic and hydrophilic substrates and at shear rates that yielded insufficient time for capillary bridges to form.

The adhesion enhancement in the presence of humidity appears to be the result of changes in the properties of the setal material (Chen and Gao 2010). Support for this hypothesis comes with the confirmation that the viscoelastic properties of the setae change with humidity (Puthoff et al. 2010; Prowse et al. 2011). These results suggested that capillarity is absent or has a limited role in the adhesion of geckos under humid conditions. Rather, humidity softens and plasticizes setal β-keratin, which increases true contact area and adhesion forces. This explanation is consistent with the fact that the change in adhesion force from low to high humidity is the same on hydrophobic and hydrophilic substrates.

Research on the role of standing water on the adhesion of geckos is ongoing. Pesika et al. (2009b) found that adhesion of isolated gecko setae is reduced in water considerably, consistent with the predictions of the Lifshiz model of van der Waals adhesion. Stark and colleagues (2012, 2013, 2015a, b) investigated the behavior of the setal arrays on a toe of a live gecko. When submerged, a plastron of air forms underneath the toe, effectively limiting the exposure of the hydrophobic setae to water. After prolonged exposure to water, the scansors are filled with water, substantially reducing adhesion on hydrophilic substrates. The effect of surface water on the clinging ability of whole animals also appears to depend on the surface chemistry of the substrate, since hydrophilic surfaces promote wetting of the setal tissue. As long at toe pads remain in the unwetted state, adhesion on hydrophilic, intermediately wetting, or hydrophobic surfaces is the same (Stark et al. 2013). However, wet hydrophilic surfaces promote toe-pad wetting, decreasing the animal's overall clinging force (Stark et al. 2013).

#### 11.2.5.3 Dominance of Geometry in VdW Interactions

The theoretical magnitude of the dispersive vdW force ($F_{vdW}$) between a planar substrate and a circular planar spatula of radius $R$ (Israelachvili 2011) is $F_{vdW} = AR^2/6D^3$, and for a planar substrate and a curved spatula of radius $R$, $F_{vdW} = AR/6D^2$, where $A$ represents the Hamaker coefficient and $D$ is the gap distance (typically 0.2 nm for solids in contact). $A$ is a function of the volume and polarizability of the molecules involved.

For materials interacting in dry air, $A \sim 10^{-19}$ J. Altering the chemical composition of one or both surfaces can alter $A$, which can be as low as one-half to one-third this value for some polymer–polymer interactions (e.g., PTFE or polystyrene) and as high as five times this value for some metal-on-metal interactions. In water, $A$ can be reduced by an order of magnitude. Nevertheless, this variation in $A$ is only about an order of magnitude, while gap distance $D$ and contact area may vary by six or more orders of magnitude without macroscopically visible changes at the interface. Moreover, the effect of the gap distance is exponential to a power of at least two. Thus, adhesive surface effects due to vdW interactions are a function primarily of geometry, not of chemistry. A van der Waals mechanism for adhesion in gecko setae suggests that continuum theory models of the mechanics of surface contact (Johnson 1985) may be applicable. Then again, since the complex structure of setae and spatulae differs dramatically from the ideal curved and planar surfaces used in contact mechanics models, one might question the validity of models based on simple geometries to the function of gecko setae.

#### 11.2.5.4 JKR Model of Spatulae

The mechanics of contact have been modeled using continuum theory and highly simplified geometries. For example, the Johnson/Kendall/Roberts (JKR) model considers the force $F_{\mathrm{JKR}}$ required to pull an elastic sphere of radius $R$ from a planar surface (Johnson et al. 1971). The predicted adhesion force is $F_{\mathrm{JKR}} = (3/2)\pi R\gamma$, where $\gamma$ is the adhesion energy determined by the materials the sphere and the surface are comprised of. Using an approximate value of $R = 100$ nm for the spatular size and taking $\gamma = 50$ mJ/m$^2$, the predicted pull off force for a gecko spatula is $F_{\mathrm{JKR}} = 23.6$ nN, approx. twice the value measured by AFM (Huber et al. 2005a) (Table 11.1).

Another test of the validity of the JKR model is to begin with the forces measured in single setae and then calculate the size of the associated JKR sphere (Autumn et al. 2002b). Adhesion is $\approx 40$ μN per seta on silicon cantilever surfaces (Table 11.1). The setal tip is approximately 43 μm$^2$ in area, and therefore, the adhesive stress ($\sigma$) is $\sigma =$ force/area $\approx 917$ kPa. If the spatulae are packed tightly, $\sigma \approx F_{\mathrm{JKR}}$/(approx. JKR contact area) $= (3/2)\pi R\gamma/\pi R^2 = (3/2)\gamma/R$. Using a typical value of $\gamma$ for dispersive vdW interactions between surfaces (50–60 mJ/m$^2$), solving for the predicted radius ($R_{\mathrm{JKR}}$) of individual spatular contacts using empirical force measurements gives $R_{\mathrm{JKR}}$ as 82–98 nm (or contacts 164–196 nm across). These values are remarkably close to empirical measurements of the size of real gecko spatulae (200 nm across) (Ruibal and Ernst 1965; Autumn et al. 2000), yet obviously spatulae are not spherical (Fig. 11.1e). Note that the preceding estimate of $R_{\mathrm{JKR}}$ differs from that of Autumn and colleagues (2002b) in that they estimated the area of one seta using setal density and arrived at a similar but somewhat lower value for $\sigma$. The confirmation that the JKR model predicts similar magnitudes of force as observed in setae suggested the extraordinary conclusion that adhesion can be enhanced simply by splitting a surface into small protrusions to increase the surface density of individual contacts (Autumn et al. 2002b) and that adhesive stress is proportional to $1/R$ (Arzt et al. 2002). This assessment was initially supported by a comparative analysis of setae in lizards and arthropods (Arzt et al. 2003) (see Sect. 11.5.1). However, scaling of spatular density as a function of body mass was rejected once phylogenetic relationships were accounted for (Peattie and Full 2007).

**Table 11.1** Scaling of adhesion and friction stresses in Tokay gecko setae

| Scale | Mode | Force | Area | Stress (kPa) |
|---|---|---|---|---|
| Single spatula (Huber et al. 2005a) | Adhesion | 10 nN | 0.02 μm$^2$ | 500 |
| JKR model predictions for single spatula | Adhesion | 24 nN | 0.031 μm$^2$ | 750 |
| Kendall peel predictions for single spatula | Adhesion | 10 nN | 0.04 μm$^2$ | 250 |
| Single seta (Autumn et al. 2000) | Adhesion | 20 μN | 43.6 μm$^2$ | 460 |
| Single seta (Autumn et al. 2002a, b) | Adhesion | 40 μN | 43.6 μm$^2$ | 920 |
| Single seta (Autumn et al. 2000) | Friction | 200 μN | 43.6 μm$^2$ | 4600 |
| (Autumn et al. 2006b) | Adhesion | 35 mN | ~1 mm$^2$ | ~35 |
| Setal array (Hansen and Autumn 2005) | Friction | 370 mN | 0.99 mm$^2$ | 370 |
| (Autumn et al. 2006b) | Friction | 75 mN | ~1 mm$^2$ | ~75 |
| Single toe (Hansen and Autumn 2005) | Friction | 4.3 N | 0.19 cm$^2$ | 230 |
| Single foot (Autumn et al. 2002a, b) | Friction | 4.6 N | 0.22 cm$^2$ | 190 |
| Two feet (Irschick et al. 1996) | Friction | 20.4 N | 2.27 cm$^2$ | 89.9 |

Shear stress decreases approximately exponentially (Fig. 11.6), or approximately linearly on a log–log scale. The Kendall peel model prediction uses a square spatula of 100 nm on a side. The JKR model prediction uses a spherical spatula with 100 nm radius. Both predictions use an adhesion energy of $\gamma = 50$ mJ/m$^2$. Note that the similarity of area between single toe and single foot is due to the use of larger geckos in the single-toe measurements

#### 11.2.5.5 Kendall Peel Model of Spatulae

Spatulae may also be modeled as nanoscale strips of adhesive tape (Huber et al. 2005a; Spolenak et al. 2005b; Hansen and Autumn 2005). Using the approach of Kendall (1975), the peel force is $F_{\text{peel}} = \gamma w/(1 - \cos\theta)$, assuming there is negligible elastic energy storage in the spatula as it is pulled off. $w$ is the width of the spatula, $\gamma$ is the adhesion energy as for the JKR model, and $\theta$ is the angle that the adhesive strips make with the substrate. Empirical measurements of vertical ($\theta = 90°$) spatular adhesion (Huber et al. 2005a) suggest that each spatula adheres with approximately 10 nN of force. Using $\gamma = 50$ mJ/m$^2$, typical of vdW interactions, the Kendall peel model predicts a spatular width of 200 nm, remarkably close to the actual dimension (Ruibal and Ernst 1965; Autumn et al. 2000).

The peel zone model of adhesion (Tian et al. 2006; Pesika et al. 2007) is an extension of the Kendall peel model that places additional emphasis on the nanometer-scale aspects of adhesion. The curvature of the peeling film over the region where intermolecular forces are most significant introduces a geometrical correction to the peeling equation: $F_{\text{peel}} \approx 2\gamma w\theta/\pi(1 - \cos\theta)\sin\theta$. When $\theta = 90° = \pi/2$ rad, the result is the same as for the Kendall solution, but as $\theta$ becomes small, $F$ increases much more rapidly as more intermolecular forces are brought into play.

Theoretical considerations suggest that generalized continuum models of spatulae as spheres or nanotape are applicable to the range of spatula size and β-keratin stiffness of setae found in reptiles and arthropods (Spolenak et al. 2005b). Interestingly, at the 100-nm size scale, the effect of shape on adhesion force may be limited (Gao and Yao 2004; Spolenak et al. 2005b). However, at sizes above

100 nm, and especially above 1 μm, Spolenak et al. (2005b) concluded that shape should have a very strong effect on adhesion force. A phylogenetic comparative analysis of attachment force in lizards and insects will be an important test of this hypothesis.

### *11.2.6 Property (5): Rate-Dependent Adhesion*

The gecko adhesion system also possesses the property of rate-dependent friction and adhesion. All friction is rate dependent, a fact typically reflected in the convention that there are separate coefficients of friction $\mu_{\mathrm{static}}$ and $\mu_{\mathrm{dynamic}}$ for the respective cases of a stationary contact and one with relative motion between the surfaces. Detailed models of friction adopt a truly dynamic coefficient of friction $\mu = \mu(v)$, where $v$ is the velocity of relative motion. Ordinarily, dry, hard materials slip more easily as they slide more rapidly. However, the gecko adhesive setae, also dry and hard, become stickier as they slide faster (Gravish et al. 2010; Puthoff et al. 2010, 2013). Additionally, gecko setae are extremely wear resistant and capable of sliding as far as 300 m without significant deterioration (Gravish et al. 2010).

The adhesion produced follows the friction according to the principles of frictional adhesion. This is a property that might have significant ramifications for the capabilities of the organism, since very strong forces can be exerted by a gecko moving rapidly (i.e., falling) with respect to a substrate. These stable tribological forces in sliding gecko setae may emerge from the stochastic stick–slip of a large population of individual fibrils with high resonant frequencies, and wear resistance may be a consequence of stick–slip motion that minimizes rubbing and involves semi-regular elastic energy release. Moreover, each seta is small enough to be affected by thermal energy, raising the possibility that rate enhancement is in part due to thermally activated kinetics (Gravish et al. 2010; Puthoff et al. 2013).

## 11.3 Antiadhesive Properties of Gecko Setae

Paradoxical as it may seem, there is growing evidence that gecko setae are strongly antiadhesive. Gecko setae do not adhere spontaneously to surfaces, but instead require a mechanical program for attachment (Autumn et al. 2000). Unlike adhesive tapes, gecko setae do not self-adhere. Pushing the setal surfaces of a gecko's feet together does not result in strong adhesion. Also unlike conventional adhesives, gecko setae do not seem to stay dirty.

### *11.3.1 Properties (6) and (7): Self-Cleaning and Anti-Self-Adhesion*

Dirt particles are common in nature (Little 1979), yet casual observation suggests that geckos' feet are quite clean (Fig. 11.1b). Sand, dust, leaf litter, pollen, and plant waxes would seem likely to contaminate gecko setae. Hairlike elements on plants accumulate micron-scale particles (Little 1979) that could come into contact with gecko feet during climbing. Indeed, insects must cope with particulate contamination that reduces the function of their adhesive pads (Gorb and Gorb 2002) and spend a significant proportion of their time grooming (Stork 1983) in order to restore function. On the other hand, geckos have not been observed to groom their feet (Russell and Rosenberg 1981), yet apparently retain the adhesive ability of their setae during the months between shed cycles. How geckos manage to keep their toes dean while walking about with sticky feet has remained a puzzle until recently (Hansen and Autumn 2005). While self-cleaning by water droplets has been shown to occur in plant (Barthlott and Neinhuis 1997) and animal (Baum et al. 2002) surfaces, no adhesive had been shown to self-clean.

Gecko setae are the first known self-cleaning adhesive (Hansen and Autumn 2005). Tokay geckos with 2.5-μm-radius microspheres applied to their feet recovered their ability to cling to vertical surfaces after only a few steps on clean glass. Hansen and Autumn contaminated toes on one side of the animal with an excess of 2.5-μm-radius silica-alumina microspheres and compared the shear stress produced to that of uncontaminated toes on the other side of the animal. Prior researchers had suggested that geckos' unique toe peeling motion (digital hyperextension) might aid in cleaning of the toe pads (Russell 1979; Bauer et al. 1996; Hu et al. 2012; Xu et al. 2015), so the geckos' toes were immobilized and applied by hand to the surface of a glass force plate to determine if self-cleaning could occur without toe peeling. After only four simulated steps on a clean glass surface, the geckos recovered enough of their setal function to support their body weight by a single toe. To test the hypothesis that self-cleaning is an intrinsic property of gecko setae and does not require a gecko, Hansen and Autumn isolated arrays of setae, glued them to plastic strips, and simulated steps using a servomanipulation system called RoboToe. The shear stress in clean setal arrays was compared to that in the same arrays with a monolayer of microspheres applied to their adhesive surfaces. Self-cleaning of microspheres occurred in arrays of setae isolated from the gecko. Again as for live gecko toes, isolated setal arrays rapidly recovered the shear force lost due to contamination by microspheres. Presumably, the microspheres were being preferentially deposited on the glass substrate and did not remain strongly attached to the setae.

Contact mechanical models suggest that it is possible that self-cleaning occurs by an energetic disequilibrium between the adhesive forces attracting a dirt particle to the substrate and those attracting the same particle to one or more spatulae (Fig. 11.3) (Hansen and Autumn 2005). The models suggest that self-cleaning may in fact require $\gamma$ of spatulae to be relatively low (equal to or less than that of the

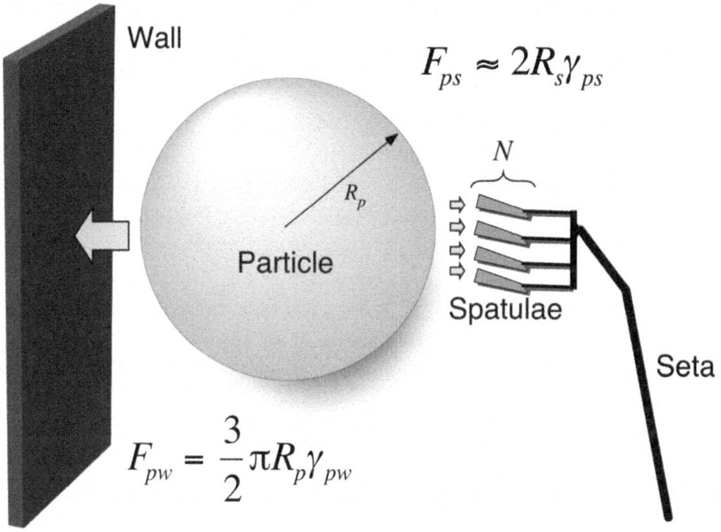


**Fig. 11.3** Model of self-cleaning in gecko setae from Hansen and Autumn (2005). If we model spatula as nanoscale strips of adhesive tape (Kendall 1975) that peel during detachment, the particle-spatula pull off force is given by $F = 2R_s\gamma_{ps}$, where $\gamma_{ps}$ is the adhesion energy at the particle-spatula interface and $2R$ is the width of the spatula, assuming negligible elastic energy storage. The pull off force of the dirt particle from a planar wall, using the JKR model (Johnson et al. 1971), is $F_{pw} = (3/2)\pi R_p\gamma_{pw}$, where $\gamma_{ps}$ is the adhesion energy of the particle to the wall. *N* represents the number of spatulae attached simultaneously to each dirt particle to achieve energetic equilibrium

wall), perhaps constraining the spatula to be made of a hydrophobic material. So, geckos may benefit by having setae made of an antiadhesive material: decreasing $\gamma$ decreases adhesion energy of each spatula, but promoting self-cleaning should increase adhesion of the array as a whole by maximizing the number of uncontaminated spatulae. If $\gamma$ were to be increased by supplementing the vdW forces with stronger intermolecular forces such as polar or H-bonding types, it is likely that self-cleaning and anti-self-fouling properties would be lost. Thus the self-cleaning and anti-self-fouling properties may represent a sweet spot in the evolutionary design space for adhesive nanostructures. Additionally, the rate dependence of friction and adhesion (Gravish et al. 2010; Puthoff et al. 2013) influences self-cleaning (Xu et al. 2015). As the sliding and detachment velocity are parameters an animal can ideally control, the capability to actively clean the scansor surfaces via a dynamical process is present.

## *11.3.2 Property (8): Nonsticky Default State*

The discovery that maximal adhesion in isolated setae requires a small push perpendicular to the surface, followed by a small parallel drag (Autumn et al. 2000), explained the load dependence and directionality of adhesion observed at the whole-animal scale by Haase (1900) and Dellit (1934) and was consistent with the structure of individual setae and spatulae (Ruibal and Ernst 1965; Hiller 1968). In their resting state, setal stalks are recurved proximally. When the toes of the gecko are planted, the setae may become bent out of this resting state, flattening the stalks between the toe and the substrate such that their tips point distally. This small preload and a micron-scale displacement of the toe or scansor proximally may serve to bring the spatulae (previously in a variety of orientations) uniformly flush with the substrate, maximizing their surface area of contact. Adhesion results and the setae are ready to bear the load of the animal's body weight.

To test the hypothesis that the default state of gecko setal arrays is to be nonsticky, Autumn and Hansen (2006) estimated the fraction of area able to make contact with a surface in setae in their unloaded state. Only less than 7 % of the area at the tip of a seta is available for initial contact with a smooth surface, and 93 % is empty air. This suggests that, initially, during a gecko's foot placement, the contact fraction of the distal region of the setal array must be very low. Yet, the dynamics of the foot must be sufficient to increase the contact fraction substantially to achieve the extraordinary values of adhesion and friction that have been measured in whole animals (Autumn et al. 2002b; Hansen and Autumn 2005; Irschick et al. 1996) and isolated setae (Autumn et al. 2000, 2002b; Hansen and Autumn 2005). Thus gecko setae may be nonsticky by default because only a very small contact fraction is possible without mechanically deforming the setal array.

How much does the contact fraction increase during attachment? While there are no empirical measurements of the number of spatulae in contact as a function of adhesion (or friction) force, it is possible to estimate from measurements of single setae. Empirical measurements and theoretical estimates of spatular adhesion (Autumn et al. 2000, 2002b; Arzt et al. 2003; Huber et al. 2005a; Spolenak et al. 2005b; Hansen and Autumn 2005) suggest that each spatula generates 10–40 nN with an approximate area of 0.02 $\mu m^2$; this gives an adhesive stress of 500–2000 kPa. A single seta on a Si MEMs cantilever can generate approximately 920 kPa (Table 11.1). The value of 10 nN of adhesion measured in single spatulae using an AFM (Huber et al. 2005a) implies that 4000 spatulae would need to be attached to equal the peak adhesion force (40 μN) measured in single setae (Autumn et al. 2002b). However, each seta contains not more than 1000 spatulae (Ruibal and Ernst 1965; Schleich and Kästle 1986). Therefore, a spatular force of 40 nN corresponds to a conservative estimate of setal contact fraction during attachment. In the case of a spatular adhesion force of 40 nN, the adhesive stress is 2000 kPa; a contact fraction of 46 % is required to yield this stress. This suggests that, unless the force of adhesion of a spatula has been greatly underestimated, the contact fraction must increase from 6 % to 46 %, or by approximately 7.5-fold, following preload and drag.

## 11.4 Modeling Adhesive Nanostructures

### *11.4.1 Effective Modulus of a Setal Array*

The gecko adhesive is a microstructure in the form of an array of millions of high-aspect-ratio shafts. The effective elastic modulus of this array ($E_{eff}$) (Persson 2003; Sitti and Fearing 2003) is much lower than Young's modulus ($E$) of β-keratin. Thus, arrays of setae should behave as a softer material than bulk β-keratin. The modulus of β-keratin in tension is approx. 2.5 GPa in bird feathers (Bonser and Purslow 1995) and 1.3–1.8 GPa in bird claws (Bonser 2000). Young's moduli of lizard β-keratins in general (Fraser and Parry 1996) and gecko β-keratins in particular (Alibardi 2003) remain unknown at present. The behavior of a setal array during compression and subsequent relaxation will depend on the mode(s) of deformation of individual setae. Bending is a likely mode of deformation (Simmermacher 1884) (Fig. 11.1d), and a simple approach is to model arrays of setae as arrays of cantilever beams (Persson 2003; Sitti and Fearing 2003; Glassmaker et al. 2004; Hui et al. 2004; Spolenak et al. 2005a; Autumn et al. 2006c; Pesika et al. 2009a). One might question the applicability of models based on a simple geometry for the complex, branched structure of the seta. However, as with the JKR and Kendall models (Sects. 11.2.5.4 and 11.2.5.5) applied to spatulae, the simple cantilever model is surprisingly well supported by empirical measurements of setal arrays (Geisler et al. 2005; Autumn et al. 2006c).

The effective modulus of an array of vertical fibers depends on parameters such as the material, geometric properties, and density of the fibers (Campolo et al. 2003; Sitti and Fearing 2003). A single setal stack is approximately a cantilever beam subject to a lateral load ($F$) at its tip. The resulting tip displacement due to bending is $\Delta = FL^3/3EI$, where $L$ is the length, $E$ is the elastic modulus of the material, and $I = \pi R^4/4$ is the area moment of inertia of the (cylindrical) cantilever (Timoshenko and Gere 1984) (Fig. 11.4a). For a cantilever at an angle $\phi$ to the substrate and under a normal load $F_n$, the resolved force lateral to the cantilever is $F = F_n \cos \phi$ (Fig. 11.4b). This results in a lateral tip displacement of $\Delta = F_n L^3 \cos(\phi)/3EI$ and a normal tip displacement of $\Delta_n = \Delta \cos \phi = F_n L^3 \cos^2(\phi)/3EI$.

For a system with an effective modulus, Hooke's law is

$$\sigma = E_{eff}\varepsilon, \tag{11.1}$$

where $\sigma$ is the stress applied perpendicularly to the setal array and $\varepsilon$ is the resulting perpendicular strain. For an array with setal density $\rho$, the stress is $F_n\rho$, and the

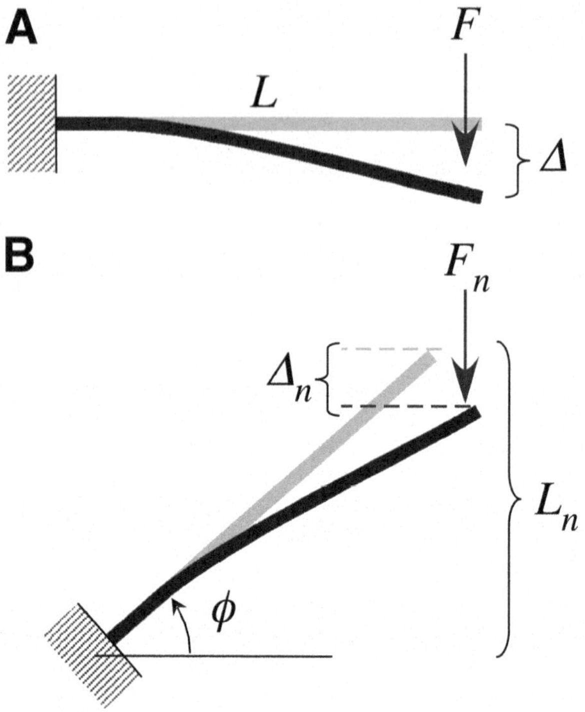


**Fig. 11.4** Free body diagram of (**a**) cantilever beam and (**b**) angled cantilever beam based on the model of Sitti and Fearing (2003). This model is similar to that of Persson (Persson 2003), who used a spring-based approach

strain is $\varepsilon = \Delta_n/(L \sin \phi)$. Expanding and substituting these equations into Eq. (11.1) gives

$$E_{\mathrm{eff}} = 3EI\rho \sin \phi / L^2 \cos^2(\phi)(1 \pm \mu \tan \phi). \tag{11.2}$$

A typical tokay setal array has approx. 14,000 setae per mm$^2$ ($\rho = 1.44 \times 10^{10}$ m$^{-2}$), so the shaft angle $\phi$ (Fig. 11.4b) required to yield an effective modulus of 100 kPa (the upper limit of Dahlquist's criterion; see Sect 11.6) is $\phi = 50°$ for $E = 1.0$ GPa and $\phi = 36.65°$ for $E = 2.0$ GPa.

We measured the forces resulting from deformation of isolated arrays of tokay gecko setae to determine $E_{\mathrm{eff}}$ and tested the validity of the cantilever model. We found that $E_{\mathrm{eff}}$ of tokay gecko setae falls near 100 kPa, close to the upper limit of Dahlquist's criterion for tack (Fig. 11.5) (Geisler et al. 2005; Autumn et al. 2006c). Additionally, we observed values of $\phi$ for tokay gecko setae near 43°, further supporting the validity of the cantilever model (Fig. 11.4).

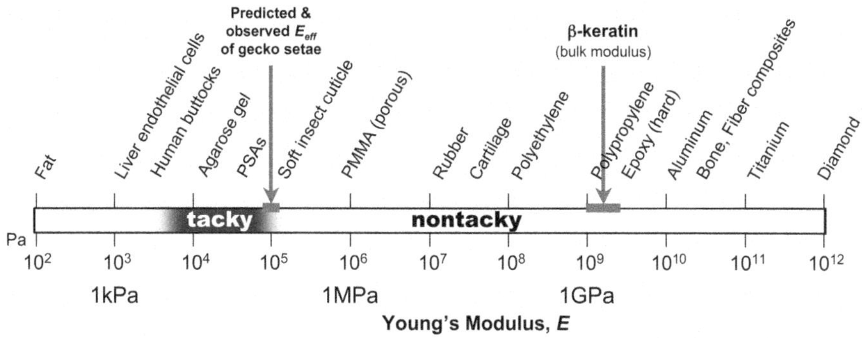


**Fig. 11.5** Young's modulus ($E$) of various materials, including the approximate values of bulk β-keratin and the effective modulus ($E_{\mathrm{eff}}$) of natural setal arrays. A value of $E \sim 100$ kPa (measured at 1 Hz) is the upper limit of the Dahlquist criterion for tack, which is based on empirical observations of PSAs (Dahlquist 1969; Pocius 2012). A cantilever beam model (Eq. 11.2) predicts a value of $E_{\mathrm{eff}}$ near 100 kPa, as observed for natural setae and PSAs. It is notable that geckos have evolved $E_{\mathrm{eff}}$ close to the limit of tack. This value of $E_{\mathrm{eff}}$ may be tuned to allow strong and rapid adhesion, yet prevent spontaneous or inappropriate attachment

## *11.4.2 Rough Surface and Antimatting Conditions*

Adhesion force has been shown to depend on the roughness of the substrate at the seta level (Huber et al. 2007) and the toe level (Pugno and Lepore 2008; Gillies et al. 2013; Stark et al. 2015b). The cantilever model predicts that a high density of setae should be selected for in increasing adhesive force of setal arrays. First, it follows from the JKR model (Autumn et al. 2002b; Arzt et al. 2003) that packing in more spatulae should increase adhesion in an array of setae. Second, the cantilever model suggests that thinner setal shafts should decrease $E_{\mathrm{eff}}$ and promote a greater contact fraction on rough surfaces (Stork 1983; Scherge and Gorb 2001; Jagota and Bennison 2002; Campolo et al. 2003; Persson 2003; Persson and Gorb 2003; Sitti and Fearing 2003; Spolenak et al. 2005a). The cantilever model also suggests that longer and softer setal shafts, and a lower shaft angle $\phi$, will result in better adhesion on rough surfaces because these parameters will reduce $E_{\mathrm{eff}}$. On a randomly rough surface, some setal shafts should be bent in compression (concave), while others will be bent in tension (convex). The total force required to pull off a setal array from a rough surface should therefore be determined by the cumulative adhesive force of all the attached spatulae, minus the sum of the forces due to elastic deformation of compressed setal shafts.

If setae mat together (Stork 1983), it is likely that adhesive function will be compromised. Interestingly, the same parameters that promote strong adhesion on rough surfaces should also cause matting of adjacent setae (Persson 2003; Sitti and Fearing 2003; Glassmaker et al. 2004; Hui et al. 2004; Spolenak et al. 2005a). The distance between setae and the stiffness of the shafts will determine the amount of force required to bring the tips together for matting to occur. It follows from the cantilever model that stiffer, shorter, and thicker stalks will allow a greater packing density without matting. As is the case for self-cleaning (Hansen and Autumn 2005), setae should be made of materials with lower surface energy to prevent self-adhesion and matting. Satisfying both antimatting and rough surface conditions may require a compromise of design parameters. Spolenak et al. (2005a) devised "design maps" for setal adhesive structures, an elegant approach to visualizing the parametric tradeoffs needed to satisfy the rough surface and antimatting conditions while at the same time maintaining structural integrity of the material.

## 11.5 Scaling

Small and large organisms are dominated by different forces (McMahon and Bonner 1983). Inertial forces usually dwarf adhesive forces in organisms gecko-size and above. Geckos are unusual among macroscale organisms in having adhesive forces dominate their world. The astonishing adaptive radiation in geckos and their unique ecologies can be seen as an emergent property of integration across seven orders of magnitude in size (Pianka and Sweet 2005)—from the nanoscale spatula and the microscale seta to the mesoscale scansors and the macroscale body (Fig. 11.1).

While it is tempting to focus on the smallest level in the gecko adhesive system, integration of multiple levels in the compliance hierarchy is needed to achieve reliable and controllable adhesion and friction. Self-cleaning adhesive nanostructures cannot adhere if they never get near the surface. Compliant scansors and the compliant adipose or vascular tissue underlying the scansors may be important in spreading the load during foot placement (Russell 1986, 2002). The complex morphology and musculature of the toes, feet, and limbs play a critical role in bringing the compliant scansors to bear upon the substrate in the appropriate manner and in detaching them without large forces (Russell 1975). Simulation studies of animal-like climbers suggest that tuning limb compliance correctly is much more important for climbing than for running. In particular, the ratios of linear and torsional compliances at the foot and ankle have an enormous effect on climbing stability and efficacy (Autumn et al. 2005).

### *11.5.1 Scaling of Pad Area and Spatular Size*

Shear force of the two front feet of pad-bearing lizards (geckos, anoles, and skinks) is highly correlated with pad area, even when the effects of body size and phylogeny are accounted for (Irschick et al. 1996). However, there is significant variation in shear force among taxa of similar size and pad area, suggesting that other factors are important in determining the strength of the setal adhesive. The JKR model (Autumn et al. 2002b; Arzt et al. 2003) predicts that larger spatulae should result in lower forces; this is supported by an inverse correlation between body mass and the size of the spatula or setal tip in lizards and arthropods (Arzt et al. 2003), but Peattie and Full (Peattie and Full 2007) rejected this correlation using a phylogenetically independent contrasts across broad sample of lizards and arthropods.

### *11.5.2 Scaling of Stress*

Amontons' first law states that the relationship of shear force (friction) to load is a constant value, $\mu$ (the coefficient of friction). Amontons' second law predicts that $\mu$ is independent of the area of contact (Bhushan 2013; Ringlein and Robbins 2004). When pulled in shear (Autumn et al. 2000, 2002b), gecko setae seem to violate Amontons' laws, as do tacky polymers where the forces of adhesion can be much greater than the external load. Shear stress in setae increases greatly with a decrease in contact area, suggesting that at larger scales, fewer spatulae are attached and/or the contact fraction within spatulae is reduced (Fig. 11.6; Table 11.1). The scaling of shear stress ($\tau$) is exponential and scales as log $\tau = 1.14 - 0.24$ log (area). It is unknown whether stress is uniformly spread across the toe or foot (Russell 2002), or if there are high stress concentrations on the setal arrays of a few scansors. The force of only 2 % of setae, and only 25 % of setal arrays, is required to yield the maximum shear stresses measured at the whole-animal level. However, at the setal level, it appears that most spatulae must be strongly attached to account for theoretical and empirical values of adhesion, suggesting that the seta is highly effective at making contact with a smooth surface. If each spatula can generate 10–40 nN, it would take 1000–4000 spatulae to yield the 40 μN of adhesion measured in single setae. However, each seta bears only 100–1000 spatulae. Clearly further work is needed to resolve this discrepancy. The relationship between adhesion and friction also demands further investigation. Existing data suggest that friction at the seta level is about two to four times the adhesion.

## 11.6 Comparison of Conventional and Gecko Adhesives

Conventional adhesives are materials that are used to join two surfaces. Typically, adhesives are liquids that are chemically compatible with both surfaces and have sufficiently low viscosity that wetting of the surfaces occurs either spontaneously or with a small amount of pressure (Baier et al. 1968; Kinloch 1987; Pocius 2012). Surface treatments are often needed to raise the interfacial energies between one or both surfaces and the adhesive. Liquid hard-set adhesives (e.g., epoxy or cyanoacrylate glues) flow easily during application, but cure to make a strong, permanent bond. Because they are stiff when cured, hard-set adhesives can resist plastic creep caused by sustained loading. However, hard-set adhesives are single use: their bonds must be broken or dissolved for removal, and once broken, hard-set adhesives do not rebond.

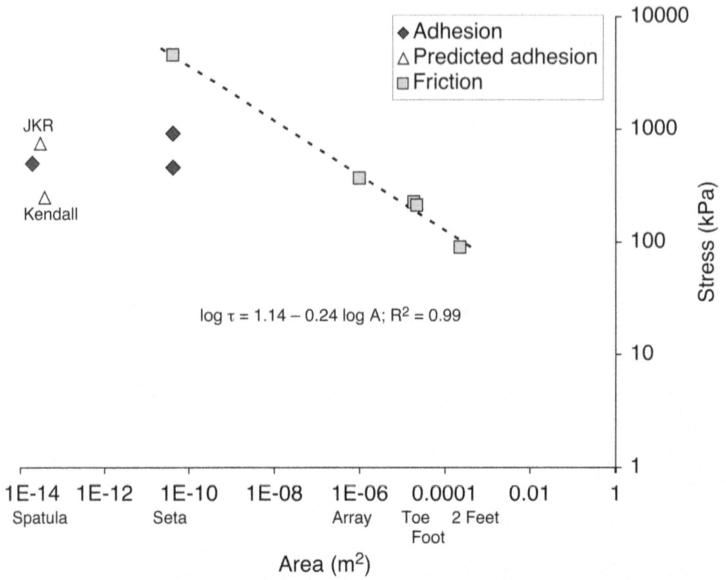


**Fig. 11.6** Stress vs. area in the gecko adhesive hierarchy (see Table 11.1 for numerical values and literature sources). JKR and Kendall model predictions for spatular adhesive stress (*triangles*) bound the measured value of Huber et al. (2005a). Text below the area axis shows the associated level in the gecko adhesive hierarchy (Fig. 11.1)

Conventional PSAs are fabricated from soft, tacky, viscoelastic materials (Gay and Leibler 1999; Gay 2002; Pocius 2012). Tacky materials are those that exhibit spontaneous plastic deformation that increases true area of contact with the surface at the molecular scale. Theoretical considerations (Creton and Liebler 1996) agree with Dahlquist's (1969; Pocius 2012) empirical observation that Young's modulus below 100 kPa (at 1 Hz) is needed to achieve a high contact fraction with the substrate. Additives known as tackifiers are commonly used to promote plastic deformation in PSAs during contact (Pocius 2012). PSAs such as masking tape or sticky notes are capable of repeated attachment and detachment cycles without residue because the dominant mechanism of adhesion is weak intermolecular forces. PSAs adhering with weak intermolecular forces can require much more energy to pull off of surfaces than do rigid adhesives relying on strong chemical bonds. As soft polymeric adhesives are pulled apart from a surface, polymer chains or bundles of polymer chains can be elongated into pillars in a process known as crazing. The total fracture energy can greatly exceed the sum of all the bond energies at the interface since work must be done on the craze as well as to break adhesive bonds at the interface. Thus the strong adhesion in polymeric adhesives results from long bonds rather than from strong bonds (Persson 2003). However, because they are soft polymeric materials, PSAs are prone to creep, degradation, self-adhesion, and fouling.

In contrast to the soft polymers of PSAs, the adhesive on the toes of geckos is made of hard protein (β-keratin) with values of $E$ four to five orders of magnitude greater than the upper limit of Dahlquist's criterion. Therefore, one would not expect a β-keratin structure to function as a PSA by readily deforming to make intimate molecular contact with a variety of surface profiles. However, since the gecko adhesive is a microstructure in the form of an array of millions of high-aspect-ratio shafts (setae), the effective elastic modulus, $E_{eff}$ (Jagota and Bennison 2002; Persson 2003; Sitti and Fearing 2003; Glassmaker et al. 2004; Hui et al. 2004; Spolenak et al. 2005a), is much lower than the value of $E$ of bulk β-keratin. The effective modulus of gecko setal arrays is close to 100 kPa (Geisler et al. 2005; Autumn et al. 2006c). Gecko setal arrays possess some of the properties of PSAs although the bulk material properties of β-keratin place it in the class of stiff, nonviscous materials (Fig. 11.5) (Dahlquist 1969; Creton and Liebler 1996; Gay and Leibler 1999; Gay 2002; Jagota and Bennison 2002; Persson 2003; Persson and Gorb 2003; Sitti and Fearing 2003; Pocius 2012).

There is emerging evidence that an array of gecko setae can act like a tacky, deformable material, while individual setae and spatulae retain the structural integrity of stiff protein fibers. This may enable the gecko adhesive to tolerate heavy, repeated use without creep or degradation. Indeed, theoretical considerations suggest that the fibrillar structure of the gecko adhesive can be thought of as a permanent craze (Jagota and Bennison 2002; Persson 2003) that has a higher fracture energy than a solid layer of adhesive material. As with polymer crazes, setal structures under stress could store energy elastically in each seta of the array, and then as setae are pulled off, elastic energy could be dissipated internally without contributing to propagation of the crack between the adhesive and substrate (Hui et al. 2004; Jagota and Bennison 2002; Persson 2003). Unlike polymer crazes, setal structures may dissipate energy primarily elastically rather than plastically. However, measurements of isolated gecko setae during detachment suggest that the high fracture toughness of the setal interface is not due to elastic losses, but rather to frictional losses (Gravish et al. 2008). As setal arrays are pulled away from a surface, micron-scale sliding increases the net detachment energy.

Gecko setae do not bond spontaneously on contact, as do PSAs. Gecko setae have a nonsticky default state (Autumn and Hansen 2006) and require mechanical deformation to initiate adhesion and friction (Autumn et al. 2000; Autumn and Peattie 2002). Again in contrast to PSAs, gecko setae are anisotropic and possess a built-in release mechanism. Setae are sticky when forces are directed with the curvature of the shaft and released when forces are directed away from the curvature of the shaft (Autumn et al. 2000; Autumn and Peattie 2002; Gao et al. 2005).

## 11.7 Gecko-Inspired Synthetic Adhesive Nanostructures

Using a nanostructure to create an adhesive is a novel and bizarre concept. It is possible that if it had not evolved, humans would never have invented it. With the inspiration of biology, materials incorporating adhesive nanostructures are being developed (Fig. 11.1f). The growing list of benchmark properties—eight of which are presented in this chapter—can be used to evaluate the degree of geckolike function of synthetic prototypes. Early synthetic setae (Autumn et al. 2002b; Geim et al. 2003; Sitti and Fearing 2003; Peressadko and Gorb 2004; Northen and Turner 2005), while functional in some sense, possessed few truly geckolike properties. For example, consider the adhesion coefficient $\mu' = F_{\mathrm{adhesion}}/F_{\mathrm{preload}}$ as a metric for geckolike adhesive function. By this criterion, the material of Geim et al. (Geim et al. 2003) is not geckolike since it required a very large preload of 50 N to yield 3 N and 0.3 atm of adhesion, yielding $\mu' = 0.06$. The synthetic setae of Northen and Turner (Northen and Turner 2005) perform significantly better with $\mu' = 0.125$, but still well below the benchmark of real gecko setae with $\mu' = 8$–16.

Since the development of these early prototype materials, implementation of numerous gecko-inspired synthetic adhesives has been documented in the literature (Kwak et al. 2011). Frequently, these materials exceed the capabilities of the natural system for a single property, such as self-cleaning capabilities (Lee and Fearing 2008; Lee and Bhushan 2012; Abusomwan and Sitti 2014) or surface roughness tolerance (Lee et al. 2009), but fall short in replicating the combination of properties to any degree. Developing materials that exhibit all eight benchmark functional properties of natural gecko adhesives will require considerable theoretical, experimental, and materials-design work.

Applications abound for a dry self-cleaning adhesive that does not rely on soft polymers or chemical bonds. Biomedical applications such as endoscopy and tissue adhesives (Pain 2000; Menciassi and Dario 2003) are one example. However, any materials chosen for synthetic setae in biomedical applications would need to be nontoxic and nonirritating (Baier et al. 1968). Other applications include MEMS switching (Decuzzi and Srolovitz 2004), wafer alignment (Slocum and Weber 2003), micromanipulation (Pain 2000; Jeong et al. 2014), and robotics (Autumn et al. 2005; Daltorio et al. 2007; Kim et al. 2008). Since a nanostructure could be applied directly to a surface, it is conceivable that geckolike structures could replace screws, glues, and interlocking tabs in many assembly applications such as automobile dashboards or mobile phones.

Sports applications such as fumble-free football gloves or rock climbing aids (Irving 1955; Hawkes et al. 2014) could be revolutionary. Using gecko technology to climb is not a new idea. In a seventeenth-century Indian legend, Shivaji and his Hindu warriors used adhesive lizards from the Deccan region as grappling devices to scale a shear rock cliff and mount a surprise attack on a Maharashtrian clifftop stronghold (Ghandi 2002).

## 11.8 Future Directions in the Study of the Gecko Adhesive System

Adhesion in geckos remains a sticky problem that is generating at least as many new questions as answers. Much of the fertility of this area stems from an integration of biology, physics, and engineering (Autumn et al. 2014). For example, the relationship between friction and adhesion is one of the most fundamental issues in surface science (Ringlein and Robbins 2004; Luan and Robbins 2005). One of the most striking properties (Table 11.2) of the gecko adhesive system is the coupling between adhesion and friction. Without a shear load, setae detach easily. Indeed, without shear loading of opposing toes or legs, a gecko could not hang from the ceiling. Integration of the macroscale system with the as yet undefined relationship between friction and adhesion at the nanoscale could yield important design principles for natural and synthetic setal structures.

Natural surfaces are rarely smooth, and an important next step will be to measure empirically the effect of surface roughness (Vanhooydonck et al. 2005) on friction and adhesion in gecko setae to test the predictions of the new generation of theoretical models for rough surface contacts with micro- and nanostructures (Persson and Gorb 2003). Under real-world conditions where surfaces are fractal (Greenwood 1992; Persson and Gorb 2003), compliance is required at each level of the gecko adhesive hierarchy: spatula, seta, lamella, toe, and leg. Models including a spatular array at the tip of a seta have not yet been developed. Similarly, models of lamellar structure will be needed to explain function on roughness above the micron scale.

Biological diversity of setal and spatular structure is high and poorly documented, though some advances in the arena have been made (Hagey et al. 2014). Basic morphological description will be required. Theory predicts that tip shape affects pull off force less at smaller sizes (Gao and Yao 2004), so it is possible that part of spatular variation is due to phylogenetic effects, but material constraints such as tensile strength of β-keratin must be considered as well (Autumn et al. 2002b; Spolenak et al. 2005a; Puthoff et al. 2010; Prowse et al. 2011). The collective behavior of the setal array will be a productive research topic (Gao and Yao 2004). Diversity of the array parameters, density, dimension, and shape is great but not well documented. In particular, the shape of setal arrays on lamellae demands further investigation. Phylogenetic analysis (Harvey and Pagel 1991; Peattie and Full 2007) of the variation in setal structure and function will be required to tease apart the combined effects of evolutionary history, material constraints, and adaptation (Autumn et al. 2002a).

The molecular structure of setae is not yet known. Setae are made primarily of β-keratin, but a histidine-rich protein or proteins may be present as well (Alibardi 2003). One possible role of non-keratin proteins is as a glue that holds the keratin fibrils together in the seta (Fig. 11.1d) (Alibardi 2003; Rizzo et al. 2006). This suggests a possible role of genes coding for histidine-rich protein(s) in tuning the

**Table 11.2** Properties, principles, and parameters of the gecko adhesive system

| Properties | Principles | Parameters |
|---|---|---|
| 1. Anisotropic attachment (Autumn et al. 2000) | Cantilever beam (Autumn et al. 2000; Sitti and Fearing 2003; Spolenak et al. 2005b) | Shaft length, radius, density (Sitti and Fearing 2003) |
| 2. High $\mu'$ (pull off/preload) (Autumn et al. 2000) | | Shaft angle (Sitti and Fearing 2003; Persson 2003) |
| 3. Low detachment force (Autumn et al. 2000) | Low effective stiffness (Sitti and Fearing 2003; Persson 2003) | Shaft modulus (Sitti and Fearing 2003)<br>Spatular shape (Persson and Gorb 2003; Spolenak et al. 2005b) |
| 4. Material independence (Autumn et al. 2002a, b; Hiller 1968, 1969) | | Spatular size (Arzt et al. 2003)<br>Spatular shape (Gao and Yao 2004; Spolenak et al. 2005b)<br>Spatular density (Arzt et al. 2003; Peattie et al. 2004) |
| 5. Rate dependence (Gravish et al. 2010; Puthoff et al. 2013) | (Persson 1995; Baumberger et al. 1999; Brörmann et al. 2013) | Dynamic friction coefficient |
| 6. Self-cleaning (Hansen and Autumn 2005) | Nanoarray | Spatular bulk modulus |
| 7. Anti-fouling | Small contact area | Particle size, shape, surface energy |
| 8. Nonsticky default state (Autumn and Hansen 2006) | Nontacky spatulae<br>Hydrophobic, vdW spatulae | Spatular size, shape, surface energy |

This table lists known properties of the gecko adhesive, proposed principles (or models) that explain the properties, and model parameters for each properly. "JKR" refers to the Johnson/Kendall/Roberts model of adhesion (Johnson et al. 1971)

material properties of the setal shaft. The outer molecular groups responsible for adhesion at the spatular surface will also be an important topic for future research.

Clearly there is a great desire to engineer a material that functions like a gecko adhesive, yet progress has been limited. A biomimetic approach of attempting to copy gecko setae blindly is unlikely to succeed due to the complexity of the system (Fig. 11.1) and the fact that evolution generally produces satisfactory rather than optimal structures. Instead, development of biologically inspired adhesive nanostructures will require careful identification and choice of design principles (Table 11.2) to yield selected geckolike functional properties. As technology and the science of gecko adhesion advance, it may become possible to tune design parameters to modify functional properties in ways that have not evolved in nature.

It is remarkable that the study of a lizard is contributing to understanding the fundamental processes underlying adhesion and friction (Fakley 2001; Urbakh et al. 2004; Autumn et al. 2014) and providing biological inspiration for the design of novel adhesives and climbing robots. Indeed, the broad relevance and applications of the study of gecko adhesion underscore the importance of basic, curiosity-based research.

**Acknowledgments** I am grateful to many colleagues for their help with this chapter, including Eduard Arzt, Sanford Autumn, Violeta Autumn, Emerson De Soma, Andrew Dittmore, Ron Fearing, Valeurie Friedman, Bob Full, Bill Geisler, Stas Gorb, Gerrit Huber, Jacob Israelachvili, Carmel Majidi, Anne Peattie, Holger Pfaff, Tony Russell, Andy Smith, and Simon Sponberg. Thanks to Stas Gorb and MPI Stuttgart for the Cryo-SEM image of a single seta. Thanks also to Carmel Majidi and Ron Fearing for Eqs. (11.1) and (11.2). Supported by DARPA N66001-03-C-8045, NSF-NIRT 0304730, DCI/NGIA HM1582-05-2022, and Johnson & Johnson Dupuy-Mitek Corp.

## References

Abusomwan UA, Sitti M (2014) Mechanics of load-drag-unload contact cleaning of gecko-inspired fibrillar adhesives. Langmuir 30(40):11913–11918

Alibardi L (2003) Ultrastructural autoradiographic and immunocytochemical analysis of setae formation and keratinization in the digital pads of the gecko Hemidactylus turcicus (Gekkonidae, Reptilia). Tissue Cell 35(4):288–296

Altevogt R (1954) Probleme eines Fuβes. Kosmos. Gesellschaft der Naturfreunde (Stuttgart) 50:428–430

Aristotle (1910) Historia animalium. Translation by D'Arcy Thompson. The Clarendon Press, Oxford

Arzt E, Enders S, Gorb S (2002) Towards a micromechanical understanding of biological surface devices. Z Met 93(5):345–351

Arzt E, Gorb S, Spolenak R (2003) From micro to nano contacts in biological attachment devices. Proc Natl Acad Sci 100(19):10603–10606

Autumn K, Hansen W (2006) Ultrahydrophobicity indicates a non-adhesive default state in gecko setae. J Comp Physiol A 192(11):1205–1212

Autumn K, Peattie AM (2002) Mechanisms of adhesion in geckos. Integr Comp Biol 42 (6):1081–1090

Autumn K, Liang YA, Tonia Hsieh S, Zesch W, Chan WP, Kenny TW, Fearing R, Full RJ (2000) Adhesive force of a single gecko foot-hair. Nature 405(6787):681–685

Autumn K, Ryan MJ, Wake DB (2002a) Integrating historical and mechanistic biology enhances the study of adaptation. Q Rev Biol 77(4):383–408

Autumn K, Sitti M, Liang YA, Peattie AM, Hansen WR, Sponberg S, Kenny TW, Fearing R, Isrealachvili JN, Full RJ (2002b) Evidence for van der Waals adhesion in gecko setae. Proc Natl Acad Sci 99(19):12252–12256

Autumn K, Buehler M, Cutkosky M, Fearing R, Full RJ, Goldman D, Groff R, Provancher W, Rizzi AA, Saranli U, Saunders A, Koditschek DE (2005) Robotics in Scansorial Environments. In: Gerhart GR, Shoemaker CM, Gage DW (eds) Proceedings of the SPIE vol. 5804: unmanned ground vehicle technology VII. SPIE, Bellingham, WA, pp 291–302

Autumn K, Dittmore A, Santos D, Spenko M, Cutkosky M (2006a) Frictional adhesion: a new angle on gecko attachment. J Exp Biol 209(18):3569–3579

Autumn K, Hsieh ST, Dudek DM, Chen J, Chitaphan C, Full RJ (2006b) Dynamics of geckos running vertically. J Exp Biol 209(2):260–272

Autumn K, Majidi C, Groff RE, Dittmore A, Fearing R (2006c) Effective elastic modulus of isolated gecko setal arrays. J Exp Biol 209(18):3558–3568

Autumn K, Niewiarowski PH, Puthoff JB (2014) Gecko adhesion as a model system for integrative biology, interdisciplinary science, and bioinspired engineering. Annu Rev Ecol Evol Syst 45:445–470

Baier RE, Shafrin EG, Zisman WA (1968) Adhesion: mechanisms that assist or impede it. Science 162(3860):1360–1368

Barthlott W, Neinhuis C (1997) Purity of the sacred lotus, or escape from contamination in biological surfaces. Planta 202(1):1–8

Bauer AM (1998) Morphology of the adhesive tail tips of carphodactyline geckos (*Reptilia: Diplodactylidae*). J Morphol 235(1):41–58

Bauer AM, Russell AP, Powell GL (1996) The evolution of locomotor morphology in *Rhoptropus* (Squamata: Gekkonidae): Functional and phylogenetic considerations. Afr J Herpetol 45 (1):8–30

Baum C, Meyer W, Stelzer R, Fleischer L-G, Siebers D (2002) Average nanorough skin surface of the pilot whale (*Globicephala melas*, Delphinidae): considerations on the self-cleaning abilities based on nanoroughness. Mar Biol 140(3):653–657

Baumberger T, Berthoud P, Caroli C (1999) Physical analysis of the state- and rate-dependent friction law. II Dynamic friction. Phys Rev B 60(6):3928–3939

Bellairs A (1970) The life of reptiles. Universe Books, New York

Bhushan B (2013) Introduction to tribology. Wiley, New York

Biewener AA, Full RJ (1992) Force platform and kinematic analysis. In: Biewener AA (ed) Biomechanics: structures and systems: a practical approach. IRL Press at Oxford University Press, Oxford, pp 45–73

Blackwall J (1845) On the means by which walk various animals on the vertical surface of polished bodies. Ann Mag Nat Hist Ser 1 15(96):115–119

Bonser RHC (2000) The Young's modulus of ostrich claw keratin. J Mater Sci Lett 19 (12):1039–1040

Bonser RHC, Purslow PP (1995) The Young's modulus of feather keratin. J Exp Biol 198 (4):1029–1033

Brainerd EL (1994) Adhesion force of ants on smooth surfaces. Am Zool 34(5):128A (Abstract from 1994 American Society of Zoologists Annual Meeting.)

Braun M (1878) Zur Bedeutung der Cuticularborsten auf den Haftlappen der Geckotiden. Arbeiten aus dem Zoologisch-Zootomischen Institut in Würzburg 4:231–237

Brörmann K, Barel I, Urbakh M, Bennewitz R (2013) Friction on a microstructured elastomer surface. Tribol Lett 50(1):3–15

Campolo D, Jones S, Fearing RS (2003) Fabrication of gecko foot-hair like nano structures and adhesion to random rough surfaces. In: Proceedings of the third IEEE conference on nanotechnology, vol. 2. IEEE, Los Alamitos, pp 856–859

Cartier O (1872) Studien über den feineren Bau der Epidermis bei den Geckotiden. Verhandlungen der Physikalisch-medincinischen Gesellschaft zu Würzburg 3:7–22

Cartier O (1874a) Studien über den feineren Bau der Haut bei Reptilien. I Die Epidermis der Geckotiden Arbeiten aus dem. Zoologisch-Zootomischen Institut in Würzburg 1:83–96

Cartier O (1874b) Studien über den feineren Bau der Haut bei Reptilien. II Ueber die Wachsthumserscheinimgen der Oberhaut von Schlangen und Eidechsen bei der Häutung Arbeiten aus dem. Zoologisch-Zootomischen Institut in Würzburg 1:239–258

Chen B, Gao H (2010) An alternative explanation of the effect of humidity in gecko adhesion: stiffness reduction enhances adhesion on a rough surface. Int J Appl Mech 2(1):1–9

Chen JJ, Peattie AM, Autumn K, Full RJ (2006) Differential leg function in a sprawled-posture quadrupedal trotter. J Exp Biol 209(2):249–259

Chen B, Wu P, Gao H (2009) Pre-tension generates strongly reversible adhesion of a spatula pad on substrate. J R Soc Interface 6(35):529–537

Chui BW, Kenny TW, Mamin HJ, Terris BD, Rugar D (1998) Independent detection of vertical and lateral forces with a sidewall-implanted dual-axis piezoresistive cantilever. Appl Phys Lett 72(11):1388–1391

Creton C, Liebler L (1996) How does tack depend on contact time and contact pressure? J Polym Sci B Polym Phys 34(3):545–554
Dahlquist CA (1969) Pressure-sensitive adhesives. In: Patrick RL (ed) Treatise on adhesion and adhesives, vol 2. Materials. Dekker, New York, pp 219–260
Daltorio KA, Gorb S, Peressadko A, Horchler AD, Wei TE, Ritzmann RE, Quinn RD (2007) Microstructured polymer adhesive feet for climbing robots. MRS Bull 32(6):504–508
Decuzzi P, Srolovitz DJ (2004) Scaling laws for opening partially adhered contacts in MEMS. J Microelectromech Syst 13(2):377–385
Delaugerre M, Alain G, Leoncini A (2015) One island, two geckos and some powder. Why and how a colonization process can fail? In: X Congresso Nazionale della Societas Herpetologica Italica. Societas Herpetologica Italica, Pavia, Italy, pp 117–121
Dellit W-D (1934) Zur anatomie und physiologie der Geckozehe. Jenaische Zeitschrift für Naturwissenschaft 68:613–656
Dewitz H (1882) Wie ist es den Stubenfliegen und vielen anderen Insecten möglich, an senkrechten Glaswänden emporzulaufen? Sitzungsberichte der Gesellschaft Naturforschender Freunde zu Berlin 17(1):5–7
Edwards JS, Tarkanian M (1970) The adhesive pads of Heteroptera: a re-examination. Proc R Entomol Soc Lond A Gen Entomol 45(1-3):1–5
Emerson SB, Diehl D (1980) Toe pad morphology and mechanisms of sticking in frogs. Biol J Linn Soc 13(3):199–216
Fakley M (2001) Smart adhesives. Chem Ind Mag:691–695
Fraser RDB, Parry DAD (1996) The molecular structure of reptilian keratin. Int J Biol Macromol 19(3):207–211
Gadow H (1901) Amphibia and reptiles. Cambridge natural history, vol 8. Macmillan & Co., Ltd., London
Gao H, Yao H (2004) Shape insensitive optimal adhesion of nanoscale fibrillar structures. Proc Natl Acad Sci 101(21):7851–7856
Gao H, Wang X, Yao H, Gorb S, Arzt E (2005) Mechanics of hierarchical adhesion structures of geckos. Mech Mater 37(2-3):275–285
Gay C (2002) Stickiness—some fundamentals of adhesion. Integr Comp Biol 42(6):1123–1126
Gay C, Leibler L (1999) Theory of tackiness. Phys Rev Lett 82(5):936–940
Geim AK, Dubonos SV, Grigorieva IV, Novoselov KS, Zhukov AA, Shapoval SY (2003) Microfabricated adhesive mimicking gecko foot-hair. Nat Mater 2(7):461–463
Geisler B, Dittmore A, Gallery B, Stratton T, Fearing R, Autumn K (2005) Deformation of isolated gecko setal arrays: bending or buckling? 2. Kinetics. Society for Integrative and Comparative Biology, San Diego (Abstract from 2005 Society for Integrative and Comparative Biology Annual Meeting)
Gennaro JG Jr (1969) The gecko grip. Nat Hist (The Journal of The American Museum of Natural History) 78(7):36–43
Ghandi M (2002) The ughly buglies. Swagat Mag Media Transasia, Bangalore
Gillett JD, Wigglesworth VB (1932) The climbing organ of an insect, *Rhodnius prolixus* (Hemiptera; Reduviidae). Proc R Soc B 111(772):364–376
Gillies AG, Henry A, Lin H, Ren A, Shiuan K, Fearing RS, Full RJ (2013) Gecko toe and lamellar shear adhesion on macroscopic, engineered rough surfaces. J Exp Biol 217(2):283–289
Glassmaker JN, Jagota A, Hui C-Y, Kim J (2004) Design of biomimetic fibrillar interfaces: 1. Making contact. J R Soc Interface 1(1):23–33
Gorb EV, Gorb SN (2002) Attachment ability of the beetle *Chrysolina fastuosa* on various plant surfaces. Entomol Exp Appl 105(1):13–28
Gravish N, Wilkinson M, Autumn K (2008) Frictional and elastic energy in gecko adhesive detachment. J R Soc Interface 5(20):339–348
Gravish N, Wilkinson M, Sponberg S, Parness A, Esparza N, Soto D, Yamaguchi T, Broide M, Cutkosky M, Creton C, Autumn K (2010) Rate-dependent frictional adhesion in natural and synthetic gecko setae. J R Soc Interface 7(43):259–269

Green DM (1981) Adhesion and the toe-pads of treefrogs. Copeia 1981(4):790–796

Greenwood JA (1992) Problems with surface roughness. In: Singer IL, Pollock HM (eds) Fundamentals of friction: macroscopic and microscopic processes, Volume 220 of the NATO ASI Series. Kluwer Academic Publishers, Dordrecht, The Netherlands, pp 57–76

Haase A (1900) Untersuchungen über den Bau und die Entwicklung der Haftlappen bei den Geckotiden. Archiv für Naturgeschichte 1–2(2):321–346

Hagey TJ, Puthoff JB, Holbrook M, Harmon LJ, Autumn K (2014) Variation in setal micromechanics and performance of two gecko species. Zoomorphology 133(2):111–126

Han D, Zhou K, Bauer AM (2004) Phylogenetic relationships among gekkotan lizards inferred from C-*mos* nuclear DNA sequences and a new classification of the Gekkota. Biol J Linn Soc 83(3):353–368

Hanna G, Jon W, Jon Barnes WP (1991) Adhesion and detachment of the toe pads of tree frogs. J Exp Biol 155(1):103–125

Hansen WR, Autumn K (2005) Evidence for self-cleaning in gecko setae. Proc Natl Acad Sci 102 (2):385–389

Harvey PH, Pagel MD (1991) The comparative method in evolutionary biology. Oxford series in ecology and evolution. Oxford University Press, Oxford, UK

Hawkes EW, Eason EV, Christensen DL, Cutkosky MR (2014) Human climbing with efficiently scaled gecko-inspired dry adhesives. J R Soc Interface 12(102):20140675

Hecht MK (1952) Natural selection in the lizard genus *Aristelliger*. Evolution 6(1):112–124

Hepworth J (1854) On the structure of the foot of the fly. J Cell Sci s1–2(7):158–163

Hiller U (1968) Untersuchungen zum Feinbau und zur Funktion der Haftborsten von Reptilien. Zeitschrift für Morphologie der Tiere 62(4):307–362

Hiller U (1969) Correlation between corona-discharge of polyethylene films and the adhering power of *Tarentola M. mauritanica* (Rept.). Forma et Functio 1:350–352

Hiller U (1971) Form und Funktion der Hautsinnesorgane bei Gekkoniden. Forma et Functio 4:240–253

Hiller U (1976) Comparative studies on the functional morphology of two gekkonid lizards. J Bombay Nat Hist Soc 73(2):278–282

Hora SL (1924) The adhesive apparatus on the toes of certain geckos and tree frogs. J Proc Asiat Soc Bengal 9:137–145

Hu C, Alex Greaney P (2014) Role of seta angle and flexibility in the gecko adhesion mechanism. J Appl Phys 116(7):074302

Hu S, Lopez S, Niewiarowski PH, Xia Z (2012) Dynamic self-cleaning in gecko setae via digital hyperextension. J R Soc Interface 9(76):2781–2790

Huber G, Gorb SN, Spolenak R, Arzt E (2005a) Resolving the nanoscale adhesion of individual gecko spatulae by atomic force microscopy. Biol Lett 1(1):2–4

Huber G, Mantz H, Spolenak R, Mecke K, Jacobs K, Gorb SN, Artz E (2005b) Evidence for capillary contributions to gecko adhesion from single spatula nanomechanical measurements. Proc Natl Acad Sci 102(45):16293–16296

Huber G, Gorb SN, Hosoda N, Spolenak R, Arzt E (2007) Influence of surface roughness on gecko adhesion. Acta Biomater 3(4):607–610

Hui C-Y, Glassmaker NJ, Tang T, Jagota A (2004) Design of biomimetic fibrillar interfaces: 2. Mechanics of enhanced adhesion. J R Soc Interface 1(1):35–48

Irschick DJ, Austin CC, Petren K, Fisher RN, Losos JB, Ellers O (1996) A comparative analysis of clinging ability among pad-bearing lizards. Biol J Linn Soc 59(1):21–35

Irving RLG (1955) A history of British mountaineering. B.T. Batsford, London

Israelachvili J (2011) Intermolecular and surface forces. Academic, San Diego, CA, USA

Jagota A, Bennison SJ (2002) Mechanics of adhesion through a fibrillar microstructure. Integr Comp Biol 42(6):1140–1145

Jeong J, Kim J, Song K, Autumn K, Lee J (2014) Geckoprinting: assembly of microelectronic devices on unconventional surfaces by transfer printing with isolated gecko setal arrays. J R Soc Interface 11(99):20140627

Johnson KL (1985) Contact mechanics. Cambridge University Press, Cambridge

Johnson KL, Kendall K, Roberts AD (1971) Surface energy and the contact of elastic solids. Proc R Soc A 324(1558):301–313

Kendall K (1975) Thin-film peeling—the elastic term. J Phys D Appl Phys 8(13):1449–1452

Kim TW, Bhushan B (2008) The adhesion model considering capillarity for gecko attachment system. J R Soc Interface 5(20):319–327

Kim S, Spenko M, Trujillo S, Heyneman B, Santos D, Cutkosky MR (2008) Smooth vertical surface climbing with directional adhesion. IEEE Trans Robot 24(1):65–74

Kinloch AJ (1987) Adhesion and adhesives: science and technology, 1st edn. Chapman & Hall, New York

Kwak MK, Pang C, Jeong H-E, Kim H-N, Yoon H, Jung H-S, Suh K-Y (2011) Towards the next level of bioinspired dry adhesives: new designs and applications. Adv Funct Mater 21 (19):3606–3616

Landmann L (1986) The skin of reptiles: epidermis and dermis. In: Bereiter-Hahn J, Matoltsy AG, Richards KS (eds) Biology of the integument, vol 2: Vertebrates. Springer Science + Business Media, Berlin, Germany

Lee H, Bhushan B (2012) Fabrication and characterization of hierarchical nanostructured smart adhesion surfaces. J Colloid Interface Sci 372(1):231–238

Lee J, Fearing RS (2008) Contact self-cleaning of synthetic gecko adhesive from polymer microfibers. Langmuir 24(19):10587–10591

Lee YI, Kogan M, Larsen JR (1986) Attachment of the potato leafhopper to soybean plant surfaces as affected by morphology of the pretarsus. Entomol Exp Appl 42(2):101–107

Lee J, Bush B, Maboudian R, Fearing RS (2009) Gecko-inspired combined lamellar and nanofibrillar array for adhesion on nonplanar surface. Langmuir 25(21):12449–12453

Lees AD, Hardie J (1988) The organs of adhesion in the Aphid *Megoura Viciae*. J Exp Biol 136 (1):209–228

Little P (1979) Particle capture by natural surfaces. Agric Aviat 20:129–144

Luan B, Robbins MO (2005) The breakdown of continuum models for mechanical contacts. Nature 435(7044):929–932

Maderson PFA (1964) Keratinized epidermal derivatives as an aid to climbing in gekkonid lizards. Nature 203(4946):780–781

Mahendra BC (1941) Contributions to the bionomics, anatomy, reproduction and development of the Indian house-gecko, *Hemidactylus flaviviridis* Rüppel. Part II. The problem of locomotion. Proc Ind Acad Sci B 13(5):288–306

McMahon TA, Bonner JT (1983) On size and life. Scientific American Books - W. H. Freeman & Co., New York

Menciassi A, Dario P (2003) Bio-inspired solutions for locomotion in the gastrointestinal tract: background and perspectives. Philos Trans R Soc A 361(1811):2287–2298

Niewiarowski PH, Lopez S, Ge L, Hagen E, Dhinojwala A (2008) Sticky gecko feet: the role of temperature and humidity. PLoS One 3(5), e2192

Northen MT, Turner KL (2005) A batch fabricated biomimetic dry adhesive. Nanotechnology 16 (8):1159–1156

Pain S (2000) Sticking power. New Sci 168(2270/2271):62–67

Peattie AM, Full RJ (2007) Phylogenetic analysis of the scaling of wet and dry biological fibrillar adhesives. Proc Natl Acad Sci 104(47):18595–18600

Peattie AM, Fearing RS, Full RJ (2004) Using a simple beam model to predict morphological variation in adhesive gecko hairs. Society for Integrative and Comparative Biology, New Orleans (Abstract from 2004 Society for Integrative and Comparative Biology Annual Meeting)

Peressadko A, Gorb SN (2004) When less is more: experimental evidence for tenacity enhancement by division of contact area. J Adhes 80(4):247–261

Persson BNJ (1995) Theory of friction: stress domains, relaxation, and creep. Phys Rev B 51 (19):13568–13585

Persson BNJ (2003) On the mechanism of adhesion in biological systems. J Chem Phys 118 (16):7614–7621

Persson BNJ, Gorb S (2003) The effect of surface roughness on the adhesion of elastic plates with application to biological systems. J Chem Phys 119(21):11437–11444

Pesika NS, Yu T, Zhao B, Rosenberg K, Zeng H, McGuiggan P, Autumn K, Israelachvili JN (2007) Peel-zone model of tape peeling based on the gecko adhesive system. J Adhes 83 (4):383–401

Pesika NS, Gravish N, Wilkinson M, Zhao B, Zeng H, Yu T, Israelachvili J, Autumn K (2009a) The crowding model as a tool to understand and fabricate gecko-inspired dry adhesives. J Adhes 85(8):512–525

Pesika NS, Zeng H, Kristiansen K, Zhao B, Yu T, Autumn K, Israelachvili J (2009b) Gecko adhesion pad: a smart surface? J Phys Condens Matter 21(46):464132

Peterson JA, Williams EE (1981) A case history in retrograde evolution: the *onca* lineage in Anoline lizards II Subdigital fine structure. Bull Museum Comp Zool 149(4):215–268

Pianka ER, Sweet SL (2005) Integrative biology of sticky feet in geckos. BioEssays 27 (6):647–652

Pocius AV (2012) Adhesion and adhesive technology. Carl Hanser Verlag, Munich, Germany

Prowse M, Puthoff JB, Wilkinson M, Autumn K (2011) Effects of humidity on the mechanical properties of gecko setae. Acta Biomater 7(2):733–738

Pugno NM, Lepore E (2008) Observation of optimal gecko's adhesion on nanorough surfaces. BioSystems 94(3):218–222

Puthoff JB, Prowse MS, Wilkinson M, Autumn K (2010) Changes in materials properties explain the effects of humidity on gecko adhesion. J Exp Biol 213(21):3699–3704

Puthoff JB, Holbrook M, Wilkinson MJ, Jin K, Pesika NS, Autumn K (2013) Dynamic friction in natural and synthetic gecko setal arrays. Soft Matter 9(19):4855–4863

Ringlein J, Robbins MO (2004) Understanding and illustrating the atomic origins of friction. Am J Phys 72(7):884–891

Rizzo NW, Gardner KH, Walls DJ, Keiper-Hrynko NM, Ganzke TS, Hallahan DL (2006) Characterization of the structure and composition of gecko adhesive setae. J R Soc Interface 3(8):441–451

Röll B (1995) Epidermal fine structure of the toe tips of *Sphaerodactylus cinereus* (Reptilia, Gekkonidae). J Zool 235(2):289–300

Rosenberg HI, Rose R (1999) Volar adhesive pads of the feathertail glider, Acrobates pygmaeus (Marsupialia; Acrobatidae). Can J Zool 77(2):233–248

Ruibal R, Ernst V (1965) The structure of the digital setae of lizards. J Morphol 117(3):271–293

Russell AP (1975) A contribution to the functional analysis of the foot of the Tokay, *Gekko gecko* (Reptilia: Gekkonidae). J Zool 176(4):437–476

Russell AP (1976) Some comments concerning the interrelationships amongst gekkonine geckos. In: Bellairs AA, Cox CB (eds) Morphology and biology of reptiles, number 3 in Linnean Society Symposium Series. Academic, London, pp 217–244

Russell AP (1979) Parallelism and integrated design in the foot structure of Gekkonine and Diplodactyline Geckos. Copeia 1979(1):1–21

Russell AP (1981) Descriptive and functional anatomy of the digital vascular system of the tokay, *Gekko gecko*. J Morphol 169(3):293–323

Russell AP (1986) The morphological basis of weight-bearing in the scansors of the tokay gecko (Reptilia: Sauria). Can J Zool 64(4):948–955

Russell AP (2002) Integrative functional morphology of the Gekkotan adhesive system (Reptilia: Gekkota). Integr Comp Biol 42(6):1154–1163

Russell AP, Bauer AM (1988) Paraphalangeal elements of gekkonid lizards: a comparative survey. J Morphol 197(2):221–240

Russell AP, Bauer AM (1990a) Digit I in pad-bearing gekkonine geckos: alternate designs and the potential constraints of phalangeal number. Mem Qld Mus 29(2):453–472

Russell AP, Bauer AM (1990b) *Oedura* and *Afroedura* (Reptilia: Gekkonidae) revisited: similarities of digital design, and constraints on the development of multiscansorial subdigital pads? Mem Qld Mus 29(2):473–486

Russell AP, Rosenberg HI (1981) Self-grooming in *Diplodactylus spinigerus* (Reptilia: Gekkonidae), with a brief review of such behaviour in reptiles. Can J Zool 59(3):564–566

Sauer RA (2009) Multiscale modelling and simulation of the deformation and adhesion of a single gecko seta. Comput Methods Biomech Biomed Engin 12(6):627–640

Scherge M, Gorb S (2001) Biological micro- and nanotribology: natures solutions nanoscience and technology series. Springer, Berlin, Germany

Schleich HH, Kästle W (1986) Ultrastrukturen an Gecko-Zehen (Reptilia: Sauria: Gekkonidae). Amphibia-Reptilia 7(2):141–166

Schmidt H (1904) Zur Anatomie und Physiologie der Geckopfote. Jenaische Zeitschrift für Naturwissenschaft 39:551–580

Simmermacher G (1884) Haftapparate bei Wirbeltieren. Der Zoologische Garten 25(10):289–301

Sitti M, Fearing RS (2003) Synthetic gecko foot-hair micro/nano-structures as dry adhesives. J Adhes Sci Technol 17(8):1055–1073

Slocum AH, Weber AC (2003) Precision passive mechanical alignment of wafers. J Microelectromech Syst 12(6):826–834

Spolenak R, Gorb S, Arzt E (2005a) Adhesion design maps for bio-inspired attachment systems. Acta Biomater 1(1):5–13

Spolenak R, Gorb S, Gao H, Arzt E (2005b) Effects of contact shape on the scaling of biological attachments. Proc R Soc A 461(2054):305–319

Stark AY, Sullivan TW, Niewiarowski PH (2012) The effect of surface water and wetting on gecko adhesion. J Exp Biol 215(17):3080–3086

Stark AY, Badge I, Wucinich NA, Sullivan TW, Niewiarowski PH, Dhinojwala A (2013) Surface wettability plays a significant role in gecko adhesion underwater. Proc Natl Acad Sci 110 (16):6340–6345

Stark AY, Ohlemacher J, Knight A, Niewiarowski PH (2015a) Run don't walk: locomotor performance of geckos on wet substrates. J Exp Biol 218(15):2435–2441

Stark AY, Palecek AM, Argenbright CW, Bernard C, Brennan AB, Niewiarowski PH, Dhinojwala A (2015b) Gecko adhesion on wet and dry patterned substrates. PLoS One 10(12), e0145756

Stewart WJ, Higham TE (2014) Passively stuck: death does not affect gecko adhesion strength. Biol Lett 10(12):20140701

Stork NE (1980) Experimental analysis of adhesion of *Chrysolina Polita* (Chrysomelidae: Coleoptera) on a variety of surfaces. J Exp Biol 88(1):91–108

Stork NE (1983) A comparison of the adhesive setae on the feet of lizards and arthropods. J Nat Hist 17(6):829–835

Sun W, Neuzil P, Kustandi TS, Sharon O, Samper VD (2005) The nature of the gecko lizard adhesive force. Biophys J 89(2):14–17

Tian Y, Pesika N, Zeng H, Rosenberg K, Zhao B, McGuiggan P, Autumn K, Israelachvili J (2006) Adhesion and friction in gecko toe attachment and detachment. Proc Natl Acad Sci 103 (51):19320–19325

Timoshenko SP, Gere SM (1984) Mechanics of materials. Thomson Brooks/Cole, Independence, KY, USA

Urbakh M, Klafter J, Gourdon D, Israelachvili J (2004) The nonlinear nature of friction. Nature 430(6999):525–528

Vanhooydonck B, Andronescu A, Herrel A, Irschick DJ (2005) Effects of substrate structure on speed and acceleration capacity in climbing geckos. Biol J Linn Soc 85(3):385–393

Vinson J, Vinson J-M (1969) The saurian fauna of the Mascarene Islands. In: Mauritius institute bulletin, vol 6. The Mauritius Institute, Port Louis, Mauritius, pp 203–320

Vitt LJ, Zani PA (1997) Ecology of the nocturnal lizard *Thecadactylus rapicauda* (Sauria: Gekkonidae) in the Amazon region. Herpetologica 53(2):165–179 von Wittich (1854) Der Mechanismus der Haftzehen von *Hyla arborea*. Archiv für Anatomie, Physiologie und Wissenschaftliche Medicin 8:170–183

Wagler J (1830) Natürliches System der Amphibien: mit vorangehender Classification der Säugethiere und Vögel: ein Beitrag zur vergleichenden Zoologie, 1st edn. J.G. Cotta'schen Buchhandlung, München, Germany

Wainwright SA, Biggs WD, Currey JD, Gosline JM (1982) Mechanical design in organisms. Princeton University Press, Princeton, NJ, USA

Weitlaner F (1902) Eine Untersuchung fiber den Haftfuβ des Gecko. Verhandlungen der Zoologisch-Botanischen Gesellschaft in Wien 52:328–332

Williams EE, Peterson JA (1982) Convergent and alternative designs in the digital adhesive pads of scincid lizards. Science 215(4539):1509–1511

Xu Q, Wan Y, Hu TS, Liu TX, Tao D, Niewiarowski PH, Yu T, Liu Y, Dai L, Yang Y, Xia Z (2015) Robust self-cleaning and micromanipulation capabilities of gecko spatulae and their bio-mimics. Nat Commun 6:8949

Yao H, Gao H (2006) Mechanics of robust and releasable adhesion in biology: Bottom-up designed hierarchical structures of gecko. J Mech Phys Solids 54(6):1120–1146

# Chapter 12
# Adhesive Secretions in Harvestmen (Arachnida: Opiliones)

**Jonas O. Wolff, Solimary García-Hernández, and Stanislav N. Gorb**

**Abstract** Opiliones, colloquially also known as harvestmen or daddy longlegs, are arachnids capable of producing and releasing a variety of secretions that are used to deter predators. The fact that a large fraction of these animals also produce efficient glues for trapping prey, gluing eggs to substrates, attaching soil particles to their body or eggs for camouflage purposes, or transferring sperm, is rather unknown. Not only the physical properties of these glues are interesting, but also the supplementary cuticular structures, that work hand in hand with the secretions to produce highly efficient adhesive mechanisms. Here we give an overview on the occurrence, properties, and associated structures of adhesive secretions in harvestmen and discuss their biological functions.

## 12.1 Introduction

Following insects, the arachnids comprise the second most diverse class of arthropods with more than 100,000 described species. Arachnids have conquered nearly all terrestrial habitats and evolved numerous fascinating life forms, which are only poorly studied. Arachnids also produce materials with impressive properties, like

J.O. Wolff (✉)
Zoological Institute: Functional Morphology and Biomechanics, Kiel University, Am Botanischen Garten 9, 24118 Kiel, Germany

Department of Biological Sciences, Macquarie University, Sydney, NSW 2109, Australia
e-mail: jonas.wolff@mq.edu.au

S. García-Hernández
Programa de Pós-graduação em Ecologia, Instituto de Biociências, Universidade de São Paulo, Rua do Matão, trav. 14, no. 321, São Paulo 05508-900, SP, Brazil
e-mail: solimarygh@usp.br

S.N. Gorb
Zoological Institute: Functional Morphology and Biomechanics, Kiel University, Am Botanischen Garten 9, 24118 Kiel, Germany
e-mail: sgorb@zoologie.uni-kiel.de

© Springer International Publishing Switzerland 2016 
A.M. Smith (ed.), *Biological Adhesives*, DOI 10.1007/978-3-319-46082-6_12

the extremely tough silk of spiders (Gosline et al. 1984) anchored to substrates by strong glue plaques (Wolff et al. 2015a); the stiff cuticle of spider fangs (Politi et al. 2012); or adhesive structures at the feet of spiders (Kesel et al. 2003), amblypygids (Wolff et al. 2015b), and mites (Mizutani et al. 2006). However, the majority of these natural materials and substances are not investigated or even not known.

Opiliones, colloquially also known as harvestmen or daddy longlegs, are the third-largest order of arachnids, composed of nearly 6500 species distributed in all continents except for Antarctica and divided in four recent suborders, namely, Cyphophthalmi, Eupnoi, Dyspnoi, and Laniatores (Machado et al. 2007; Kury 2012) (Fig. 12.1). Harvestmen are known to produce a variety of volatile secretions in specialized scent glands that are used to deter predators (Holmberg 1986; Machado et al. 2005; Raspotnig 2012) or for intraspecific communication (Holmberg 1986; Machado et al. 2002). The fact that a large fraction of these animals also produce efficient glues used for trapping prey (Rimsky-Korsakow 1924; Wachmann 1970; Wolff et al. 2014), gluing eggs to substrates (Juberthie 1964; Martens 1993) or attaching soil particles to their body for camouflage purposes (Schwangart 1907; Roewer 1923; Schönhofer 2013) is rather unknown. In some cases they also play a role in sperm transfer (Karaman 2005; Novak 2005). Not only the physical properties of these glues but also the supplementary cuticular structures working hand in hand with the secretions, in order to produce highly efficient adhesive mechanisms, are very interesting. Here we give an overview on the occurrence, properties, and associated structures of the adhesive secretions in harvestmen and discuss their biological functions.

## 12.2 Types of Adhesive Secretions and Their Distribution Among Harvestmen

Glue (an adhesive) is a substance that wets surfaces and builds chemical or physical bonds with them, resulting in strong adhesive forces (Kinloch 1987). Efficient glue shows a good wetting behavior on the target surface, strong adhesion (i.e., by building hydrogen or covalent bonds with the substrate surface), and a similarly strong cohesion (inner strength). In most cases, there is a switch in cohesion after the application of the glue to the surface: The initial cohesion should be preferably much lower than the adhesion to the target surface, to provide a good wetting behavior, but should then quickly increase to a value of at least as high as the adhesion, to provide a strong bonding. High strength can be achieved in very different ways, depending on whether durable or quick-and-reversible adhesion is needed for a particular application or biological function.

Two different types of adhesive secretions occur among harvestmen. The first type is the glue that solidifies and creates a permanent bonding. Such solidifying adhesives recruit strong cohesion by the hardening process, which may be due to

**Fig. 12.1** Analogy of hairlike structures supplying viscid glue for prey capture. (**a**, **b**) *Drosera capensis*, an insectivorous plant with trichomes secreting sticky mucilage (Photos: J. Uribe). (**c**) Nymph of the harvestman *Mitostoma chrysomelas*, with its pedipalps densely covered in glue-secreting setae (*arrowhead*) (Photo: J. Pageler). (**d**) Adhesion test with an ablated pedipalp. The viscous adhesive stretches into long threads

polymerization or the evaporation of a solvent (usually water) after emergence (Nachtigall 1974). In harvestmen, solidifying glues are used to attach soil particles to the body to blend the animal in its environment (*soil camouflage*), to attach eggs to substrates, or as a defense mechanism against egg predators.

The second type of adhesive secretions stays in a fluid condition and exhibits shear-thickening properties. This means that if a droplet of such secretion is brought into contact with an object and then pulled apart quick enough, the flow will be hindered due to viscous or viscoelastic forces. This results in a temporarily high inner strength, which depends on the rate of pulling. Such *viscid glue* is a common means of prey capture, since it offers a very quick and reversible bonding mechanism (Betz and Kölsch 2004). It is well known from carnivorous plants, such as the sundew *Drosera*, which exposes its glue in small droplets attached to the globular tips of trichomes (Fig. 12.1a, b), representing an effective insect trap. The shear-thickening mechanism of the sundew glue was recently explained by its macromolecular structure: It is based on a hygroscopic polysaccharide network that is highly compliant under pressure, but stretches into long, tough nano-fibers under tension (Huang et al. 2015). Similar, although significantly smaller, hairlike structures (setae) carrying droplets of a sticky substance have been described from the pedipalps (the second pair of limbs in arachnids) of small soil-dwelling harvestmen, the Nemastomatidae (Rimsky-Korsakow 1924) (Fig. 12.1c, d). Wolff et al. (2014) were curious about the actual function of these setae and studied the prey-capture behavior and the rheological properties of the adhesive fluid in these animals. In a subsequent study, Wolff et al. (2016) used a broad comparative approach to answer the question how widespread this mechanism is among harvestmen. They found that glandular setae are distributed among the suborders Dyspnoi and Eupnoi, presumably with its origin in a common ancestor (Fig. 12.2). This shows that not only the small cryptic Nemastomatidae are active predators that use glue to immobilize their prey but, among others, also the Phalangiidae, which comprise the well-known long-legged harvestmen widespread in temperate parts of the Northern Hemisphere. However, the structure of the glue-carrying setae and the properties of their secretions vary and are distinguished between so-called plumose setae (or "bottle-brush setae") and clavate setae (or "drumstick-like setae") (see Sect. 12.3).

Besides, viscid glue of pedipalpal glands has presumably independently evolved in the suborder Laniatores, which we discovered in nymphs of Cosmetidae and Gonyleptidae. In particular, Cosmetidae nymphs show modifications of the pedipalps, especially of the pretarsus, which is covered by numerous gland openings (see Sect. 12.4). The glue does not assemble at the tip of setae, but forms a continuous coat at the tip of the pedipalp. Although we could not find direct evidence, we assume that this glue is used during prey capture in a same manner like in Nemastomatidae.

## 12.3 Glandular Setae in Palpatores

Harvestmen were long time believed to be scavengers that are not capable to capture and overwhelm agile prey. Presumably, because of their fragile appearance and clumsy movements, they do not make an impression to be effective predators. However, there are observations of harvestmen catching walking, jumping, and even flying insects with surprising ease (Rühm 1926; Bristowe 1949; Immel 1955;

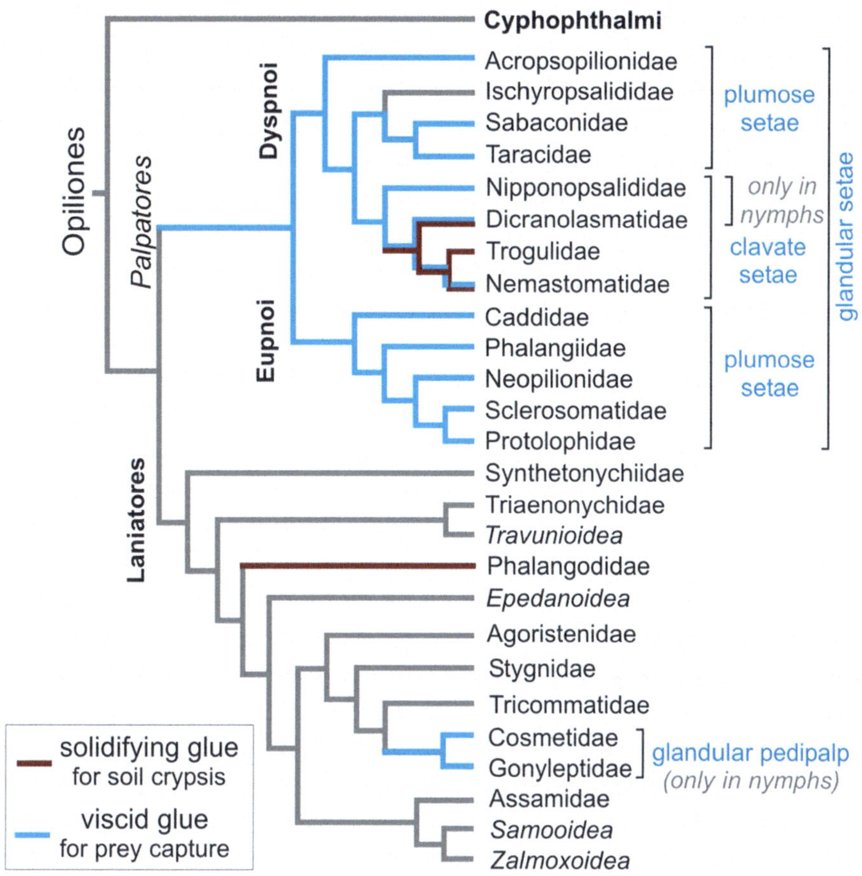


**Fig. 12.2** Phylogenetic distribution of adhesive secretions in harvestmen (Opiliones). Compiled cladogram based on Shultz and Regier (2001), Sharma and Giribet (2011), Hedin et al. (2012), Schönhofer (2013), Groh and Giribet (2014), Pinto-da-Rocha et al. (2014), Giribet and Sharma (2015). Characters may be inhomogeneously distributed within lineages (not displayed). The distribution of the glandular setae follows Wolff et al. (2016). Note that there is no data on pedipalpal glue in Laniatores families other than Cosmetidae, Gonyleptidae, and Agoristenidae, so the character distribution should be handled as a hypothesis. Distribution of egg coating secretions is not displayed here since it is not studied thoroughly

Phillipson 1960; Gruber 1993; Santos and Gnaspini 2002). In feeding trials with *Mitostoma chrysomelas* (Nemastomatidae), Wolff et al. (2014) observed that fast-moving springtails up to a weight that exceeds that of the harvestman are effectively caught with the help of the thin and non-armored pedipalps only (Fig. 12.3a, b). Gut content analysis had previously shown that springtails are the primary diet of these animals (Adams 1984). Close inspection of the pedipalps reveals the glandular setae, each with a small droplet of glue at the tip. These setae occur from the femur to the tarsus with the highest density and the largest droplets at the distal dorsal side of the tibia. The setae are 20–30 μm long and hence barely discernable without a

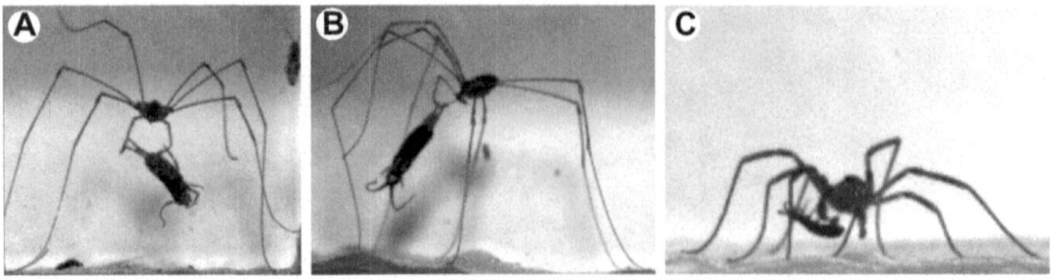


**Fig. 12.3** Use of the sticky pedipalps during prey capture. Single frames of high-speed video sequences (1000 frames per second) of the prey capture of springtails by two different species of harvestmen. (**a**, **b**) *Mitostoma chrysomelas* (Nemastomatidae), with a prey attached to its pedipalps. The springtail sticks so strongly that it cannot free itself by strong pushes of its furca. (**c**) *Sabacon* sp. (Sabaconidae) with a springtail attached to its pedipalps. This species exhibits strongly modified pedipalps bearing plumose setae

magnification lens. Despite of their small size, these setae have a highly elaborate structure and multiple functions. They serve as structures, which supply the glue and anchor the droplets at an exposed position, while additionally working both as chemo- and presumably mechanoreceptors that provide direct feedback about the prey and the success of agglutination.

The structure of these setae is of high significance for a proper function of the glue. Below we discuss the role of different features of these highly specialized setae.

**Microtrichia** Setae are basically thin cylindrical structures that taper toward the tip. If such objects are covered by a reasonable amount of a viscoelastic fluid, droplets form due to the interplay of adhesion and surface tension (so-called Rayleigh instability) (Rayleigh 1878, 1892; Kalliadasis and Chang 1994). A well-known example for this phenomenon is the capture threads of orb web spiders, in which the glue coating forms a stable beads-on-a-string (BOAS) structure along the silk fiber (Edmonds and Vollrath 1992; Sahni et al. 2011; Torres et al. 2014). If a target substrate touches a glue droplet on such a cylindrical object and is pulled apart, the resulting forces highly depend on the pull-off angle. When it is pulled vertically from the cylinder, the forces are high, because the flow is hindered. However, when the substrate is moved in parallel to the cylinder, the fluid can easily flow along the surface of the cylinder, and the substrate is easily shifted. In the case of a tapering seta, this means that the droplet can easily be shifted toward the tip, where its adhesion to the seta diminishes because the surface area becomes smaller toward the tip. Eventually the droplet may totally get lost, which means the target substrate will be released from the seta. In orb web spiders, there are special tapered setae on the tarsi, which actually use this effect to prevent agglutination when walking on sticky capture threads (Briceño and Eberhard 2012). However, for a predator that aims to agglutinate a prey item, this effect renders the best glue useless. In harvestmen, unique anchoring structures have been evolved that stop the flow along the setal shaft, such that even under high load, the fluid is not detached from the seta (see Fig. 12.1d). These structures are microtrichia ("plumes") that branch off vertically from the shaft. This structure repeats on the nano-level, since the distal tips of these microtrichia are split into branchlets that, again, build a rectangle with the microtrichial shaft. The result of such a geometry is a highly effective flow stopper that keeps the droplet attached even under high tensile load.

Plumose and clavate setae differ in the arrangement of the microtrichia on the setal shaft. In plumose setae, the microtrichia are distributed along the distal section (the distal ¼ to ½) (Fig. 12.4d). The arrangement is often anisotropic with more microtrichia on one side (Fig. 12.4d). The usually two branchlets at the tip of microtrichia are often knob-like swollen at their ends. The glue droplet of a plumose seta has an oval shape.

The clavate setae are derived structures, which presumably have a higher adhesive performance. The microtrichia are concentrated in a globule underneath the setal tip, such that they point in nearly every direction (Figs. 12.4b, c and 12.5). This means that the droplet pull-off resistance is nearly independent on the pull-off angle. Again, this structure is repeated at the nanoscale, since there are three regularly arranged branchlets at the tip of the microtrichia (Fig. 12.4c, e). The glue droplet of a clavate seta is of a globular shape, with a diameter of about 10 μm (Fig. 12.4a). Another feature of clavate setae are distinct slit-like openings of the secretory channels (Figs. 12.4c and 12.5), which are lacking in most plumose setae and may facilitate the quick emergence of the adhesive fluid.

**Material Gradient** Arthropod setae are usually rather compliant, unless they are spinelike modified to serve defensive or prey-capture purposes. In contrast, the glandular setae of harvestman pedipalps are rather stiff, and it is nearly impossible to bend the shaft without breakage. This is due to a high degree of sclerotization of the cuticular material, which is easily seen by its dark color. However, the distal part which carries the microtrichia appears translucent under a light microscope. This part is compliant and may absorb some tensile stress acting on the glue droplet in contact situation with a prey. This raises the question, why not the whole seta is compliant, since this would prevent stress concentration in the tip, which may lead to breakage. The reason may be that the glue droplet must stay in a defined position to avoid contact with other setae or the rest of the pedipalp. Similar material gradient are known from adhesive setae of beetle feet, used for attachment during locomotion and while resting on surfaces (Peisker et al. 2013). Here the comparably higher stiffness of the shaft is crucial to avoid matting between neighboring setae in the dense pad, whereas the soft tip ensures a proper contact formation to various potential substrates.

Interestingly, in clavate setae, their tapered tips protrude through the glue droplet and are usually not wetted (Fig. 12.4a). This part is rather stiff and prevents agglutination of a vertically touching surface, which we frequently observed during adhesion measurements using a glass surface as substrate. Only high load could press the tip down such that the glue could contact and spread onto the glass surface. This may be crucial to prevent agglutination to objects in the environment by

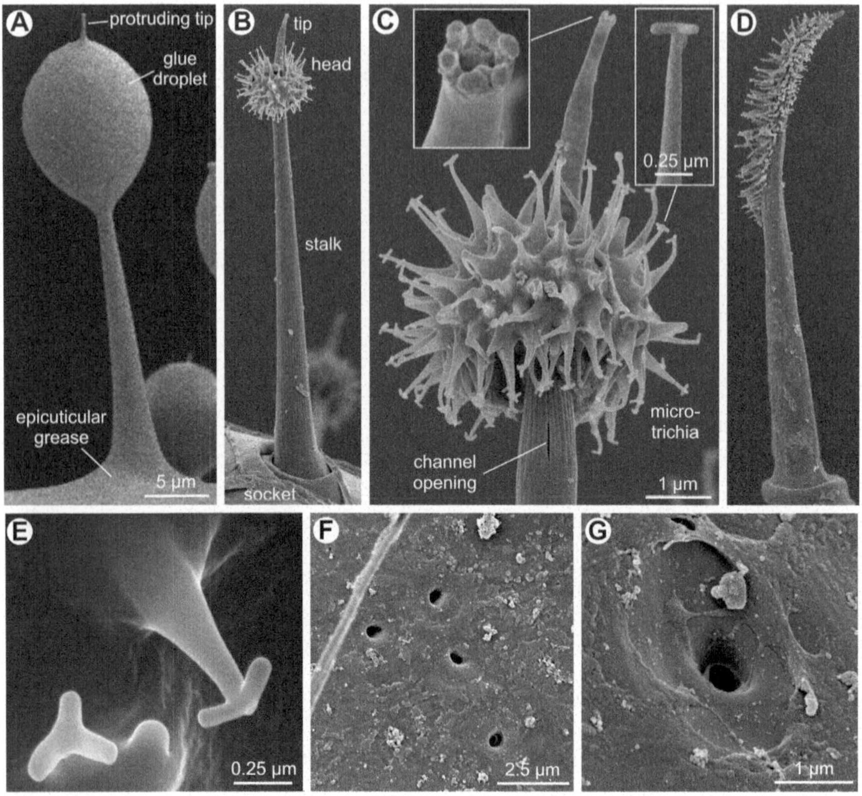


**Fig. 12.4** Fine structure of pedipalpal glandular setae. (**a**) Cryo-scanning electron micrograph (Cryo-SEM) of a clavate seta of *Mitostoma chrysomelas*. Note the presence of epicuticular fluids that are washed off in (**b**), and the setal tip protruding the glue droplet. (**b**, **c**) Same after washing with acetone, followed by critical point drying, revealing the ultrastructural details of this specialized type of seta. (**c**) Detail of the microtrichious head and tip. The adhesive fluid emerges from slit-like openings underneath the head. Insets demonstrate details of the tip (apical view) with pores of the chemosensor, and the branching tip of a microtrichium. (**d**) Plumose seta of *Thrasychirus gulosus* (Neopilionidae). Note the anisotropic distribution of microtrichia in the upper half of the seta. (**e**) Details of the apical branchlets of the trifid microtrichia of a plumose seta of *M. chrysomelas*. (**f**,**g**) Pores in the cuticula between the glandular setae of *Mediostoma stussineri* (Nemastomatidae). These may supply an epicuticular grease, as seen in (**a**) that may have an anti-adhesive effect to prevent self-agglutination

accident, since these animals often dwell in highly structured microhabitats, such as leaf litter. On the other hand, the body structures of the prey, such as setae, can attach immediately, because they protrude between the clavate setae where they get in touch with the glue droplets.

**Properties of the Adhesive Fluid** In a previous study, we measured the pull-off forces of the glue droplets of single clavate setae of *Nemastoma lugubre* and *Mitostoma chrysomelas* after bringing them in contact with glass surfaces (Wolff

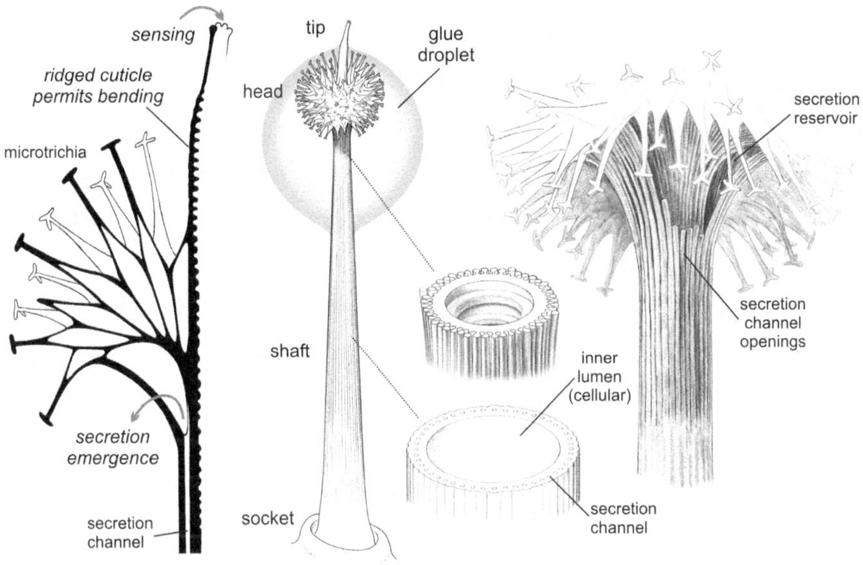


**Fig. 12.5** Schematic illustration of the structure of a clavate seta (Drawings by H. Wijnhoven)

et al. 2014). The pull-off force increases with pull-off velocity (Fig. 12.6). At a rate of 1 μm/s, it is on average 1.2 μN, whereas at 1 mm/s it is up to 7 μN. For comparison, the weight force of an average springtail is 1–2 μN (Christian 1979; Verhoef and Witteveen 1980). This means that one seta alone can theoretically be sufficient to hold the prey. Rate dependence is caused by the non-Newtonian properties of the adhesive fluid: At high rates it is stretched into long filaments (up to 40 times of the initial droplet diameter) that remain in a nearly cylindrical shape with an even stress distribution. At low speed, the filament breaks up and develops a beads-on-a-string morphology due to its viscoelastic properties. When droplets formation begins, the pull-off force drops although the filament can be pulled to a substantial length until it fails. Such so-called pressure sensitive adhesives (PSA) are perfect for the capture of agile prey, because the prey will usually struggle wildly to come free, which causes even stronger adhesion.

Further, we estimated the elastic modulus of the secretion by integrating strength and extensibility of the glue during our pull-off tests. It is in an estimated range of 9–15 kPa, which is in agreement with Dahlquist's criterion for PSAs (<100 kPa) (Dahlquist 1969; Creton 2003).

The adhesive fluid of clavate setae also shows a good wetting behavior. It easily spreads on the complex micro- and nanostructures of springtails, which are known for their omniphobic (fluid-repellent) properties (Helbig et al. 2011).

The chemical composition of the adhesive fluid is not studied thoroughly. Rimsky-Korsakow (1924) did solution and staining tests on the clavate setal secretion of *N. lugubre*. He came to the conclusion that the glue is based on

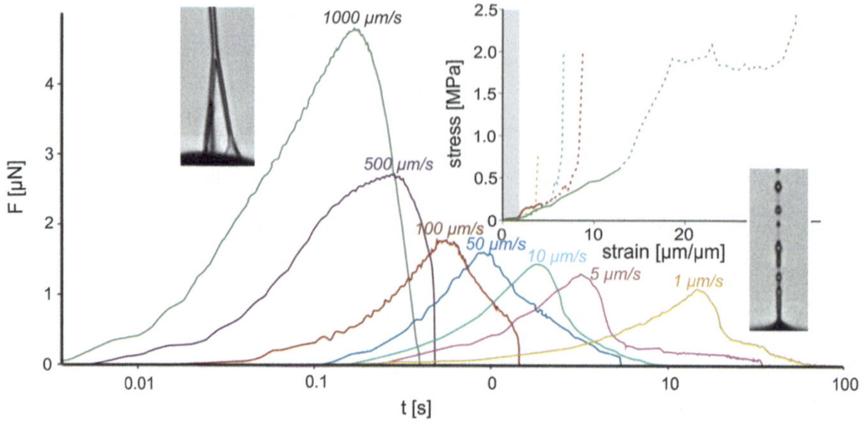


**Fig. 12.6** Time-force and strain–stress curves of single clavate setae of *Mitostoma chrysomelas* pulled from a glass surface at different speeds. The *gray* area in the strain–stress diagram marks the initial elastic phase used to calculate Young's modulus. Inset images show micromorphology of the glue thread at high and low pulling rates (adapted from Wolff et al. 2014)

triglycerides with oleic fatty acids. This seems plausible since the cuticle of some harvestmen, like *M. chrysomelas*, is covered with secretions of wax crystals or oils (J.O. Wolff, unpublished data). Comparable oils in plants have recently successfully been used as a raw material for the synthesis of a biocompatible PSA (Wool and Bunker 2007), which underlines the suitability of such substances as adhesives. From a biological point of view, it would be interesting to analyze the composition of the glue in harvestmen of different evolutionary lineages. In specimens preserved in ethanol, there are distinct differences in the amount of secretion remains between harvestmen with plumose and clavate setae. The plumose setae are often clean, which means that the glue has totally dissolved over time. When air-drying freshly ablated pedipalps, the glue droplets significantly shrink, which may indicate a high water content. In contrast, air-dried pedipalps of species with clavate setae keep their glue droplets over a long period of time (several months). Specimens, stored in 70 % ethanol for years usually carry remains of the secretion. To remove the glue from the setae, in order to visualize the ultrastructure like in Fig. 12.4c, the specimen has to be rinsed with acetone for at least half an hour. These differences in glue composition are likely accompanied by differences in the mechanical properties, which are subject for future studies.

**Anti-adhesive Fluids** A concluding question is how the harvestmen with such powerful adhesives manage not to conglutinate their own body parts by accident. When comparing Cryo-scanning electron images (Cryo-SEM) of fresh shock-frozen samples of *N. lugubre* pedipalps with washed and critical point dried samples, we found a liquid film covering the cuticle at the bases and shafts of the setae (Fig. 12.4a). In ethanol-preserved specimens of some nemastomatids, we found numerous small pore openings (Fig. 12.4f, g) between the setae, which may extrude a fluid that presumably does not mix with the clavate setal secretion. Such mechanism has been reported from orb web spiders, which get stuck to their own webs, when the epicuticular lipids are washed off (Kropf et al. 2012; Briceño and Eberhard 2012).

## 12.4 Pedipalpal Adhesives in Laniatores Nymphs

Laniatores is the largest suborder of Opiliones, with over 4000 described species worldwide, the majority of which are found in the tropics (Machado et al. 2007; Kury 2012). These harvestmen usually possess raptorial pedipalps with long spines. Until now it was unknown that some of them may utilize viscid glue for prey capture similar to various Palpatores.

In Cosmetidae, there is a distinct dimorphism in the pedipalps between nymphs and (pre-)matures (Goodnight and Goodnight 1976; Rodriguez et al. 2014). Nymphs have elongated and modified pedipalps, which lack any spines and have a bulbous pretarsus (Fig. 12.7d). Ultrastructural examination of the pedipalp in a nymph of *Cynortellana quadrimaculata* from Cuba revealed that the distal patella, the tibia, the tarsus, and the modified pretarsus are densely covered with 1.5 μm wide tubelike gland openings with remains of some amorphous secretion (Fig. 12.7e). Recently, it was observed in some nymphs of unknown Cosmetidae from Costa Rica that the pedipalps are covered with a sticky substance (G. Machado, pers. comm.). We presume that it is an adhesive that is used for prey capture like in nemastomatids. An unusual brush of bristles on the pretarsus may help to assemble a larger amount of glue at the tip, which may be used then to pick up prey arthropods. However, the feeding biology of these harvestmen is entirely unknown to date.

Further observations were made with nymphs of the closely related Gonyleptidae. Although these harvestmen exhibit the raptorial pedipalps typical for Laniatores, we found them similarly coated with translucent substance. In order to prove the stickiness of the secretion, we did simple test with living individuals in the field. The pedipalps were slightly touched with forceps or a finger and slowly retracted. If visible threads of a viscous substance formed, or if the pedipalp stayed attached, we concluded the presence of a sticky secretion. In order to determine the species, we collected adult individuals found at the same sites. In total we did 62 tests with eight species of Laniatores:

16 nymphs of *Iporangaia pustulosa* (Gonyleptidae: Progonyleptoidellinae)
2 nymphs of *Neosadocus maximus* (Gonyleptidae: Gonyleptinae)
1 nymph of *Sodreana leprevosti* (Gonyleptidae: Sodreaninae)
1 nymph of *Ampheres leucopheus* (Gonyleptidae: Caelopyginae)
10 nymphs of *Serracutisoma proximum* (Gonyleptidae: Goniosomatinae), all of which from Atlantic Forest of State of São Paulo, Brazil
2 nymphs of *Mitogoniella taquara* (Gonyleptidae: Goniosomatinae), from Atlantic Forest of State of Rio de Janeiro, Brazil

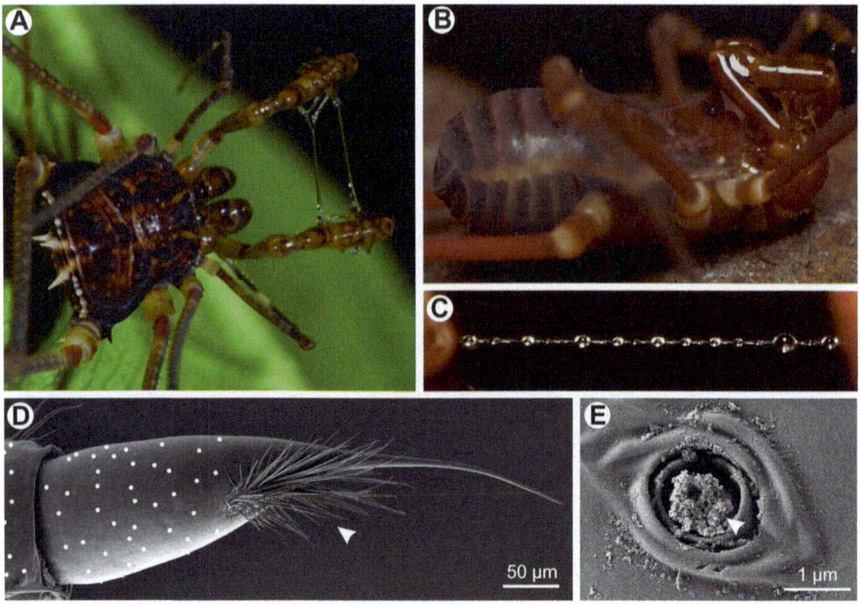


**Fig. 12.7** Pedipalpal glue in Laniatores nymphs. (**a**) Nymph of an undetermined species of Cranainae (Gonyleptidae), sitting in an ambushing position. Note the viscid glue on the pedipalps (Photo: A. Anker). (**b**) Nymph of *Phalangodus briareos*. Note the much higher reflectance of the pedipalps indicating the glue cover (Photo: J. Uribe). (**c**) The glue on the pedipalps can be pulled into long threads that form BOAS structures over time (Photo: J. Uribe). (**d**) Scanning electron micrograph of the modified pedipalp pretarsus of a *Cynortellana quadrimaculata* nymph. *White dots* mark the position of gland openings, *arrowhead* points to a brush of bristles, which is very unusual for an arthropod pretarsus. (**e**) Detail of a gland opening, with remains of an amorphous secretion

25 nymphs of *Phalangodus briareos* (Gonyleptidae: Cranainae) (Fig. 12.7b)
5 nymphs of *Avima* sp. (Agoristenidae: Leiosteninae), from an Andean cave in Santander, Colombia

We found evidence for sticky secretions in all tested species of Gonyleptidae. The highest amounts of glue were observed in Cranainae. In these, the secretions could be pulled into long threads, which formed BOAS structures, indicating their viscoelastic properties (Fig. 12.7c). In individuals of *P. briareos*, which were collected in dry plastic boxes and taken to the lab, we observed a significant loss of the sticky secretion, suggesting high water content. The pedipalps of *Avima* sp. (Agoristenidae) are not sticky and do not exhibit any fluids. We could not verify the presence or absence of sticky secretions in other families related to Gonyleptidae and Cosmetidae due to the lack of material. The mechanism might be more widespread, although we assume it is most developed in Cosmetidae and Cranainae.

Due to their nocturnal activity, the feeding biology of nymphs of all these species is unknown. However, adults of *P. briareos* (Gonyleptidae: Cranainae) prey on insects and other invertebrates using their raptorial pedipalps (S. García-

Hernández, unpublished data). A photo by A. Anker taken at night in the rain forest of Puerto Nariño (Colombian Amazon) shows a nymph of Cranainae with large amount of the glue on its pedipalps, sitting in an ambushing position, which resembles that of adults (Fig. 12.7a). It remains to be investigated which kind of prey is caught and how the glue is utilized during prey capture.

In all the above cases the amount of secretion is reduced in subadult instars and totally absent in adults. A similar ontogenetic change is found in the dyspnoid Dicranolasmatidae (Gruber 1993) and Nipponopsalididae (Miyoshi 1942). Gruber (1993) did observations on the feeding biology of *Dicranolasma pauper* and found that the nymphs are predominantly feeding on living arthropods, which are captured with the help of the pedipalps, whereas the adults prefer feeding on dead invertebrates and slowly moving prey. It is likely that the nutrition requirements differ between nymphs and adults, with the nymphs relying more on fresh protein for growth. This may explain the switch in feeding biology, correlated with the differences in glue production.

## 12.5 Soil Camouflage

Some harvestmen can blend in their environment, such that they are nearly not discernable, because of a cover with soil particles (*soil crypsis*). This is particularly known for the Trogulidae (Schwangart 1907) (Fig. 12.8a, b), but also from Dicranolasmatidae (Gruber 1993) (Fig. 12.8g, h), some Nemastomatidae (*Ortholasma*, *Dendrolasma*, *Vestiferum*) (Schönhofer 2013), some Sclerosomatidae (*Sclerosoma* spp.) (Gnaspini and Hara 2007), some Triaenonychidae (Gnaspini and Hara 2007), some Phalangodidae (Roewer 1923), some Podoctidae (Martens 1993) (Fig. 12.9d) and some Gonyleptidae (Hernandariinae, Ampycinae, some Pachylinae) (DaSilva and Pinto-da-Rocha 2010). The particles are attached via a solidifying (presumably tanning) secretion. In *Trogulus* spp., the secretion is stored in bubble-like cuticular protuberances (*Drüsenwärzchen* = glandular warts) and assembles between them after extrusion (Schwangart 1907) (Fig. 12.8c–e).

We examined a freshly molted individual of *Trogulus martensi* by means of scanning electron microscopy. There were no discernable pores in the warts nor between them, hence the secretion is presumably released by diffusion through their thinned cuticular walls. The hardened secretion forms a very robust crust with embedded sand grains and detritus, which can hardly be removed. Short, hook-like curved setae on some warts (Fig. 12.8e) additionally help to catch and retain soil particles. The tarsi, eyes, and mouth parts (pedipalps and chelicerae) are the only clean body parts.

In *Dicranolasma* spp., no glandular warts are present. The secretion assembles as a film on the cuticle and forms clumps on the short body setae (Fig. 12.8i). In nymphs, the crust is comparably easy to remove, presumably because of a lower amount of setae on the dorsum that interlock with the solidified secretion crust, or

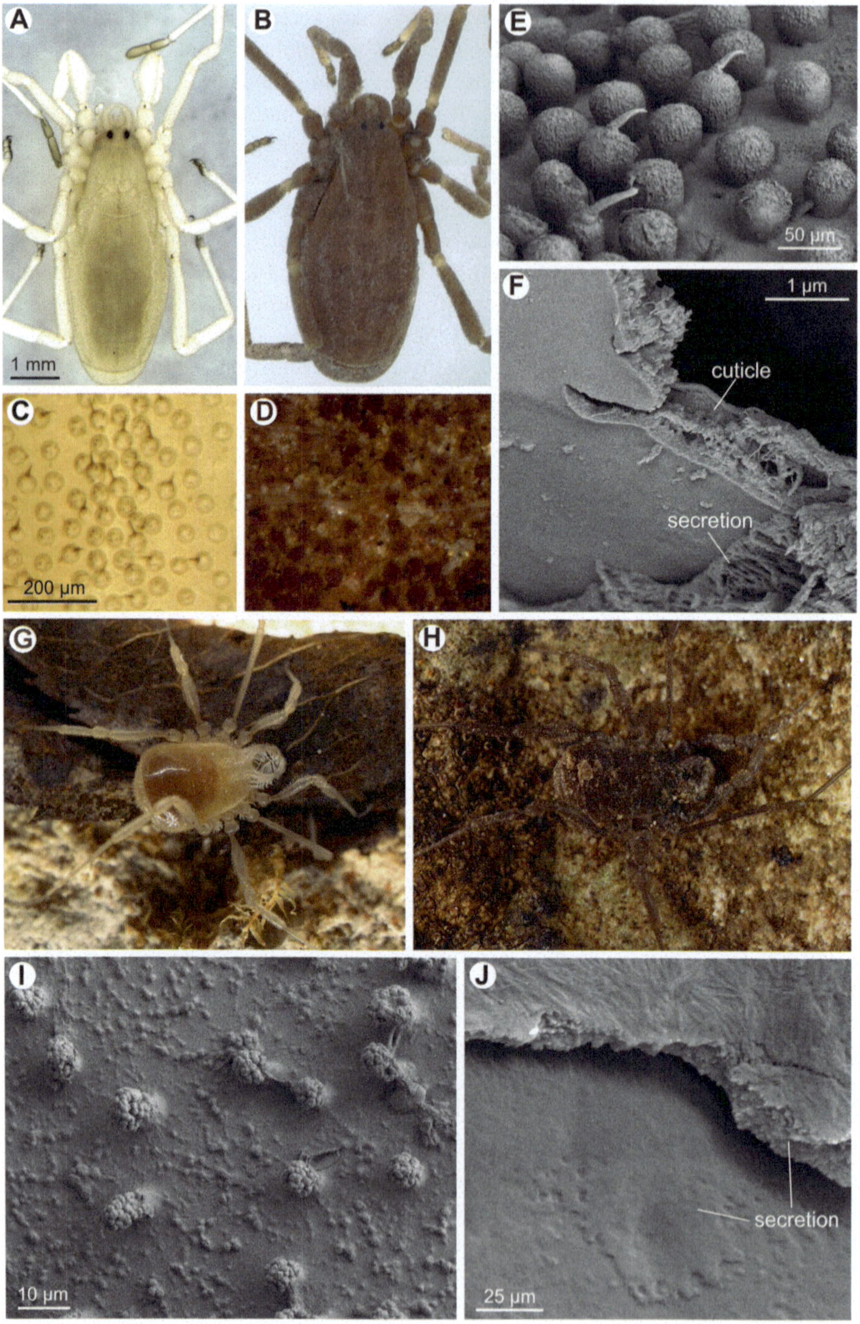


**Fig. 12.8** Soil camouflage in Dyspnoi. (**a–f**) *Trogulus martensi* (Trogulidae). (**a**, **c**, **e**, **f**) Freshly molted. (**b**, **d**) Usual habitus with dried crust of secretion and soil particles. (**c–e**) Detail of dorsal integument with glandular warts. (**c**, **e**) After molting the warts are still exposed and clearly visible. (**d**) A secretion emerges and assembles between the warts, catches soil particles, and dries forming a crust that is hardly removable [same magnification and section as in (**c**)]. (**e**) Scanning electron

due to lower amounts of secretion. Nonetheless, often large clumps of soil material are assembled on the dorsum.

The composition and properties of these secretions are not known. Our ultrastructural analysis shows that there are different components, of which one is fibrous (Fig. 12.8f, j). In other animals, permanent glues are primarily based on self-assembling proteins or polypeptides that may cross-link and build large polymer complexes after their application to the surface (Naldrett 1993; Silverman and Roberto 2007; Hennebert et al. 2008; Dickinson et al. 2009; Voigt and Gorb 2010). However, since harvestmen often have a cover of lipids, these might also be waxlike substances.

A similar soil camouflage behavior is known from other arthropods; however, mechanisms of soil particle adhesion differ. In spiders of the genera *Sicarius* (Sicariidae) and *Homalonychus* (Homalonychiidae), for instance, there are specialized setae with fine microtrichia that adhere to sand particles via van der Waals forces (Duncan et al. 2007). In others, like *Cryptothele* spp. (Zodariidae), there are barbs that mechanically interlock with detritus material (Ramírez et al. 2014). A solidifying secretion, encrusting soil particles like in harvestmen, is present in predatory mites of the family Caeculidae (J.O. Wolff, unpublished data).

Although defense from predators seems to be the plausible function for soil camouflage, it is not experimentally studied in arachnids. The soil cover may not only hide the animal from visually guided predators but might also represent a hard armor (Schaider and Raspotnig 2009) and a mechanical barrier against parasites. In reduviid bugs, which exhibit a similar mechanism, a defensive effect was shown (Brandt and Mahsberg 2002), but the coating is even beneficial in the absence of predators, presumably because it assists thermoregulation (Ramírez et al. 2013).

## 12.6 Egg Coatings and Spermatophores

Some harvestmen may produce specialized adhesive secretions for egg protection. In some Laniatores, females glue their eggs on substrates and hide them by attaching soil particles, such as sand grains or pieces of plants or wood, picked up from the substrate (Machado and Raimundo 2001; Willemart 2001; Stanley 2011) (Fig. 12.9a). Egg hiding behavior is generally interpreted as a way of camouflage

**Fig. 12.8** (continued) micrograph (SEM) of the glandular warts. Note the short, curved setae on some warts that assist in the capture and retention of soil particles. (**f**) Cuticle of glandular wart, partly cut with a scalpel. The cuticle is smooth; hence the roughness seen is already emerged secretion. There are no discernable pores; hence the secretion may emerge by diffusion through the cuticle. (**g–j**) *Dicranolasma* sp. (Dicranolasmatidae) (Photos **g**, **h**: S. Huber). (**g**) Freshly molted individual. (**h**) Usual habit with soil cover, photographed in its natural environment. (**i**, **j**) SEM of the dorsal integument of a nymph. (**i**) Crust of dried secretion. The nubs are clumps of the secretion adhering to setae. Glandular warts are absent in this family. (**j**) Fracture of the secretion crust. The underlying smooth cuticle is visible. The secretion seems to contain nano-fibrils

Fig. 12.9 Adhesive egg coatings in Laniatores. (**a**) Female of *Longiperna zonata* picking up debris and attaching particles to the egg. In the inset, two eggs with agglutinated pieces of plants are visible (Photos: B. A. Buzatto). (**b**) Mucus-covered egg clutch of *Iporangaia pustulosa*. Note that the mucus coat remains stable and sticky over a long period, since the eggs already show a late developmental stage. (Photo: G. Requena). (**c**) Eggs of *Phalangodus briareos* glued onto a wet rock surface with means of a solidifying secretion (*arrowheads* point to plaques of the adhesive) (Photo: J. Uribe). (**d**) Male of *Leytpodoctis oviger*, carrying eggs glued to the right femur of the leg IV. Also note the soil incrustation on the dorsum (Photo: J. Martens & J. Trautner)

targeting against egg predators that use visual or tactile cues (Canals 1936; Juberthie 1972; Cokendolpher and Jones 1991; Willemart 2001). Although the egg-covering behavior seems to be common, the composition and properties of the secretions used to attach particles to the egg surface are unknown. In some species of Laniatores, Dyspnoi and Eupnoi, the clutches are covered by mucus produced by females and deposited on the eggs during or just after oviposition (Juberthie 1964; Witaliński and Żuwała 1981; Machado et al. 2004; Requena et al. 2009) (Fig. 12.9b). Mucus is a viscous substance based on glycoproteins or polysaccharides, as it has been shown for the sticky egg coating of *Leiobunum rotundum* (Sclerosomatidae) (Witaliński and Żuwała 1981). In many species, it forms a tough hydrogel that swells after the eggs are coated, due to hydroscopic components (Juberthie 1964). Experiments with *Iporangaia pustulosa* revealed that this mucus coat works as an efficient physical barrier against egg predators (Requena et al. 2009). Presumably, the same mucus coat helps this species to attach its eggs to the surface of leaves (Fig. 12.9b). The gonyleptid *Phalangodus briareos* glues its eggs onto humid rock surfaces in caves (Fig. 12.9c), where they have to stay attached for about 60 days until the nymphs hatch (S. García-Hernández, unpublished data). This secretion seems not to be hygroscopic and hardens after deposition. A bizarre usage of egg coating secretions has been reported from *Leytpodoctis oviger* (Podoctidae), in which the eggs are glued onto the male's leg by the female (Martens 1993) (Fig. 12.9d). Martens (1993) remarked the strong attachment of the eggs and that "the position of the eggs did not change after preservation in 70 % alcohol and remained constant for at least 2 years" although the specimens were manipulated for taxonomic studies. Hence, this substance seems to solidify after secretion. These examples show that egg coating adhesives vary greatly in their material properties. Adhesive egg coatings are widespread among harvestmen; however, they are sparsely studied. For instance, an open question is, if the adaptation to specific oviposition substrates is reflected by difference in both glue composition and properties.

Finally, there might be some cases, where sticky secretions assist sperm transfer. In contrast to all other harvestmen, the basal Cyphophthalmi perform an indirect sperm transfer via a spermatophore (Karaman 2005). Karaman (2005) and Novak (2005) reported that sticky substances coming from ovipositor glands help to attach the spermatophore to the genital opening and chelicerae of the female.

## 12.7 Conclusion

The present review on adhesives in harvestmen shows a range of rarely known and only sparsely studied adaptations that are accompanied by fascinating structures and behaviors. It shows that spiders are not the only arachnid group that use viscid glue for prey capture. The glandular setae found on the pedipalps of Palpatores are perhaps one of the most complex cuticular microstructures associated with glue production. Due to their droplet-retaining properties, these structures are worth a closer look for biomimetic approaches to solve problems of a minute glue droplet application.

The finding of adhesive secretions in the pedipalps of some Laniatores nymphs underlines how little we know about this cryptic, yet abundant, group of arachnids. A thorough study of the biochemistry and rheology of harvestmen cuticular secretions could help to understand the evolution of viscid glues adapted for prey capture.

Soil catching and egg coating adhesives are even less understood and represent an interesting field for future biochemical and experimental studies.

**Acknowledgements** We thank Glauco Machado for sharing his discovery of glue in cosmetid nymphs. We like to thank John Uribe, Jörg Pageler, Arthur Anker, Siegfried Huber, Bruno Buzatto, Gustavo Requena, and Jochen Martens for their enthusiastic help and permission to reproduce their photographs. Thanks to Hay Wijnhoven for his kind permission to reuse his excellent drawings of clavate setae. J. O. Wolff was supported by a PhD scholarship of the German National Merit Foundation and a Macquarie Research Fellowship (MQRF) of Macquarie University, Sydney, Australia. S. García-Hernández was supported by a master student fellowship from Fundação de Amparo à Pesquisa do Estado de São Paulo, Brazil (FAPESP 2012/23135-6).

## References

Adams J (1984) The habitat and feeding ecology of woodland harvestmen (Opiliones) in England. Oikos 42(3):361–370

Betz O, Kölsch G (2004) The role of adhesion in prey capture and predator defence in arthropods. Arthropod Struct Dev 33(1):3–30. doi:10.1016/j.asd.2003.10.002

Brandt M, Mahsberg D (2002) Bugs with a backpack: the function of nymphal camouflage in the West African assassin bugs *Paredocla* and *Acanthaspis* spp. Anim Behav 63(2):277–284

Briceño R, Eberhard W (2012) Spiders avoid sticking to their webs: clever leg movements, branched drip-tip setae, and anti-adhesive surfaces. Naturwissenschaften 99(4):337–341

Bristowe W (1949) The distribution of harvestmen (Phalangida) in Great Britain and Ireland, with notes on their names, enemies and food. J Anim Ecol 18(1):100–114

Canals J (1936) Observaciones biológicas em arácnidos del orden Opiliones. Revista de Chilena de Historia Natural 40:61–63

Christian E (1979) Der Sprung der Collembolen. Zool Jahrb Abt Syst Oekol Geogr Tiere 83 (4):457–490

Cokendolpher JC, Jones SR (1991) Karyotype and notes on the male reproductive system and natural history of the harvestman Vonones sayi (Simon) (Opiliones, Cosmetidae). Proc Entomol Soc Washington 93:86–91

Creton C (2003) Pressure-sensitive adhesives: an introductory course. MRS Bull 28(06):434–439

Dahlquist CA (1969) Pressure-sensitive adhesives. In: Patrick RL (ed) Treatise on adhesion and adhesives, vol 2. Dekker, New York, pp 219–260

DaSilva MB, Pinto-da-Rocha R (2010) Systematic review and cladistic analysis of the Hernandariinae (Opiliones: Gonyleptidae). Zoologia (Curitiba) 27(4):577–642

Dickinson GH, Vega IE, Wahl KJ, Orihuela B, Beyley V, Rodriguez EN, Everett RK, Bonaventura J, Rittschof D (2009) Barnacle cement: a polymerization model based on evolutionary concepts. J Exp Biol 212(21):3499–3510

Duncan RP, Autumn K, Binford GJ (2007) Convergent setal morphology in sand-covering spiders suggests a design principle for particle capture. Proc R Soc B Biol Sci 274(1629):3049–3057

Edmonds DT, Vollrath F (1992) The contribution of atmospheric water vapour to the formation and efficiency of a spider's capture web. Proc R Soc Lond B Biol Sci 248(1322):145–148

Giribet G, Sharma PP (2015) Evolutionary Biology of Harvestmen (Arachnida, Opiliones). Annu Rev Entomol 60:157–175

Gnaspini P, Hara MR (2007) Defense mechanisms. In: Pinto-da-Rocha R, Machado G, Giribet G (eds) Harvestmen: the biology of Opiliones. Harvard University Press, Cambridge, pp 374–399

Goodnight ML, Goodnight CJ (1976) Observations on the systematics, development, and habits of *Erginulus clavotibialis* (Opiliones: Cosmetidae). Trans Am Microsc Soc 95:654–664

Gosline JM, Denny MW, DeMont ME (1984) Spider silk as rubber. Nature 309(5968):551–552

Groh S, Giribet G (2014) Polyphyly of Caddoidea, reinstatement of the family Acropsopilionidae in Dyspnoi, and a revised classification system of Palpatores (Arachnida, Opiliones). Cladistics 31(3):277–290

Gruber J (1993) Beobachtungen zur Ökologie und Biologie von *Dicranolasma scabrum* (HERBST) (Arachnida: Opiliones) Teil I. Ann Naturhist Mus Wien 94:393–426

Hedin M, Tsurusaki N, Macías-Ordóñez R, Shultz JW (2012) Molecular systematics of sclerosomatid harvestmen (Opiliones, Phalangioidea, Sclerosomatidae): geography is better than taxonomy in predicting phylogeny. Mol Phylogenet Evol 62(1):224–236

Helbig R, Nickerl J, Neinhuis C, Werner C (2011) Smart skin patterns protect springtails. PLoS One 6(9), e25105

Hennebert E, Viville P, Lazzaroni R, Flammang P (2008) Micro-and nanostructure of the adhesive material secreted by the tube feet of the sea star *Asterias rubens*. J Struct Biol 164(1):108–118

Holmberg RG (1986) The scent glands of Opiliones: a review of their function. In: Eberhard WG, Lubin YD, Robinson BC (eds) Proceedings of the ninth international congress of arachnology, Panama, 1983. Smithsonian Institution Press, pp 131–133

Huang Y, Wang Y, Sun L, Agrawal R, Zhang M (2015) Sundew adhesive: a naturally occurring hydrogel. J R Soc Interface 12:20150226. doi:10.1098/rsif.2015.0226

Immel V (1955) Einige Bemerkungen zur Biologie von *Platybunus bucephalus* (Opiliones, Eupnoi). Zool Jahrb Abt Syst 83:475–484

Juberthie C (1964) Recherches sur la biologie des Opilions. Ann Spéléol 19:5–238

Juberthie C (1972) Reproduction et développement d'un opilion Cosmetidae, *Cynorta cubana* (Banks) de Cuba. Ann Spéléol 27:773–785

Kalliadasis S, Chang H-C (1994) Drop formation during coating of vertical fibres. J Fluid Mech 261:135–168

Karaman IM (2005) Evidence of spermatophores in Cyphophthalmi (Arachnida, Opiliones). Rev Suisse Zool 112:3–11

Kesel A, Martin A, Seidl T (2003) Adhesion measurements on the attachment devices of the jumping spider Evarcha arcuata. J Exp Biol 206(16):2733–2738

Kinloch AJ (1987) Adhesion and adhesives: science and technology. Springer Science & Business Media, The Netherlands, 441 pp

Kropf C, Bauer D, Schläppi T, Jacob A (2012) An organic coating keeps orb-weaving spiders (Araneae, Araneoidea, Araneidae) from sticking to their own capture threads. J Zool Syst Evol Res 50(1):14–18

Kury A (2012) A synopsis of catalogs and checklists of harvestmen (Arachnida, Opiliones). Zootaxa 3184:35–58

Machado G, Raimundo RLG (2001) Parental investment and the evolution of subsocial behaviour in harvestmen (Arachnida Opiliones). Ethol Ecol Evol 13:133–150

Machado G, Bonato V, Oliveira PS (2002) Alarm communication: a new function for the scent-gland secretion in harvestmen (Arachnida: Opiliones). Naturwissenschaften 89(8):357–360

Machado G, Requena GS, Buzatto BA, Osses F, Rossetto LM (2004) Five new cases of paternal care in harvestmen (Arachnida: Opiliones): implications for the evolution of male guarding in the Neotropical family Gonyleptidae. Sociobiology 44:577–598

Machado G, Carrera PC, Pomini AM, Marsaioli AJ (2005) Chemical defense in harvestmen (Arachnida, Opiliones): do benzoquinone secretions deter invertebrate and vertebrate predators? J Chem Ecol 31(11):2519–2539

Machado G, Pinto-da-Rocha R, Giribet G (2007) What are harvestmen? In: Pinto-da-Rocha R, Machado G, Giribet G (eds) Harvestmen: the biology of Opiliones. Harvard University Press, Cambridge, MA, pp 1–13

Martens J (1993) Further cases of paternal care in Opiliones (Arachnida). Trop Zool 6:97–107

Miyoshi Y (1942) Post-embryonic development of *Ischyropsalis abei* Sato et Suzuki (orig.: 三好保徳: サスマタアゴザトウムシ *Ischyropsalis abei* Sato et Suzuki の生長に伴ふ形態の變化). Acta Arachnol 7(3–4):109–120

Mizutani K, Egashira K, Toukai T, Ogushi J (2006) Adhesive force of a spider mite, *Tetranychus urticae*, to a flat smooth surface. JSME Int J Ser C 49(2):539–544

Nachtigall W (1974) Biological mechanisms of attachment. Springer, Berlin, Heidelberg, New York

Naldrett MJ (1993) The importance of sulphur cross-links and hydrophobic interactions in the polymerization of barnacle cement. J Mar Biol Assoc UK 73(03):689–702
Novak T (2005) Notes on spermatophores in *Cyphopthalmus duricorius* Joseph (Arachnida: Opiliones:Sironidae). Ser Hist Nat 15:277–280
Peisker H, Michels J, Gorb SN (2013) Evidence for a material gradient in the adhesive tarsal setae of the ladybird beetle *Coccinella septempunctata*. Nat Comm 4:1661
Phillipson J (1960) A contribution to the feeding biology of *Mitopus morio* (F) (Phalangida). J Anim Ecol 29(1):35–43
Pinto-da-Rocha R, Bragagnolo C, Marques FPL, Antunes J (2014) Phylogeny of harvestmen family Gonyleptidae inferred from a multilocus approach (Arachnida: Opiliones). Cladistics 30:519–539
Politi Y, Priewasser M, Pippel E, Zaslansky P, Hartmann J, Siegel S, Li C, Barth FG, Fratzl P (2012) A Spider's Fang: how to design an injection needle using chitin-based composite material. Adv Funct Mater 22(12):2519–2528
Ramírez PA, González A, Botto-Mahan C (2013) Masking behavior by *Mepraia spinolai* (Hemiptera: Reduviidae): anti-predator defense and life history trade-offs. J Insect Behav 26 (4):592–602
Ramírez MJ, Grismado CJ, Labarque FM, Izquierdo MA, Ledford JM, Miller JA, Haddad CR, Griswold CE (2014) The morphology and relationships of the walking mud spiders of the genus *Cryptothele* (Araneae: Zodariidae). Zool Anz 253(5):382–393
Raspotnig G (2012) Scent gland chemistry and chemosystematics in harvestmen. Biol Serbica 34 (1–2):5–18
Rayleigh Lord SRS (1878) On the instability of jets. Proc Lond Math Soc 10:4–13
Rayleigh Lord SRS (1892) XIX. On the instability of cylindrical fluid surfaces. Philos Mag Ser 5 34(207):177–180
Requena GS, Buzatto BA, Munguia-Steyer R, Machado G (2009) Efficiency of uniparental male and female care against egg predators in two closely related syntopic harvestmen. Anim Behav 78:1169–1176
Rimsky-Korsakow AP (1924) Die Kugelhaare von *Nemastoma lugubre*, Müll. Zool Anz 60:1–16
Rodriguez AL, Townsend VR, Johnson MB, White TB (2014) Interspecific variation in the microanatomy of cosmetid harvestmen (Arachnida, Opiliones, Laniatores). J Morphol 275 (12):1386–1405
Roewer CF (1923) Die Weberknechte der Erde. Gustav Fischer, Jena
Rühm J (1926) Über die Nahrung von *Phalangium* L. Zool Anz 68:154–158
Sahni V, Labhasetwar DV, Dhinojwala A (2011) Spider silk inspired functional microthreads. Langmuir 28(4):2206–2210
Santos FH, Gnaspini P (2002) Notes on the foraging behavior of the Brazilian cave harvestman *Goniosoma spelaeum* (Opiliones, Gonyleptidae). J Arachnol 30(1):177–180
Schaider M, Raspotnig G (2009) Unusual organization of scent glands in *Trogulus tricarinatus* (Opiliones, Trogulidae): Evidence for a non-defensive role. J Arachnol 37(1):78–83
Schönhofer AL (2013) A taxonomic catalogue of the Dyspnoi Hansen and Sørensen, 1904 (Arachnida: Opiliones). Zootaxa 3679(1):1–68
Schwangart F (1907) Beiträge zur Morphologie und Systematik der Opilioniden: 1. Über das Integument der Troguloidae. Zool Anz 31:161–183
Sharma PP, Giribet G (2011) The evolutionary and biogeographic history of the armoured harvestmen–Laniatores phylogeny based on ten molecular markers, with the description of two new families of Opiliones (Arachnida). Invertebr Syst 25(2):106–142
Shultz JW, Regier JC (2001) Phylogenetic analysis of Phalangida (Arachnida, Opiliones) using two nuclear protein-encoding genes supports monophyly of Palpatores. J Arachnol 29 (2):189–200
Silverman HG, Roberto FF (2007) Understanding marine mussel adhesion. Mar Biotechnol 9 (6):661–681

Stanley E (2011) Egg hiding in four harvestman species from Uruguay (Opiliones: Gonyleptidae). J Arachnol 39:495–496

Torres FG, Troncoso OP, Cavalie F (2014) Physical characterization of the liquid adhesive from orb-weaving spiders. Mater Sci Eng C Mater Biol Appl 34:341–344

Verhoef H, Witteveen J (1980) Water balance in Collembola and its relation to habitat selection; cuticular water loss and water uptake. J Insect Physiol 26(3):201–208

Voigt D, Gorb SN (2010) Egg attachment of the asparagus beetle *Crioceris asparagi* to the crystalline waxy surface of *Asparagus officinalis*. Proc R Soc B 277(1683):895–903

Wachmann E (1970) Der Feinbau der sog. Kugelhaare der Fadenkanker (Opiliones, Nemastomatidae). Z Zellforsch 103:518–525

Willemart RH (2001) Egg covering behavior of the neotropical harvestman *Promitobates ornatus* (Opiliones, Gonyleptidae). J Arachnol 28:249–252

Witaliński W, Žuwała K (1981) Ultrastructural studies of egg envelopes in harvestmen (Chelicerata, Opiliones). Int J Inver Rep Dev 4(2):95–106

Wolff JO, Schönhofer AL, Schaber CF, Gorb SN (2014) Gluing the 'unwettable': soil-dwelling harvestmen use viscoelastic fluids for capturing springtails. J Exp Biol 217(19):3535–3544

Wolff JO, Grawe I, Wirth M, Karstedt A, Gorb SN (2015a) Spider's super-glue: thread anchors are composite adhesives with synergistic hierarchical organization. Soft Matter 11(12):2394–2403

Wolff JO, Seiter M, Gorb SN (2015b) Functional anatomy of the pretarsus in whip spiders (Arachnida, Amblypygi). Arthropod Struct Dev 44(6):524–540

Wolff JO, Schönhofer AL, Martens J, Wijnhoven H, Taylor CK, Gorb SN (2016) The evolution of pedipalps and glandular hairs as predatory devices in harvestmen (Arachnida, Opiliones). Zool J Linn Soc 177(3):558–601

Wool RP, Bunker SP (2007) Polymer-solid interface connectivity and adhesion: Design of a bio-based pressure sensitive adhesive. J Adhes 83(10):907–926

# Chapter 13
# Unraveling the Design Principles of Black Widow's Gumfoot Glue

**Dharamdeep Jain, Todd A. Blackledge, Toshikazu Miyoshi, and Ali Dhinojwala**

**Abstract** Prey capture adhesives produced by web-building spiders have intrigued humans for many years and provide important insights to develop adhesives that work in humid environments. These humidity-responsive glues are laid down by spiders in various types of webs, primarily orb webs and cobwebs. The formation and function of viscid glue in the capture spirals of orb webs is well-studied compared to the vertically aligned gumfoot glue strands in cobwebs. While the glue droplets in cobwebs contain some peptides, they act as viscoelastic liquids, rather than viscoelastic solids, and the cause of glue stickiness is poorly understood. However, the recent discovery of glycoproteins and hygroscopic salts in the gumfoot adhesives brings a new perspective to explain the mechanism of adhesion of these microscopic droplets. In this chapter, we summarize the current state of our understanding of the chemical composition, morphology, and mechanism of adhesion of gumfoot glue threads. Additionally, we present molecular evidence that both salts and glycoproteins are important for strong adhesion in a humid environment and show how understanding the mechanism of cobweb spider adhesives will help in designing materials that are active and functional in high humidity.

## 13.1 Introduction

Webs are an integral part of any spiders' life cycles and serve multiple functions such as shelter, prey capture, reproduction, and predator defense (Foelix 1982; Blackledge et al. 2009; Sahni et al. 2011a). These silk-laden constructs evolved from ancestral sheet structures on the ground consisting of dry silks that might

D. Jain • T. Miyoshi • A. Dhinojwala (✉)
Department of Polymer Science, University of Akron, Akron, OH 44325-3909, USA
e-mail: ali4@uakron.edu

T.A. Blackledge
Department of Biology, Integrated Biosciences Program, University of Akron, Akron, OH 44325-3908, USA

© Springer International Publishing Switzerland 2016 
A.M. Smith (ed.), *Biological Adhesives*, DOI 10.1007/978-3-319-46082-6_13

**Fig. 13.1** Adult female western black widow spider (*Latrodectus hesperus*)

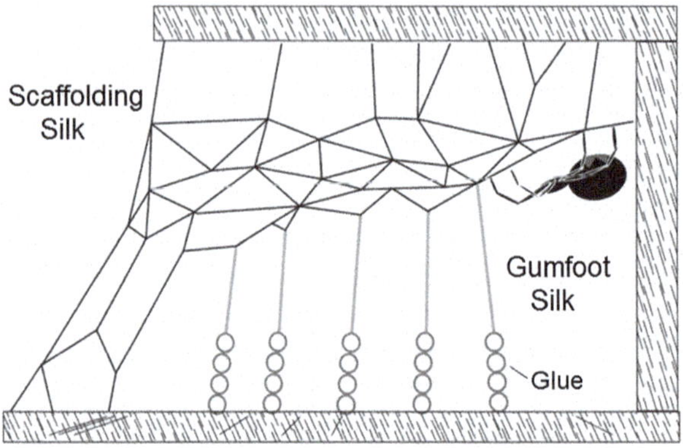


**Fig. 13.2** Model structure of the three-dimensional cobweb built by *Latrodectus* spiders depicting the two regions: dry scaffolding silk (*upper part*) and gumfoot silk (*lower part*) containing glue droplets

physically entangle prey into highly complex orb web and cobweb structures coated with sticky capture silk from the aggregate gland (Vollrath and Selden 2007; Blackledge et al. 2009). The evolution of three-dimensional cobweb structures from more ancient two-dimensional orb web architectures was a major event in the evolution of spider diversity that is associated with an increase in the number and diversity of spider species (Griswold et al. 1998) and enhanced protection against predators (Blackledge et al. 2003). The evolution of cobwebs also resulted in novel uses of aggregate glue silk (Griswold et al. 1998; Blackledge et al. 2003, 2005a, 2009; Blackledge and Zevenbergen 2007).

Cobwebs are constructed by the spider family Theridiidae that includes the famous widow spiders in the genus *Latrodectus* (Fig. 13.1) (Chamberlin and Ivie 1935; d'Amour et al. 1936; Garb et al. 2004). While the architectures of cobwebs are diverse across species and often behaviorally plastic (Eberhard et al. 2008), black widows construct webs with two distinct regions (Fig. 13.2) (Benjamin and Zschokke 2003; Argintean et al. 2006; Blackledge et al. 2005a, b; Sahni et al. 2011a). The upper part consists of numerous dry major ampullate (MA) silk fibers, termed "scaffolding silk," that form a sheet structure suspended by supporting threads in a maze-like geometry. The lower part known consists of vertical MA silk, termed "gumfoot silk," coated with ~100 μm diameter adhesive glue droplets extending 0.5–2 cm up from the threads' attachment at the base of the web (Fig. 13.3). Scaffolding silk mechanically supports the web, knocks down

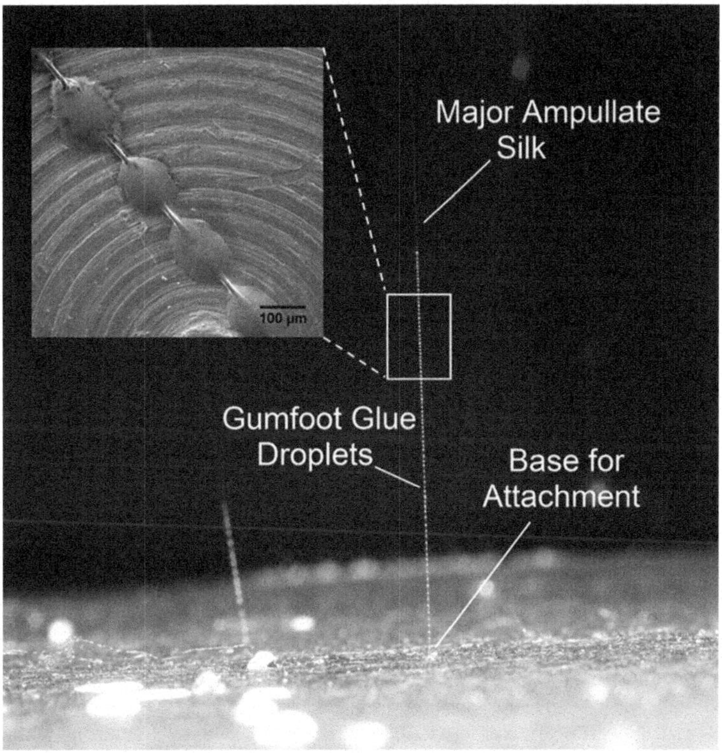


**Fig. 13.3** Arrangement of gumfoot glue. The real-time image depicts the structural arrangement of gumfoot silk in a cobweb. The strands are attached to the base and comprise of axial major ampullate silk thread coated with glue droplets. The *inset* is a SEM micrograph of a strand of gumfoot silk

flying prey, and provides an avenue on which the spider maneuvers. Gumfoot silk threads capture walking prey when the glue droplets adhere to the prey and the threads detach from their connection to the substrate—tension in the threads then pulls struggling prey up into the air (Benjamin and Zschokke 2003; Argintean et al. 2006; Blackledge et al. 2005a, b; Sahni et al. 2011a, 2012).

Black widow spiders are found in many different habitats (Chamberlin and Ivie 1935; Garb et al. 2004) and are popular species to study because of the toxicity of their venom. Widow venom generated by them contains a neurotoxin called α-latrotoxin that is active against vertebrate nervous systems and therefore holds biomedical potential (Finkelstein et al. 1976; Frontali et al. 1976; Meldolesi et al. 1986; Ushkaryov et al. 2004). Widow spiders are also model species for the investigation of silk because they are easy to work with in the laboratory, the chemistry and structure of many of their silks are well-characterized, and their silk gene expression is described at the genomic and transcriptomic levels (Casem et al. 1999; Moore and Tran 1999; Lawrence et al. 2004; Hu et al. 2005a, b, 2007; Vasanthavada et al. 2007, 2012; Blasingame et al. 2009; Sahni et al. 2011b, 2012;

Jenkins et al. 2013; Xu et al. 2014; Jain et al. 2015). Most of this research focuses on MA silk, known for its excellent mechanical properties (Blackledge et al. 2005a, b; Swanson et al. 2006), with the goal of designing high-toughness synthetic fibers. The chemistry and mechanism of adhesion of the gumfoot silk has only been a focus of investigation recently (Hu et al. 2007; Sahni et al. 2011b; Jain et al. 2015), but widow spider glue shows unusual performance properties compared to better-studied orb web spiders, and hold promise for application to synthetic adhesives.

## 13.2 Gumfoot Glue

### *13.2.1 Secretion Mechanism*

The aggregate glands produce viscid glues in spiders (Foelix 1982). There are two pairs of such glands in the cobweb spiders: typical and atypical. The sticky secretions on the silk strands are produced from the anterior pair of nodular "typical" aggregate glands, which resemble the set of similar glands in orb web spiders (Kovoor 1977; Kelly 1989; Townley and Tillinghast 2013). During cobweb construction, the glue is secreted from the typical gland after the spider attaches the first MA thread of the gumfoot thread to the substrate and begins moving back up to the sheet while spinning a second MA thread (Blackledge et al. 2005a, b; Sahni et al. 2011a). The other "atypical" aggregate gland is used in defenses and prey capture, producing a secretion that is combed out by setae on the tarsi of the fourth legs and subsequently hardens over a few seconds (Kelly 1989; Townley and Tillinghast 2013). Our focus here is specifically on the glue produced by the typical aggregate gland.

### *13.2.2 Material Properties*

#### 13.2.2.1 Viscoelasticity

The glycoproteins in viscid glue act as viscoelastic adhesives, and this makes a key contribution to how spider webs capture prey (Sahni et al. 2010, 2011b). In the case of orb webs, the glycoproteins act as viscoelastic solids—keeping prey trapped for longer times at slow extension rates while producing stronger adhesion forces at the fast extension rates generated by the impact of fast-moving insects (Sahni et al. 2010). Gumfoot glue has a strikingly different behavior because the glue droplets behave as viscoelastic liquids (Fig. 13.4). When these glue droplets are subjected to a constant deformation, the stress relaxes nearly to zero, indicating significantly reduced cross-linking in the gumfoot glue droplets and the absence of an anchoring core or granule in the droplet structure. The viscoelastic liquid nature

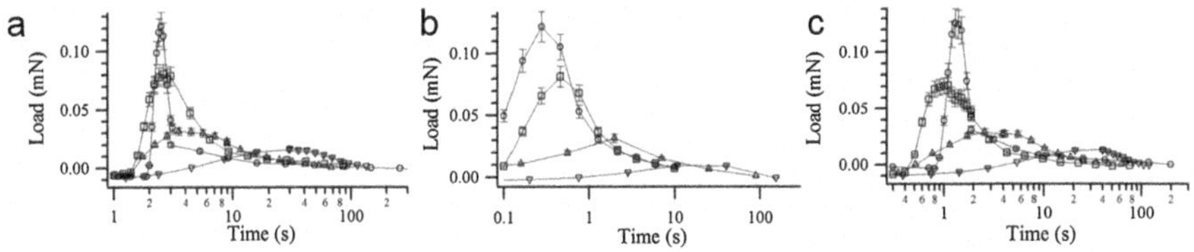


**Fig. 13.4** Viscoelastic nature of gumfoot glue. (**a–c**) are the load relaxation plots of single gumfoot glue droplet stretched by a constant length at 15 % RH, 40 % RH, and 90 % RH, respectively. The different symbols represent pull off rates: 1 μm/s (*inverted triangles*), 10 μm/s (*upright triangles*), 50 μm/s (*squares*), and 100 μm/s (*circles*). The stress in the droplets plateaus out near zero indicating the absence of cross-linking and establishes the viscoelastic liquid nature of the droplets (Reproduced from Sahni, V, Blackledge, TA, Dhinojwala A (2011b) Changes in the Adhesive Properties of Spider Aggregate Glue During the Evolution of Cobwebs. Sci Rep 1:41)

of gumfoot glue is further supported by observations of the droplets coalescing, sliding, or spreading over the fiber at high relative humidity (Sahni et al. 2011b).

#### 13.2.2.2 Chemical Composition

Gumfoot glue is a mixture of several components in an aqueous solution. The glue has both water-soluble (~56 %) and water-insoluble (~44 %) fractions (Jain et al. 2015) as depicted in Fig. 13.5a. Water-soluble components consist of peptides (Hu et al. 2007; Jain et al. 2015) and hygroscopic salts (Townley and Tillinghast 2013; Jain et al. 2015), while glycoproteins (Jain et al. 2015) constitute the water-insoluble component. Details about each constituent are discussed in the following subsections.

##### Hygroscopic Salts

Salts present in capture silk are often termed low molecular mass (LMM) components (Townley and Tillinghast 2013). Broadly classified as organic and inorganic, these salts are water-soluble hygroscopic compounds that take up moisture from the atmosphere and help in making the glue tacky enough to stick (Vollrath et al. 1990; Townley et al. 1991, 2012; Townley and Tillinghast 2013). Organic salts (~60 wt. % of water-soluble mass) are the major component in capture glue and consist of amino acid-like compounds such as choline, glycine, betaine, alanine, and GABamide (Townley and Tillinghast 2013). Inorganic salts form 10–20 wt. % of water-soluble mass and include potassium, sodium, nitrate, and dihydrogen phosphate moieties. Viscid glue droplets show diverse variation in the composition of these compounds across species (Vollrath et al. 1990; Townley et al. 1991, 2012; Townley and Tillinghast 2013).

Many of these compounds are also found in cobweb glues. Atypical secretions from *Latrodectus mactans* and *Latrodectus hesperus* contain GABA, GABamide,

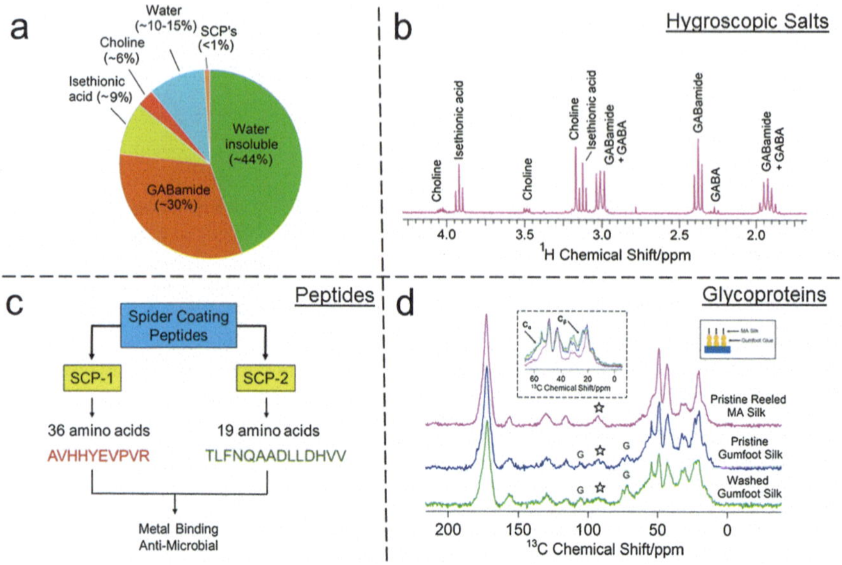


**Fig. 13.5** Chemical composition of gumfoot glue. The figure is a representation of major components in gumfoot glue. (**a**) The overall summary of components in the form of a pie chart. (**b**) The $^{1}$H NMR solution-state NMR for water-soluble extract highlighting the presence of hygroscopic salts. (**c**) The types of peptides present in the glue, and (**d**) depicts the solid-state NMR evidence for the existence of glycoproteins (G) in gumfoot glue (Figures **b** and **d** reprinted with permission from Jain D, Zhang C, Cool LR, Blackledge TA, Wesdemiotis C, Miyoshi T, Dhinojwala A (2015) Composition and Function of Spider Glues Maintained During the Evolution of Cobwebs. Biomacromolecules 16:3373–80. Copyright (2015) American Chemical Society)

choline, proline, glycine, and isethionic acid (Kelly 1989; Townley and Tillinghast 2013). Recent detailed efforts in understanding salt compositions in gumfoot glue present in the webs of *Latrodectus hesperus* using solution-state NMR (Fig. 13.5b) show the presence of three distinct salts: GABamide (~30 %), isethionic acid (~9 %), and choline (~6 %), in addition to small traces of GABA (Jain et al. 2015). Water uptake and interaction with proteins to facilitate adhesion are likely the primary function of these salts, but they may also impart wettability to the prey cuticle and act as toxic compound antimicrobials in the glue droplets (Townley et al. 2012; Townley and Tillinghast 2013).

### Water-Soluble Peptides

Spider coating peptides (SCPs) are present in the water-soluble extracts from the gumfoot glue (Hu et al. 2007; Jain et al. 2015). The peptides are classified as SCP-1 (36 amino acids) and SCP-2 (19 amino acids) (Hu et al. 2007) (Fig. 13.5c) and are present in small quantities (less than 1 %) compared to the hygroscopic salts in the water-soluble fraction (Jain et al. 2015). While SCPs do not appear to be in the silk protein family, detailed analysis of the peptide sequences using MALDI-ToF revealed sequences, AVHHYEVPVR and TLFNQAADLLDHVV in SCP-1 and SCP-2, respectively, that have also been detected in the water-soluble fractions of egg case silks and scaffolding connection (pyriform) joints. The roles of these peptides in the glue are not clear, although SCP-1 exhibits metal-ion binding capabilities. Other hypothesized roles include participating in oxidation–reduction reactions, acting as antimicrobial agents and affecting the conformation of proteins (Hu et al. 2007). Although SCPs are present in small quantities in the gumfoot glue (Jain et al. 2015), it is also possible that they influence the adhesion of the glue droplets.

Glycoproteins

Proteinaceous compounds have been detected in the combed secretions of *Latrodectus* spiders (Kovoor 1977, 1987; Kelly 1989; Townley and Tillinghast 2013). However, they lack the visible core seen in the centers of adhesive glycoproteins in orb-weaver glue droplets. The adhesive proteins were only recently detected in large quantities in the gumfoot glue droplets of *Latrodectus hesperus* (Jain et al. 2015). An insoluble residue was clearly visible after washing the salts from the glue droplets with water, and the presence of glycoproteins was confirmed by using histochemical and spectroscopy analysis. Both gumfoot glue and a positive control, viscid glue from orb web of *Larinioides cornutus*, showed strong magenta color with periodic acid-Schiff (PAS) stain, indicating glycoproteins. In addition, $^{13}$C CP/MAS NMR (cross polarization/magic-angle spinning nuclear magnetic resonance spectroscopy) spectra of pristine gumfoot silk and washed gumfoot silk showed the presence of glycoprotein peaks around 75 ppm and 105 ppm that were absent in a negative control—major ampullate silk (Fig. 13.5d) (Jain et al. 2015). These results support previous histochemical evidence of glycoproteins in the typical aggregate glands of *Latrodectus* spiders (Kovoor 1977, 1987). The discovery of glycoproteins in gumfoot glue clears up the confusion about how these droplets stick. However, the chemical nature of the glycoprotein, which appears to be structurally and visually different from its viscid silk counterpart, stills need to be investigated.

Lipids

Lipid-based molecules such as fatty acids and long-chain methyl esters (1-methoxy alkanes) are in the nonpolar solvent washings of webs and in glue droplets of some orb web spiders, including *Nephila clavipes* (Schulz 2001; Salles et al. 2006). Their origin from the aggregate glands is still not clear and they seem to be absent in other orb species (Schulz 1997, 1999). Similar compounds like 1-methoxy alkanes have been detected in cobwebs of *Latrodectus*, but again their presence in gumfoot droplets is debatable due to lack of evidence in the aggregate glands (Schulz

1999). Nevertheless, it is suggested that lipids in the glue droplets may have antimicrobial properties (Schulz 1997, 2001; Salles et al. 2006) and repel ants (Salles et al. 2006). In addition, lipids may increase the spreading of glue on hydrophobic prey epicuticles and make prey more susceptible to toxic glue compounds by increasing cuticle permeability (Salles et al. 2006; Townley and Tillinghast 2013).

### 13.2.3 Humidity-Mediated Adhesion

The adhesion of spider glue droplets is humidity responsive (Sahni et al. 2010, 2011b; Opell et al. 2011, 2013; Amarpuri et al. 2015) due to the presence of hygroscopic salts (Vollrath et al. 1990; Townley et al. 1991, 2012; Townley and Tillinghast 2013). The interaction of the glue with water influences key adhesion parameters such as viscosity and droplet extensibility so that they adhere better as humidity increases. Glue viscosity decreases as humidity increases, making the droplets stretch further before they detach from the surface. This increase the work done in peeling until the glue modulus decreases at some very high humidity that little work is done during peeling (Opell et al. 2011; Amarpuri et al. 2015). In overly dry conditions, droplets behave as stiff or rigid materials and are unable to spread and make firm adhesive contact. The volume of the glue droplet also changes with humidity (Sahni et al. 2011b; Opell et al. 2013). With increase in humidity, the droplets absorb water and the volume of the glue droplet increases, leading to better spreading and contact on the surface. However, over lubrication in some cases can cause adhesion loss.

The humidity response of cobweb glue droplets is notably different from orb weavers. With increase in humidity, droplets of gumfoot glue, unlike viscid glue, coalesce and form bigger droplets that can flow/spread across the axial dragline fiber, but the overall volume remains largely constant (Fig. 13.6a, insets a-d). The lack of a visible proteinaceous core and the presence of less viscous glycoproteins, compared to orb spiders' viscid glue, likely enable the droplets to coalesce together. Cobweb glue adhesion is also largely invariant to humidity. The effect of humidity on adhesion of gumfoot glue has been tested in two different ways: single drop adhesion (Sahni et al. 2011b) and whole thread adhesion (Jain et al. 2015). Single droplet adhesion showed both adhesion force and adhesion energy to be constant with humidity (Fig. 13.6b), in contrast to orb spiders' glues (Sahni et al. 2011b). The whole thread adhesion results (Fig. 13.7) were similar in that adhesion did not depend on humidity above 30 % RH (relative humidity). However, at 10 % RH, the glue behaves solid-like and fails to make good adhesive contact with the surface (Fig. 13.7a). Like in orb spiders, the salts take up water as humidity increases and make the glue stick, but this occurs at very low RH. After washing the threads with water to remove the salts, the adhesion was weak at all humidities, indicating that in the absence of salts, glycoproteins alone cannot absorb sufficient water (Fig. 13.7b) (Jain et al. 2015). Thus cobweb spider glue still utilizes salts to uptake water from

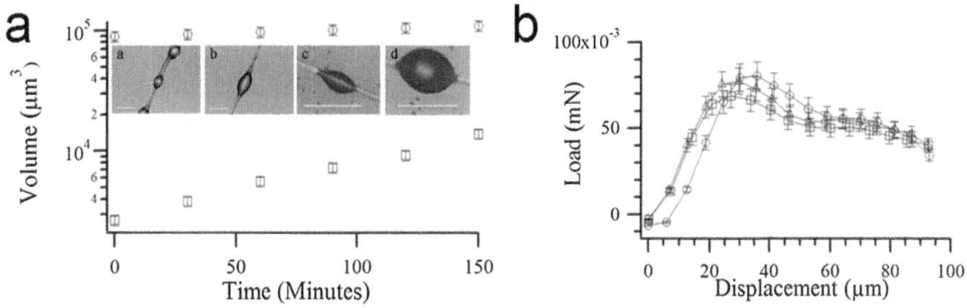


**Fig. 13.6** Effect of humidity on volume and single glue droplet adhesion. (**a**) The comparison between gumfoot glue (*circles*) and viscid glue (*squares*) droplets for change in volume with time under high humidity. Insets a,b depicts the gumfoot glue droplets and c,d depicts the viscid glue droplet conditioned at 0 % RH (a,c) and 100 % RH (b,d). Scale bar is 100 μm. (**b**) The single droplet load vs. extension at 50 μm/s at 15 % RH (*circles*), 40 % RH (*squares*), and 90 % RH (*triangles*). The adhesion force is constant over different humidity conditions (Reproduced from Sahni, V, Blackledge, TA, Dhinojwala A (2011b) Changes in the Adhesive Properties of Spider Aggregate Glue During the Evolution of Cobwebs. Sci Rep 1:41)

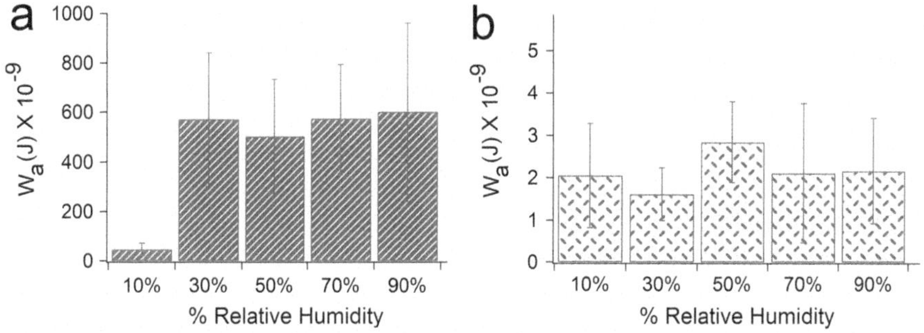


**Fig. 13.7** Whole thread adhesion. The work of adhesion of pristine (**a**) and washed (**b**) gumfoot silk threads, respectively, pulled off from glass substrates at various humidity conditions (Reprinted with permission from Jain D, Zhang C, Cool LR, Blackledge TA, Wesdemiotis C, Miyoshi T, Dhinojwala A (2015) Composition and Function of Spider Glues Maintained During the Evolution of Cobwebs. Biomacromolecules 16:3373–80. Copyright (2015) American Chemical Society)

the air, but its adhesion is much more broadly tuned compared to orb spiders, where individual species maximize their adhesion at very different, relatively narrow windows of humidity.

The functional implications for cobwebs of the independence of the adhesive strength of gumfoot glue to relative humidity from 30 % to 90 % RH is unknown, but the behavior is strikingly different from orb spiders. *Latrodectus* are found in many habitats across many different climatic zones (Garb et al. 2004) so broadly tuned glue may preserve adhesion in varying humidity conditions in which they reside for efficient prey capture (Sahni et al. 2011b). However, many orb spiders also need to cope with variable habitats. Instead, the key may lie in the differences in web architectures and silk compositions. The flagelliform silk fibers in the core of viscid threads of orb weavers are highly extensible and rubbery in part because they are supercontracted by the water in the glue due to the fibers' high proline content (Liu et al. 2008). The major ampullate silk in a gumfoot thread has very little proline (Liu et al. 2008) and is instead 47 wt. % of β-sheet domains (alanine, glycine, and serine) (Jenkins et al. 2013). This could prevent the "local supercontraction" in the gumfooted lines and maintain the integrity of the capture region of the cobweb (Sahni et al. 2011b). The lack of local supercontraction may also relate to the glue droplets present on the silk taking in less water, leading to the weak response towards adhesion in humidity environments. These ideas are largely speculations, so that the question remains why did gumfoot glue evolve to become relatively independent of RH when orb-weaver viscid glues optimize stickiness at very particular humidities?

## *13.2.4 Molecular Mechanism of Adhesion*

As described in Sects. 13.2.2 and 13.2.3, the major components of gumfoot glue are organic salts and glycoproteins, and the presence of both is critical in adhesion in presence of humidity. Here, we present novel molecular evidence that the organic salts in the glue absorb water and become mobile with increase in RH and affect the hydration behavior of the glue proteins. We directly measure the molecular level mobility of the gumfoot silk strands produced by the western black widow *Latrodectus hesperus* using solid-state nuclear magnetic spectroscopy (method details described in Sahni et al. 2014; Jain et al. 2015). CP/MAS and direct polarization/magic-angle spinning (DP/MAS) were performed on pristine and washed gumfoot silks exposed to different humidity environments (10 % RH, 60 % RH, and 90 % RH) to understand the dynamic behavior of the glue constituents (salts and glycoproteins) in the presence of humidity.

### 13.2.4.1 Detecting Gumfoot Glue Peaks

The samples used for the solid-state NMR analysis were gumfoot silk strands that combined both the axial major ampullate thread and the gumfoot glue. To identify the peaks specifically related to glue (salts and glycoproteins), we measured the $^{13}C$ CP/MAS spectra for forcibly reeled major ampullate silk threads as a control and compared it with the pristine gumfoot silk spectrum (Fig. 13.8). The peaks corresponding to the major ampullate silk match with the earlier reported assignments (Creager et al. 2010; Jenkins et al. 2013). Pristine gumfoot silk spectra, apart from the major ampullate peaks, show additional peaks related to the organic salts, specifically at 23 ppm (GABamide, choline), 33 ppm (GABamide), 40 ppm (GABamide), 53 ppm (isethionic acid, choline), and 55–58 ppm (isethionic acid, choline). These peak assignments were confirmed by studying the $^{13}C$ solution-state NMR spectra for pure organic salts (results not shown). Glycoproteins in the gumfoot glue have been detected previously (Jain et al. 2015) and described in Sect. 13.2.2.2. Additionally, as mentioned in Sect. 13.2.2.2, salts form a major

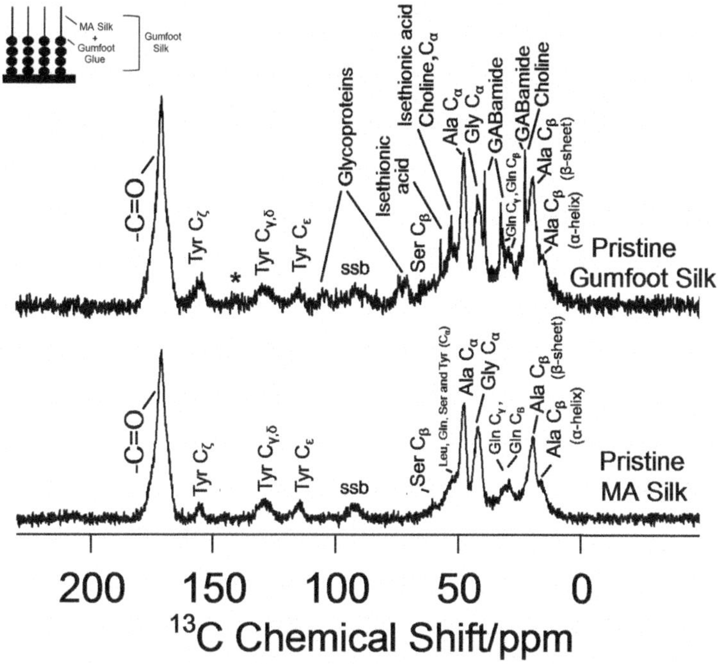


**Fig. 13.8** Identifying components of gumfoot silk. The comparative CP/MAS spectra of pristine gumfoot silk and major ampullate silk. Salts and glycoproteins comprising the glue are visible in gumfoot silk spectrum and absent in control MA silk spectrum. *Inset* depicts the gumfoot silk arrangement. All spectra were recorded at MAS frequency of 6000 Hz, 60 % RH, and 25 °C. Starred peak is unidentifiable component and ssb refers to the spinning sideband

component of the gumfoot glue; the observations of similar salt peaks (GABamide, isethionic acid, and choline) in the solid-state NMR analysis confirm that these peaks must originate from the gumfoot glue.

#### 13.2.4.2 Effect of Salts on the Humidification of Glue Proteins

Figure 13.9 shows the CP/MAS spectra for pristine (Fig. 13.9a) and washed (Fig. 13.9b) gumfoot silk at different RH (10 %, 60 %, and 90 %). Poor signal to noise ratio (due to lack of $^{13}C$ labeled and small sample) limits our efficiency in glycoprotein analysis. However, in the case of pristine silk, increased humidity softens the silk. The salt peaks (GABamide, isethionic acid, and choline) are visible at 10 % RH and 60 % RH, but they become lower in intensity/disappear at higher RH (90 %), indicating higher mobility. However, the glycoprotein peaks at 75 ppm and 105 ppm do not show a dramatic change in intensity with as humidity increases, in contrast to previous observations with viscid glues (Sahni et al. 2014). The reason behind this behavior is not clear, but this observation may explain why gumfoot glues are largely inert to humidity. Other protein moieties in the gumfoot silk, such

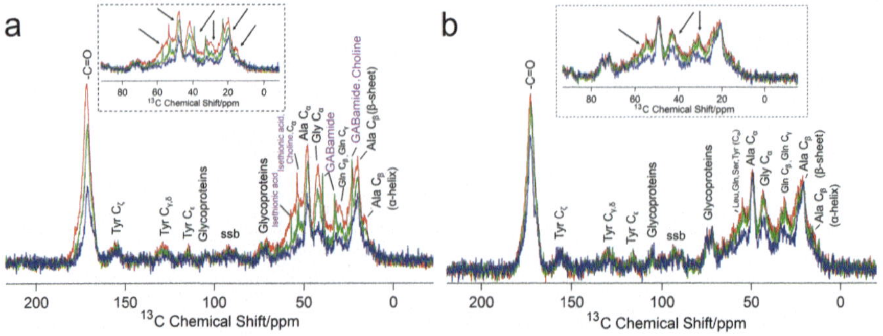


**Fig. 13.9** Role of salts in humidification of glue proteins. (**a**, **b**) The CP/MAS spectra of pristine and washed gumfoot silk spectra, respectively, at 10 % RH (*red*), 60 % RH (*green*), and 90 % RH (*blue*). The pristine silk gets affected by humidity and softens as compared to washed silk. *Insets* in each spectrum show the peaks affected in aliphatic region with humidity exposure. $C_\alpha$ region near 55 ppm (**a**) comprises of Leu $C_\alpha$, Ser $C_\alpha$, Gln $C_\alpha$, and Tyr $C_\alpha$ signatures. All spectra were recorded at MAS frequency of 6000 Hz and 25 °C. ssb refers to the spinning sideband

as amino acids constituting major ampullate silk (Creager et al. 2010) and also hypothesized proteins present in glue (Jain et al. 2015), are affected by water. Decrease in the intensity (Fig. 13.9a, inset) of the peaks related to amino acids of these proteins, as the humidity increases, indicates enhanced protein backbone and side-chain mobility (Creager et al. 2010). This effect is especially visible in the aliphatic $C_\alpha$ region with amino acid signatures—Leu $C_\alpha$, Ser $C_\alpha$, Gln $C_\alpha$, Tyr $C_\alpha$ (shoulder near 55 ppm), Ala $C_\alpha$ (49 ppm), and Gly $C_\alpha$ (40 ppm)—showing lower peak intensities with increase in RH. Other aliphatic peaks such as Gln $C_{\gamma,\beta}$ (30 ppm) and Ala $C_\beta$ (random coil, 17 ppm) are also affected. The peak intensity of more hydrophobic Ala $C_\beta$ (22 ppm) segments shows no effect because these amino acids are mostly in the antiparallel β-sheet conformation (Creager et al. 2010), which does not get perturbed by water. In the case of washed gumfoot silk, it is evident that the salts are not present and their removal decreases the overall water uptake and mobility of the gumfoot silk. The change in intensity for regions corresponding the amino acids of MA and glue proteins is not dramatic in contrast to pristine silk. The glycoprotein peak at 75 ppm shows increased intensity as compared to pristine silk indicating rigidity in the absence of salts. Overall this result suggests that the presence of salts is important for the water uptake by the proteins in glue and MA silk.

#### 13.2.4.3 Effect of Humidity on Salt Mobility

To identify the mobile components in gumfoot glue, DP/MAS experiments were carried out for pristine (Fig. 13.10a) and washed (Fig. 13.10b) gumfoot silks at different RH (10 %, 60 %, and 90 %). For the pristine silk (Fig. 13.10a), NMR peaks are identified for three different salts, GABamide (24, 35, and 40 ppm), isethionic

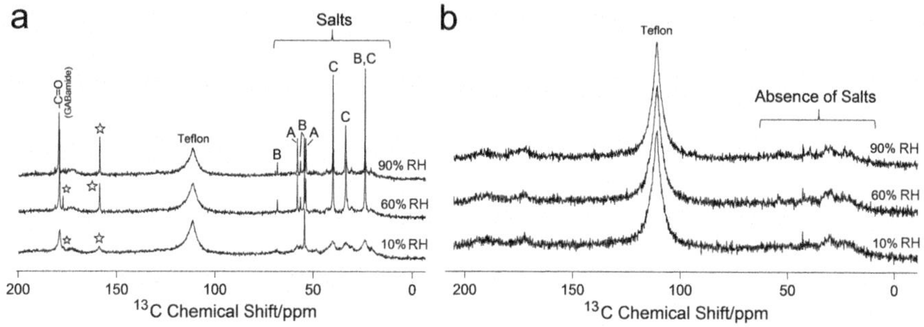


**Fig. 13.10** Effect of humidity on salt mobility. (**a**, **b**) The DP/MAS spectra of pristine and washed gumfoot silk at 10 % RH, 60 % RH, and 90 % RH. In the case of pristine silk, salts (A, isethionic acid; B, choline; and C, GABamide) become mobile with increase in humidity. Washed silk does not get affected by humidity due to lack of salts. Starred peaks in (**a**) are unidentified peaks. All spectra were recorded at MAS frequency of 6000 Hz and 25 °C

acid (55 and 59 ppm), and choline (the signal at 23 ppm, which overlaps with one at 24 ppm from GABamide, 53 ppm, 55 ppm, and 67 ppm). The carbonyl resonance from GABamide shows a peak at 182 ppm. Peaks related to glycoproteins (70 and 105 ppm) were not observed in the aliphatic region spectra over the range of RH studied here, indicating that they are not as mobile as the organic salts. Also, there were two unidentified peaks (156 and 177 ppm) in the aromatic/carbonyl region. Clearly, these peaks are not from the major ampullate silk present in the gumfoot silk and possibly belong to the glycoprotein glue. The peak width decreases as the RH increases and this suggests that the salts are absorbing water and becoming more mobile. Second, choline is the only salt among the three that is hygroscopic at low RH (Vollrath et al. 1990; Townley et al. 1991), and it shows a clear single sharp resonance at 55 ppm at 10 % RH. All of the salt peaks are absent in washed silk, as expected. These results show salts to be the mobile component in the glue that is responsible for absorbing moisture from the environment with increase in RH.

#### 13.2.4.4 Implications of Molecular-Level Findings Toward Macro-Level Adhesion

Solid-state NMR provides direct molecular information of how gumfoot silk components interact with humidity. At the macro-level, pristine silk (Fig. 13.7a) adheres while washed silk (devoid of salts) shows near zero adhesion (Fig. 13.7b). The removal of salts makes the silk inert to humidity such that it fails to absorb the water necessary to make the silk sticky (Jain et al. 2015). NMR findings directly correlate with these macro-level observations. CP/MAS results for pristine silk demonstrate that the glue responds to humidity and softens due to water uptake by salts with increase in RH (Fig. 13.9a) Thus, at the macro-level, the glue is able to spread and adhere to the substrate. On the other hand, washed silk shows the inability of the glue to absorb water without the presence of salts (Figs. 13.9b and 13.10b), ultimately rendering it nonsticky. The low adhesion at 10 % RH (Fig. 13.7a) points out that the salts are dry and the glue does not have sufficient water to plasticize and make it flow and spread on the substrate. Direct molecular evidence of this behavior can be seen in DP/MAS spectrum of pristine silk (Fig. 13.10a), where salts at 10 % RH exhibit broad peaks indicating rigidity. Increases in RH lead to increase in water uptake of salts and sharpening of peaks, again indicative of enhanced adhesion when RH is greater than 10 %.

### *13.2.5 Summary*

Gumfoot adhesives, present at the base of cobwebs, target pedestrian insects. These glues are deposited in droplets on an axial dragline thread and consist of soluble peptides, hygroscopic salts, and glycoproteins. The droplets are viscoelastic liquids by nature and show adhesion that is independent of RH (30 % to 90 %) and therefore works in diverse moisture environments. However, water is necessary for the function of the glue so that adhesion depends on both salts and glycoproteins. Salts, in particular, keep the glue proteins moist and mobile and are important for adhesion in the presence of humidity. Clearly the humidity response of gumfoot glues is an intriguing observation from the perspective of designing materials that perform well under broad ranges of humidity. Efforts in this direction have been limited and need to be undertaken for designing novel adhesives that can work at high RH.

**Acknowledgments** The authors would like to express gratitude to the National Science Foundation (NSF) for funding the NMR studies on gumfoot silk, Bill Hsuing for the help in the collection of major ampullate silk from silk glands of Black Widow, Sarah Han and Dr. Matjaz Gregoric for pictures in Fig.13.1 and Fig.13.3 respectively and Dr. Wei Chen for the assistance in solid-state NMR experiments.

## References

Amarpuri G, Zhang C, Diaz C, Opell BD, Blackledge TA, Dhinojwala A (2015) Spiders tune glue viscosity to maximize adhesion. ACS Nano 9:11472–11478

Argintean S, Chen J, Kim M, Moore AMF (2006) Resilient silk captures prey in black widow cobwebs. Appl Phys A 82:235–241

Benjamin SP, Zschokke S (2003) Webs of theridiid spiders: construction, structure and evolution. Biol J Linn Soc 78:293–305

Blackledge TA, Zevenbergen JM (2007) Condition-dependent spider web architecture in the western black widow, *Latrodectus hesperus*. Anim Behav 73:855–864

Blackledge TA, Coddington J, Gillespie R (2003) Are three-dimensional spider webs defensive adaptations? Ecol Lett 6:13–18

Blackledge TA, Summers AP, Hayashi CY (2005a) Gumfooted lines in black widow cobwebs and the mechanical properties of spider capture silk. Zoology (Jena) 108:41–46

Blackledge TA, Swindeman JE, Hayashi CY (2005b) Quasistatic and continuous dynamic characterization of the mechanical properties of silk from the cobweb of the black widow spider *Latrodectus hesperus*. J Exp Biol 208:1937–1949

Blackledge TA, Scharff N, Coddington JA, Szüts T, Wenzel JW, Hayashi CY, Agnarsson I (2009) Reconstructing web evolution and spider diversification in the molecular era. Proc Natl Acad Sci USA 106:5229–5234

Blasingame E, Tuton-Blasingame T, Larkin L, Falick AM, Zhao L, Fong J, Vaidyanathan V, Visperas A, Geurts P, Hu X, La Mattina C, Vierra C (2009) Pyriform spidroin 1, a novel member of the silk gene family that anchors dragline silk fibers in attachment discs of the black widow spider, *Latrodectus hesperus*. J Biol Chem 284:29097–29108

Casem M, Turner D, Houchin K (1999) Protein and amino acid composition of silks from the cob weaver, *Latrodectus hesperus* (black widow). Int J Biol Macromol 24:103–108

Chamberlin R, Ivie W (1935) The black widow spider and its varieties in the United States. Bull Univ Utah 25:1–29

Creager MS, Jenkins JE, Thagard-Yeaman LA, Brooks AE, Jones JA, Lewis RV, Holland GP, Yarger JL (2010) Solid-state NMR comparison of various spiders' dragline silk fiber. Biomacromolecules 11:2039–2043

d'Amour F, Becker F, van Riper W (1936) The black widow spider. Q Rev Biol 11:123–160

Eberhard WG, Agnarsson I, Levi HW (2008) Web forms and the phylogeny of theridiid spiders (Araneae: Theridiidae): chaos from order. Syst Biodivers 6:415–475

Finkelstein A, Rubin LL, Tzeng MC (1976) Black widow spider venom: effect of purified toxin on lipid bilayer membranes. Science 193:1009–1011

Foelix RF (1982) Biology of spiders. Harvard University Press, Harvard

Frontali N, Ceccarelli B, Gorio A, Mauro A, Siekevitz P, Tzeng MC, Hurlbut WP (1976) Purification from black widow spider venom of a protein factor causing the depletion of synaptic vesicles at neuromuscular junctions. J Cell Biol 68:462–479

Garb JE, González A, Gillespie RG (2004) The black widow spider genus *Latrodectus* (Araneae: Theridiidae): phylogeny, biogeography, and invasion history. Mol Phylogenet Evol 31:1127–1142

Griswold CE, Coddington JA, Hormiga G, Scharff N (1998) Phylogeny of the orb-web building spiders (Araneae, Orbiculariae: Deinopoidea, Araneoidea). Zool J Linn Soc 123:1–99

Hu X, Kohler K, Falick AM, Moore AMF, Jones PR, Sparkman OD, Vierra C (2005a) Egg case protein-1. A new class of silk proteins with fibroin-like properties from the spider *Latrodectus hesperus*. J Biol Chem 280:21220–21230

Hu X, Lawrence B, Kohler K, Falick AM, Moore AMF, McMullen E, Jones PR, Vierra C (2005b) Araneoid egg case silk: a fibroin with novel ensemble repeat units from the black widow spider, *Latrodectus hesperus*. Biochemistry 44:10020–10027

Hu X, Yuan J, Wang X, Vasanthavada K, Falick AM, Jones PR, La Mattina C, Vierra CA (2007) Analysis of aqueous glue coating proteins on the silk fibers of the cob weaver, *Latrodectus hesperus*. Biochemistry 46:3294–3303

Jain D, Zhang C, Cool LR, Blackledge TA, Wesdemiotis C, Miyoshi T, Dhinojwala A (2015) Composition and function of spider glues maintained during the evolution of cobwebs. Biomacromolecules 16:3373–3380

Jenkins J, Sampath S, Butler E, Kim J, Henning RW, Holland GP, Yarger JL (2013) Characterizing the secondary protein structure of black widow dragline silk using solid-state NMR and X-ray diffraction. Biomacromolecules 14:3472–3483

Kelly S (1989) The chemical composition of the defensive secretion of the spider *Latrodectus mactans* (Fabricius). MS Thesis, University of New Hampshire

Kovoor J (1977) Données histochimiques sur les glandes séricigènes de la veuve noire *Latrodectus mactans* Fabr. (Araneae, Theridiidae). Ann Sci Nat Zool Biol Anim 12[e] Sér 19:63–87

Kovoor J (1987) Comparative structure and histochemistry of silk-producing organs in arachnids. In: Nentwig W (ed) Ecophysiology of spiders. Springer, Berlin
Lawrence BA, Vierra CA, Moore AMF (2004) Molecular and mechanical properties of major ampullate silk of the black widow spider, *Latrodectus hesperus*. Biomacromolecules 5:689–695
Liu Y, Sponner A, Porter D, Vollrath F (2008) Proline and processing of spider silks. Biomacromolecules 9:116–121
Meldolesi J, Scheer H, Madeddu L, Wanke E (1986) Mechanism of action of α-latrotoxin: the presynaptic stimulatory toxin of the black widow spider venom. Trends Pharmacol Sci 7:151–155
Moore AMF, Tran K (1999) Material properties of cobweb silk from the black widow spider *Latrodectus hesperus*. Int J Biol Macromol 24:277–282
Opell BD, Karinshak SE, Sigler MA (2011) Humidity affects the extensibility of an orb-weaving spider's viscous thread droplets. J Exp Biol 214:2988–2993
Opell BD, Karinshak SE, Sigler MA (2013) Environmental response and adaptation of glycoprotein glue within the droplets of viscous prey capture threads from araneoid spider orb-webs. J Exp Biol 216:3023–3034
Sahni V, Blackledge TA, Dhinojwala A (2010) Viscoelastic solids explain spider web stickiness. Nat Commun 1:19
Sahni V, Blackledge TA, Dhinojwala A (2011a) A review on spider silk adhesion. J Adhes 87:595–614
Sahni V, Blackledge TA, Dhinojwala A (2011b) Changes in the adhesive properties of spider aggregate glue during the evolution of cobwebs. Sci Rep 1:41
Sahni V, Harris J, Blackledge TA, Dhinojwala A (2012) Cobweb-weaving spiders produce different attachment discs for locomotion and prey capture. Nat Commun 3:1106
Sahni V, Miyoshi T, Chen K, Jain D, Blamires SJ, Blackledge TA, Dhinojwala A (2014) Direct solvation of glycoproteins by salts in spider silk glues enhances adhesion and helps to explain the evolution of modern spider orb webs. Biomacromolecules 15:1225–1232
Salles HC, Volsi ECFR, Marques MR, Souza BM, dos Santos LD, Tormena CF, Mendes MA, Palma MS (2006) The venomous secrets of the web droplets from the viscid spiral of the orb-weaver spider *Nephila clavipes* (Araneae, Tetragnathidae). Chem Biodivers 3:727–741
Schulz S (1997) The chemistry of spider toxins and spider silk. Angew Chem Int Ed Engl 36:314–326
Schulz S (1999) Structural diversity of surface lipids from spiders. In: Diederichsen U, Lindhorst TK, Westermann B, Wessjohann LA (eds) Bioorganic chemistry: highlights and new aspects. Wiley-VCH, Weinheim
Schulz S (2001) Composition of the silk lipids of the spider *Nephila clavipes*. Lipids 36:637–647
Swanson BO, Blackledge TA, Beltrán J, Hayashi CY (2006) Variation in the material properties of spider dragline silk across species. Appl Phys A 82:213–218
Townley MA, Tillinghast EK (2013) Aggregate silk gland secretions of araneoid spiders. In: Nentwig W (ed) Spider ecophysiology. Springer, Berlin, pp 283–302
Townley MA, Bernstein DT, Gallagher KS, Tillinghast EK (1991) Comparative study of orb web hygroscopicity and adhesive spiral composition in three araneid spiders. J Exp Zool 259:154–165
Townley MA, Pu Q, Zercher CK, Neefus CD, Tillinghast EK (2012) Small organic solutes in sticky droplets from orb webs of the spider *Zygiella atrica* (Araneae; Araneidae): β-alaninamide is a novel and abundant component. Chem Biodivers 9:2159–2174
Ushkaryov YA, Volynski KE, Ashton AC (2004) The multiple actions of black widow spider toxins and their selective use in neurosecretion studies. Toxicon 43:527–542
Vasanthavada K, Hu X, Falick AM, La Mattina C, Moore AMF, Jones PR, Yee R, Reza R, Tuton T, Vierra C (2007) Aciniform spidroin, a constituent of egg case sacs and wrapping silk fibers from the black widow spider *Latrodectus hesperus*. J Biol Chem 282:35088–35097
Vasanthavada K, Hu X, Tuton-Blasingame T, Hsia Y, Sampath S, Pacheco R, Freeark J, Falick AM, Tang S, Fong J, Kohler K, La Mattina-Hawkins C, Vierra C (2012) Spider glue proteins have distinct architectures compared with traditional spidroin family members. J Biol Chem 287:35986–35999
Vollrath F, Selden P (2007) The role of behavior in the evolution of spiders, silks and webs. Annu Rev Ecol Evol Syst 38:819–846

Vollrath F, Fairbrother WJ, Williams RJ, Tillinghast EK, Bernstein DT, Gallagher KS, Townley MA (1990) Compounds in the droplets of the orb spider's viscid spiral. Nature 345:526–527
Xu D, Yarger JL, Holland GP (2014) Exploring the backbone dynamics of native spider silk proteins in Black Widow silk glands with solution-state NMR spectroscopy. Polymer 55:3879–3885

# Chapter 14
# High-Strength Adhesive Exuded from the Adventitious Roots of English Ivy

**Yujian Huang and Mingjun Zhang**

**Abstract** As a trending topic in recent years, tremendous efforts in the exploration of molecular bases for a variety of adhesive events in diverse biological organisms have considerably improved our understanding of relevant principles capable of being implemented as guidelines for directing the design and development of adhesive biomaterials and devices with expected functionalities. In this chapter, we focus on describing the recent advance in the exploration of a high-strength bioadhesive derived from the adventitious roots of English ivy (*Hedera helix*), which is a root climber that possesses strong capacity to cling vertical surfaces. The molecular mechanisms underlying this high-strength adhesive, especially the intriguing roles of bulk glycoprotein-rich spherical nanoparticles in favoring the generation of strong adhesion strength within the adhesive substances, are discussed in detail. Relevant progress in the development of ivy-mimetic and ivy-inspired adhesive composites is also illustrated.

## 14.1 Introduction

While an adhesive is commonly regarded as a substance capable of bridging two surfaces, a bioadhesive is appropriately defined as unmodified natural glue or mucilage derived from diverse biological species for attachment (Favi et al. 2014; Park and Park 1990). Throughout millions of years' evolution, the refined development of these bioadhesives among a variety of biological organisms has allowed

Y. Huang • M. Zhang, Ph.D., D.Sc. (✉)
Department of Biomedical Engineering, College of Engineering, The Ohio State University, Columbus, OH 43210, USA

Dorothy M. Davis Heart & Lung Research Institute, The Ohio State University Wexner Medical Center, Columbus, OH 43210, USA

Interdisciplinary Biophysics Graduate Program, The Ohio State University, Columbus, OH 43210, USA
e-mail: zhang.4882@osu.edu

© Springer International Publishing Switzerland 2016 
A.M. Smith (ed.), *Biological Adhesives*, DOI 10.1007/978-3-319-46082-6_14

them to be involved in numerous physiological events essential for the survival and growth of corresponding hosts (Bar-Cohen 2005). Analogous to conventional adhesives, bioadhesives are expected for their capacity to spread across the substrates, fill the irregularities on the surfaces, and connect dissimilar materials by taking advantage of the strong tensile and shear strengths generated upon curing (Petrie 2007). Surprisingly, in most cases, bioadhesives also possess exceptional features that are unmatched in engineered synthetic counterparts, including but not limited to the rapid reversible adhesion behavior that allows insects to climb surface vertically, the strong subaqueous adhesive action that favors the adsorption of mussels on the hulls of ships, and the extraordinary weather-resistant adhesive activity that ensures the attachment of ivy toward multiple substrates.

Plants and animals have evolved a series of strategies regulating their attachment and climb over corresponding surfaces. A representative instance of the dry adhesion is the gecko, which applies a sophisticated hierarchical architecture constituted with spatula (100–200 nm), setae (5–10 μm), branches (1–2 μm), and lamellae (1–2 mm), to generate sufficient contact area and thus allow the gecko to climb vertical surfaces irreversibly (Hansen and Autumn 2005; Tian et al. 2006; Autumn et al. 2000; Bhushan 2007). Distinguished from the dry adhesion, the bioadhesive named byssus that is secreted by the byssal glands developed in the feet of mussel represents another approach exploited by nature for surface attachment (Cha et al. 2008; Waite and Tanzer 1981; Lee et al. 2006).

In addition to the extensive exploration of the molecular mechanisms controlling diverse adhesive behaviors in nature, researchers have sought to summarize the principles underlying these molecular events and exploit them as guidelines for inspiring the design and development of the new generation of engineered synthetic glues in recent years. These synthetic adhesive composites, aiming to mimic and realize similar forms and functions comparable to the templates offered by respective bioadhesives, have been designated as bioinspired adhesives (Favi et al. 2014). As a result, fundamental concepts of bioinspired engineering have been implemented throughout the fabrication of these bioinspired adhesives, intended to create high-strength, quickly bonded and eco-friendly engineered synthetic glues with desirable functionalities, such as the repeated usage or underwater performance, but without noticeable toxic activities (Peattie 2009).

As an emerging field, bioadhesives derived from plants have drawn increasing interest owing to their exceptional performance and nearly endless diversity of the substances that constitute the adhesive composites. Analogous to animals, plants have developed efficient strategies for attachment after more than 400 million years of stringent evolution (Favi et al. 2014). Different types of organs functioning for attachment have evolved independently in various climbing plants. While the whole stems are employed by twining plants to climb host trees and other substrates (Silk and Hubbard 1991; Isnard et al. 2009), specialized organs including hooks, adventitious roots, attachment pads, and tendrils are exploited by other climbing plants to attach to their corresponding supports (Rowe et al. 2006; Burnham and Revilla-Minaya 2011). These specialized organs are commonly evolved from roots, branches, and leaves (Seidelmann et al. 2012). In particular, root climbers are defined as self-clinging plants that develop adventitious climbing roots allowing them to attach onto multiple substrates such as tree bark or rocks (Melzer et al. 2008; Groot et al. 2003; Seidelmann et al. 2012). Even though the detailed attachment strategies utilized by each root climber are diverse (Seidelmann et al. 2012; Melzer et al. 2010), to date, sticky mucilage is observed to be exuded by all species that have been investigated (Darwin 1865; Groot et al. 2003; Melzer et al. 2010; Seidelmann et al. 2012; Bowling and Vaughn 2008a). These adhesive exudates undoubtedly aid in the attachment of root climbers on vertical substrates independent of the overall size and shape of the support. However, due to the evolutionary pressures and physiological differences, the mechanisms controlling the attachment of root climbers toward the substrates are dramatically different from their counterparts in the animal kingdom. In this chapter, we focus on describing recent advances in the exploration of molecular bases for a bioadhesive derived from the adventitious roots of English ivy (*Hedera helix*), which is a typical root climber that possesses common morphological and physiological traits of congener plants. Additionally, relevant progress in the development of ivy-mimetic and ivy-inspired adhesive composites is also illustrated.

## 14.2 Uniqueness and Importance of the Bioadhesive Derived from English Ivy

*H. helix* (English ivy), an evergreen plant, is a typical root climber characterized by its exceptional capacity to climb along vertical substrates, including but not limited to trees, rocks, fences, and brick walls, up to a height of 30 m (Metcalfe 2005). English ivy has been investigated in detail in terms of its ecological implications (Metcalfe 2005), physiological features (Oberhuber and Bauer 1991), pharmaceutical potentials (Majester-Savornin et al. 1991), and morphogenesis (Rogler and Hackett 1975). Throughout the juvenile phase of the ontogenesis of English ivy, clusters of adventitious roots originate from shoots (Tobler 1912) and develop into unbranched attachment roots in a length of ~1–15 mm under appropriate moisture conditions (Melzer et al. 2010). Root hairs, indicating protrusion of single cells, are observed on the attachment roots (Melzer et al. 2010; Lenaghan and Zhang 2012). In 1876, Charles Darwin described his observation that the adventitious roots of English ivy are capable of secreting yellowish mucilage while climbing surfaces. During the climbing process, ivy uses adhering flat disks developed from the adventitious roots to cling to the surface, and an individual adhesive disk is estimated to possess the ability to support a weight of approximately two pounds, despite the weight of the disk itself, which is only ~0.5 mg (Darwin 1865; Zhang et al. 2008b). This information suggests here that the adhesive disk formed by English ivy is capable of generating a tensile force 1.8 million times greater than its own weight.

### 14.2.1 Hierarchical Attachment Strategies of English Ivy

Despite the noticeable adhesion strength, little progress has been made in elucidating the detailed attachment procedures of English ivy toward substrates until recent years (Melzer et al. 2010). A four-phase model for the attachment mechanisms of English ivy was proposed by Melzer et al. in light of the morphological observations, i.e., (1) physical contact, (2) closure of the root with the substrate, (3) chemical adhesion, and (4) shape changes in the root hairs (Melzer et al. 2010). Different strategies are thought to work together to favor the attachment of English ivy toward various supports. Two hypothetical patterns of initial contact between the adventitious roots of English ivy and corresponding substrates are proposed, but direct evidence to support either case is still weak until the exact signal that triggers the physical contact could be identified in the future. The second phase is believed to be induced by the initial physical contact, regarding the structural adaptation of the adventitious roots to the topology of the substrates attached. A two-step reaction is involved in this phase. The contact area between the adventitious roots and corresponding substrates is enlarged during the first reaction, as a result of the enhanced form closure. Upon the second reaction, the increase in diameter of the adventitious roots improves the enduring connections, resulting in a strong mechanical interlocking. The chemical adhesion described in the third phase is also thought to be triggered by the physical contact. During the chemical adhesion, bundles of "threads" of glue were captured between the root hairs and the attached substrates. A passive change in shape of the root hairs represents the fourth phase of the attachment, a process driven by the water evaporation (Melzer et al. 2010).

### 14.2.2 Spherical Nanoparticles Observed in the Ivy-Derived Adhesive

In light of the proposed hierarchical attachment procedures of English ivy toward corresponding supports as described above, it is apparent that the mucilage exuded by the adventitious roots undoubtedly exerts vital roles in driving the clinging of English ivy and supporting the climbing of this plant on vertical surfaces. To further explore the physicochemical cues implied in this bioadhesive, atomic force microscopy (AFM) was employed to characterize the supramolecular structures within this adhesive substance, aiming to elucidate the molecular bases controlling the generation of high-strength adhesion (Zhang et al. 2008b). For such a purpose, a 3-in. silicon wafer and a piece of mica were placed to the end of an ivy branch. English ivy climbed across the silicon wafer and mica, and after a week, the branches were removed and the traces left on the surfaces were examined using AFM. Abundant spherical nanoparticles were observed to be secreted from the tendrils of the adhering flat disks (Fig. 14.1) (Zhang et al. 2008b). The particles detected are fairly uniform, with an average size of ~70 nm in diameter as measured

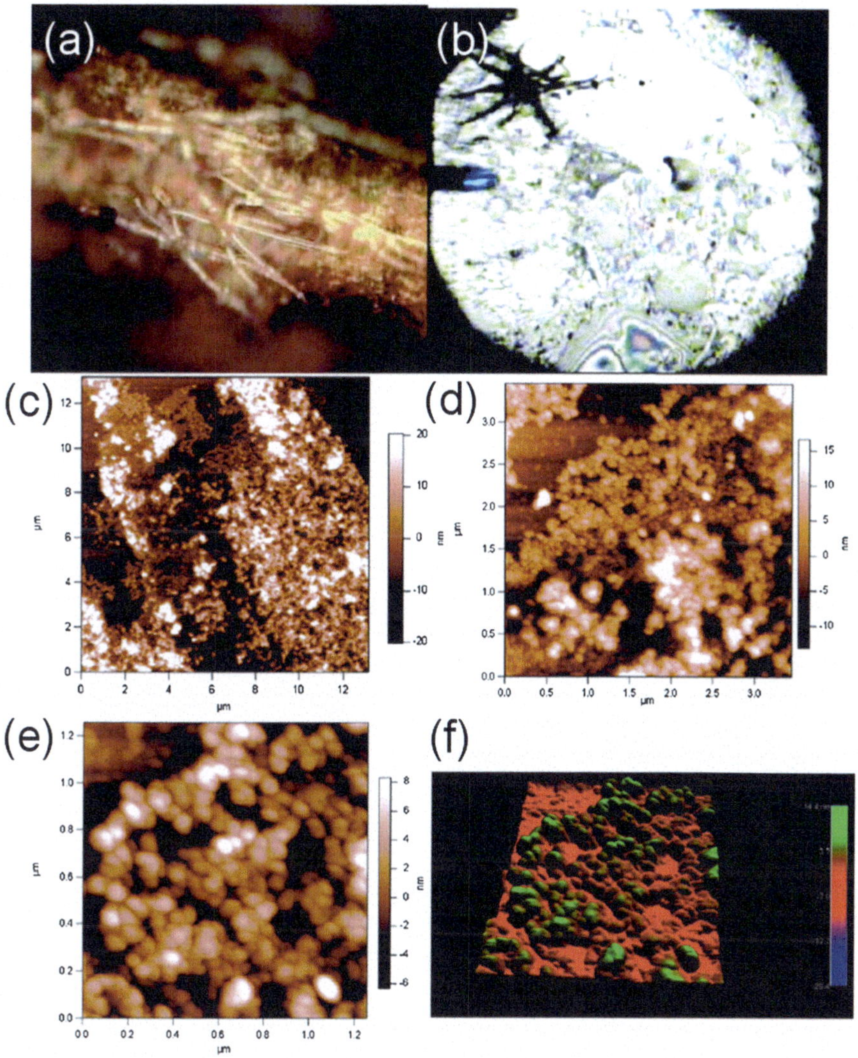


**Fig. 14.1** Nanoparticles secreted from the adventitious roots of English ivy. (**a**) An optical microscopic image of the ivy adhering disks with protruding fingers developed on an adventitious root. (**b**) An optical microscopic image of an adhesive disk left by the adventitious roots on a silicon surface. Each finger is approximately 250–350 µm. (**c–e**) AFM topographic images of scan size of 14 × 14 µm, 3.5 × 3.5 µm, and 1.25 × 1.25 µm, respectively. (**f**) A 3D view of (**e**). Images are from (Zhang et al. 2008b)

from cursor profiles of the AFM images and a mean height of approximately 20–30 nm. The difference in the particle sizes measured either vertically or horizontally is attributed to the deformation caused by the AFM tips. In the aqueous suspension, the hydrodynamic size of the ivy nanoparticles is $108.8 \pm 3.1$ nm in diameter, with a negatively charged surface under neutral condition (zeta potential = $-28.5 \pm 3.2$ mV), as measured by dynamic light scattering (DLS) and electrophoretic light scattering (ELS) at room temperature (Huang et al. 2015). Intuitively, these spherical nanoparticles are hypothesized to play a direct and significant role in the surface clinging of English ivy (Zhang et al. 2008b).

Accompanied by the secretion of spherical nanoparticles from the adventitious roots, the yellowish substances mentioned by Charles Darwin were also observed to be gradually exuded in a gel stage in the early stage and become dry subsequently. The adhesive exudate was observed to be attached tightly toward the substrates after the substances were completely dried on the surface, a process promoted by water evaporation (Zhang et al. 2008b; Darwin 1865).

### *14.2.3 Monitoring the Secretion of the Nanocomposite-Rich Adhesive in Real Time*

To explore the specific functions of the spherical nanoparticles within the ivy-derived adhesive and the manner in which these nanoparticles are involved during the formation of strong adhesion strength at the interface, the secretory procedures of the mucilage from the adventitious roots of English ivy were real-time monitored using a novel video microscopy apparatus (Lenaghan and Zhang 2012). In brief, a Meiji EMZ-13TR stereoscope was mounted horizontally onto a drill press stand. A Handycam HDV high-definition video camera was installed onto the stereoscope, by means of a trinocular mount. Owing to the tiny size of individual root hair, ~10 μm in diameter, imaging of these structures perpendicular to an attaching substrate was not effective. As a result, to eliminate such problem, a Gold Seal 25 mm × 25 mm cover glass was fixed between the adventitious roots and the camera system, providing a transparent surface and thus allowing the attachment to be directly visualized. This setup served the purpose of tracing the secretion of the ivy-derived adhesive and the attachment process. The cover glass was then handled with forceps for subsequent attachment studies. After the adventitious roots contacted the glass and the secretion of adhesive was monitored for at least 8 h, the cover glass was removed and examined using AFM. After the analysis of approximately 150 h of video, the key stages involved in the attachment of English ivy were capable of being discerned (Lenaghan and Zhang 2012). It is concluded that the root tip is the key structure involved in providing the signal to attach, a significant breakthrough which forwards the progress toward answering the question raised by Melzer et al., regarding specific signals required for triggering the cascade reactions in the first stage of attachment (Melzer et al. 2010). Prior to the contact, millimeter-sized adventitious roots elongate with the tips pointing in diverse directions. After contact with the surface, there is an increase in the number of micrometer-sized root hairs, showing an accelerated growth. It is also determined that this initial contact stimulates the production of adventitious roots above the next node. In a series of trials throughout a month, it is also observed that without such contact stimuli, adventitious roots do not arise above the point of contact and exhibit a considerably slow growth. This suggests that the tip of the adventitious root serves as a pressure sensor signaling not only for the attaching root but also inducing increased production of adventitious roots at posterior nodes (Lenaghan and Zhang 2012). The second stage of attachment consists of parallel bending of the adventitious root to the surfaces, giving rise to a closer contact between the root hairs and the substrates. It is observed that the root hairs start to exude mucilage at this stage, even if they are not in contact with the corresponding surfaces. This information here indicates that shear force is not essential for deposition of the adhesive substances. It is also found that the adhesive droplets from several root hairs combined to form bulky droplets that contact the surface, owing to the close proximity of the root hairs with one another. The entire process of secretion commonly takes 4–6 h after contact, prior to cessation. Additionally, the correlations between the exuded adhesives and the uniform spherical nanoparticles within the exudates are established by AFM studies (Lenaghan and Zhang 2012). These spheroidal nanoparticles are observed to be abundantly dispersed in the adhesive substances (Lenaghan and Zhang 2012).

## 14.3 Identification of the Chemical Constituents of the Spheroidal Nanoparticles

### *14.3.1 Massive Harvest of the Purified Ivy Nanoparticles*

In order to identify the chemical composition of the ivy-derived spherical nanoparticles and the exuded adhesives, it is necessary to develop a platform capable of producing sufficient amount of materials for physical and chemical analyses and to establish a purification system for cost-effectively harvesting bulk ivy nanoparticles. Accordingly, a rooting chamber for the fabrication of the adventitious roots of English ivy was designed and constructed (Burris et al. 2012). Indole-3-butyric acid (IBA) was applied for inducing the generation of the adventitious roots and the biosynthesis of ivy nanoparticles. It was observed that the adventitious roots of English ivy elongated to a length up to 30 mm in the chamber after the induction of IBA-K, whereas under natural conditions, the adventitious roots can merely grow unbranched to a length of 1–15 mm, due to the difference in humidity. Under this high humidity condition in the chamber, the adventitious roots grow unabated for at least 1 month. The maximum yield of the adventitious roots was reached by a 4 h induction of 1 mg/ml IBA-K to juvenile shoot segments of English ivy cultured in vessels. After 2-week growth, the adventitious roots were harvested and flash frozen in liquid nitrogen. To obtain the purified ivy nanoparticles, macerated roots were squeezed using a glass dounce tissue grinder, and only the liquid from the tissue was collected. After centrifugation and filtration as detailed in (Burris et al. 2012), the nanoparticles obtained were further purified by dialysis against DI water and size-exclusion-chromatography-high-performance liquid chromatography (SEC-HPLC) (Lenaghan et al. 2013). Successful isolation and purification of the ivy-derived nanoparticles were examined and validated by AFM characterization and dynamic light scattering evaluation, as described in (Burris et al. 2012). Approximately 90 mg of ivy-derived nanoparticles per 12 g of the adventitious roots was obtained via this procedure.

### 14.3.2 Physicochemical Characterization

To identify the chemical constituents of the ivy-derived nanoparticles, the physiochemical properties of these spheroidal nanoparticles were preliminarily characterized. The first step in chemical analysis of the ivy nanoparticles is to confirm the organic nature of these nanoparticles and test if metal ions are contained in these nanostructures. This is especially important considering the bulk metallic nanoparticles that can be formed naturally from heavy metal substrates. A series of studies have demonstrated the potential of plants, including English ivy, to manufacture nanoparticles from $HAuCl_4$, $AgNO_3$, $FeCl_3$, and many others (Chandran et al. 2006; Shankar et al. 2003). To rule out the possibility of the presence of metallic components in the ivy nanoparticles, inductively coupled plasma mass spectrometry (ICP-MS) was utilized to examine the purified ivy nanoparticles. Given that any type of metal ions detected in the purified ivy nanoparticles was below respective trace levels, it is concluded that the ivy nanoparticles are organic nanoarchitectures (Lenaghan et al. 2013). After evidencing the organic nature of the ivy nanoparticles, the next step was to determine the exact types of molecules that may be associated with the formation of the nanostructures. In the earlier studies of analogous climbing plants, such as Boston ivy (*Parthenocissus tricuspidata*) and Virginia creeper (*Parthenocissus quinquefolia*), the vast majority of the components that have been identified in the secreted adhesives, via immunocytochemical analyses, are mucilaginous pectic polysaccharides, callose, tanniferous substances, and acid mucopolysaccharides (Bowling and Vaughn 2008a; Endress and Thomson 1976). Unfortunately, similar spherical nanostructures were not detected in any of these studies, presumably owning to the technical limitations previously. In the meanwhile, proteins are considered to be the predominant building blocks leading to the generation of strong adhesion strength in other biological organisms, including mussels (*Mytilus edulis*) and polychaetes (*Phragmatopoma californica*) (Fant et al. 2002; Ninan et al. 2007). The bioadhesives derived from these organisms have shown the presence of nanoparticles, mainly arising from the interactions with divalent cations (Stevens et al. 2007; Lenaghan et al. 2013). The elemental analysis by ICP-MS was conducted to determine the amounts of carbon, nitrogen, and sulfur present in the ivy nanoparticles. Since up to 51.77 % carbon and 4.72 % nitrogen are identified in the purified ivy nanoparticles, the relatively high carbon-to-nitrogen ratio indicates that biomacromolecules, such as protein or nucleic acids, are presumably the predominant constituents essential for the construction of the overall architectures of the ivy nanoparticles (Lenaghan et al. 2013). Additionally, 0.32 % sulfur was identified in the ivy-derived nanoparticles, another hint for the existence of biomacromolecules, such as proteins, within the purified ivy nanoparticles (Sevier and Kaiser 2002). These results are consistent with those obtained from the chemical composition analysis by means of LC/MS (Zhang et al. 2008b). Considering that the surfaces to which the nanoparticles commonly attach are inorganic or at least polar substrates, involving rocks, bricks, tree barks, and many others, the putative compositions of the ivy-derived nanoparticles suggest that these nanostructures exert functions at the interface relying on hydrogen bonds (Zhang et al. 2008b).

While the evidence strongly supports the presence of proteins in the ivy-derived nanoparticles, the possibility of the presence of other biomolecules, involving polysaccharides and lipids, could not be excluded. As such, Fourier transform infrared (FTIR) spectroscopy was carried out on lyophilized ivy nanoparticles to gain further chemical information of these nanostructures. The FTIR spectra of the ivy nanoparticles were recorded and compared to those originated from a typical protein sample, bovine serum albumin (BSA) and also a common polysaccharide extensively exploited in the preparation of nanomaterials, chitosan (Lenaghan et al. 2013). In light of the FTIR data, in combination with the elemental information obtained from the ICP-MS analyses, it is thought that the most likely component of the ivy-derived nanoparticles is glycoprotein, as a result of the mutually shared amide II band present in the BSA spectra and the shared CO–C band present in the chitosan spectra. Additionally, the shift in the broadband at 3407 $cm^{-1}$ indicated that O–H bonds are present, further suggesting the existence of carbohydrate structures within the purified ivy nanoparticles. According to all the information gained from above physicochemical analyses, it is concluded that the ivy-derived nanoparticles are comprised of proteins and/or glycoproteins (Lenaghan et al. 2013).

### *14.3.3 Glycoprotein Nature of the Ivy Nanoparticles*

To further identify the chemical constituents of the ivy-derived nanoparticles in detail, gel electrophoresis was performed by Lenaghan et al. The hypothesis that the proteins and/or glycoproteins could be detected from the purified ivy nanoparticles was evaluated by several different gel electrophoresis analyses. It was determined that sodium dodecyl sulfate polyacrylamide gel electrophoresis (SDS-PAGE) with a 5 % stacking gel and 10 % resolving gel yielded the ideal results for assessing the purified ivy nanoparticles (Lenaghan et al. 2013). Additionally, the sample was pretreated with 2 M thiourea, 8 M urea, and 3 % SDS, to completely denature and solubilize the ivy nanoparticles, in order to reduce the background staining. After electrophoresis for 4 h at 180 V, respective gels were either silver-stained or evaluated with a glycoprotein stain. For reproducibility, the ivy-derived nanoparticles were isolated and examined by two independent groups (Lenaghan et al. 2013). Surprisingly, only one band was observed in all of the ivy nanoparticle samples, with a high MW above 460 kDa, under the harsh denaturing conditions. This high MW band stained positive for protein, as assessed by silver stain, while also positive for glycoprotein, as tested by Pro-Q glycoprotein stain (Lenaghan et al. 2013). In contrast, it was observed that the glycoprotein stain did not cross-react with the non-glycosylated proteins present in the protein ladder. To eliminate potential contamination and ensure that the presence of the glycoprotein band was indeed associated with the ivy nanoparticles, three separate isolations were carried out using distinct batches of the adventitious roots harvested. As expected, the band was consistent across all three tests. This result indicates that the ivy nanoparticles consist of at least one, if not several glycoproteins (Lenaghan et al. 2013).

In light of the fact that the presence of glycoproteins has been determined in the ivy nanoparticles as detailed above, logically, it is subsequently hypothesized that the ivy nanoparticles may consist of typical glycoproteins or proteoglycans present in the extracellular matrix (ECM) of plant cells. Among them, one subfamily of hydroxyproline (Hyp)-rich glycoproteins (HRGPs), arabinogalactan proteins (AGPs), seems to be the ideal candidate due to its vital role in supporting the morphogenesis and function of root hairs (Showalter 2001; Cannon et al. 2008; Callow and Callow 2006; Velasquez et al. 2011; Seifert and Roberts 2007). More importantly, the tertiary structure of the AGPs has theoretically been predicted to be spheroidal according to a "wattle-blossom" model established by analyzing the constituents and flexibility of the AGPs as a whole (Fincher et al. 1983; Showalter 2001; Ellis et al. 2010; Baldwin et al. 1993).

To test if AGPs are contained in the ivy nanoparticles, Yariv phenylglycoside dye, a reagent also called β-glucosyl Yariv (β-GlcY), which selectively binds to the AGPs via recognizing both given protein moieties and β-1, 3-galactan chains with greater than five residues (Kitazawa et al. 2013; Showalter 2001; Sardar et al. 2006; Ellis et al. 2010; Du et al. 1994; Poon et al. 2012), was applied for the identification. Characteristic AGP-like smeared bands in a high MW range were observed on respective SDS-PAGE gels stained with either Coomassie brilliant blue or 0.2 % (w/v) β-GlcY dye (Xu et al. 2007; Poon et al. 2012; Huang et al. 2016). Consistent with the gel information, the presence of AGPs in the purified ivy nanoparticles was further evidenced by Western blotting analysis using two monoclonal antibodies (mAbs), JIM13 and JIM14, which specifically recognize glycan epitopes of typical AGPs (Huang et al. 2016). In addition, FTIR measurements also indirectly validated the existence of AGPs in the ivy nanoparticles, in comparison of the IR spectra of the ivy nanoparticles with those of a standard reference AGP, gum arabic (Huang et al. 2016). Furthermore, it was also observed that the ivy nanoparticles exhibited similar photoelectron spectra to the gum arabic, as measured by X-ray photoelectron spectroscopy (XPS). All the information described above demonstrates that as expected, the AGPs indeed exist in the purified ivy nanoparticles (Huang et al. 2016).

To further determine the proportion that AGPs account for in the purified ivy nanoparticles, reversed-phase (RP)-HPLC was carried out to segregate each non-covalently bonded structural domain (Huang et al. 2016). Upon gradient elution at a flow rate of 1 ml/min, a series of fractions were obtained, including one major and four minor ones. Apart from the solvent peak designated as fraction 1, all remnant fractions were loaded onto SDS-PAGE gels with equal amounts of samples (30 μg per lane). Among them, only the major fraction designated as fraction 4 was detected by Coomassie brilliant blue and β-GlcY simultaneously. Additionally, via weighing each lyophilized fraction, it was found that the fraction 4 occupied up to 94 % (w/w) of the entire amount, indicating that the vast majority of the components in the ivy nanoparticles consist of AGP-rich fraction. Spheroidal nanoparticles, with an average diameter that resembled the purified ivy nanoparticles derived from SEC-HPLC, were observed in fraction 4 as examined by AFM, whereas analogous nanoparticles were not detected in any other fractions, further suggesting that the minor fractions might be nonessential constituents for the assembly of the overall architecture of the ivy nanoparticles. Given that structural domains with loosely non-covalent binding should be separated during the gradient elution in RP-HPLC, these data suggest that the purified ivy nanoparticles should be regarded as the minimal units, i.e., individual molecules, rather than clusters of multiple molecules. To date, this architecture is thought to be the most regular spherical nanostructure obtained in AGP-rich molecules and also in agreement with the "wattle-blossom" model (Fincher et al. 1983).

Given that the AGPs have been identified to be the predominant constituent in the ivy nanoparticles, an exploration of the detailed structures of the AGP molecules, including the monosaccharide composition, linkages, uronic acid content, and protein backbones, may contribute to the elucidation of the precise functions of the AGP-rich ivy nanoparticles involved in the generation of the strong adhesion strength. The glycosyl composition and linkages of sugar residues obtained from fraction 4 were quantitatively determined at the Complex Carbohydrate Research Center (CCRC) of the University of Georgia, GA (Huang et al. 2016). It was observed that the fraction 4 was rich in Gal and Ara, with respective proportions of up to 28.1 % and 36.1 % of the total monosaccharides, while 8.1 % Rha, 9.6 % GalA, 5.8 % GlcA, 2.6 % Xyl, 5.3 % Glc, and 4.4 % Man were also identified among the entire monosaccharides. Moreover, the existence of type II arabinogalactans (AGs) in the fraction 4 was also validated by glycosyl linkage analysis, due to the presence of terminal Gal, as well as 3-, 6-, and 3, 6-Gal residues. In general, the ivy nanoparticles exhibit glycan structures homologous to typical AGPs in terms of the monosaccharide composition and linkages. In the meanwhile, the detected GalA and pectic glycosyl linkages, including 2-, 2, 4-Rha, and 4-GalA*p*, also indicate the presence of rhamnogalacturonan-I (RG-I) and possible homogalacturonan (HG) in the ivy nanoparticles. In combination with information gained from 1D $^{1}$H NMR spectra which substantiated the existence of α-GalA*p*, α-Rha*p*, α-Ara*f*, and β-Gal*p* residues in the fraction 4, evidence is sufficient to support the concept that the ivy nanoparticles are primarily composed of pectic AGPs, a bulky proteoglycan architecture similar to the arabinoxylan pectin arabinogalactan protein 1 (APAP1) and an RG-I-enriched fraction named Ara101P that are both identified in *Arabidopsis thaliana* (Tan et al. 2013). Notably, even though Glc, Man, and Xyl are not commonly regarded as characteristic monosaccharides in AGPs and pectin, they accounted for a proportion of greater than 12 % (mol%) of the total monosaccharides as a whole in the fraction 4 (Huang et al. 2016).

Typically, AGPs consist of core proteins and diverse *O*-glycans mainly comprising type II arabinogalactans (AGs) and short oligoarabinosides (Tan et al. 2004; Showalter 2001; Kitazawa et al. 2013; Ellis et al. 2010; Seifert and Roberts 2007). It has been demonstrated that numerous AGP molecules possess negative surface charge that is thought to be associated with the presence of uronic acid substituents, i.e., GlcA and GalA, at the termini of type II AGs (Showalter 2001; Fincher et al. 1983; Ellis et al. 2010). In this respect, likewise, the negative surface charge exhibited by the ivy nanoparticles as described in Sect. 14.2.2 presumably arises from the deprotonation of GlcA*p* and GalA*p* residues in solution.

In addition to the monosaccharide composition and linkages, the core protein that constitutes the framework for the construction of the overall architecture of the ivy nanoparticle was also explored (Huang et al. 2016). For such a purpose, the AGP-rich fraction gathered from RP-HPLC, i.e., fraction 4, was deglycosylated, and a short segment of amino acid sequence at the *N*-terminus of the deglycosylated protein, Ala–Hyp–Hyp–Hyp–Thr–Asp–Ala, was determined via Edman degradation. According to this *N*-terminal sequence, degenerate primers were designed, and the nucleotide sequence for the full-length cDNA encoding the core protein was then determined by 5′- and 3′-RACE cloning. Moreover, the identified full protein sequence, designated as IVY ARABINOGALACTAN PROTEIN (IAGP), demonstrated a moderate similarity to four other HRGPs derived from *Arabidopsis* in a sequence alignment. Notably, in a BLASTp search in the genome of the *A. thaliana*, the IAGP identified from *H. helix* also exhibited a moderate extent of similarity to several cytochrome c oxidase subunits, implying potential homology between these two types of proteins. Accordingly, a phylogenetic analysis was carried out to determine the relationship of the IAGP with these analogous proteins obtained from BLASTp. Among them, the IAGP and the AGP9 were the closest relatives. The identity of the IAGP obtained from the RACE cloning was further validated by MALDI-TOF MS analysis by determining the masses of the tryptic peptides of the deglycosylated fraction 4 obtained from RP-HPLC. Several peptides with respective masses detected by the MS matched the predicted molecular masses of expected tryptic peptides. Additionally, the hydroxylation degree and the distribution of the Hyps in other estimated tryptic peptides were deduced from relevant molecular masses gained from the MS test. Given that greater than 90 % of the prolines within the IAGP appear in the form of Hyp, which commonly bears type II AGs and short oligoarabinosides in typical AGPs, the IAGP should be highly *O*-glycosylated in the ivy nanoparticles, similar to other AGPs identified from *A. thaliana*.

## 14.4 Molecular Basis for the Ivy-Derived Adhesive

### 14.4.1 *Preferable Surface Wetting Favored by the Ivy Nanoparticles*

As one of the most intricate glycoprotein families present in the extracellular space of plant cells (Showalter 2001; Crawford and Yanofsky 2008; Crawford et al. 2007; Ellis et al. 2010), the diversity of the protein backbones and the complexity of the anchored *O*-glycans allow the AGPs to be involved in numerous aspects of growth and development. In particular, it has been demonstrated that the AGPs are capable of aiding in the formation of pollen tubes, facilitating cellular communication, inducing and adjusting the release of regulatory factors, and participating in many other cellular events (Showalter 2001). In addition, the AGP molecules are also thought to exert adhesive function on the stigma surface, acting as adherent base for capturing pollens, as evidenced in the earlier studies (Clarke et al. 1979; Crawford et al. 2007; Lennon et al. 1998; Ge et al. 2010). As such, it is hypothesized that the AGP-rich nanoparticles exuded by the adventitious roots of English ivy may be involved in similar molecular events to those observed on the stigma surface. In this respect, the efforts in the identification of the roles of the AGP-rich ivy nanoparticles in the generation of strong adhesion strength may considerably improve our understanding of a series of adhesive behaviors in plant kingdom. Furthermore, in terms of the engineering application, given that one typical AGP-rich molecule, gum arabic, has been extensively developed as commercial glues in the stamp industry (Showalter 2001), the insight into the principles controlling the molecular interactions within the ivy adhesive may have broader impacts on offering guidelines for directing the design and fabrication of engineered or biomimetic glues.

To explore the potential molecular mechanisms that allow the AGP-rich nanoparticles derived from English ivy to intensify the surface adhesion, the intrinsic viscosity of the purified ivy nanoparticles was examined (Huang et al. 2016). As expected, similar to a variety of typical AGPs, AGP-rich ivy nanoparticles exhibited an exceedingly low intrinsic viscosity in solution ($[\eta]_{\text{ivy NPs}} = 30.4 \pm 1.9$ ml/g), whereas approximately 3.7-fold higher intrinsic viscosity was obtained in gum arabic and ~ 8- to 20-fold higher viscosities were observed in pectin and sodium alginate, respectively. Given that the low intrinsic viscosity is commonly regarded as the most exceptional physical trait with respect to the AGPs (Fincher et al. 1983; Ellis et al. 2010; Showalter 2001), and "wattle-blossom" theoretical model proposed previously has associated this low intrinsic viscosity with the anticipated spheroidal appearance of the majority of the AGPs (Fincher et al. 1983; Ellis et al. 2010; Showalter 2001), it is logical to expect the potential correlations between this physical property and the physiological activities of the AGP-rich ivy nanoparticles within the adhesive exudates, even though the implication of this physical property during the growth and development of plants has not been illustrated in the earlier studies (Ellis et al. 2010; Showalter 2001; Fincher et al. 1983). Accordingly, it is important to elucidate the physiological meanings relevant to the low intrinsic viscosity of the AGP-rich ivy nanoparticles in *H. helix*.

In most cases, to realize ideal adhesive action, an engineered synthetic glue should be designed and optimized to ensure preferable functionalities in two aspects: a favorable wetting behavior on the surface and an effective strategy to originate sufficient adhesion strength during the subsequent curing (hardening) procedure (Comyn 1997; Ferguson 2000; Damico 2005; Xie et al. 2015; Mehdizadeh et al. 2012; Mehdizadeh and Yang 2013; Callow and Callow 2006). Analogous to a series of conventional glues, to accomplish adhesion at the interface, the sticky exudates derived from English ivy should initially wet the surface, a spreading motion capable of driving an intimate and extensive contact over the substrates. In light of the established wetting theory, the liquid adhesives with lower viscosities typically exhibit a better wetting performance than the highly viscous glues (Comyn 1997; Ferguson 2000; Zhu et al. 2013; Lee et al. 2000; Duncan et al. 2005). Thereby, in the case of the ivy adhesive, the comparatively low intrinsic viscosity of the AGP-rich ivy nanoparticles in solution is beneficial for the wetting activity that can be achieved by the ivy adhesive. Additionally, owing to their tiny scale, these nano-sized spheroidal particles are also exceptional in penetrating surface irregularities present on most substrates, further promoting intimate interactions between the ivy-derived adhesive and corresponding substrates.

Apart from the wetting activity favored by the low intrinsic viscosity of the AGP-rich ivy nanoparticles, as a typical structural adhesive, in principle, a curing step is requisite for the ivy-derived adhesive to create sufficient cohesive strength within the exudate, aiding in the formation of strong adhesion strength at the interface (Comyn 1997; Ferguson 2000; Damico 2005; Mehdizadeh et al. 2012; Mehdizadeh and Yang 2013; Xie et al. 2015; Callow and Callow 2006). Commonly, curing (hardening) of adhesives arises from water evaporation and/or chemical cross-linking (Comyn 1997; Ferguson 2000; Damico 2005; Mehdizadeh et al. 2012; Mehdizadeh and Yang 2013; Xie et al. 2015; Callow and Callow 2006). Notably, in the earlier studies, morphological descriptions with respect to the drying process involved in the curing of botanic adhesives derived from the adventitious roots of English ivy (Zhang et al. 2008b), and the climbing organs of analogous plants (Groot et al. 2003), have substantially advanced our understanding of the relevant molecular events regulating the adhesive action in plant kingdom.

### 14.4.2 Chemical Components Within the Ivy-Derived Adhesive

In order to investigate the hardening of the ivy-derived adhesive in detail and the role of the AGP-rich ivy nanoparticles in promoting this process, a cultivation platform was developed to collect and observe the sticky exudate in situ, as detailed in Sect. 14.2.3. Throughout the secretory and curing progress, traces left on the surface of the fixed silicon wafers were monitored using SEM (Huang et al. 2016). It was observed that considerable spheroidal nanoparticles were secreted toward the silicon wafers, present around the imprints of the adventitious roots of English ivy, demonstrating a similar average size to those extracted and purified in vitro. In addition, the ivy nanoparticles remaining on the silicon wafers exhibited a strong tendency to cluster, forming bulky pads at the interface. Given that the AGPs are characterized by their capacity to agglomerate, as described in (Showalter 2001), it is reasonable to propose that the aggregation of the ivy nanoparticles within the sticky exudate is driven by the physicochemical interactions of the AGPs. Consistently, in the case of the AGP-rich nanoparticles isolated and purified in vitro by means of SEC-HPLC, a similar agglomerate fashion of nanoparticles over time was also observed (Huang et al. 2016). Furthermore, the agglomerate progress of the nanoparticles toward bulky pads demonstrated a trend in the formation of a compact film at the interface, implying the pivotal roles of the spheroidal nanoparticles within the exudates during the curing procedures.

In addition to the abundant spherical nanoparticles observed in the periphery of the imprints of the root hairs, a translucent gel-like porous network comprising tightly cross-linked spheres in nanoscale was also captured in the adhesive remnant on the silicon wafers by SEM observation (Huang et al. 2016). To determine the chemical constituents of this matrix, the remnant substances exuded from the adventitious roots on the silicon wafers were resuspended in PBS and examined using ELISA screening test, with 38 mAbs raised against the vast majority of the polysaccharides present in the plant cell wall (Huang et al. 2016). Accompanied by arabinogalactans, diverse pectic epitopes are also richly distributed in the adhesive secretions. Moreover, subsequent glycosyl composition assay identified that 4.09 % (mol%) Gal*Ap* residues were contained in the adhesive substances recovered from the remnant on the silicon wafers, further suggesting the existence of pectic polysaccharides in the mucilage derived from the adventitious roots of English ivy. Meanwhile, from the monosaccharide composition analyses, it is also noteworthy that the proportion of Glc, Man, and Xyl as a whole is greater than 65 % (mol%) of the total monosaccharides, suggesting the presence of cellulose and/or hemicellulose in the imprints remnant on the silicon wafers. These substances presumably arise from the encapsulation of partial components of the plant cell wall within the cured adhesive, a phenomenon that has been detailed in Zhang et al. (2008b) and Groot et al. (2003). Given that arabinogalactans and pectic substances have been identified to be two of the predominant components in the majority of botanic adhesives, including those obtained from the climbing organs of Virginia creeper, Boston ivy, and *Ficus pumila*, as shown in the previous cytochemical analyses (Groot et al. 2003; Bowling and Vaughn 2008a, b, 2009; Young et al. 2008; Bowling et al. 2008; Meloche et al. 2007), it is logical to expect that these two acidic polysaccharides possess exceptional capacity to effectively support the adhesive function of the sticky exudates at the interface. In particular, for the mucilage secreted by the root hairs of English ivy, the pectic acids may exist alone and/or be covalently bonded to the AGPs within the ivy nanoparticles as interpreted above.

### 14.4.3 Calcium-Driven Cross-Linking

More evidence regarding the coexistence of the AGPs and the pectic polysaccharides within the extracellular matrix of ivy root cells was obtained from the subsequent immunohistochemical assessment, further implying the functional correlations between these two components (Huang et al. 2016). Kevin C. Vaughn et al. have appropriately described the pectins and the AGPs in the botanic adhesive as "mucilaginous molecules that are spread across the surface of the structure to be attached, filling in the gaps" (Vaughn and Bowling 2011). In particular, "arabinans and AGPs appear to be an even more mobile component of the adhesive, filling in spaces between the papillate epidermal cells and even moving into small cracks in the structure that is attached" (Vaughn and Bowling 2011). In this respect, the effort in the exploration of the potential interactions among these acidic polysaccharides/glycoproteins may substantially improve our understanding of the molecular events controlling the generation of strong adhesion strength within the ivy-derived bioadhesive. For such a purpose, a dot blotting test and a fluorescent combination assay were carried out to evaluate the hypothetical binding between the AGP-rich ivy nanoparticles and the pectic polysaccharides, apart from the aforementioned covalent connections of the AGPs and the pectic fraction within the ivy nanoparticles (Huang et al. 2016). In both tests, FITC-conjugated ivy nanoparticles are prone to being arrested by the adsorbed pectin (20–34 % esterified, extracted from citrus fruit) in the presence of calcium ions, while this combination is partially inhibited by the $Ca^{2+}$-chelating agent, EGTA (Huang et al. 2016). By means of these tests, calcium ion-driven electrostatic interaction is determined to be the predominant force aiding in the combination of the AGP-rich ivy nanoparticles and the acidic pectic polysaccharides. This electrostatic interaction is displayed by calcium ions in facilitating the cross-linking among carboxyl groups of the uronic acid residues within the AGPs and the pectic acids. It is noteworthy that this $Ca^{2+}$-driven event has been frequently discussed in the earlier reports (Showalter 2001; Baldwin et al. 1993; Immerzeel et al. 2006). Meanwhile, given that $Ca^{2+}$ is one of the richest and most physiologically vital ions present in the extracellular space of plant cells (O'Neill et al. 2004; Demarty et al. 1984; Jarvis 1984), it is reasonable to conclude that the $Ca^{2+}$-regulated cross-linking among these acidic polysaccharides/glycoproteins undoubtedly renders potent driving force, effectively promoting the curing (hardening) progress of the sticky exudate derived from the adventitious roots of English ivy. In particular, under natural conditions, AGPs have shown a trend in raising the porosity of the native pectic gel, functioning as "pectin plasticizers" as described by several authors (Seifert and Roberts 2007; Ellis et al. 2010; Lamport et al. 2006). In this respect, it seems reasonable to propose here that in the ivy-derived adhesive, uronic acid-rich AGPs and pectic polysaccharides may cross-link to form a porous network upon hardening, an architecture shown in the SEM observation as mentioned above.

## *14.4.4 Putative Model of the Ivy-Derived Bioadhesive*

In light of the information gained, the molecular basis for the ivy-derived adhesive at the interface is envisaged and summarized (Fig. 14.2) (Huang et al. 2016). Initially, for the attachment, the AGP-rich ivy nanoparticles are secreted toward the extracellular space of root cells upon contact with corresponding substrates in a manner that has been detailed in Sect. 14.2.3. These spheroidal nanoparticles are concentrated during evaporation, and the highly structural flexibility of the protein backbone as well as the anchored bulky AG branches of the AGPs allows these macromolecules to be tightly packed (Fincher et al. 1983), reaching an intimate connection between adjacent nanoparticles. Subsequent $Ca^{2+}$-driven cross-linking among carboxyl groups of the uronic acid residues within the AGPs and the pectic acids in the extracellular space favors the cohesion of the adjacent AGP-rich nanoparticles, gives rise to the generation of an adhesive film (Showalter 2001; Tan et al. 2004; Baldwin et al. 1993), further aids in the curing progress of the exuded adhesive, and thus realizes the adhesive function at the interface by restraining the relative movement of the adventitious roots and the corresponding substrates. Throughout these procedures, the AGP-rich ivy nanoparticles also possess the capacity to permeate irregularities present on the substrates, resulting in a strong mechanical interlocking at the interface and further ensuring an ideal

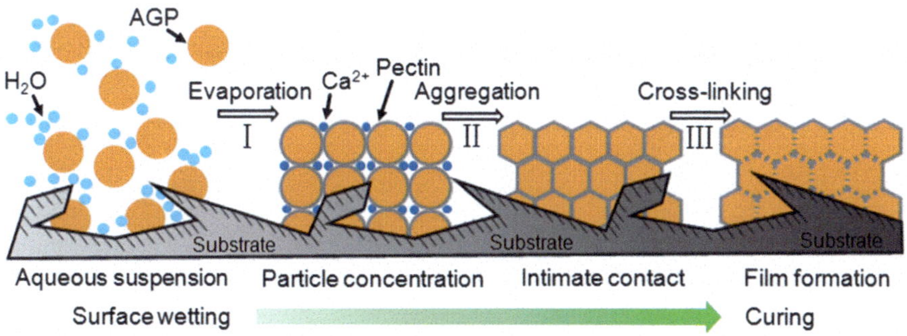


**Fig. 14.2** A schematic drawing of the molecular basis for the ivy-derived adhesive. Adhesive substances are exuded toward the extracellular spaces of root cells, containing AGP-rich nanoparticles, in a manner that has been detailed in Sect. 14.2.3. The spheroidal shape of these nanoparticles allows them to spread at the interface and permeate into the substrates, as a result of their low intrinsic viscosity. Upon evaporation, nanoparticles are concentrated and packed to reach tight contact, giving rise to the formation of a film. Further calcium-dependent cross-linking among carboxyl groups of the AGPs and the pectic polysaccharides elevates the cohesive strength of this film, and the intimate contact of the nanoparticles with the corresponding substrates causes an effective mechanical interlocking at the interface. Image is cited from Huang et al. (2016)

adhesive action (Comyn 1997; Ferguson 2000; Mehdizadeh et al. 2012; Mehdizadeh and Yang 2013; Xie et al. 2015; Callow and Callow 2006; Zhu et al. 2013; Lee et al. 2000).

The manner in which nano-sized particles are involved during the formation of an adhesive film between two adherends has been well elucidated in the case of the poly(vinyl acetate) adhesive, which is a conventional glue prepared by means of emulsion polymerization (Budhlall et al. 2003; Ferguson 2000; Zhang et al. 2008a; Chen et al. 2011a, b; Cai et al. 2003). Typically, commercially available poly(vinyl acetate) glues are comprised of synthetic particles in nanoscale dispersed in aqueous suspension (Zhang et al. 2008a; Chen et al. 2011a, b; Cai et al. 2003). After being applied to a substrate, a three-step strategy was employed by this glue to accomplish adhesive activity at the interface, including (1) tight packing of the particles upon evaporation, (2) deformation of the particles toward intimate interactions, and (3) coalescence (cross-linking) between adjacent particles to create a cohesively strong solid (Zhang et al. 2008a; Chen et al. 2011a, b; Cai et al. 2003). Analogously, these engineered synthetic nanoparticles boost the mechanical interlocking of the glue at the interface in a similar pattern to the spheroidal nanoparticles observed in the ivy-derived adhesive, as documented in the previous reports (Budhlall et al. 2003; Ferguson 2000; Winnik and Yekta 1997; Men 2012; Comyn 1997). In this respect, in contrast to the consecutive molecular events in which the AGP-rich ivy nanoparticles are involved as detailed above, it is logical to propose here that these two types of polymeric nanoparticles share considerable mutual principles underlying their respective adhesive activities.

## 14.5 Adhesion Strength

### *14.5.1 Adhesion Force of the Ivy-derived Adhesive Characterized by AFM*

An AFM-based approach, relying on the principles of tensile testing, was developed to quantitatively evaluate the strength of the ivy-derived adhesive. Force curves were plotted versus displacement by the AFM system during a 24-h test. A strong adhesion force up to 298 nN and a high elasticity were obtained from the adhesive substances (Xia et al. 2011).

### *14.5.2 Adhesion Strength of the Ivy-Mimetic Adhesive Composites*

In light of the uncovered molecular basis for the ivy-derived bioadhesive, a reconstructed biomimetic adhesive was subsequently developed by integrating the purified ivy nanoparticles with pectin in the presence of 2 mM calcium ions, to offer an engineering instance that might further evidence the putative adhesion mechanisms summarized above (Huang et al. 2016). To evaluate the behavior of the prepared adhesive composites, the adhesion strength of this reconstructed adhesive was examined by lap joint shear strength test, an approach that has been extensively applied to quantitatively assess bioadhesives (Rose et al. 2014; Matos-Pérez et al. 2012; Zhang et al. 2014; Mehdizadeh et al. 2012; Xie et al. 2015). The variation in shear strength of the prepared adhesive constructs was monitored over time to reflect the curing progress (Huang et al. 2016). In general, the shear strength of the ivy-mimetic composites at failure was elevated with increasing time and reached a plateau of maximum value in approximately 3 days. To explore the specific roles of each component within the developed material, the shear strengths of the $Ca^{2+}$-free and EGTA-containing adhesive constructs, as well as those of the individual ivy nanoparticles/pectin incorporated with calcium ions, were also traced throughout the test under the same condition. Similarly, the maximal shear strengths of the respective control groups could not be approached until approximately 3 days. However, a significant difference in the shear strength values that could be reached after 3 days was observed among distinct adhesive composites. On day 7, a shear strength up to $0.53 \pm 0.033$ MPa was achieved by the ivy-mimetic adhesive construct prepared in the presence of $Ca^{2+}$ and in the absence of EGTA, substantially greater than those of the $Ca^{2+}$-free or EGTA-containing counterparts, indicating the significance of the calcium ions in developing an ivy-mimetic adhesive composite with the expected level of performance. Meanwhile, the shear strengths of the composites consisting of 2 mM $Ca^{2+}$ integrated with either ivy nanoparticles or pectin alone were approximately $0.12 \pm 0.036$ and $0.40 \pm 0.017$ MPa, respectively, still markedly weaker than that of the ivy-mimetic adhesive constructs 7 days after preparation.

## 14.6 Conclusions and Future Prospects

The spherical nanoparticles detected in the mucilage exuded by the adventitious roots of English ivy are identified to be predominantly composed of AGPs, a superfamily of HRGPs present in the plant cell wall. It has been proposed that the genes encoding the HRGPs may originate from a "superfamily" of ancient genes relevant to diverse adhesive events in both animals and plants (Callow and Callow 2006). In comparing the cell walls of plants to the corresponding extracellular matrices of animals, with respective frameworks built with HRGP extensin family and collagen, obviously, considerable commonalities are shared by these two types of polymeric networks (Callow and Callow 2006). In particular, repeat motifs dominated by Hyp are observed in both matrices, defining and determining the helical conformation during the complicated assembly and cross-linking process (Ferris et al. 2001). Meanwhile, the helical conformation is thought to be stabilized by the arabinosyl/galactosyl modification of the Hyp within both networks (Callow and Callow 2006; Ferris et al. 2001). Notably, a similar Hyp-rich motif is also identified in the adhesive proteins isolated from mussel (Waite and Tanzer 1981; Lin et al. 2007), suggesting the potential evolutionary homology between the adhesive proteins derived from animals and plants (Callow and Callow 2006). As another typical subfamily of HRGPs, the AGPs identified in the ivy nanoparticles are also rich in Hyp. In this respect, there may be still some ubiquitous principles between these two types of bioadhesives waiting for us to explore.

To explore the feasibilities of developing ivy-inspired biomaterials or devices with desirable functionalities by implementing the molecular mechanisms unlocked from the ivy-derived bioadhesive, an initial attempt was made in tissue engineering. In particular, the AGP-rich ivy nanoparticles were exploited as nano-fillers to develop fibrous scaffolds incorporated with collagen, via mimicking the porous matrix where the ivy-derived nanoparticles are embedded under natural condition (Huang et al. 2015). Enhanced adhesion of smooth muscle cells and accelerated proliferation of mouse aortic SMCs are observed in this newly constructed scaffold (Huang et al. 2015). In addition, these spherical nanoparticles were also evaluated as drug-loaded nano-carriers and sunscreen fillers for corresponding biomedical applications, due to their biocompatible nature and exceptional physical (optical), chemical, and biological properties (Huang et al. 2013, 2015).

To summarize, the AGP-rich ivy nanoparticles are essential components in the mucilage exuded by the adventitious roots of English ivy, capable of favoring the generation of strong adhesion strength which aids in surface clinging. The advance in unlocking the molecular mechanisms underlying the ivy-derived adhesive not only forwards our understanding of other botanic bioadhesives but may also open new frontiers for the development of ivy-mimetic and ivy-inspired high-strength adhesive composites eventually, to expand their engineering applications.

## References

Autumn K, Liang YA, Hsieh ST, Zesch W, Chan WP, Kenny TW, Fearing R, Full RJ (2000) Adhesive force of a single gecko foot-hair. Nature 405(6787):681–685

Baldwin TC, McCann MC, Roberts K (1993) A novel hydroxyproline-deficient arabinogalactan protein secreted by suspension-cultured cells of *Daucus carota* (purification and partial characterization). Plant Physiol 103(1):115–123

Bar-Cohen Y (2005) Biomimetics: biologically inspired technologies. CRC Press, Boca Raton, FL

Bhushan B (2007) Adhesion of multi-level hierarchical attachment systems in gecko feet. J Adhes Sci Technol 21(12–13):1213–1258

Bowling A, Vaughn K (2008a) Structural and immunocytochemical characterization of the adhesive tendril of Virginia creeper (*Parthenocissus quinquefolia* [L.] Planch.). Protoplasma 232(3-4):153–163

Bowling AJ, Vaughn KC (2008b) Immunocytochemical characterization of tension wood: gelatinous fibers contain more than just cellulose. Am J Bot 95(6):655–663

Bowling AJ, Vaughn KC (2009) Gelatinous fibers are widespread in coiling tendrils and twining vines. Am J Bot 96(4):719–727

Bowling AJ, Maxwell HB, Vaughn KC (2008) Unusual trichome structure and composition in mericarps of catchweed bedstraw (*Galium aparine*). Protoplasma 233(3-4):223–230
Budhlall B, Shaffer O, Sudol E, Dimonie V, El-Aasser M (2003) Atomic force microscopy studies of the film surface characteristics of poly (vinyl acetate) latexes prepared with poly (vinyl alcohol). Langmuir 19(23):9968–9972
Burnham RJ, Revilla-Minaya C (2011) Phylogenetic influence on twining chirality in lianas from Amazonian Peru 1. Ann MO Bot Gard 98(2):196–205
Burris JN, Lenaghan SC, Zhang M, Stewart CN (2012) Nanoparticle biofabrication using English ivy (*Hedera helix*). J Nanobiotechnol 10:41
Cai HH, Li SD, Tian GR, Wang HB, Wang JH (2003) Reinforcement of natural rubber latex film by ultrafine calcium carbonate. J Appl Polym Sci 87(6):982–985
Callow JA, Callow ME (2006) The Ulva spore adhesive system. In: Biological adhesives. Springer, Heidelberg, pp 63–78
Cannon MC, Terneus K, Hall Q, Tan L, Wang Y, Wegenhart BL, Chen L, Lamport DT, Chen Y, Kieliszewski MJ (2008) Self-assembly of the plant cell wall requires an extensin scaffold. Proc Natl Acad Sci USA 105(6):2226–2231
Cha HJ, Hwang DS, Lim S (2008) Development of bioadhesives from marine mussels. Biotechnol J 3(5):631
Chandran SP, Chaudhary M, Pasricha R, Ahmad A, Sastry M (2006) Synthesis of gold nanotriangles and silver nanoparticles using *Aloe vera* plant extract. Biotechnol Prog 22 (2):577–583
Chen X, Fischer S, Men Y (2011a) Temperature and relative humidity dependency of film formation of polymeric latex dispersions. Langmuir 27(21):12807–12814
Chen X, Fischer S, Yi Z, Boyko V, Terrenoire A, Reinhold F, Rieger J, Li X, Men Y (2011b) Structural reorganization of a polymeric latex film during dry sintering at elevated temperatures. Langmuir 27(13):8458–8463
Clarke A, Gleeson P, Harrison S, Knox RB (1979) Pollen-stigma interactions: identification and characterization of surface components with recognition potential. Proc Natl Acad Sci USA 76 (7):3358–3362
Comyn J (1997) Adhesion science, vol 13. Royal Society of Chemistry, London
Crawford BCW, Yanofsky MF (2008) The formation and function of the female reproductive tract in flowering plants. Curr Biol 18(20):R972–R978
Crawford BC, Ditta G, Yanofsky MF (2007) The NTT gene is required for transmitting-tract development in carpels of *Arabidopsis thaliana*. Curr Biol 17(13):1101–1108
Damico DJ (2005) Advances in adhesives, adhesion science, and testing, vol 1463. ASTM International, West Conshohocken, PA
Darwin C (1865) On the movements and habits of climbing plants. J Linnean Soc Lond Bot 9 (33-34):1–118
Demarty M, Morvan C, Thellier M (1984) Calcium and the cell wall. Plant Cell Environ 7 (6):441–448
Du H, Simpson RJ, Moritz RL, Clarke AE, Bacic A (1994) Isolation of the protein backbone of an arabinogalactan-protein from the styles of *Nicotiana alata* and characterization of a corresponding cDNA. Plant Cell Online 6(11):1643–1653
Duncan B, Mera R, Leatherdale D, Taylor M, Musgrove R (2005) NPL Report
Ellis M, Egelund J, Schultz CJ, Bacic A (2010) Arabinogalactan-proteins: key regulators at the cell surface? Plant Physiol 153(2):403–419
Endress AG, Thomson WW (1976) Ultrastructural and cytochemical studies on the developing adhesive disc of Boston ivy tendrils. Protoplasma 88(2–4):315–331
Fant C, Elwing H, Höök F (2002) The influence of cross-linking on protein-protein interactions in a marine adhesive: the case of two byssus plaque proteins from the blue mussel. Biomacromolecules 3(4):732–741
Favi PM, Yi S, Lenaghan SC, Xia L, Zhang M (2014) Inspiration from the natural world: from bio-adhesives to bio-inspired adhesives. J Adhes Sci Technol 28(3-4):290–319

Ferguson CJ (2000) Core-shell polymers from styrene and vinyl acetate for use as wood adhesives. A thesis submitted in partial fulfilment of the requirements for the degree of doctor of philosophy in chemistry at the University of Canterbury, Christchurch. University of Canterbury, New Zealand

Ferris PJ, Woessner JP, Waffenschmidt S, Kilz S, Drees J, Goodenough UW (2001) Glycosylated polyproline II rods with kinks as a structural motif in plant hydroxyproline-rich glycoproteins. Biochemistry (Mosc) 40(9):2978–2987

Fincher GB, Stone BA, Clarke AE (1983) Arabinogalactan-proteins: structure, biosynthesis, and function. Annu Rev Plant Physiol 34(1):47–70

Ge X, Chang F, Ma H (2010) Signaling and transcriptional control of reproductive development in Arabidopsis. Curr Biol 20(22):R988–R997

Groot E, Sweeney E, Rost T (2003) Development of the adhesive pad on climbing fig (*Ficus pumila*) stems from clusters of adventitious roots. Plant Soil 248(1–2):85–96

Hansen W, Autumn K (2005) Evidence for self-cleaning in gecko setae. Proc Natl Acad Sci USA 102(2):385–389

Huang Y, Lenaghan SC, Xia L, Burris JN, Stewart CN Jr, Zhang M (2013) Characterization of physicochemical properties of ivy nanoparticles for cosmetic application. J Nanobiotechnol 11 (1):1–12

Huang Y, Wang Y-J, Wang Y, Yi S, Fan Z, Sun L, Lin D, Anreddy N, Zhu H, Schmidt M (2015) Exploring naturally occurring ivy nanoparticles as an alternative biomaterial. Acta Biomater 25:268–283

Huang Y, Wang Y, Tan L, Sun L, Petrosino J, Cui M-Z, Hao F, Zhang M (2016) Nano-spherical arabinogalactan protein: a key component in the high-strength adhesive secreted by English ivy. Proc Natl Acad Sci USA 113:E3193–E3202

Immerzeel P, Eppink MM, De Vries SC, Schols HA, Voragen AG (2006) Carrot arabinogalactan proteins are interlinked with pectins. Physiol Plant 128(1):18–28

Isnard S, Cobb AR, Holbrook NM, Zwieniecki M, Dumais J (2009) Tensioning the helix: a mechanism for force generation in twining plants. Proc R Soc Lond B Biol Sci 276:2643–2650

Jarvis MC (1984) Structure and properties of pectin gels in plant cell walls. Plant Cell Environ 7 (3):153–164

Kitazawa K, Tryfona T, Yoshimi Y, Hayashi Y, Kawauchi S, Antonov L, Tanaka H, Takahashi T, Kaneko S, Dupree P (2013) β-Galactosyl yariv reagent binds to the β-1, 3-galactan of arabinogalactan proteins. Plant Physiol 161(3):1117–1126

Lamport DT, Kieliszewski MJ, Showalter AM (2006) Salt stress upregulates periplasmic arabinogalactan proteins: using salt stress to analyse AGP function. New Phytol 169 (3):479–492

Lee JW, Park JH, Robinson JR (2000) Bioadhesive-based dosage forms: the next generation. J Pharm Sci 89(7):850–866

Lee H, Scherer NF, Messersmith PB (2006) Single-molecule mechanics of mussel adhesion. Proc Natl Acad Sci USA 103(35):12999–13003

Lenaghan SC, Zhang M (2012) Real-time observation of the secretion of a nanocomposite adhesive from English ivy (*Hedera helix*). Plant Sci (Amsterdam, Neth) 183:206–211

Lenaghan SC, Burris JN, Chourey K, Huang Y, Xia L, Lady B, Sharma R, Pan C, LeJeune Z, Foister S (2013) Isolation and chemical analysis of nanoparticles from English ivy (*Hedera helix* L.). J R Soc Interface 10(87):20130392

Lennon KA, Roy S, Hepler PK, Lord EM (1998) The structure of the transmitting tissue of *Arabidopsis thaliana* (L.) and the path of pollen tube growth. Sex Plant Reprod 11(1):49–59

Lin Q, Gourdon D, Sun C, Holten-Andersen N, Anderson TH, Waite JH, Israelachvili JN (2007) Adhesion mechanisms of the mussel foot proteins mfp-1 and mfp-3. Proc Natl Acad Sci USA 104(10):3782–3786

Majester-Savornin B, Elias R, Diaz-Lanza A, Balansard G, Gasquet M, Delmas F (1991) Saponins of the ivy plant, *Hedera helix*, and their leishmanicidic activity. Planta Med 57(3):260–262

Matos-Pérez CR, White JD, Wilker JJ (2012) Polymer composition and substrate influences on the adhesive bonding of a biomimetic, cross-linking polymer. J Am Chem Soc 134(22):9498–9505
Mehdizadeh M, Yang J (2013) Design strategies and applications of tissue bioadhesives. Macromol Biosci 13(3):271–288
Mehdizadeh M, Weng H, Gyawali D, Tang L, Yang J (2012) Injectable citrate-based mussel-inspired tissue bioadhesives with high wet strength for sutureless wound closure. Biomaterials 33(32):7972–7983
Meloche CG, Knox JP, Vaughn KC (2007) A cortical band of gelatinous fibers causes the coiling of redvine tendrils: a model based upon cytochemical and immunocytochemical studies. Planta 225(2):485–498
Melzer B, Steinbrecher T, Kraft O, Schwaiger R, Speck T (2008) Anhaftungsmechanismen von Efeu (*Hedera helix* L.): Erste Ergebnisse zu Struktur und Funktionsweise. In: Bionik: Patente aus der Natur, 4. Bionik Kongress, pp 284–289
Melzer B, Steinbrecher T, Seidel R, Kraft O, Schwaiger R, Speck T (2010) The attachment strategy of English ivy: a complex mechanism acting on several hierarchical levels. J R Soc Interface 7(50):1383–1389
Men Y (2012) Crystallographic deformation in mechanically soft colloidal crystals derived from polymeric latex dispersions. Soft Matter 8(21):5723–5727
Metcalfe DJ (2005) *Hedera helix* L. J Ecol 93(3):632–648
Ninan L, Stroshine R, Wilker J, Shi R (2007) Adhesive strength and curing rate of marine mussel protein extracts on porcine small intestinal submucosa. Acta Biomater 3(5):687–694
Oberhuber W, Bauer H (1991) Photoinhibition of photosynthesis under natural conditions in ivy (*Hedera helix* L.) growing in an understory of deciduous trees. Planta 185(4):545–553
O'Neill MA, Ishii T, Albersheim P, Darvill AG (2004) Rhamnogalacturonan II: structure and function of a borate cross-linked cell wall pectic polysaccharide. Annu Rev Plant Biol 55:109–139
Park K, Park H (1990) Test methods of bioadhesion. In: Bioadhesive drug delivery systems. CRC Press, Boca Raton, FL, pp 43–64
Peattie AM (2009) Functional demands of dynamic biological adhesion: an integrative approach. J Comp Physiol B 179(3):231–239
Petrie E (2007) Handbook of adhesives and sealants. McGraw-Hill Education, London
Poon S, Heath RL, Clarke AE (2012) A chimeric arabinogalactan protein promotes somatic embryogenesis in cotton cell culture. Plant Physiol 160(2):684–695
Rogler CE, Hackett WP (1975) Phase change in *Hedera helix*: stabilization of the mature form with abscisic acid and growth retardants. Physiol Plant 34(2):148–152
Rose S, Prevoteau A, Elzière P, Hourdet D, Marcellan A, Leibler L (2014) Nanoparticle solutions as adhesives for gels and biological tissues. Nature 505(7483):382–385
Rowe NP, Isnard S, Gallenmüller F, Speck T (2006) Diversity of mechanical architectures in climbing plants: an ecological perspective. In: Ecology and biomechanics: a mechanical approach to the ecology of animals and plants. Taylor & Francis, Boca Raton, FL, pp 35–59
Sardar HS, Yang J, Showalter AM (2006) Molecular interactions of arabinogalactan proteins with cortical microtubules and F-actin in Bright Yellow-2 tobacco cultured cells. Plant Physiol 142 (4):1469–1479
Seidelmann K, Melzer B, Speck T (2012) The complex leaves of the monkey's comb (*Amphilophium crucigerum*, Bignoniaceae): a climbing strategy without glue. Am J Bot 99 (11):1737–1744
Seifert GJ, Roberts K (2007) The biology of arabinogalactan proteins. Annu Rev Plant Biol 58:137–161
Sevier CS, Kaiser CA (2002) Formation and transfer of disulphide bonds in living cells. Nat Rev Mol Cell Biol 3(11):836–847
Shankar SS, Ahmad A, Pasricha R, Sastry M (2003) Bioreduction of chloroaurate ions by geranium leaves and its endophytic fungus yields gold nanoparticles of different shapes. J Mater Chem 13(7):1822–1826

Showalter A (2001) Arabinogalactan-proteins: structure, expression and function. Cell Mol Life Sci 58(10):1399–1417

Silk WK, Hubbard M (1991) Axial forces and normal distributed loads in twining stems of morning glory. J Biomech 24(7):599–606

Stevens MJ, Steren RE, Hlady V, Stewart RJ (2007) Multiscale structure of the underwater adhesive of *Phragmatopoma californica*: a nanostructured latex with a steep microporosity gradient. Langmuir 23(9):5045–5049

Tan L, Qiu F, Lamport DTA, Kieliszewski MJ (2004) Structure of a hydroxyproline (Hyp)-arabinogalactan polysaccharide from repetitive Ala-Hyp expressed in transgenic *Nicotiana tabacum*. J Biol Chem 279(13):13156–13165

Tan L, Eberhard S, Pattathil S, Warder C, Glushka J, Yuan C, Hao Z, Zhu X, Avci U, Miller JS (2013) An Arabidopsis cell wall proteoglycan consists of pectin and arabinoxylan covalently linked to an arabinogalactan protein. Plant Cell Online 25(1):270–287

Tian Y, Pesika N, Zeng H, Rosenberg K, Zhao B, McGuiggan P, Autumn K, Israelachvili J (2006) Adhesion and friction in gecko toe attachment and detachment. Proc Natl Acad Sci USA 103 (51):19320–19325

Tobler F (1912) Die Gattung Hedera: Studien über Gestalt und Leben des Efeus, seine Arten und Geschichte. Fischer, Berlin

Vaughn KC, Bowling AJ (2011) Biology and physiology of vines. Hortic Rev 38:1–21

Velasquez SM, Ricardi MM, Dorosz JG, Fernandez PV, Nadra AD, Pol-Fachin L, Egelund J, Gille S, Harholt J, Ciancia M, Verli H, Pauly M, Bacic A, Olsen CE, Ulvskov P, Petersen BL, Somerville C, Iusem ND, Estevez JM (2011) O-glycosylated cell wall proteins are essential in root hair growth. Science 332(6036):1401–1403

Waite JH, Tanzer ML (1981) Polyphenolic substance of *Mytilus edulis*: novel adhesive containing L-dopa and hydroxyproline. Science 212(4498):1038–1040

Winnik MA, Yekta A (1997) Associative polymers in aqueous solution. Curr Opin Colloid Interface Sci 2(4):424–436

Xia L, Lenaghan SC, Zhang M, Wu Y, Zhao X, Burris JN, Stewart CN Jr (2011) Characterization of English ivy (*Hedera helix*) adhesion force and imaging using atomic force microscopy. J Nanopart Res 13(3):1029–1037

Xie D, Guo J, Mehdizadeh MR, Tran RT, Chen R, Sun D, Qian G, Jin D, Bai X, Yang J (2015) Development of injectable citrate-based bioadhesive bone implants. J Mater Chem B 3 (3):387–398

Xu J, Tan L, Goodrum KJ, Kieliszewski MJ (2007) High-yields and extended serum half-life of human interferon α2b expressed in tobacco cells as arabinogalactan-protein fusions. Biotechnol Bioeng 97(5):997–1008

Young RE, McFarlane HE, Hahn MG, Western TL, Haughn GW, Samuels AL (2008) Analysis of the Golgi apparatus in Arabidopsis seed coat cells during polarized secretion of pectin-rich mucilage. Plant Cell 20(6):1623–1638

Zhang J, Hu S, Rieger J, Roth SV, Gehrke R, Men Y (2008a) Effect of annealing on the deformation mechanism of a styrene/n-butyl acrylate copolymer latex film investigated by synchrotron small-angle X-ray scattering. Macromolecules 41(12):4353–4357

Zhang M, Liu M, Prest H, Fischer S (2008b) Nanoparticles secreted from ivy rootlets for surface climbing. Nano Lett 8(5):1277–1280

Zhang H, Bré LP, Zhao T, Zheng Y, Newland B, Wang W (2014) Mussel-inspired hyperbranched poly (amino ester) polymer as strong wet tissue adhesive. Biomaterials 35(2):711–719

Zhu Z, Zhai Y, Zhang N, Leng D, Ding P (2013) The development of polycarbophil as a bioadhesive material in pharmacy. Asian J Pharm Sci 8(4):218–227

# Chapter 15
# Biomimetic Adhesives and Coatings Based on Mussel Adhesive Proteins

**Yuan Liu, Hao Meng, Phillip B. Messersmith, Bruce P. Lee, and Jeffrey L. Dalsin**

**Abstract** Nature provides many outstanding examples of adhesive strategies from which chemists and material scientists can draw inspiration in their pursuit of new adhesive materials. Mussels secrete adhesive proteins, which enable these organisms to bind tightly to various surfaces under water. One of the key structure component of mussel adhesive protein (MAP) is the presence of a catecholic amino acid, 3,4-dihydroxyphenylalanine (DOPA), which plays an important role in the curing and interfacial binding of MAP. The catechol side chain is capable of undergoing various reversible and irreversible interactions with both organic and inorganic substrates. Modification of inert polymer systems with DOPA and other catechol derivatives have imparted these materials with water-resistant adhesive properties and the ability to cure rapidly. This chapter focuses on the various strategies used in developing biomimetic adhesives and coatings, as well as recent developments of self-healing and smart materials that employ MAP chemistry.

## 15.1 Introduction

Nature provides many outstanding examples of adhesive strategies from which chemists and material scientists can draw inspiration in their pursuit of new adhesive materials. As described in other chapters of this book, detailed studies of the adhesive mechanisms of geckos, mussels, and other organisms during the

Y. Liu • H. Meng • B.P. Lee (✉)
Department of Biomedical Engineering, Michigan Technological University, 1400 Townsend Drive, Houghton, MI 49931, USA
e-mail: bplee@mtu.edu

P.B. Messersmith
Departments of Materials Science and Engineering and Bioengineering, University of California, Berkeley, 210 Hearst Mining Building, Berkeley, CA 94720, USA

J.L. Dalsin
Department of Biomedical Engineering, Northwestern University, Evanston, IL, USA

© Springer International Publishing Switzerland 2016

A.M. Smith (ed.), *Biological Adhesives*, DOI 10.1007/978-3-319-46082-6_15

past several decades have enhanced our understanding of the underlying physico-chemical principles to the extent that direct translation of this knowledge into biomimetic strategies for synthesizing new practical adhesives is now possible. Although new biomimetic adhesives have the potential for impact in many areas of technology, one of the more compelling outlets for these materials is in healthcare delivery, which will be the focus of this chapter. Obvious parallels exist between the marine and human physiologic environment, and a strategy that works well in one context may be useful in the other.

Efforts to develop biomimetic adhesives are most effective when guided by detailed understanding of the key features and mechanisms of natural adhesives. A simple example of this is given by recent mimicry of gecko adhesive in photolithographically micropatterned polymers (Geim et al. 2003), an approach made possible by earlier studies elaborating the fine morphological detail and adhesive force of individual gecko foot setae (Autumn et al. 2000). In the case of secreted adhesives such as those employed by marine organisms, biomimetic efforts are only possible when there is basic understanding of the key macromolecular components and their compositions. Mussel adhesive proteins (MAPs) have been extensively studied, providing much guidance to biomimetic efforts. We begin with a brief review of the unique chemical features of mussel adhesive proteins, a subject that is discussed in greater detail in several review papers (Waite et al. 2005; Lee et al. 2011) and previous chapter. Several approaches to MAP biomimetic are being pursued in labs throughout the world, including extraction of native proteins, expression of genetically engineered MAPs, and chemical synthesis of MAP-mimetic polymers. The emphasis of this review will be on synthetic mussel-inspired material for adhesive construction and surface coating. Finally, we describe the use of mussel-inspired chemistry for designing novel materials in both medical and nonmedical fields.

## 15.2 Mussel Adhesive Proteins and DOPA

The attachment apparatus of mussels is called the byssus, which is a bundle of threads extending from within the shell of the mussel, each of which terminates in an adhesive plaque attached to a substrate (Fig. 15.1) (Waite 1991, 1999). An entire byssal thread and adhesive plaque is formed by secretion of liquid protein from within glands of the mussel foot in a process that resembles polymer injection molding. Solidification of the liquid occurs rapidly to provide a stable attachment, and the process is sequentially repeated many times to deposit each byssal thread and plaque. The proteins of the byssal threads resemble collagen and elastin and will not be further discussed (Waite 1991). The adhesive plaque located at the distal end of each thread will be emphasized here, because this is where adhesion to the substrate is enforced.

At least six *Mytilus edulis* foot proteins (Mefps) have been identified (Waite 1999; Rzepecki et al. 1992; Zhao and Waite 2006). Mefp-1 is the key protein of the

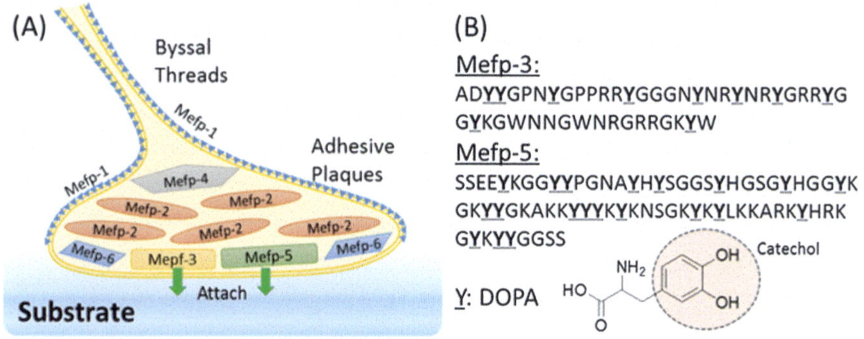


**Fig. 15.1** Schematic illustration of a byssal thread and adhesive plaque (**a**). The sequences of the interfacial proteins Mefp-3 and Mefp-5 are shown along with the chemical structure of DOPA (**b**) (adapted from Lee et al. 2011)

byssal cuticle, while Mefp-2 through 6 are found within the adhesive plaque. These proteins all share a common distinguishing feature: the presence of 3,4-dihydroxyphenylalanine (DOPA), a residue formed by posttransitional hydroxylation of tyrosine. It is the 3,4-dihydroxyphenyl (catechol) side chain that gives DOPA its unique properties related to adhesion. The DOPA content of Mefps ranges from a few percent to well above 20 % (Waite et al. 2005). Mefp-3 and Mefp-5 are predominately found near the interface between the adhesive plaque and substratum (Waite and Qin 2001; Warner and Waite 1999) and may function as mediators of adhesion at the interface between the plaque and the foreign surface.

Mefp-3 and Mefp-5 have the highest DOPA contents of the foot proteins at 20 and 27 mol %, respectively, and also contain a large number of basic residues (arginine in Mefp-3 and lysine in Mefp-5) (Papov et al. 1995; Waite and Qin 2001). Mefp-5 also contains phosphoserine residues in a N-terminal sequence reminiscent of acidic mineral-binding motifs that appear in statherin and osteopontin (Waite and Qin 2001). Little is known about the role of phosphoserine residues, although it is interesting to note that phosphoserines are known to bind to calcareous mineral surfaces, which are common substrates of mussels in the marine environment (Long et al. 1998; Meisel and Olieman 1998).

## 15.3 Catechol Chemistry

The DOPA residues found in the mussel adhesive proteins are widely believed to fulfill two important roles: cohesion and adhesion. Cohesion, as we use the term here, refers to the bulk elastic properties of the adhesive, whereas adhesion refers very specifically to those physicochemical interactions occurring at the interface between adhesive pad and substrate. Both cohesive and adhesive properties of MAPs are important in achieving the final outcome of secure, water-resistant

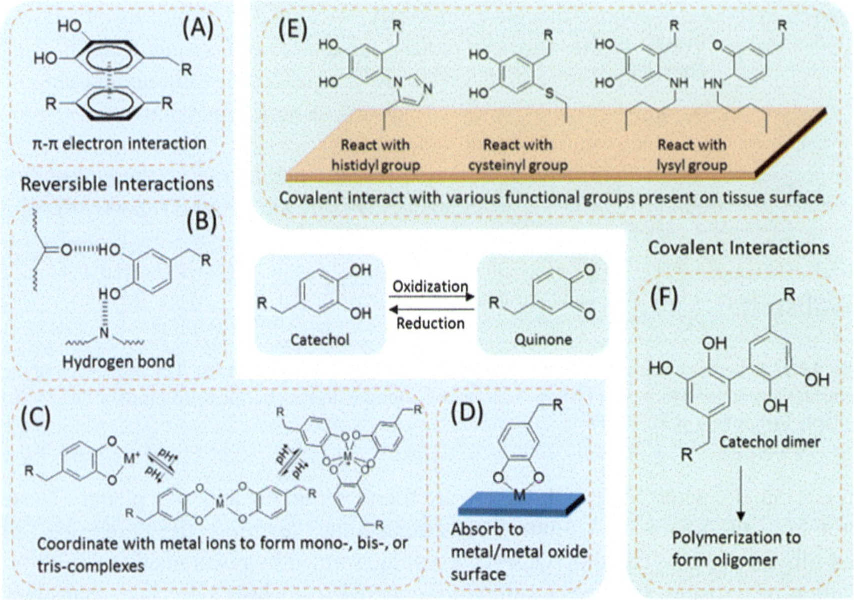


**Fig. 15.2** Catechol side chain chemistry. Center: the catechol–quinone equilibrium is affected by pH and the presence of redox-active species. (**a**) Catechol forms π–π electron interactions with benzene rings. (**b**) Catechol forms hydrogen bonds through phenol hydroxyl groups. (**c**) Catechol chelates metal ions to form complexes with different stoichiometry depending on the solution pH. (**d**) Catechol forms coordination bonds with metal and metal oxide surface. (**e**) Quinone covalently reacts with nucleophilic groups present on tissue surface through Michael addition or Schiff base substitution resulting in adhesion to tissue substrates. (**f**) Catechol and quinone polymerize to form oligomers

attachment to surfaces. The catechol side chain of DOPA is capable of various catechol–catechol and catechol–surface interactions, leading to the curing of the catechol-containing adhesive and strong interfacial binding (Fig. 15.2). Additionally, catechol is a unique molecule capable of forming strong bonds to both inorganic and organic substrates while utilizing either reversible physical or irreversible covalent crosslinks.

### *15.3.1 Reversible Interactions*

The benzene ring of catechol can form π–π interaction with another benzyl moiety (Fig. 15.2a), allowing the catechol-containing material to be able to bind to surfaces rich in aromatic compounds such as polystyrene (Baty et al. 1997). The hydroxyl groups of catechol forms extensive hydrogen bonds (Fig. 15.2b), which allows catechol to compete with water for hydrogen bonding sites and absorb onto mucosal tissues (Deacon et al. 1998; Huang et al. 2002; Schnurrer and Lehr 1996; Chirdon et al. 2003). This interaction also contributes to the cohesive properties of the catechol-containing materials (Waite 1987, 1999).

Catechol is capable of forming strong complexes with metal ions ($Fe^{3+}$, $Ca^{2+}$, $Cu^{2+}$, $Ti^{3+}$, $Ti^{4+}$, $Mn^{2+}$, $Mn^{3+}$, $Zn^{2+}$) with different stoichiometry (i.e., mono-, bis-, and tris-complexes; Fig. 15.2c) (Martin et al. 1996; Rodriguez et al. 1996; Sever and Wilker 2004; Taylor et al. 1996). Strong and reversible catechol–metal ion complexation is responsible for the wear resistance properties, high extensibility, and elevated hardness of mussel byssal cuticles (Holten-Andersen et al. 2009; Sun and Waite 2005). Metal ions account for as much as 1 % by weight in Mefp-1 and its removal resulted in a 50 % reduction in cuticle hardness. Similarly, catechol forms reversible complex with boronic acid (Pizer and Babcock 1977; Fang et al. 2004). The formation of catechol–boronate complexation has been utilized as a temporary protecting group for the synthesis of catechol-modified monomers and polymers (Lee et al. 2007a, b).

The affinity of catechol for metal is not limited to soluble metal ions but is also extended to metal and metal oxide surfaces (Fig. 15.2d). Catechol can be absorbed onto various metal (i.e., Ti alloys and stainless steel) and metal oxides (i.e., $Au_2O_3$, $Al_2O_3$, $Fe_3O_4$, $SiO_2$, $TiO_2$, ZnO) (Kummert and Stumm 1980; Soriaga and Hubbard 1982; Hansen et al. 1995; Lee et al. 2006a; Araujo et al. 2005; Creutz and Chou 2008; Dalsin et al. 2003, 2005; Kawabata et al. 1996; Anderson et al. 2010; Wei et al. 2013; Amstad et al. 2009). Single molecule pull-off tests have demonstrated that the force required to break a metal–catechol bond (0.8 nN) averaged around 40 % that of a covalent bond (2 nN) (Lee et al. 2006b). This interaction is reversible and is one of the strongest reversible bonds involving a biological molecule reported to date.

## 15.3.2 *Oxidization-Induced Covalent Interactions*

In the presence of oxidizing reagent (i.e., $IO_4^-$, $H_2O_2$, enzyme, etc.), catechol is oxidized to its quinone form (Yamamoto and Tatehata 1995; Lee et al. 2001). Catechol can also autoxidize in a slightly basic aqueous solution (Burzio and Waite 2000; Herlinger et al. 1995; Yu et al. 1999). Quinone is highly reactive and can form covalent crosslinks with various nucleophilic functional groups present on tissue surface (i.e., $-NH_2$, –SH, imidazole from lysine, cysteine, and histidine residues, respectively) through either Michael addition or Schiff base substitution (Fig. 15.2e), contributing to forming interfacial bonds with tissue substrates (Burke et al. 2007; Brodie et al. 2011). Additionally, quinone can polymerize to form oligomers consisting of up to six DOPA residues (Fig. 15.2f), leading to the rapid hardening of catechol-containing adhesives (Lee et al. 2002).

Crosslinking of catechol is highly dependent on pH. Under mild acidic conditions (pH 5.7–6.7), catechol crosslinks at a slower rate as a result of increased stability of transient oxidation intermediates (i.e., quinone methide), which reduce the rate of subsequent crosslinking reactions (Cencer et al. 2014; Li and Christensen 1994). On the other hand, a fast crosslinking rate was observed at a neutral to basic pH (pH 7.4–8). Interfacial covalent binding to tissue substrates is also limited at a lower pH due to the protonation of nucleophilic functional groups (e.g., p$K_a$ of ε-lysine ~10) (Cencer et al. 2014). These reported results provide guides for bioadhesive design due to variable physiological pH levels (pH 4–6 for skin (Ohman and Vahlquist 1994), pH 6.7–7.1 for subcutaneous tissue (Soller et al. 1996), pH 7 for dysoxic tissue due to extensive hemorrhage or ischemia (Soller et al. 2003)).

Catechol that contains a free amine group (i.e., dopamine) undergoes intramolecular cyclization between the catechol side chain and the amine group to form a leukochrome, which can continue to polymerize in a process similar to melanin formation (Lee et al. 2002; Kuang et al. 2014). Application of polydopamine as versatile coating and grafting platform is introduced in a later section.

### *15.3.3 ROS Generation During Catechol Oxidization*

During the oxidation process of the catechol group, reactive oxygen species (ROS) such as superoxide anions ($O_2^{\cdot -}$) and hydrogen peroxide ($H_2O_2$) are generated (Fig. 15.3) (Clement et al. 2002; Meng et al. 2015). Micromolar level of $H_2O_2$ is released from polymer network-bound catechol over a period of 48 h under typical cell culture environment (pH 7.4, 37 °C), which drastically decreases cell viability (Meng et al. 2015). Introduction of catalase, a common enzyme that decomposes $H_2O_2$ into $H_2O$ and $O_2$ (Chelikani et al. 2004), could counteract the cytotoxicity effect of the generated $H_2O_2$ (Meng et al. 2015), indicating that the release of $H_2O_2$ is the main source of cytotoxicity in culture media. Release of $H_2O_2$ also stimulates acute inflammatory response in vivo and recruits macrophages to the biomaterial–tissue interface (Meng et al. 2015). The release of ROS can be both advantageous [i.e., promote wound healing (Arul et al. 2012), antimicrobial effect (Brian et al. 2012), and cell signaling (Anderson et al. 2008)] and disadvantageous (i.e., chronic inflammation (Brian et al. 2012)) for a given biomedical application. Therefore, ROS concentration needs to be rigorously regulated for catechol-containing biomaterials.

### *15.3.4 Chemical Modification to Catechol*

Chemical modification of catechol with electron withdrawing chloro and nitro groups (Fig. 15.4a, b, respectively) greatly enhances the interfacial binding strength and reactivity of the catechol group. These modifications lower the dissociation constants (p$K_a$) of the catechol –OH groups and allow the catechol to form complexes with metal ions at a higher stoichiometry or at a reduced pH (Amstad et al. 2011; Menyo et al. 2013). Nitro group functionalized catechol binds more tightly to inorganic surfaces, exhibits increased stability at elevated temperature,

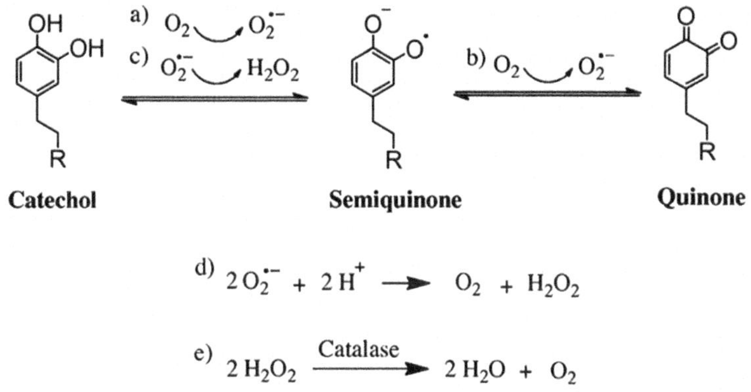


**Fig. 15.3** Mechanism of $H_2O_2$ generated from catechol composed hydrogel during autoxidation process. (**a**) $O_2$ oxidizes catechol to semiquinone and itself transformed into superoxide $O_2^{\cdot -}$. (**b**) $O_2$ further oxidizes semiquinone into quinone form and generates more $O_2^{\cdot -}$. (**c**) $O_2^{\cdot -}$ is quickly consumed and reacts with catechol to form $H_2O_2$. (**d**) $O_2^{\cdot -}$ also reacts with proton ions to form $H_2O_2$. (**e**) Catalase is an enzyme to catalyze $H_2O_2$ into $H_2O$ and $O_2$ (reproduced from Meng et al. 2015)

and demonstrates higher resistance against oxidation under a more basic condition (Amstad et al. 2009, 2011; Ding et al. 2015). The nitro group functionalization also results in a faster crosslinking and degradation rates, and the interfacial crosslinking could occur at a wider pH range when compared to unmodified catechol (Cencer et al. 2015; Anderson et al. 2015). Additionally, nitrocatechol is photolabile (Shafiq et al. 2012), while chlorocatechol prevented biofilm formation and exhibited antimicrobial properties (Garcia-Fernandez et al. 2013). Similarly, pyridine-modified catecholic quinone (Fig. 15.4c) also exhibits strong ability to form metal ion complexes (Menyo et al. 2013) with improved interfacial binding strength to inorganic surfaces (Amstad et al. 2009, 2011).

## 15.4 Mussel-Inspired Adhesive Polymers

### *15.4.1 Extraction and Expression of MAPs*

The purification and chemical analysis of MAPs were first reported by Waite and Andersen (1978). Since then, numerous adhesive proteins from *M. edulis* as well as other species of mussels have been reported (Pardo et al. 1990; Papov et al. 1995; Waite and Qin 2001). The ultimate utility of isolation and purification of MAPs for commercial use remains debatable, however, as large numbers of mussels are needed for even a modest amount of purified protein (Burzio et al. 1997). A possible exception has been reported in the case of *Aulacomya atra*, where yield of purified adhesive protein can be 15–20 times more than other marine mussel species

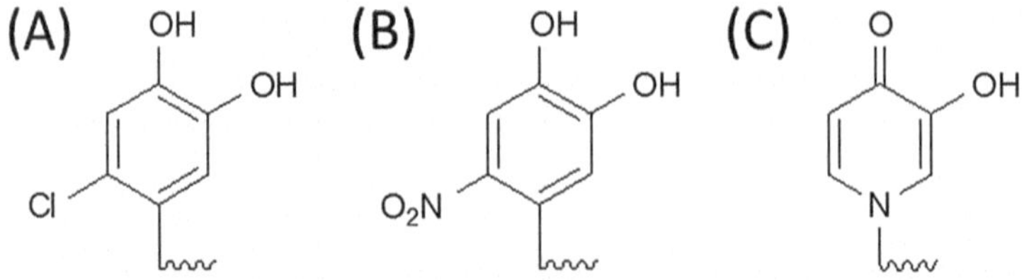


**Fig. 15.4** Chemical structure of chemically modified catechol. The proton (–H) at the para position of the catechol benzene ring was substituted with chloro (**a**) and nitro group (**b**). (**c**) Benzene ring was substituted with a pyridine group

(Characklis 1990). In vitro cell culture and in vivo transplantation experiments showed that purified MAPs appeared to be nontoxic and well tolerated by biological systems (Pitman et al. 1989; Olivieri et al. 1990); however, biological response to these proteins at the cell, tissue, and systemic levels needs to be further investigated.

Although purified MAPs have been suggested for use in a variety of medical contexts, there exist only a handful of literature reports with sufficient experimental detail to allow adequate analysis and interpretation of the results (Green et al. 1987; Grande and Pitman 1988; Ninan et al. 2003; Schnurrer and Lehr 1996). Early studies focused on the use of MAPs for facilitating cell adhesion to substrate or tissue surfaces (Grande and Pitman 1988; Green et al. 1987). Mucoadhesive properties of MAPs have been explored in thin film (Schnurrer and Lehr 1996) and soluble form (Deacon et al. 1998). MAPs were highly mucoadhesive in both forms, suggesting possible future use of MAP-mimetic polymers for adhesion and drug delivery to the mucosal surfaces of the eye, gastrointestinal tract, and reproductive tract.

Purified MAPs have been investigated as potential tissue adhesives (Ninan et al. 2003), where a highly concentrated paste of MAPs was applied to pig skin tissue surfaces and the bond strength determined mechanically. The results of this study suggest potential use as an adhesive, although the curing time of the MAP glue was too slow for possible clinical use and under most conditions the adhesive strength of MAP bonded tissue was similar to or lower than fibrin glue. It is possible that chemical or enzymatic crosslinking reagents could improve the performance of the MAP glue.

Microbial production of marine adhesive proteins has been investigated using recombinant DNA technology. cDNA encoding the adhesive proteins from different species of mussels has been isolated, and microbial hosts such as *E. coli*, *S. cerevisiae*, and *B. subtilis* have been used to express the desired sequences (Maugh et al. 1988; Strausberg et al. 1989; Filpula et al. 1990; Hwang et al. 2004, 2005, 2007; Kim et al. 2008; Choi et al. 2011; Silverman and Roberto 2006). Initially, the purified proteins lack the desired water-resistant adhesion and cohesive strength due to the absence of DOPA functionality as this is a posttranslational modification (Strausberg et al. 1989). Monophenol catecholase such as tyrosinase was used to transform tyrosine residues into DOPA, and a conversion of up to 70 % of tyrosine residues was reported (Yamamoto 1995; Marumo and Waite 1986). Successful adhesion to various substrates including polystyrene, glass, hydrogel, and collagen using recombinant adhesive proteins has been reported (Strausberg and Link 1990). Recently, protein sequences from *E. coli* fiber-forming amyloid and mussel adhesive protein were combined to form a chimeric peptide capable of self-assembling into higher-ordered, adhesive nanofibers (Zhong et al. 2014). This approach may potentially lead to future designs of genetically engineered adhesives that can self-assemble into nano-domains and unique hierarchical structures.

### 15.4.2 Synthetic Mussel-Inspired Polymer Adhesive

Chemical synthesis of peptide analogues of MAP has been extensively investigated by Yamamoto and colleagues (Tatehata et al. 2000; Yamamoto 1987; Yamamoto et al. 1995), who utilized solution methods to synthesize consensus repeat sequences of MAP that were then chemically coupled to yield MAP-like polypeptides. The adhesive strength to iron, alumina, glass, nylon, polyethylene, Teflon, and pig skin was measured mechanically (Tatehata et al. 2000; Yamamoto 1987), or work of adhesion measured by wettability (contact angle) experiments (Yamamoto et al. 1995). Although these MAP-mimetic polypeptides were generally found to be adhesive, the contribution of DOPA to adhesion was not entirely clear from these studies, as similar adhesive values were obtained for DOPA-free polypeptides (Yamamoto 1987; Yamamoto et al. 1995). In one notable case, however, in vitro tissue adhesion to pig skin was clearly enhanced in polypeptides containing tyrosine, which was converted to DOPA and presumably further oxidized during curing of the adhesive (Tatehata et al. 2000). An in vivo pig skin tissue adhesion study was also performed using these polypeptides; however, no mechanical performance data was reported (Tatehata et al. 2001).

Deming and coworkers synthesized MAP-mimetic polypeptides (Fig. 15.5a) by polymerization of *N*-carboxyanhydride monomers of lysine and DOPA to form DOPA-containing peptide (Deming 1997; Yu and Deming 1998) and tested adhesion to steel, aluminum, glass, and several synthetic polymer surfaces (Yu et al. 1999; Yu and Deming 1998). Copolypeptides containing 10 mol % DOPA were crosslinked via oxidation and adhered approximately ten times stronger to aluminum than a polylysine control polymer. Although suggestive of an adhesive role for DOPA, the results are difficult to interpret in view of the potential effects of DOPA crosslinking on cohesive strength of the polymer adhesive. A subsequent study of metal substrate adhesion by the same group using model peptide compounds and similar polypeptides further defined the role of non-oxidized DOPA as adhesive, and that of oxidized DOPA species as contributing primarily to cohesion (Yu et al. 1999). Another example of DOPA-containing adhesive polymer is DOPA/styrene copolymer system (Fig. 15.5b) (Westwood et al. 2007; Matos-Pérez et al. 2012; Meredith et al. 2014). The incorporation of

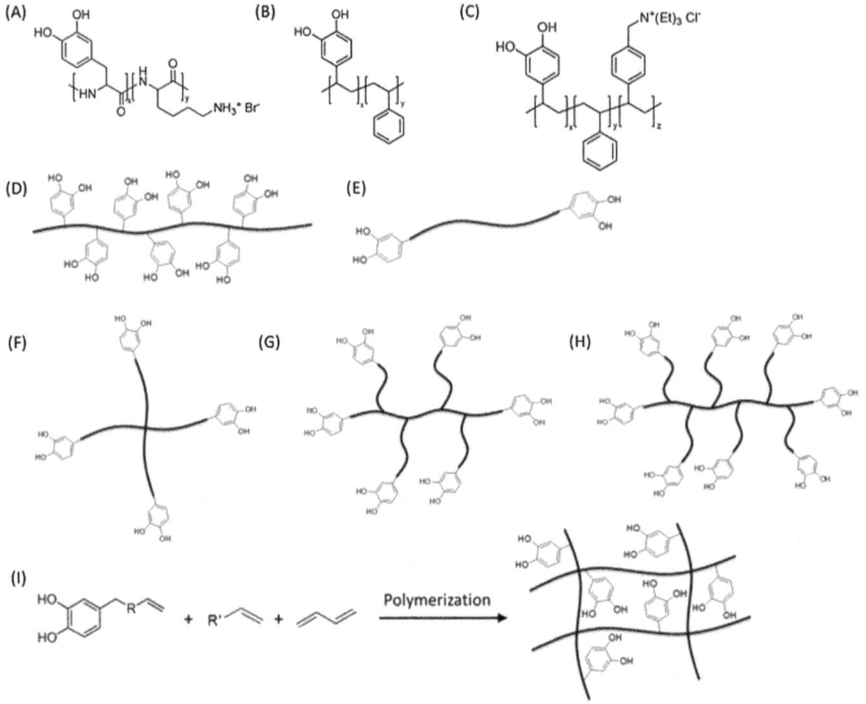


**Fig. 15.5** Chemical structures of mussel-inspired polymers. (**a**) Copolymer of DOPA and lysine. (**b**) Copolymer of catechol and styrene. (**c**) Copolymer containing catechol, styrene, and cationic side chains. (**d**) Macromolecule with side chain modification of catechol. (**e–h**) Linear and 4-, 6-, 8-armed PEG end-modified with catechol. (**i**) Schematic of catechol-containing polymer network preparation via free radical polymerization

styrene has demonstrated extremely high adhesive strength, up to 11 MPa adhesive strength was achieved in aluminum surface lap shear adhesion testing. And copolymerization of DOPA with the cationic composition exhibited enhanced binding property to rock surface in wet environment (Fig. 15.5c) (White and Wilker 2011; Yu et al. 1999).

Incorporation of catechol groups into polymer side chain (Fig. 15.5d) usually aims for promoting the adhesion to tissue and substrate (Lee et al. 2007a, b, 2013b; You et al. 2011; Matos-Perez and Wilker 2012) or providing the functional groups for cohesive crosslinking based on reversible/irreversible catechol interactions (Zhang et al. 2012; Skelton et al. 2013; Ryu et al. 2011; Lee et al. 2010c). Catechol-modified polyacrylamide main chain was blended with a nano-silicate to construct nanocomposite hydrogel with enhanced material properties through strong interfacial interactions between the catechol and the incorporated nanoparticles (Ding et al. 2015). A small amount of incorporated catechol (~0.1 wt% in swollen hydrogel) could significantly increase the network crosslinking density over a wide pH range (pH = 3.0–9.0). Similarly, increased stability was observed in calcium crosslinked catechol-modified alginate gel through catechol–catechol dimer formation (Hong et al. 2015). Additionally, this catechol modification strategy has been widely applied to biopolymers which have desirable degradation rate and bioactivity. Polysaccharide-based bioadhesive polymers have been created by direct coupling of either amine or acid-functionalized catechols onto alginate (Lee et al. 2013a; Kastrup et al. 2012), chitosan (Kim et al. 2010, 2013), dextran (Shazly et al. 2010; Park et al. 2013), gelatin (Choi et al. 2014; Yang et al. 2014), and hyaluronic acid (Neto et al. 2014; Hong et al. 2013).

Linear (Fig. 15.5e) and multi-armed (Fig. 15.5f–h) poly(ethylene glycols) (PEGs) end-capped with catechol have been employed to prepare fast-curing, injectable tissue adhesive and sealant (Bilic et al. 2010; Lee et al. 2002). PEG is a hydrophilic and biocompatible polymer that is widely used for biomedical application (Harris et al. 1991). Crosslinking of catechol at polymer end chain through oxidation reaction or catechol–metal ion coordination gives rise to the formation of 3D polymeric network in situ. Adhesives fabricated with PEG–catechol have demonstrated cytocompatibility, have comparable adhesive properties with commercial fibrin glue (Catron et al. 2006), and have successfully sealed human fetal membrane defects (Bilic et al. 2010; Haller et al. 2012). The adhesive and degradation properties of PEG–catechol adhesive can be controlled by PEG architecture (i.e., linear, 4, 6, 8 arms), molecular weight (2–20 kDa), and PEG–catechol linkage (i.e., ester, urethane, amide) (Harris et al. 1991; Dalsin et al. 2012; Messersmith et al. 2014).

Similar to the PEG–catechol polymers described above, poly(ethylene oxide)–poly(propylene oxide)–poly(ethylene oxide) (PEO–PPO–PEO) block copolymers end-capped with catechol increase adhesive properties of a thermal-responsive polymer system (Huang et al. 2002). The PEO–PPO–PEO block copolymer undergoes temperature-induced aggregation phenomena as a result of the hydrophobic nature of the PPO block (Alexandridis 1997). At low temperature and concentration, PEO–PPO–PEO block copolymers exist in solution as dissolved monomers, but self-assemble at higher concentrations and temperatures into block copolymer micelles that form under conditions defined by the critical micelle concentration (CMC, at constant temperature) and the critical micelle temperature (CMT, at constant concentration). Highly concentrated solutions of certain PEO–PPO–PEO block copolymers, such as Pluronic F127 ($PEO_{100}$–$PPO_{65}$–$PEO_{100}$) and F68 ($PEO_{78}$–$PPO_{30}$–$PEO_{78}$), exhibit solgel transitions when heated from ambient to body temperature (Huang et al. 2002). This physical behavior provides a basis for future use of these polymers for inducing gel formation in medical applications. In addition to linear PEO–PPO–PEO block copolymers, catechol-modified 4-armed PEO–PPO block copolymers exhibited a volumetric reduction and mechanical toughening when the temperature is increased to physiological condition (Barrett et al. 2013). The ability to reduce swelling for the PEO–PPO-based adhesive is advantageous as excessive swelling could lead to severe complications such as nerve compression (Mulder et al. 2009; Blackburn and Smyth 2007).

Another advance in enhancing the mussel-inspired adhesives property is the modification of amphiphilic multiblock copolymer films with the adhesive catechol (Murphy et al. 2010; Lee et al. 2010a, b, 2012). This type of adhesive polymer contains PEG segments which allow the polymer to maintain good "wetting" with the tissue substrates and hydrophobic polyester segments such as polycaprolactone (PCL) to increase cohesive strength due to its ability to self-assemble in the presence of water. This PEG–PCL–catechol polymer can be casted as a thin film (50–120 μm) on biologic scaffolds (i.e., small intestinal submucosa, dermal tissues, and pericardium) (Murphy et al. 2010) or synthetic meshes (Lee et al. 2010a, 2012). These adhesive-coated constructs exhibited 20-fold increase in adhesive strength compared to fibrin glue, reached almost 60 % that of cyanoacrylate-based adhesive, and demonstrated the ability of augmenting primary suture repair of transected porcine Achilles tendons (Brodie et al. 2011).

Preparation of an oxidation-free crosslinking of catechol-containing polymers relies on free radical copolymerization of a catechol-containing monomer, backbone monomer, and crosslinker (Fig. 15.5i) (Lee et al. 2004). The effect of catechol-containing monomers on gelation time, gel conversion, and elastic modulus of the photocured hydrogels was investigated. While the presence of catechol-containing monomers did not prevent polymerization, longer irradiation time was needed to achieve gelation. Gel conversion, extent of catechol incorporation, and elastic modulus also decreased with increasing catechol concentration, suggesting a retarding effect on free radical polymerization. Despite this effect, catechol was successfully incorporated into hydrogels with elastic moduli (ca. 50 kPa) suitable for many biomedical applications. To counteract the retardation effect of catechol on free radical initiated polymerization, catechol-modified triblock copolymers consisting of polylactic acid (PLA) and PEG have been used to segregate catechol from polymerizable methacrylate group to achieve rapid photo-initiated curing (Catron et al. 2006). Similarly, removal of molecular oxygen has been used to create hydrogel networks with minimal reduction in mechanical properties (Anderson et al. 2015; Liu et al. 2013).

## 15.5 Mussel-Inspired Material Surface Functionalization

### *15.5.1 Antifouling Coating Mediated by Catechol-Interface Interactions*

Control of initial nonspecific adsorption of proteins at bio-interfaces is fundamental to studying the biological response to materials and is important in dictating clinical outcomes of implanted devices. In some applications it is important to limit nonspecific fouling of device surfaces, as this can affect device performance and lead to adverse biological responses. Cell and bacterial colonization of surfaces is generally mediated by adsorbed proteins; thus, surfaces which most effectively limit protein adsorption are less likely to support cell and bacterial adhesion.

A common method to reduce nonspecific cell and protein fouling is immobilizing the antifouling polymers on biomaterial surfaces (Dalsin and Messersmith 2005; Nath et al. 2004). In the absence of active release of antifouling agents from the surface, experimental observations from many laboratories make an overwhelming case for the dependence of resistance to protein adsorption on antifouling polymer surface density (Kenausis et al. 2000; Pasche et al. 2003; Sofia et al. 1998; Malmsten et al. 1998; Dalsin et al. 2005), with high polymer surface density providing better fouling resistance than low-density coatings. Thus, the general goal in designing antifouling polymer strategies is to achieve a high density of stably immobilized antifouling polymer on the substrate of interest (Nath et al. 2004). This can be achieved through manipulation of different parameters such as polymer chain length, anchoring chemistry, and antifouling polymer composition as well as the modification of the coating process.

The design of MAP-mimetic polymers for this purpose is conceptually simple: one end of an antifouling polymer is conjugated to a DOPA residue or short peptide (Fig. 15.6), and the antifouling polymer is immobilized onto the substrate surface via the adhesive end group. A typical example of this strategy consists of simple constructs of linear PEGs end-functionalized with DOPA-containing peptide sequence that mimics Mefp-1 (Fig. 15.6a), as well as peptides consisting of 1–3 DOPA residues (Fig. 15.6b), or copeptide consisting of lysine and DOPA residues (Fig. 15.6c) (Dalsin et al. 2003, 2005, 2006). PEG (Fig. 15.6f) is chosen as the antifouling polymer given its ability to effectively reduce the adsorption of cells (Zhang et al. 1998; Kamath et al. 1996; Dalsin et al. 2003, 2006), microbes (Bridgett et al. 1992; Desai et al. 1992; Harris et al. 2004; Ko et al. 2008), and macromolecules (Pasche et al. 2003; Dalsin et al. 2005).

A material of great interest to the medical community is Ti, since its alloys are commonly employed in medical devices due to their excellent biocompatibility, corrosion resistance, and high strength (Brunette et al. 2001). XPS analysis of Ti modified with antifouling polymer provided evidence of a potential mechanism for catechol–Ti binding. Quantitative analysis of raw XPS spectra suggested the occurrence of catechol–Ti charge-transfer complexes (Rodriguez et al. 1996) as evidenced by increasing depletion of surface hydroxyl groups on the substrates with longer DOPA peptides (Dalsin et al. 2005). Remarkably, an anchor composed of only one catechol group was sufficient to confer excellent short-term antifouling properties on Ti, reducing cell adhesion by as much as 98 % compared to control. Control polymers having an identical structure but with tyrosine as anchor showed no reduction in cell attachment, which was attributed to differences in adhesive ability of the catechol functionality of catechol and the phenol side chain of Tyr, providing a clear demonstration of the importance of posttranslational modification of Tyr in adhesion.

It is also clear from the cell attachment results on Au and Ti that catechol does not bind equally well to all surfaces—a single catechol group was clearly sufficient for good anchorage of PEG onto Ti but insufficient on Au (Dalsin and Messersmith

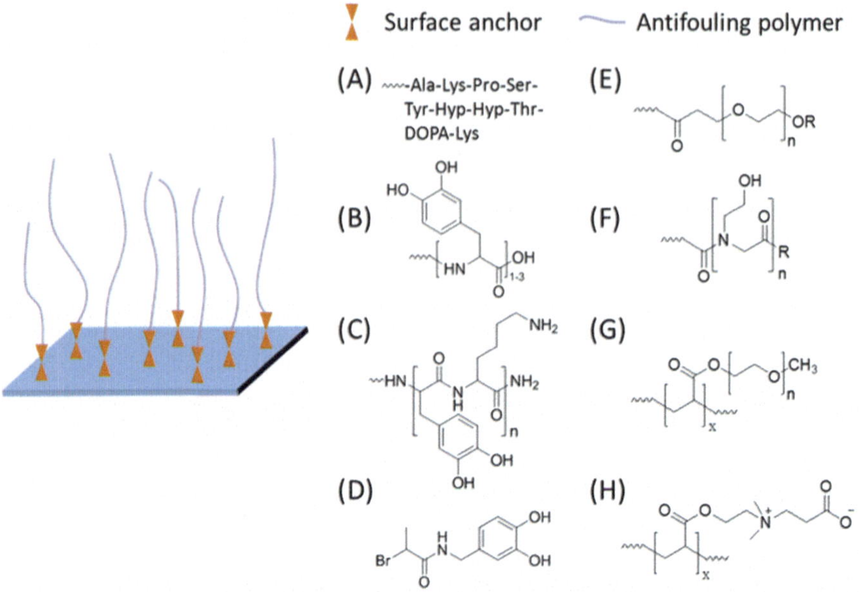


**Fig. 15.6** Representative of antifouling polymer anchor to surface. Mussel-inspired surface anchor for "graft-to" (**a–c**) and "graft-from" (**d**) strategy. (**e–h**) Various types of antifouling polymers

2003; Dalsin et al. 2005), where only the decapeptide and oligomeric poly(DOPA)-terminated PEG performed well in cell attachment assays. This result perhaps suggests that other amino acid residues play important adhesive roles on certain substrates. Given the observation many years ago that mussels attach to a great variety of substrates (Crisp et al. 1985; Young and Crisp 1982), additional experiments aimed at elucidating the strength of interactions between catechol and a variety of substrates should be an important goal for the field.

In addition to PEG-based antifouling polymers, other hydrophilic polymers such as ionic and zwitterionic polymers have demonstrated promising antifouling properties (Lee 2009, 2011; Gong et al. 2012; Braulta et al. 2010). Peptoid synthesized with oligomeric ethylene glycol (Statz et al. 2005, 2008a) and zwitterionic (Lau et al. 2015) side chains have demonstrated the ability to repel nonspecific binding of proteins, cells, and bacteria. Messersmith and colleagues (Statz et al. 2005, 2008a) have extended this general surface modification concept with the development of a new peptide mimetic, antifouling polymer consisting of a *N*-methoxyethylglycine (peptoid) oligomer (Fig. 15.6f) coupled to a Mefp-5 mimetic pentapeptide anchor consisting of alternating lysine and DOPA residues (Statz et al. 2005). A peptoid has a peptide-like polyamide backbone with side chain substitution on the amide nitrogen instead of the α-carbon. This change makes peptoids more protease-resistant than peptides, likely due to the spatial misalignment of the side chains

(Miller et al. 1995). Additionally, peptoid containing antimicrobial sequences exhibited the ability to active bactericidal properties (Statz et al. 2008b).

Another design of antifouling mussel-inspired polymer coatings is formed by so-called "*graft-from*" approaches, in which antifouling polymers are grown directly from surfaces adsorbed using surface-initiated atom transfer radical polymerization (SI-ATRP) (Fan et al. 2005, 2006) or ring-opening metathesis polymerization (SI-ROMP) (Ye et al. 2010). These approaches have the advantage of achieving thicker and higher density layers of surface-bound polymer owing to a high density of initiation sites and growing chain ends. The basic requirement for this approach is a bifunctional molecule containing an initiating functional group coupled to a moiety capable of physical or chemical adsorption to the surface of interest. A biomimetic example of this approach that is given by molecule (Fig. 15.6d) contains a catechol for adsorption to surfaces and a 2-bromoamide for initiation of atom transfer radical polymerization (ATRP). Adsorption of the initiator on Ti and ATRP of oligo(ethylene glycol) methyl ether methacrylate (OEGMEMA) (Fig. 15.6g) (Fan et al. 2005, 2006), zwitterionic carboxybetaine (Fig. 15.6h) (Gao et al. 2010; Kuang and Messersmith 2012; Li et al. 2008, 2010; Braulta et al. 2010), and thermo-responsive *N*-isopropylacrylamide (Wang et al. 2011) have been used to produce a brush-type grafted polymer. This surface-initiated polymerization approach is not limited to tethering polymers onto planar substrates. Antifouling polymers can be grafted onto substrates of various geometries and sizes (i.e., nanoparticles (Fan et al. 2005; Wang et al. 2011), nanowires (Ye et al. 2010), etc.) through anchoring the initiator to these substrates, which further expands the application of this polymer anchoring technology.

### *15.5.2 Polydopamine Coating*

Dopamine consists of a catechol and an amine group, which mimics the side chain functional groups of DOPA and lysine, respectively. Both amino acids are found in large abundance in Mefp-5, which is found predominately at the interface between adhesive plaque and substrate surface and is believed to function as a primer (Waite and Qin 2001). Oxidation of dopamine leads to intramolecular cyclization to form a leukochrome which can further polymerize through reaction pathways similar to melanin formation (Lee et al. 2007a, b). The polymerized form of dopamine (polydopamine) readily attaches to many types of surfaces (i.e., noble metals, metal oxides, semiconductor, polymers, and ceramics) (Lee et al. 2007a, b) with a wide range of geometry (i.e., particle, cube, tube) and sizes (i.e., nano-, micro-) (Cheng et al. 2012; Ren et al. 2011; Fei et al. 2008). Similar coatings can be achieved through the oxidation and polymerization of DOPA to form a poly (DOPA) coating (Kuang et al. 2014).

Polydopamine coating remains reactive and can be further modified with polymers, initiators, proteins, polysaccharides, oligonucleotides, nanoparticles, growth factors, and cells, containing suitable functional groups (i.e., $-NH_2$, –SH, etc.) or metal ions (Fig. 15.7) (Lee et al. 2007a, b, 2009; Yang et al. 2011; Kang et al. 2012; Ren et al. 2011; Ham et al. 2011). Grafting PEG onto polydopamine coating forms an antifouling surface which is able to prevent nonspecific binding of cells (Lee et al. 2007a, b). Hyaluronic acid was added onto the polydopamine surface to effectively promote bone marrow and stem cell adhesion to the surface (Lee et al. 2007a, b). Poly(DOPA) was shown to entrap antibiotics (i.e., cationic aminoglycoside gentamicin), yielding surfaces that were effective in contact killing of *S. aureus* (Kuang et al. 2014). Additionally, the catechol side chain within the polydopamine coating can effectively reduce silver ions to form silver nanoparticle forming a silver ion releasing coating with antibacterial properties (Sileika et al. 2011). Drug particles and enzymes have also been immobilized in the polydopamine coating that was further tethered with temperature or pH-sensitive polymer brushes, which demonstrated the ability to release the bioactive agents in response to changes in the temperature and pH (Ma et al. 2013).

Polydopamine can be coated directly onto nanoparticle templates. Subsequent removal of the template resulted in the formation of polydopamine capsules which can be used to encapsulate drug particles (Cui et al. 2010; Ma et al. 2013). The release rate of the encapsulated drug particles can be easily tuned by modulating the thickness of the polydopamine shell (Cui et al. 2010). Additionally, polydopamine shells treated with $Fe^{3+}$ ion solution enhanced the stability of the shell (Kim et al. 2014).

Although the formation of polydopamine and poly(DOPA) represents a simple and convenient approach for surface modification, the high price of catechol starting materials and the dark color of these compounds may limit their applications. Foods (i.e., grapes and chocolates) and beverages (i.e., coffee, tea, and red wine) are rich in polyphenols which have been known to have health benefits (i.e., antiaging) (Pandey and Rizvi 2009). Tannic acid (TA) and pyrogallol (PG), both derived from polyphenol-rich food extracts, were recently demonstrated to form a colorless coating that further enabled subsequent surface modification similar to the function of the polydopamine coating (Sileika et al. 2013). Additionally, TA- and PG-coated surfaces could support 3T3 fibroblasts growth and inhibit bacterial growth. TA-modified surface also exhibited antioxidant property. A follow-up survey of nearly 20 plant-derived phenols/polyphenols showed that many of these compounds are capable of spontaneously forming functional coatings in this manner (Barrett et al. 2014). Some of these polyphenols represent lower-cost and colorless alternatives for surface modification when compared to polydopamine- and poly(DOPA)-based coating.

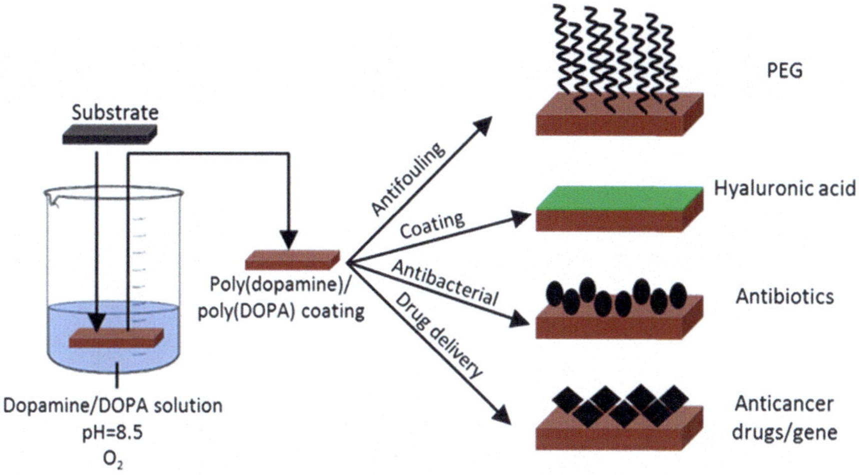


**Fig. 15.7** Polydopamine coating and secondary functional modification. Substrate immersed into pH 8.5, oxygenated dopamine/DOPA solution for several hours to form a polydopamine/poly (DOPA) coating. This coating readily underwent secondary modification for functional property, such as antifouling, adhesive coating, antibacterial, and drug delivery

## 15.6 Advanced Material Design Based on Catechol Chemistry

The ability for catechols to participate in various types of chemical interactions has inspired the development of new materials with new and improved properties. This section highlights recent development of advanced materials using mussel-mimetic chemistry which include nano-featured adhesive with the capability to bind repeatedly underwater, novel nanocomposite adhesive, targeted drug delivery, hydrogel actuator, and self-healing materials.

### *15.6.1 Gecko- and Mussel-Inspired Hybrid Adhesive*

The hierarchical nanostructure of gecko foot pads allows these organisms to cling onto vertical and even inverted surfaces reversibly (Tian et al. 2006). Inspired by this phenomenon, scientists have recreated these nanoscaled features to mimic the adhesive property of gecko foot pads (Geim et al. 2003; Yurdumakan et al. 2005). However, gecko-mimetic adhesion demonstrated drastically reduced adhesive strength to wet surfaces (Huber et al. 2005; Sun et al. 2005). To enhance the wet adhesion, a copolymer consisting of dopamine methacrylamide and 2-methoxyethyl acrylate (p(DMA-co-MEA)) was coated onto gecko-mimetic nano-pillars (Fig. 15.8) (Lee et al. 2007a, b). This hybrid biomimetic adhesive demonstrated almost 15-fold increase in adhesive properties in the presence of

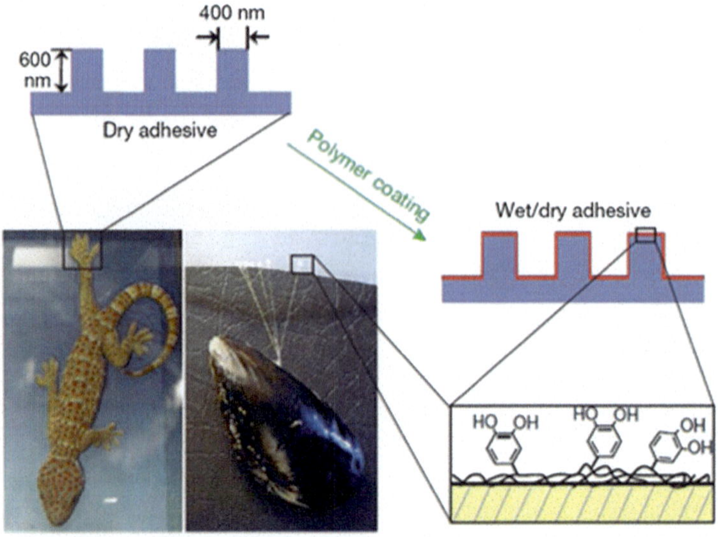


**Fig. 15.8** Gecko- and mussel-inspired hybrid nanostructure material shows robust adhesive properties both under dry and wet conditions (reproduced from Lee et al. 2007a, b)

moisture while maintaining its adhesive performance for over 1000 contact cycles in both dry and wet environments. The performance of the adhesive could be further enhanced through tuning the geometry of the nanostructures (Glass et al. 2009) as well as chemical crosslinking of the p(DMA-co-MEA) coating (Glass et al. 2010). This uniquely combination may prove useful for reversible adhesion to a variety of surfaces in both wet and dry environments.

### *15.6.2 Nanocomposite Adhesive Hydrogel*

Although various catechol-modified PEG adhesives have demonstrated excellent bioadhesive properties and rapid curing capability, PEG-based adhesives are mechanically weak due to extensive swelling. Additionally, PEG is bioinert and lacks the ability to promote cellular attachment and infiltration which are important to rapid tissue repair and regeneration (Shin et al. 2003). To simultaneously improve the mechanical properties and the bioactivity of PEG-based adhesive, a synthetic nano-silicate (Laponite, $Na^{0.7+}(Mg_{5.5}Li_{0.3}Si_8)O_{20}\,(OH)_4)^{0.7-}$) was incorporated into a PEG–catechol adhesive (Fig. 15.9) (Liu et al. 2014). Laponite has similar chemical composition as bioactive glass and mimics some of its biological properties (Hoppe et al. 2011; Hench 2009; Hench and Paschall 1973) and degrades into nontoxic products ($Na^+$, $Si(OH)_4$, $Mg^{2+}$, $Li^+$) (Thompson and Butterworth 1992). One of the degradation products is orthosilicic acid ($Si(OH)_4$), which is naturally found in numerous human tissues and organs (e.g., bone, tendon, liver,

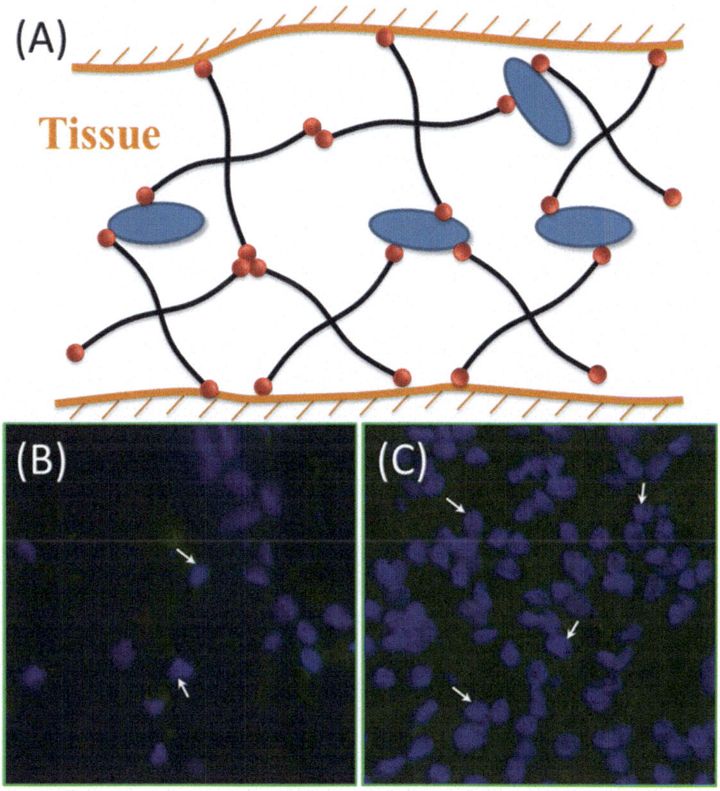


**Fig. 15.9** Injectable nanocomposite adhesive hydrogel formation and in vivo cell infiltration. (**a**) Scheme of hydrogel formation, catechol covalently crosslink and attach to tissue, physically bind to Laponite. *Line*, PEG chain; *circle*, catechol; *ellipse*, Laponite. (**b**, **c**) Immunohistochemical staining images of cells infiltrated into adhesive hydrogel containing 0 wt% and 2 wt% Laponite, respectively, after 8 weeks subcutaneous implantation in rat. Cell nuclei were stained by DAPI (*white arrows*) (reproduced from Liu et al. 2014)

and kidney tissues) (Carlisle 1970), and it has been demonstrated to promote synthesis of type I collagen and osteoblast differentiation in human osteosarcoma cells in vitro (Reffitt et al. 2003). The incorporation of Laponite increased the bulk material properties and the adhesive properties of PEG–catechol adhesive. Most importantly, incorporated Laponite promoted cellular infiltration into the nanocomposite adhesive in vivo (Fig. 15.9c).

### *15.6.3 Drug Delivery*

The pH dependency of catechol–boronate complexation was exploited to deliver an anticancer drug, bortezomib (BTZ), for its targeted delivery to cancer cells

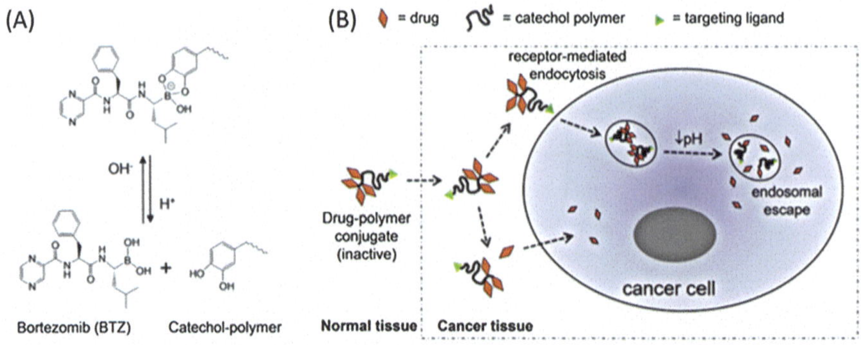


**Fig. 15.10** pH-responsive polymer drug conjugates based on catechol–boronate complexation. (**a**) Polymer formed complexation with boronic acid structure in BTZ via catechol group at neutral and alkaline pH and dissociated to release the drug in acidic environments. (**b**) The catechol polymer–BTZ conjugate dissociated in response to a mildly acidic cancer tissue microenvironment to liberate the free drug, which can be uptake by cancer cells (reproduced from Su et al. 2011)

(Fig. 15.10) (Su et al. 2011). BTZ is a dipeptide functionalized with boronic acid and has previously demonstrated inhibition of cancer cell proteasome through directly binding the boronic acid group to threonine residues in the active sites of several proteases (Adams et al. 1998; Groll et al. 2006). The drug carrier consists of PEG modified with catechol at one end and biotin at the other end (Su et al. 2011). The catechol group binds to the borate group found on BTZ at neutral and mildly basic pH through catechol–boronate complexation. Biotin functionalization enables the drug carrier to target cancer cell-targeting ligands to facilitate the entry of BTZ-loaded polymer carriers into cancer cells through endocytosis. The acidic environment of the endosome resulted in the dissociation of BTZ from PEG–catechol to release the drug. This cancer cell-targeting polymer carrier effectively promotes cellular drug uptake, proteasome inhibition, and cytotoxicity toward breast carcinoma cells.

### *15.6.4 Hydrogel Actuator*

Hydrogels that can change their shapes and physical properties in respond to environmental stimuli (i.e., temperature, pH, electricity) have found applications as soft robotic components, biosensors, artificial muscle tissues, and drug carriers (Gracias 2013; Lim et al. 2014; Ionov 2013). Recently, Lee et al. (Lee and Konst 2014) reported a hydrogel actuator that is based on catechol–metal ion coordination chemistry (Fig. 15.11). Hydrogels modified with catechol were locally printed with ferric ($Fe^{3+}$) ions by applying electric potential (2.5–25 V) using an iron electrode as the anode and an aluminum foil as the cathode electrode. Locally deposited metal ions were captured by the network-bound catechol (dark bands in Fig. 15.11a). When the hydrogels were submerged in a basic solution (pH 9.5), the formation of

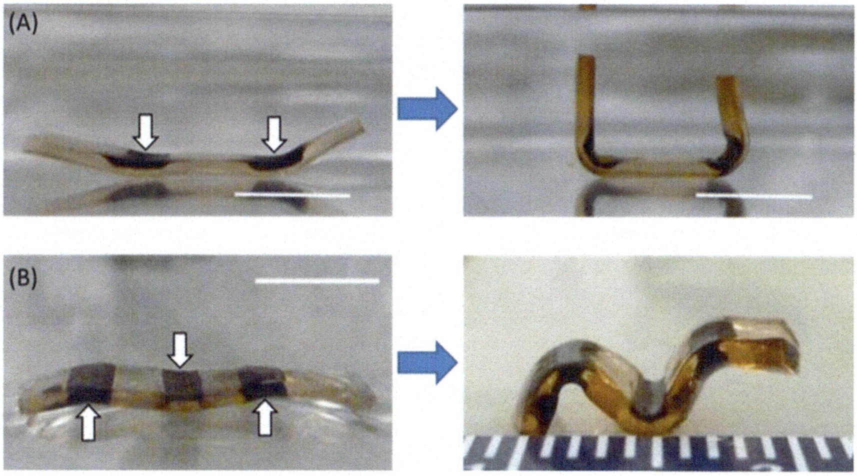


**Fig. 15.11** Catechol-containing hydrogel ionoprinted with $Fe^{3+}$ ions and incubate in pH 9.5 solution to induce hydrogel bending and forming a "U" (**a**) and "M" (**b**) shape before (*left panels*) and after (*right panels*) due to the change of hydrogel local crosslinking density. The *white arrows* indicate the direction and location of ionoprinting. Scale bars, 5 mm (reproduced from Lee and Konst 2014)

the tris-catechol-$Fe^{3+}$ complex increased local crosslinking density at the site of ionoprinting and resulted in hydrogel bending (Fig. 15.11a). When the hydrogels were subsequently submerged in an acidic solution (12 mM HCl), catechol formed mono-complex with the imprinted ion, resulting in reduced crosslinking density and allowing the hydrogel to return to its original shape (Lee and Konst 2014; Lee et al. 2014, 2015). $Fe^{3+}$ ions in hydrogel can be removed by ethylenediaminetetraacetic acid (EDTA) treatment and reintroduced in a new pattern, so that the hydrogel can be designed to bend to a different shape (Lee and Konst 2014). The rate and extent of hydrogel actuation can be effectively tuned by catechol and metal ion content, pH, hydrogel thickness, and metal ion type (Lee et al. 2015, 2016).

## 15.6.5 *Self-Healing Hydrogel*

Strong physical and chemical interactions of catechol have been used to design self-healing hydrogels (Fig. 15.12). The ability of catechol to form complexes with various types of metal ions (i.e., $Fe^{3+}$, $Zn^{2+}$, $Al^{3+}$ $Ga^{3+}$, $In^{3+}$) under basic condition have been used to create self-healing hydrogels by mixing catechol-containing polymer and metal ions (Holten-Andersen et al. 2011, 2014; Krogsgaard et al. 2014; Wang et al. 2015; Grindy et al. 2015). The modulus of the self-healing hydrogel composed by 4-armed PEG end-modified with catechol and $Fe^{3+}$ ions

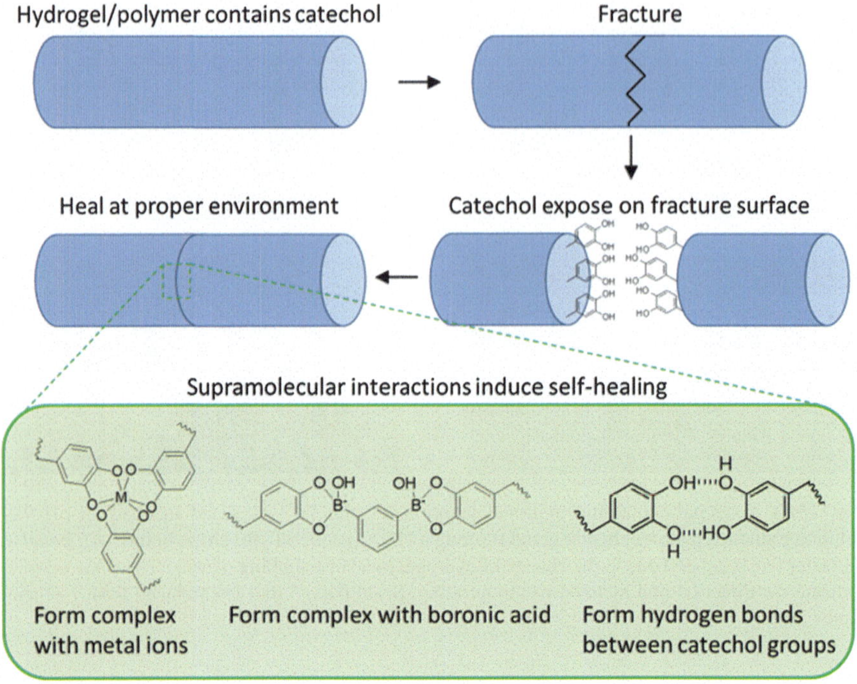


**Fig. 15.12** Schematic representations of catechol-containing polymer self-healing process based on catechol chemistry

approached those of covalently crosslinked hydrogels (Holten-Andersen et al. 2011). Similarly, pH-responsive self-healing hydrogels based on catechol–boronate complexation were reported (He et al. 2011; Zhang et al. 2013). Under basic condition (pH ~ 9), a tetrahedral boronate ester was established between 1,3-benzenediboronic acid and sodium tetrahydroborate with polymer-bound catechol to form three-dimensional polymer networks with viscoelastic behavior. Recently, a mussel-inspired polymer self-healed in acidic environment was reported (Ahn et al. 2014). Polyacrylates and polymethacrylates functionalized with triethylsilyl-protected catechol were cut and rejoined together through slight compression in acidic solution (pH = 3). The triethylsilyl-protecting group was removed under acidic aqueous environment and the exposed catechol groups formed hydrogen bonds, facilitating the polymer healing.

## 15.7 Summary and Future Outlook

Mussel adhesive proteins have the remarkable ability to secure adhesion to both organic and inorganic surfaces in the presence of water. These basic properties are highly desired for applications in both medical and nonmedical settings. One conceivable approach to exploiting the adhesive properties of MAPs is to utilize full sequence MAPs themselves, either isolated and purified from mussel adhesive glands or generated in microbial expression systems. These approaches have the theoretical advantage of utilizing the adhesive proteins in their native state without modification. However, these approaches do have their disadvantages such as biological products regulation, complexity of purification, and high cost.

Development of synthetic mussel-inspired polymers has accelerated greatly in the past years. Like the natural proteins that inspire them, catechol has been incorporated into a variety of synthetic polymers to take advantage of its adhesive and cohesive properties. Applications for such biomimetic polymers in the future will not be restricted to surgical adhesives and surface coating, as new mussel-inspired polymers are being developed as drug delivery carrier, hydrogel actuator, self-healing hydrogel, etc.

It is important to note that these synthetic efforts have benefited greatly from a rather advanced understanding of the unique composition and chemical behavior of natural MAP. Most mussel-inspired biomimetic studies have understandably focused on the catechol group, since its unique chemical characteristics and unusually high presence in MAP suggest an important role in mussel adhesion, but it is not always the case that one can easily identify a few key functional groups or design motifs around which to build a biomimetic strategy. However, MAPs found at the interface of the adhesive plaques contain other amino acid residues (i.e., lysine, arginine, phosphoserine, etc.), which likely contribute to interfacial binding (Papov et al. 1995; Waite and Qin 2001). Recent findings indicated that the composition of the adhesive (e.g., concentration of catechol, cationic and anionic side chain, polar and nonpolar residues) coexist to maintain a balance between hydrophobic and electrostatic interactions for optimizing cohesive and adhesive properties (Seo et al. 2015; Ahn et al. 2015). These studies indicate the importance of other functional groups besides the catechol is critical for future adhesive design.

**Acknowledgments** The authors would like to acknowledge NIH (R15GM104846 and R15GM112082 for BPL, YL, and HM; R37DE014193 and R01EB005772 for PBM) for financial support of this work.

## References

Adams J, Behnke M, Chen S, Cruickshank AA, Dick LR, Grenier L, Klunder JM, Ma YT, Plamondon L, Stein RL (1998) Potent and selective inhibitors of the proteasome: dipeptidyl boronic acids. Bioorg Med Chem Lett 8(4):333–338

Ahn BK, Lee DW, Israelachvili JN, Waite JH (2014) Surface-initiated self-healing of polymers in aqueous media. Nat Mat 13(9):867–872. doi:10.1038/NMAT4037

Ahn BK, Das S, Linstadt R, Kaufman Y, Martinez-Rodriguez NR, Mirshafian R, Kesselman E, Talmon Y, Lipshutz BH, Israelachvili JN, Waite JH (2015) High-performance mussel-inspired adhesives of reduced complexity. Nat Commun 6:8663. doi:10.1038/ncomms9663

Alexandridis P (1997) Poly(ethylene oxide)poly(propylene oxide) block copolymer surfactants. Curr Opin Colloid Interface Sci 2:478

Amstad E, Gillich T, Bilecka I, Textor M, Reimhult E (2009) Ultrastable iron oxide nanoparticle colloidal suspensions using dispersants with catechol-derived anchor groups. Nano Lett 9 (12):4042–4048. doi:10.1021/nl902212q

Amstad E, Gehring AU, Fischer H, Nagaiyanallur VV, Hahner G, Textor M, Reimhult E (2011) Influence of Electronegative Substituents on the Binding Affinity of Catechol-Derived Anchors to Fe(3)O(4) Nanoparticles. J Phys Chem C 115(3):683–691. doi:10.1021/Jp1109306

Anderson JM, Rodriguez A, Chang DT (2008) Foreign body reaction to biomaterials. Semin Immun 20(2):86–100. doi:10.1016/j.smim.2007.11.004

Anderson TH, Yu J, Estrada A, Hammer MU, Waite JH, Israelachvili JN (2010) The contribution of DOPA to substrate-peptide adhesion and internal cohesion of mussel-inspired synthetic peptide films. Adv Funct Mat 20(23):4196–4205. doi:10.1002/adfm.201000932

Anderson J, Lin M-H, Privette C, Flowers M, Murley M, Lee BP, Ong KG (2015) Wireless magnetoelastic sensors for tracking degradation profiles of nitrodopamine-modified poly(ethylene glycol). Scijet 4:80

Araujo PZ, Morando PJ, Blesa MA (2005) Interaction of catechol and gallic acid with titanium dioxide in aqueous suspensions. 1. Equilibrium studies. Langmuir 21(8):3470–3474. doi:10.1021/la0476985

Arul V, Masilamoni JG, Jesudason EP, Jaji PJ, Inayathullah M, Dicky John DG, Vignesh S, Jayakumar R (2012) Glucose oxidase incorporated collagen matrices for dermal wound repair in diabetic rat models: a biochemical study. J Biomat Appl 26(8):917–938

Autumn K, Liang YA, Hsieh ST, Wzesch W, Chan WP, Kenny TW, Fearing R, Full RJ (2000) Adhesive force of a single gecko foot-hair. Nature 405:681–685

Barrett DG, Bushnell GG, Messersmith PB (2013) Mechanically robust, negative-swelling, mussel-inspired tissue adhesives. Adv Healthcare Mater 2(5):745–755

Barrett DG, Sileika TS, Messersmith PB (2014) Molecular diversity in phenolic and polyphenolic precursors of tannin-inspired nanocoatings. Chem Commun 50(55):7265–7268. doi:10.1039/c4cc02961e

Baty AM, Leavitt PK, Siedlecki CA, Tyler BJ, Suci PA, Marchant RE, Geesey GG (1997) Adsorption of adhesive proteins from the marine mussel, mytilus edulis, on polymer films in the hydrated state using angle dependent X-ray photoelectron spectroscopy and atomic force microscopy. Langmuir 13(21):5702–5710

Bilic G, Brubaker C, Messersmith PB, Mallik AS, Quinn TM, Haller C, Done E, Gucciardo L, Zeisberger SM, Zimmermann R, Deprest J, Zisch AH (2010) Injectable candidate sealants for fetal membrane repair: bonding and toxicity *in vitro*. Am J Obstet Gynecol 202:85.e81–85.e89

Blackburn SL, Smyth MD (2007) Hydrogel-induced cervicomedullary compression after posterior fossa decompression for Chiari malformation—case report. J Neurosurg 106(4):302–304. doi:10.3171/ped.2007.106.4.302

Braulta ND, Gaoa C, Xuea H, Piliarik M, Jr H, Jianga S, Yua Q (2010) Ultra-low fouling and functionalizable zwitterionic coatings grafted onto $SiO_2$ via a biomimetic adhesive group for sensing and detection in complex media. Biosens Bioelectron 25:2276–2282

Brian N, Ahswin H, Smart N, Bayon Y, Wohlert S, Hunt JA (2012) Reactive oxygen species (ROS)-a family of fate deciding molecules pivotal in constructive inflammation and wound healing. Eur Cell Mater 24:249–265

Bridgett MJ, Davies MC, Denyer SP (1992) Control of staphylococcal adhesion to polystyrene surfaces by polymer surface modification with surfactants. Biomaterials 13(7):411–416

Brodie M, Vollenweider L, Murphy JL, Xu F, Lyman A, Lew WD, Lee BP (2011) Biomechanical properties of achilles tendon repair augmented with bioadhesive-coated scaffold. Biomed Mater 6(1):015014

Brunette DM, Tengvall P, Textor M, Thomsen P (2001) Titanium in medicine. Springer, New York

Burke SA, Ritter-Jones M, Lee BP, Messersmith PB (2007) Thermal gelation and tissue adhesion of biomimetic hydrogels. Biomed Mater 2(4):203–210. doi:10.1088/1748-6041/2/4/001

Burzio LA, Waite JH (2000) Cross-linking in adhesive quinoproteins: studies with model decapeptides. Biochemistry 39(36):11147–11153

Burzio LO, Burzio VA, Silva T, Burzio LA, Pardo J (1997) Environmental bioadhesion: themes and applications. Curr Opin Biotechnol 8(3):309–312

Carlisle EM (1970) Silicon: a possible factor in bone calcification. Science 167(3916):279–280. doi:10.1126/science.167.3916.279

Catron ND, Lee H, Messersmith PB (2006) Enhancement of poly (ethylene glycol) mucoadsorption by biomimetic end group functionalization. Biointerphases 1(4):134–141

Cencer MM, Liu Y, Winter A, Murley M, Meng H, Lee BP (2014) Effect of pH on the rate of curing and bioadhesive properties of dopamine functionalized poly(ethylene glycol) hydrogels. Biomacromolecules 15(8):2861–2869. doi:10.1021/bm500701u

Cencer M, Murley M, Liu Y, Lee BP (2015) Effect of nitro-functionalization on the cross-linking and bioadhesion of biomimetic adhesive moiety. Biomacromolecules 16(1):404–410. doi:10.1021/bm5016333

Characklis KC (1990) Biofilm Processes. In: Characklis WG, Marshall KC (eds) Biofilms. Wiley, New York, pp 195–231

Chelikani P, Fita I, Loewen PC (2004) Diversity of structures and properties among catalases. Cell Mol Life Sci 61(2):192–208. doi:10.1007/s00018-003-3206-5

Cheng C, Li S, Nie S, Zhao W, Yang H, Sun S, Zhao C (2012) General and biomimetic approach to biopolymer-functionalized graphene oxide nanosheet through adhesive dopamine. Biomacromolecules 13(12):4236–4246. doi:10.1021/bm3014999

Chirdon WM, O'Brien WJ, Robertson RE (2003) Adsorption of catechol and comparative solutes on hydroxyapatite. J Biomed Mater Res Part B: Appl Biomater 66B(2):532–538

Choi YS, Kang DG, Lim S, Yang YJ, Kim CS, Cha HJ (2011) Recombinant mussel adhesive protein fp-5 (MAP fp-5) as a bulk bioadhesive and surface coating material. Biofouling 27(7):729–737. doi:10.1080/08927014.2011.600830

Choi YC, Choi JS, Jung YJ, Cho YW (2014) Human gelatin tissue-adhesive hydrogels prepared by enzyme-mediated biosynthesis of DOPA and $Fe3^{+}$ ion crosslinking. J Mater Chem B 2(2):201–209. doi:10.1039/c3tb20696c

Clement M-V, Long LH, Ramalingam J, Halliwell B (2002) The cytotoxicity of dopamine may be an artefact of cell culture. J Neurochem 81(3):414–421. doi:10.1046/j.1471-4159.2002.00802.x

Creutz C, Chou MH (2008) Binding of catechols to mononuclear titanium(IV) and to 1- and 5-nm $TiO_2$ nanoparticles. Inorg Chem 47(9):3509–3514. doi:10.1021/ic701687k

Crisp DJ, Walker G, Young GA, Yule AB (1985) Adhesion and substrate choice in mussels and barnacles. J Colloid Interface Sci 104(1):40–50

Cui J, Wang Y, Postma A, Hao J, Hosta-Rigau L, Caruso F (2010) Monodisperse polymer capsules: tailoring size, shell thickness, and hydrophobic cargo loading via emulsion templating. Adv Funct Mater 20(10):1625–1631. doi:10.1002/adfm.201000209

Dalsin JL, Messersmith PB (2003) Surface modification for protein resistance using a biomimetic approach. In Thomas JL, Kiick KL, Gower LA (ed) Materials inspired by biology. Materials research society symposium series, vol 774. pp 75–80

Dalsin JL, Messersmith P (2005) Bioinspired antifouling polymers. Mater Today 8(9):38–46

Dalsin JL, Hu B, Lee BP, Messersmith PB (2003) Mussel adhesive protein mimetic polymers for the preparation of nonfouling surfaces. J Am Chem Soc 125:6

Dalsin JL, Lin L, Tosatti S, Voeroes J, Textor M, Messersmith PB (2005) Protein resistance of titanium oxide surfaces modified by biologically inspired mPEG-DOPA. Langmuir 21 (2):640–646

Dalsin JL, Sherman DL, Lee BP, Messersmith PB (2006) Versatile mussel adhesive-inspired biomimetic antifouling polymers. PMSE Prepr 94:854–855

Dalsin JL, Lee BP, Vollenweider L, Silvary S, Murphy JL, Xu F, Spitz A, Lyman A (2012) Multi-armed Catechol compound blends. US Patent 8,119,742

Deacon MP, Davis SS, Waite JH, Harding SE (1998) Structure and mucoadhesion of mussel glue protein in dilute solution. Biochemistry 37(40):14108–14112

Deming TJ (1997) Facile synthesis of block copolypeptides of defined architecture. Nature 390 (6658):386–389

Desai NP, Hossainy SFA, Hubbell JA (1992) Surface-immobilized polyethylene oxide for bacterial repellence. Biomaterials 13(7):417–420

Ding XC, Vegesna GK, Meng H, Winter A, Lee BP (2015) Nitro-group functionalization of dopamine and its contribution to the viscoelastic properties of catechol-containing nanocomposite hydrogels. Macromol Chem Phys 216(10):1109–1119. doi:10.1002/macp.201500010

Fan X, Lin L, Dalsin JL, Messersmith PB (2005) Biomimetic anchor for surface-initiated polymerization from metal substrates. J Am Chem Soc 127(45):15843–15847. doi:10.1021/ja0532638

Fan X, Lin L, Messersmith PB (2006) Cell fouling resistance of polymer brushes grafted from Ti substrates by surface-initiated polymerization: effect of ethylene glycol side chain length. Biomacromolecules 7(8):2443–2448

Fang H, Kaur G, Wang BH (2004) Progress in boronic acid-based fluorescent glucose sensors. J Fluoresc 14(5):481–489. doi:10.1023/B:Jofl.0000039336.51399.3b

Fei B, Qian BT, Yang ZY, Wang RH, Liu WC, Mak CL, Xin JH (2008) Coating carbon nanotubes by spontaneous oxidative polymerization of dopamine. Carbon 46(13):1795–1797. doi:10.1016/j.carbon.2008.06.049

Filpula DR, Lee SM, Link RP, Strausberg SL, Strausberg RL (1990) Structural and functional repetition in a marine mussel adhesive protein. Biotechnol Prog 6(3):171–177

Gao C, Li G, Xue H, Yang W, Zhang F, Jiang S (2010) Functionalizable and ultra-low fouling zwitterionic surfaces via adhesive mussel mimetic linkages. Biomaterials 31(7):1486–1492

Garcia-Fernandez L, Cui J, Serrano C, Shafiq Z, Gropeanu RA, Miguel VS, Ramos JI, Wang M, Auernhammer GK, Ritz S, Golriz AA, Berger R, Wagner M, del Campo A (2013) Antibacterial strategies from the sea: polymer-bound cl-catechols for prevention of biofilm formation. Adv Mater 25(4):529–533. doi:10.1002/adma.201203362

Geim AK, Dubonos SV, Grigorieva IV, Novoselov KS, Zhukov AA, Shapoval SY (2003) Microfabricated adhesive mimicking gecko foot-hair. Nat Mater 2:461–463

Glass P, Chung H, Washburn NR, Sitti M (2009) Enhanced reversible adhesion of dopamine methacrylamide-coated elastomer microfibrillar structures under wet conditions. Langmuir 25 (12):6607–6612. doi:10.1021/la9009114

Glass P, Chung H, Washburn NR, Sitti M (2010) Enhanced wet adhesion and shear of elastomeric micro-fiber arrays with mushroom tip geometry and a photopolymerized p(DMA-co-MEA) tip coating. Langmuir 26(22):17357–17362. doi:10.1021/la1029245

Gong Y-K, Liu L-P, Messersmith PB (2012) Doubly biomimetic catecholic phosphorylcholine copolymer: a platform strategy for fabricating antifouling surfaces. Macromol Biosci 12 (7):979–985. doi:10.1002/mabi.201200074

Gracias DH (2013) Stimuli responsive self-folding using thin polymer films. Curr Opin Chem Eng 2(1):112–119. doi:10.1016/j.coche.2012.10.003

Grande DA, Pitman MI (1988) The use of adhesives in chondrocyte transplantation surgery. Preliminary studies. Bull Hosp Jt Dis Orthop Inst 48(2):140–148

Green K, Berdecia R, Cheeks L (1987) Mussel adhesive protein: permeability characteristics when used as a basement membrane. Curr Eye Res 6(6):835–837

Grindy SC, Learsch R, Mozhdehi D, Cheng J, Barrett DG, Guan ZB, Messersmith PB, Holten-Andersen N (2015) Control of hierarchical polymer mechanics with bioinspired metal-coordination dynamics. Nat Mater 14(12):1210–1216

Groll M, Berkers CR, Ploegh HL, Ovaa H (2006) Crystal structure of the boronic acid-based proteasome inhibitor bortezomib in complex with the yeast 20S proteasome. Structure 14 (3):451–456. doi:10.1016/j.str.2005.11.019

Haller CM, Buerzle W, Kivelio A, Perrini M, Brubaker CE, Gubeli RJ, Mallik AS, Weber W, Messersmith PB, Mazza E, Ochsenbein-Koelble N, Zimmermann R, Ehrbar M (2012) Mussel-mimetic tissue adhesive for fetal membrane repair: an ex vivo evaluation. Acta Biomater 8 (12):4365–4370. doi:10.1016/j.actbio.2012.07.047

Ham HO, Liu Z, Lau KHA, Lee H, Messersmith PB (2011) Facile DNA immobilization on Surfaces through a catecholamine polymer. Angew Chem Int Ed 50(3):732–736. doi:10.1002/anie.201005001

Hansen DC, Dexter SC, Waite JH (1995) The inhibition of corrosion of S30403 stainless steel by a naturally occurring catecholic polymer. Corros Sci 37(9):1423–1441

Harris JM, Dust JM, McGill RA, Harris PA, Edgell MJ, Sedaghat-Herati RM, Karr LJ, Donnelly DL (1991) New polyethylene glycols for biomedical applications. Water-soluble polymer, ACS symposium series, vol 467. pp 418–429

Harris LG, Tosatti S, Wieland M, Textor M, Richards RG (2004) Staphylococcus aureus adhesion to titanium oxide surfaces coated with non-functionalized and peptide-functionalized poly (L-lysine)-grafted-poly(ethylene glycol) copolymers. Biomaterials 25(18):4135–4148

He LH, Fullenkamp DE, Rivera JG, Messersmith PB (2011) pH responsive self-healing hydrogels formed by boronate-catechol complexation. Chem Commun 47(26):7497–7499. doi:10.1039/C1cc11928a

Hench LL (2009) Genetic design of bioactive glass. J Eur Ceram Soc 29(7):1257–1265. doi:10.1016/j.jeurceramsoc.2008.08.002

Hench LL, Paschall HA (1973) Direct chemical bond of bioactive glass-ceramic materials to bone and muscle. J Biomed Mater Res 7(3):25–42. doi:10.1002/jbm.820070304

Herlinger E, Jameson RF, Linert W (1995) Spontaneous autoxidation of dopamine. J Chem Soc Perk T 2(2):259–263. doi:10.1039/P29950000259

Holten-Andersen N, Mates TE, Toprak MS, Stucky GD, Zok FW, Waite JH (2009) Metals and the integrity of a biological coating: the cuticle of mussel byssus. Langmuir 25(6):3323–3326. doi:10.1021/la8027012

Holten-Andersen N, Harrington MJ, Birkedal H, Lee BP, Messersmith PB, Lee KYC, Waite JH (2011) pH-induced mussel metal-ligand crosslinks yield self-healing polymer networks with near-covalent elastic moduli. Proc Natl Acad Sci 15:2651–2655

Holten-Andersen N, Jaishankar A, Harrington MJ, Fullenkamp DE, DiMarco G, He LH, McKinley GH, Messersmith PB, Leei KYC (2014) Metal-coordination: using one of nature's tricks to control soft material mechanics. J Mater Chem B 2(17):2467–2472. doi:10.1039/c3tb21374a

Hong S, Yang K, Kang B, Lee C, Song IT, Byun E, Park KI, Cho SW, Lee H (2013) Hyaluronic acid catechol: a biopolymer exhibiting a pH-dependent adhesive or cohesive property for human neural stem cell engineering. Adv Funct Mater 23(14):1774–1780. doi:10.1002/adfm.201202365

Hong SH, Shin M, Lee J, Ryu JH, Lee S, Yang JW, Kim WD, Lee H (2015) STAPLE: stable alginate gel prepared by linkage exchange from ionic to covalent bonds. Adv Healthcare Mater. doi:10.1002/adhm.201400833

Hoppe A, Güldal NS, Boccaccini AR (2011) A review of the biological response to ionic dissolution products from bioactive glasses and glass-ceramics. Biomaterials 32 (11):2757–2774. doi:10.1016/j.biomaterials.2011.01.004

Huang K, Lee BP, Ingram D, Messersmith PB (2002) Synthesis and characterization of self-assembling block copolymers containing bioadhesive end groups. Biomacromolecules 3 (2):397–406

Huber G, Mantz H, Spolenak R, Mecke K, Jacobs K, Gorb SN, Arzt E (2005) Evidence for capillary contributions to gecko adhesion from single spatula nanomechanical measurements. Proc Natl Acad Sci U S A 102:16293–16296

Hwang DS, Yoo HJ, Jun JH, Moon WK, Cha HJ (2004) Expression of functional recombinant mussel adhesive protein Mgfp-5 in Escherichia coli. Appl Environ Microbiol 70(6):3352–3359

Hwang DS, Gim Y, Cha HJ (2005) Expression of functinoal recombinant mussel adhesive protein type 3A in eschereschia coli. Biotechnol Prog 21:965–970

Hwang DS, Gim Y, Kang DG, Kim YK, Cha HJ (2007) Recombinant mussel adhesive protein Mgfp-5 as cell adhesion biomaterial. J Biotechnol 127(4):727–735. doi:10.1016/j.jbiotec.2006.08.005

Ionov L (2013) Biomimetic hydrogel-based actuating systems. Adv Funct Mater 23 (36):4555–4570. doi:10.1002/adfm.201203692

Kamath KR, Danilich MJ, Marchant RE, Park K (1996) Platelet interactions with plasma-polymerized ethylene oxide and N-vinyl-2-pyrrolidone films and linear poly(ethylene oxide) layer. J Biomater Sci Polym Ed 7(11):977–988

Kang SM, Hwang NS, Yeom J, Park SY, Messersmith PB, Choi IS, Langer R, Anderson DG, Lee H (2012) One-step multipurpose surface functionalization by adhesive catecholamine. Adv Funct Mater 22(14):2949–2955. doi:10.1002/adfm.201200177

Kastrup CJ, Nahrendorf M, Figueiredo JL, Lee H, Kambhampati S, Lee T, Cho S-W, Gorbatov R, Iwamoto Y, Dang TT, Dutta P, Yeon JH, Cheng H, Pritchard CD, Vegas AJ, Siegel CD, MacDougall S, Okonkwo M, Thai A, Stone JR, Coury AJ, Weissleder R, Langer R, Anderson DG (2012) Painting blood vessels and atherosclerotic plaques with an adhesive drug depot. Proc Natl Acad Sci U S A 109(52):21444–21449. doi:10.1073/pnas.1217972110

Kawabata T, Schepkin V, Haramaki N, Phadke RS, Packer L (1996) Iron coordination by catechol derivative antioxidants. Biochem Pharmacol 51(11):1569–1577

Kenausis GL, Voros J, Elbert DL, Huang N, Hofer R, Ruiz-Taylor L, Textor M, Hubbell JA, Spencer ND (2000) Poly(L-lysine)-g-poly(ethylene glycol) layers on metal oxide surfaces: attachment mechanism and effects of polymer architecture on resistance to protein adsorption. J Phys Chem B 104(14):3298–3309

Kim D, Hwang DS, Kang DG, Kim JYH, Cha HJ (2008) Enhancement of mussel adhesive protein production in Escherichia coli by co-expression of bacterial hemoglobin. Biotechnol Prog 24 (3):663–666. doi:10.1021/bp0703477

Kim E, Liu Y, Shi XW, Yang XH, Bentley WE, Payne GF (2010) Biomimetic approach to confer redox activity to thin chitosan films. Adv Funct Mater 20(16):2683–2694. doi:10.1002/adfm.200902428

Kim K, Ryu JH, Lee DY, Lee H (2013) Bio-inspired catechol conjugation converts water-insoluble chitosan into a highly water-soluble, adhesive chitosan derivative for hydrogels and LbL assembly. Biomater Sci-UK 1(7):783–790. doi:10.1039/C3bm00004d

Kim S, Gim T, Kang SM (2014) Stability-enhanced polydopamine coatings on solid substrates by iron(III) coordination. Prog Org Coat 77(8):1336–1339. doi:10.1016/j.porgcoat.2014.04.011

Ko R, Cadieux PA, Dalsin JL, Lee BP, Elwood CN, Razvi H (2008) Novel uropathogen-resistant coatings inspired by marine mussels. J Endourol 22(6):1153–1160

Krogsgaard M, Hansen MR, Birkedal H (2014) Metals & polymers in the mix: fine-tuning the mechanical properties & color of self-healing mussel-inspired hydrogels. J Mater Chem B 2 (47):8292–8297. doi:10.1039/c4tb01503g

Kuang J, Messersmith PB (2012) Universal surface-initiated polymerization of antifouling zwitterionic brushes using a mussel-mimetic peptide initiator. Langmuir 28(18):7258–7266. doi:10.1021/la300738e

Kuang J, Guo JL, Messersmith PB (2014) High ionic strength formation of DOPA-melanin coating for loading and release of cationic antimicrobial compounds. Adv Mater Interfaces 1(6). doi:10.1002/admi.201400145

Kummert R, Stumm W (1980) The surface complexation of organic acids on hydrous g-alumina. J Colloid Interface Sci 75(2):373–385

Lau KHA, Sileika TS, Park SH, Sousa AML, Burch P, Szleifer I, Messersmith PB (2015) Molecular design of antifouling polymer brushes using sequence-specific peptoids. Adv Mater Interfaces 2 (1):1400225. doi:10.1002/admi.201400225
Lee BP (2009) Biomimetic compounds and synthetic methods therefor. US Patent 7,622,533
Lee BP (2011) Biomimetic compounds and synthetic methods therefor. US Patent 8,030,413
Lee BP, Konst S (2014) Novel hydrogel actuator inspired by reversible mussel adhesive protein chemistry. Adv Mater 26(21):3415–3419
Lee BP, Dalsin JL, Messersmith PB (2001) Enzymatic and non-enzymatic pathways to formation of DOPA-modified PEG hydrogels. Polym Prepr 42:151–152
Lee BP, Dalsin JL, Messersmith PB (2002) Synthesis and gelation of DOPA-modified poly (ethylene glycol) hydrogels. Biomacromolecules 3(5):1038–1047
Lee BP, Huang K, Nunalee N, Shull K, Messersmith PB (2004) Synthesis of 3,4-dihydroxyphenylalanine (DOPA) containing monomers and their copolymerization with PEG-diacrylate to form hydrogels. J Biomater Sci Polym Ed 15(4):449–464
Lee BP, Chao C-Y, Nunalee FN, Motan E, Shull KR, Messersmith PB (2006a) Rapid photocurable of amphiphilic block copolymers hydrogels with high DOPA contents. Maclomolecules 39:1740–1748
Lee H, Scherer NF, Messersmith PB (2006b) Single molecule mechanics of mussel adhesion. Proc Natl Acad Sci 103:12999–13003
Lee H, Dellatore SM, Miller WM, Messersmith PB (2007a) Mussel-inspired surface chemistry for multifunctional coatings. Science 318(5849):426–430
Lee H, Lee BP, Messersmith PB (2007) A reversible wet/dry adhesive inspired by mussels and geckos. Nature 448 (7151):338-U334. doi:10.1038/Nature05968
Lee H, Rho J, Messersmith PB (2009) Facile conjugation of biomolecules onto surfaces via mussel adhesive protein inspired coatings. Adv Mater 21(4):431–434. doi:10.1002/adma.200801222
Lee Y, Chung HJ, Yeo S, Ahn CH, Lee H, Messersmith PB, Park TG (2010a) Thermo-sensitive, injectable, and tissue adhesive sol-gel transition hyaluronic acid/pluronic composite hydrogels prepared from bio-inspired catechol-thiol reaction. Soft Matter 6(5):977–983. doi:10.1039/b919944f
Lee BP, Dalsin JL, Vollenweider L, Murphy JL, Xu F, Virosco J, Lew W, White J (2010a) Bioadhesive construct. US20100137902 Patent
Lee BP, Murphy JL, Silvary S (2010b) Multibranched bioadhesive compounds and synthetic methods therefor. US Patent Application 20100197868
Lee BP, Messersmith PB, Israelachvili JN, Waite JH (2011) Mussel-inspired adhesives and coatings. Annu Rev Mater Res 41:99–132. doi:10.1146/annurev-matsci-062910-100429
Lee BP, Dalsin JL, Vollenweider L, Murphy JL, Xu F, Virosco J, Lew W, White J (2012) Bioadhesive construct sith polymer blends. US20120003888 Patent
Lee C, Shin J, Lee JS, Byun E, Ryu JH, Um SH, Kim DI, Lee H, Cho SW (2013a) Bioinspired, calcium-free alginate hydrogels with tunable physical and mechanical properties and improved biocompatibility. Biomacromolecules 14(6):2004–2013. doi:10.1021/Bm400352d
Lee K, Oh MH, Lee MS, Nam YS, Park TG, Jeong JH (2013b) Stabilized calcium phosphate nano-aggregates using a dopa-chitosan conjugate for gene delivery. Int J Pharm 445(1–2):196–202. doi:10.1016/j.ijpharm.2013.01.014
Lee BP, Liu Y, Konst S (2014) Novel hydrogel actuator based on biomimetic chemistry. MRS Proc 1710:mrss14-1710-XX1708-1701. doi:10.1557/opl.2014.511
Lee BP, Lin M-H, Narkar A, Konst S, Wilharm R (2015) Modulating the movement of hydrogel actuator based on catechol-iron ion coordination chemistry. Sens Actuators B Chem 206:456–462. doi:10.1016/j.snb.2014.09.089
Lee BP, Narkar A, Wilharm R (2016) Effect of metal ion type on the movement of hydrogel actuator based on catechol-metal ion coordination chemistry. Sensors Actuators B Chem 227:248–254
Li J, Christensen BM (1994) Effect of pH on the oxidation pathway of dopamine and dopa. J Electroanal Chem 375(1–2):219–231. doi:10.1016/0022-0728(94)03389-7

Li G, Cheng G, Xue H, Chen S, Zhang F, Jiang S (2008) Ultra low fouling zwitterionic polymers with a biomimetic adhesive group. Biomaterials 29(35):4592–4597. doi:10.1016/j.biomaterials.2008.08.021

Li G, Xue H, Gao C, Zhang F, Jiang AS (2010) Nonfouling Polyampholytes from an Ion-Pair Comonomer with Biomimetic Adhesive Groups. Maclomolecules 43:14–1

Lim HL, Hwang Y, Kar M, Varghese S (2014) Smart hydrogels as functional biomimetic systems. Biomater Sci 2(5):603–618. doi:10.1039/C3bm60288e

Liu Y, Zhan H, Skelton S, Lee BP Marine adhesive containing nanocomposite hydrogel with enhanced materials and bioadhesive properties. In: MRS Proceedings, 2013. Cambridge University Press, pp 33–38

Liu Y, Meng H, Konst S, Sarmiento R, Rajachar R, Lee BP (2014) Injectable dopamine-modified poly(ethylene glycol) nanocomposite hydrogel with enhanced adhesive property and bioactivity. ACS Appl Mater Interfaces 6(19):16982–16992. doi:10.1021/am504566v

Long JR, Dindot JL, Zebroski HK, Suzanne CRH, Campbell AA, Stayton PS, Drobny GP (1998) A peptide that inhibits hydroxyapatite growth is in an extended conformation on the crystal surface. Proc Natl Acad Sci U S A 95(21):12083–12087

Ma Z, Jia X, Hu J, Zhang G, Zhou F, Liu Z, Wang H (2013) Dual-responsive capsules with tunable low critical solution temperatures and their loading and release behavior. Langmuir 29 (19):5631–5637. doi:10.1021/la400025j

Malmsten M, Emoto K, Van Alstine JM (1998) Effect of chain density on inhibition of protein adsorption by poly(ethylene glycol) based coatings. J Colloid Interface Sci 202(2):507–517

Martin ST, Kesselman JM, Park DS, Lewis NS, Hoffmann MR (1996) Surface structures of 4-chlorocatechol adsorbed on titanium dioxide. Environ Sci Technol 30:8

Marumo K, Waite JH (1986) Optimization of hydroxylation of tyrosine and tyrosine-containing peptides by mushroom tyrosinase. Biochim Biophys Acta 872(1-2):98–103

Matos-Perez CR, Wilker JJ (2012) Ambivalent adhesives: combining biomimetic cross-linking with antiadhesive oligo(ethylene glycol). Macromolecules 45(16):6634–6639. doi:10.1021/Ma300962d

Matos-Pérez CR, White JD, Wilker JJ (2012) Polymer composition and substrate influences on the adhesive bonding of a biomimetic, cross-linking polymer. J Am Chem Soc 134(22):9498–9505

Maugh KJ, Anderson DM, Strausberg R, Strausberg SL, Mccandliss R, Wei T, Filpula D (1988) Recombinant bioadhesive proteins of marine animals and their use in adhesive compositions. USA Patent 87-US3048

Meisel H, Olieman C (1998) Estimation of calcium-binding constants of casein phosphopeptides by capillary zone electrophoresis. Anal Chim Acta 372(1–2):291–297

Meng H, Li Y, Faust M, Konst S, Lee BP (2015) Hydrogen peroxide generation and biocompatibility of hydrogel-bound mussel adhesive moiety. Acta Biomater 17:160–169. doi:10.1016/j.actbio.2015.02.002

Menyo MS, Hawker CJ, Waite JH (2013) Versatile tuning of supramolecular hydrogels through metal complexation of oxidation-resistant catechol-inspired ligands. Soft Matter. doi:10.1039/c3sm51824h

Meredith HJ, Jenkins CL, Wilker JJ (2014) Enhancing the adhesion of a biomimetic polymer yields performance rivaling commercial glues. Adv Funct Mater 24(21):3259–3267

Messersmith PB, Dalsin JL, Lee BP, Burke SA (2014) DOPA-functionalized, branched, poly (aklylene oxide) adhesives. US Patent 8,673,286

Miller SM, Simon RJ, Ng S, Zuckermann RN, Kerr JM, Moos WH (1995) Comparison of the proteolytic susceptibilities of homologous L-amino acid, D-amino acid, and N-substituted glycine peptide and peptoid oligomers. Drug Dev Res 35(1):20–32

Mulder M, Crosier J, Dunn R (2009) Cauda equina compression by hydrogel dural sealant after a laminotomy and discectomy case report. Spine 34(4):E144–E148. doi:10.1097/Brs.0b013e31818d5427

Murphy JL, Vollenweider L, Xu F, Lee BP (2010) Adhesive performance of biomimetic adhesive-coated biologic scaffolds. Biomacromolecules 11:2976–2984

Nath N, Hyun J, Ma H, Chilkoti A (2004) Surface engineering strategies for control of protein and cell interactions. Surf Sci 570(1–2):98–110

Neto AI, Cibrao AC, Correia CR, Carvalho RR, Luz GM, Ferrer GG, Botelho G, Picart C, Alves NM, Mano JF (2014) Nanostructured polymeric coatings based on chitosan and dopamine-modified hyaluronic acid for biomedical applications. Small 10(12):2459–2469. doi: 10.1002/smll.201303568

Ninan L, Monahan J, Stroshine RL, Wilker JJ, Shi R (2003) Adhesive strength of marine mussel extracts on porcine skin. Biomaterials 24(22):4091–4099

Ohman H, Vahlquist A (1994) In vivo studies concerning a pH gradient in human stratum corneum and upper epidermis. Acta Derm Venereol 74(5):375–379

Olivieri MP, Loomis RE, Meyer AE, Baier RE (1990) Surface characterization of mussel adhesive protein films. J Adhes Sci Technol 4(3):197–204

Pandey KB, Rizvi SI (2009) Plant polyphenols as dietary antioxidants in human health and disease. Oxidative Med Cell Longev 2(5):270–278

Papov VV, Diamond TV, Biemann K, Waite JH (1995) Hydroxyarginine-containing polyphenolic proteins in the adhesive plaques of the marine mussel *Mytilus edulis*. J Biol Chem 270 (34):20183–20192

Pardo J, Gutierrez E, Saez C, Brito M, Burzio LO (1990) Purification of adhesive proteins from mussels. Protein Expr Purif 1(2):147–150

Park JY, Yeom J, Kim JS, Lee M, Lee H, Nam YS (2013) Cell-repellant dextran coatings of porous titania using mussel adhesion chemistry. Macromol Biosci 13(11):1511–1519. doi:10.1002/mabi.201300224

Pasche S, De Paul SM, Voros J, Spencer ND, Textor M (2003) Poly(L-lysine)-graft-poly(ethylene glycol) assembled monolayers on niobium oxide surfaces: a quantitative study of the influence of polymer interfacial architecture on resistance to protein adsorption by ToF-SIMS and in situ OWLS. Langmuir 19(22):9216–9225

Pitman MI, Menche D, Song EK, Ben-Yishay A, Gilbert D, Grande DA (1989) The use of adhesives in chondrocyte transplantation surgery: in-vivo studies. Bull Hosp Jt Dis Orthop Inst 49(2):213–220

Pizer R, Babcock L (1977) Mechanism of complexation of boron acids with catechol and substituted catechols. Inorg Chem 16(7):1677–1681. doi:10.1021/Ic50173a021

Reffitt DM, Ogston N, Jugdaohsingh R, Cheung HFJ, Evans BAJ, Thompson RPH, Powell JJ, Hampson GN (2003) Orthosilicic acid stimulates collagen type 1 synthesis and osteoblastic differentiation in human osteoblast-like cells in vitro. Bone 32(2):127–135. doi:10.1016/s8756-3282(02)00950-x

Ren Y, Rivera JG, He L, Kulkarni H, Lee DK, Messersmith PB (2011) Facile, high efficiency immobilization of lipase enzyme on magnetic iron oxide nanoparticles via a biomimetic coating. BMC Biotechnol 11:63. doi:10.1186/1472-6750-11-63

Rodriguez R, Blesa MA, Regazzoni AE (1996) Surface Complexation at the $TiO_2$ (anatase)/aqueous solution interface: chemisorption of catechol. J Colloid Interface Sci 177

Ryu JH, Lee Y, Kong WH, Kim TG, Park TG, Lee H (2011) Catechol-functionalized chitosan/pluronic hydrogels for tissue adhesives and hemostatic materials. Biomacromolecules 12 (7):2653–2659. doi:10.1021/bm200464x

Rzepecki LM, Hansen KM, Waite JH (1992) Characterization of a cystine-rich polyphenolic protein family from the blue mussel *Mytilus edulis*. Biol Bull 183(1):123–137

Schnurrer J, Lehr C-M (1996) Mucoadhesive properties of the mussel adhesive protein. Int J Pharm 141(1,2):251–256

Seo S, Das S, Zalicki PJ, Mirshafian R, Eisenbach CD, Israelachvili JN, Waite JH, Ahn BK (2015) Microphase behavior and enhanced wet-cohesion of synthetic copolyampholytes inspired by a mussel foot protein. J Am Chem Soc 137(29):9214–9217. doi:10.1021/jacs.5b03827

Sever MJ, Wilker JJ (2004) Visible absorption spectra of metal-catecholate and metal-tironate complexes. Dalton Trans 7:1061–1072. doi:10.1039/b315811j

Shafiq Z, Cui J, Pastor-Pérez L, San Miguel V, Gropeanu RA, Serrano C, del Campo A (2012) Bioinspired Underwater Bonding and Debonding on Demand. Angew Chem Int Ed 51 (18):4332–4335. doi:10.1002/anie.201108629

Shazly TM, Baker AB, Naber JR, Bon A, Van Vliet KJ, Edelman ER (2010) Augmentation of postswelling surgical sealant potential of adhesive hydrogels. J Biomed Mater Res Part A 95A (4):1159–1169. doi:10.1002/Jbm.A.32942

Shin H, Ruhe PQ, Mikos AG, Jansen JA (2003) In vivo bone and soft tissue response to injectable, biodegradable oligo (poly (ethylene glycol) fumarate) hydrogels. Biomaterials 24 (19):3201–3211

Sileika TS, Kim H-D, Maniak P, Messersmith PB (2011) Antibacterial performance of polydopamine-modified polymer surfaces containing passive and active components. ACS Appl Mater Interfaces 3(12):4602–4610. doi:10.1021/am200978h

Sileika TS, Barrett DG, Zhang R, Lau KHA, Messersmith PB (2013) Colorless multifunctional coatings inspired by polyphenols found in tea, chocolate, and wine. Angew Chem Int Ed 52 (41):10766–10770. doi:10.1002/anie.201304922

Silverman HG, Roberto FF (2006) Cloning and expression of recombinant adhesive protein Mefp-1 of the blue mussel, Mytilus edulis. Google Patents

Skelton S, Bostwick M, O'Connor K, Konst S, Casey S, Lee BP (2013) Biomimetic adhesive containing nanocomposite hydrogel with enhanced materials properties. Soft Matter 9 (14):3825–3833

Sofia SJ, Premnath V, Merrill EW (1998) Poly(ethylene oxide) grafted to silicon surfaces: grafting density and protein adsorption. Macromolecules 31:5059–5070

Soller BR, Micheels RH, Coen J, Parikh B, Chu L, Hsi C (1996) Feasibility of non-invasive measurement of tissue pH using near-infrared reflectance spectroscopy. J Clin Monit Comput 12(5):387–395. doi:10.1007/bf02077636

Soller BR, Khan T, Favreau J, Hsi C, Puyana JC, Heard SO (2003) Investigation of muscle pH as an indicator of liver pH and injury from hemorrhagic shock. J Surg Res 114(2):195–201. doi:10.1016/S0022-4804(03)00251-8

Soriaga MP, Hubbard AT (1982) Determination of the orientation of aromatic molecules adsorbed on platinum electrodes. The effect of solute concentration. J Am Chem Soc 104 (14):3937–3945

Statz AR, Meagher RJ, Barron AE, Messersmith PB (2005) New peptidomimetic polymers for antifouling surfaces. J Am Chem Soc 127(22):7972–7973

Statz AR, Barron AE, Messersmith PB (2008a) Protein, cell and bacterial fouling resistance of polypeptoid-modified surfaces: effect of side-chain chemistry. Soft Matter 4:131–139

Statz AR, Park JP, Chongsiriwatana NP, Barron AE, Messersmith PB (2008b) Surface-immobilized antimicrobial peptoids. Biofouling 24(6):439–448

Strausberg RL, Link RP (1990) Protein-based medical adhesives. Trends Biotechnol 8(2):53–57

Strausberg RL, Anderson DM, Filpula D, Finkelman M, Link R, McCandliss R, Orndorff SA, Strausberg SL, Wei T (1989) Development of a microbial system for production of mussel adhesive protein. In: Adhesives from renewable resources, ACS symposium series, vol 385. pp 453–464

Su J, Chen F, Cryns VL, Messersmith PB (2011) Catechol polymers for pH-responsive, targeted drug delivery to cancer cells. J Am Chem Soc 133(31):11850–11853. doi:10.1021/ja203077x

Sun C, Waite JH (2005) Mapping chemical gradients within and along a fibrous structural tissue, mussel byssal threads. J Biol Chem 280(47):39332–39336. doi:10.1074/jbc.M508674200

Sun W, Neuzil P, Kustandi TS, Oh S, Samper VD (2005) The nature of the gecko lizard adhesive force. Biophys J 89:L14–16

Tatehata H, Mochizuki A, Kawashima T, Yamashita S, Yamamoto H (2000) Model polypeptide of mussel adhesive protein. I. Synthesis and adhesive studies of sequential polypeptides (X-Tyr-Lys)n and (Y-Lys)n. J Appl Polym Sci 76(6):929–937

Tatehata H, Mochizuki A, Ohkawa K, Yamada M, Yamamoto H (2001) Tissue adhesive using synthetic model adhesive proteins inspired by the marine mussel. J Adhes Sci Technol 15 (9):1003–1013

Taylor SW, Chase DB, Emptage MH, Nelson MJ, Waite JH (1996) Ferric ion complexes of a DOPA-containing adhesive protein from *Mytilus edulis*. Inorg Chem 35:6

Thompson DW, Butterworth JT (1992) The nature of laponite and its aqueous dispersions. J Colloid Interface Sci 151(1):236–243. doi:10.1016/0021-9797(92)90254-j

Tian Y, Pesika N, Zeng HB, Rosenberg K, Zhao BX, McGuiggan P, Autumn K, Israelachvili J (2006) Adhesion and friction in gecko toe attachment and detachment. Proc Natl Acad Sci U S A 103(51):19320–19325. doi:10.1073/pnas.0608841103

Waite JH (1987) Nature's underwater adhesive specialist. Int J Adhes Adhes 7(1):9–14

Waite JH (1991) Mussel beards: a coming of age. Chem Ind 2(September):607–611

Waite JH (1999) Reverse engineering of bioadhesion in marine mussels. Ann N Y Acad Sci 875:301–309

Waite JH, Andersen SO (1978) 3,4-dihydroxyphenylalanine in an insoluble shell protein of *Mytilus edulis*. Biochim Biophys Acta 541(1):107–114

Waite JH, Qin X (2001) Polyphosphoprotein from the adhesive pads of Mytilus edulis. Biochemistry 40(9):2887–2893

Waite JH, Anderson NH, Jewhurst S, Sun C (2005) Mussell adhesion: finding the tricks worth mimicking. J Adhes 81:1–21

Wang X, Ye Q, Gao T, Liu J, Zhou F (2011) Self-assembly of catecholic macroinitiator on various substrates and surface-initiated polymerization. Langmuir 28(5):2574–2581. doi:10.1021/la204568d

Wang WN, Xu YS, Li A, Li T, Liu MM, von Klitzing R, Ober CK, Kayitmazer AB, Li L, Guo XH (2015) Zinc induced polyelectrolyte coacervate bioadhesive and its transition to a self-healing hydrogel. RSC Adv 5(82):66871–66878. doi:10.1039/c5ra11915d

Warner SC, Waite JH (1999) Expression of multiple forms of an adhesive plaque protein in an individual mussel, Mytilus edulis. Mar Biol 134(4):729–734

Wei YF, Kong JH, Yang LP, Ke L, Tan HR, Liu H, Huang YZ, Sun XW, Lu XH, Du HJ (2013) Polydopamine-assisted decoration of ZnO nanorods with Ag nanoparticles: an improved photoelectrochemical anode. J Mater Chem A 1(16):5045–5052. doi:10.1039/C3ta10499k

Westwood G, Horton TN, Wilker JJ (2007) Simplified polymer mimics of cross-linking adhesive proteins. Macromolecules 40(11):3960–3964

White JD, Wilker JJ (2011) Underwater bonding with charged polymer mimics of marine mussel adhesive proteins. Macromolecules 44(13):5085–5088

Yamamoto H (1987) Synthesis and adhesive studies of marine polypeptides. J Chem Soc Perkin Trans 1(3):613–618

Yamamoto H (1995) Marine adhesive proteins and some biotechnological applications. Biotechnol Genet Eng Rev 13:133–165

Yamamoto H, Tatehata H (1995) Oxidation of synthetic precursors of adhesive proteins from mytilid bivalves using tyrosinase. J Mar Biotechnol 2(2):95–100

Yamamoto H, Ogawa T, Ohkawa K (1995) Wettability and adhesion of synthetic marine adhesive proteins and related model compounds. J Colloid Interface Sci 176(1):111–116

Yang SH, Kang SM, Lee KB, Chung TD, Lee H, Choi IS (2011) Mussel-inspired encapsulation and functionalization of individual yeast cells. J Am Chem Soc 133(9):2795–2797. doi:10.1021/ja1100189

Yang X, Zhu LP, Tada S, Zhou D, Kitajima T, Isoshima T, Yoshida Y, Nakamura M, Yan WQ, Ito Y (2014) Mussel-inspired human gelatin nanocoating for creating biologically adhesive surfaces. Int J Nanomed 9:2753–2765. doi:10.2147/Ijn.S60624

Ye Q, Wang X, Li S, Zhou F (2010) Surface-initiated ring-opening metathesis polymerization of pentadecafluorooctyl-5-norbornene-2-carboxylate from variable substrates modified with sticky biomimic initiator. Macromolecules 43(13):5554–5560. doi:10.1021/ma100479x

You I, Kang SM, Byun Y, Lee H (2011) Enhancement of blood compatibility of poly(urethane) substrates by mussel-inspired adhesive heparin coating. Bioconjug Chem 22(7):1264–1269. doi:10.1021/Bc2000534

Young GA, Crisp DJ (1982) Marine animals and adhesion. In: Allen KW (ed) Adhesion-6. Applied Science Publishers, Barkling, UK, pp 279–313

Yu M, Deming TJ (1998) Synthetic polypeptide mimics of marine adhesives. Macromolecules 31 (15):4739–4745

Yu M, Hwang J, Deming TJ (1999) Role of L-3,4-dihydroxyphenylanine in mussel adhesive proteins. J Am Chem Soc 121(24):5825–5826

Yurdumakan B, Raravikar NR, Ajayan PM, Dhinojwala A (2005) Synthetic gecko foot-hairs from multiwalled carbon nanotubes. Chem Commun 30:3799–3801

Zhang MQ, Desai T, Ferrari M (1998) Proteins and cells on PEG immobilized silicon surfaces. Biomaterials 19(10):953–960

Zhang YC, Thomas Y, Kim E, Payne GF (2012) pH- and Voltage-responsive chitosan hydrogel through covalent cross-linking with catechol. J Phys Chem B 116(5):1579–1585. doi:10.1021/Jp210043w

Zhang M, Gao B, Varnoosfaderani S, Hebard A, Yao Y, Inyang M (2013) Preparation and characterization of a novel magnetic biochar for arsenic removal. Bioresour Technol 130:457–462. doi:10.1016/j.biortech.2012.11.132

Zhao H, Waite JH (2006) Linking adhesive and structural proteins in the attachment plaque of Mytilus californianus. J Biol Chem 281(36):26150–26158. doi:10.1074/jbc.M604357200

Zhong C, Gurry T, Cheng AA, Downey J, Deng Z, Stultz CM, Lu TK (2014) Strong underwater adhesives made by self-assembling multi-protein nanofibres. Nat Nano 9(10):858–866. doi:10.1038/nnano.2014.199, http://www.nature.com/nnano/journal/v9/n10/abs/nnano.2014.199.html#supplementary-information

The manufacturer's authorised representative in the EU is Springer Nature Customer Service Centre GmbH, Europaplatz 3, 69115 Heidelberg, Germany. If you have any concerns regarding our products, please contact ProductSafety@springernature.com

Printed and bound by CPI Group (UK) Ltd, Croydon, CR0 4YY
30/09/2026
02225231-0002